METAL-LIGAND MULTIPLE BONDS

METAL-LIGAND MULTIPLE BONDS
The Chemistry of Transition Metal Complexes Containing Oxo, Nitrido, Imido, Alkylidene, or Alkylidyne Ligands

WILLIAM A. NUGENT
Central Research and Development Department
E. I. duPont de Nemours and Company
Wilmington, Delaware

JAMES M. MAYER
Department of Chemistry
University of Washington
Seattle, Washington

WILEY

A WILEY-INTERSCIENCE PUBLICATION
JOHN WILEY & SONS
New York · Chichester · Brisbane · Toronto · Singapore

Copyright © 1988 by John Wiley & Sons, Inc.

All rights reserved. Published simultaneously in Canada.

Reproduction or translation of any part of this work beyond that permitted by Section 107 or 108 of the 1976 United States Copyright Act without the permission of the copyright owner is unlawful. Requests for permission or further information should be addressed to the Permissions Department, John Wiley & Sons, Inc.

Library of Congress Cataloging-in-Publication Data

Nugent, William A., 1947–
  Metal-Ligand Multiple Bonds.

  "A Wiley-Interscience publication."
  Bibliography: p.
  Includes index.
  I. Complex compounds. 2. Transition metal compounds.
3. Ligands. I. Mayer, James M., 1958–   . II. Title.
QD474.N84 1988   546'.345   88-233
ISBN 0-471-85440-9

Printed in the United States of America

10 9 8 7 6 5 4 3 2 1

To Sharon and to Faith

PREFACE

This book had its origin in a 1986 discussion between the authors. We found ourselves sympathizing with the plight of the student or beginning researcher with an interest in the field of metal–ligand multiple bonds. To understand the present state of knowledge of this subject currently requires familiarity with a body of some 2000 key publications. Moreover, half of the relevant research has been published since 1980. The various types of multiply bonded ligands show striking similarities—and also some remarkable differences—in terms of synthetic routes, structure, spectroscopy, and chemistry, all of which can be unifiedly understood on the basis of their electronic structure. Although a few dated reviews covering some aspects of the topic were available, there existed no single source that provided this "big picture." It was to meet these needs that we undertook this volume.

Our discussion is directed at the graduate student or the chemist who has not previously been involved in research on multiply bonded ligands. Nevertheless, we believe that the established researcher in the field will find this book invaluable both for the many new tabulations of data and as a source of ideas for research directions. We further believe that this material would constitute an excellent and timely subject for a "special topics" course in inorganic chemistry.

We have attempted to organize this book in a way that stresses the unified nature of the field. For each subtopic we discuss relevant information for all the subject ligands–oxo, imido, alkylidene, and so forth. Initially, this may prove distressing to the reader with a particular question about a single ligand type, who might prefer separate chapters for oxygen-, nitrogen-, and carbon-bound ligands. To these individuals we point out that the flow within each subtopic is consistently oxygen → nitrogen → carbon. We feel any disadvantages in this approach are far outweighed by the benefits, for this organization serves not only to stress what is known about each topic but also what is not known. Can the oxo-mediated rearrangements of allylic alcohols (Section 7.1.2) be extended to alkylidene complexes to provide a powerful tool for stereospecific C—C bond formation? Does the α cleavage reaction of low-valent methoxides (Section 3.2.3) provide insight into enzymatic N—N cleavage effected by nitrogenase? The authors find that virtually every subtopic suggests interesting questions for future research. If other readers are similarly affected, we will have achieved one of our principal goals for this book.

One disappointment has been the need to exclude from coverage the subject of Fischer carbenes. Although this excision can be defended on chemical grounds, our decision to omit this extensive body of research largely reflects the availability of the recent monograph on this topic by Dötz et al. Nevertheless, we have made comparisons with the literature on Fischer carbenes whenever it appeared useful. (Given unlimited space, it also would have been interesting to have included the chemistry of sulfido complexes.)

The authors owe a debt of gratitude to several individuals who kindly read and criticized portions of the manuscript. Our heartfelt thanks go to Dr. R. Thomas Baker, Dr. Mark Burk, Dr. Nancy M. Doherty, Dr. Roald Hoffman, Dr. Richard F. Jordan, Dr. Thomas J. Meyer, Dr. Jeffrey S. Thompson, and Dr. David L. Thorn. We are also grateful to Dr. Frank Weigert for help in setting up the Datatrieve data base that we used extensively and to Dr. Zarah Ainbinder for conducting on-line literature searches. References in this book were obtained from the Chemical Abstracts Service CAS ONLINE files available from STN International, the scientific and technical information network. Much of the structural data in Chapter 5 were gathered using the Cambridge Structural Data Base. We thank the following individuals for skillful assistance in putting together the final manuscript: Mrs. E. Jayne Allen, Mrs. Debra A. Huie, Mr. David M. Lattomus, and Ms. Cheryl Meredith.

Finally we thank our wives and families for their unflagging support and for their patience during the long hours required to complete the writing.

WILLIAM A. NUGENT
JAMES M. MAYER

Wilmington, Delaware
Seattle, Washington
December, 1987

CONTENTS

LIGAND ABBREVIATIONS   xi

1 LIGAND TYPES AND OVERVIEW   1
     1.1 Coverage in This Book, 1
     1.2 Oxo Ligands, 3
     1.3 Nitrido Complexes, 6
     1.4 Imido Complexes, 8
     1.5 Hydrazido(2-) and Related Ligands, 9
     1.6 "Fischer-Type" Carbenes, 11
     1.7 "Schrock-Type" Alkylidene Ligands, 12
     1.8 Alkylidyne Ligands, 15
     1.9 Summary, 16

2 ELECTRONIC STRUCTURE   21
     2.1 Nature of the Metal–Ligand Multiple Bond, 21
     2.2 Periodic Trends, 26
     2.3 Ligand Field Descriptions, 33
     2.4 *Ab Initio* Calculations, 39
     2.5 Theoretical Studies of Specific Systems, 41

3 REACTIONS RESULTING IN THE FORMATION OF MULTIPLY BONDED LIGANDS   52
     3.1 General Considerations, 52
     3.2 By Cleavage of α Bond of Precursor, 52
     3.3 By Formation of a Bond to the α Atom, 90
     3.4 By Formation of a Bond to the β Atom, 93
     3.5 Concluding Remarks, 98

4 VIBRATIONAL AND NMR SPECTROSCOPY OF MULTIPLY BONDED LIGANDS   112
     4.1 IR Spectra of Terminal Oxo and Nitrido Complexes, 112
     4.2 Vibrational Spectra of Bridging Nitrides, 121
     4.3 IR Spectra of Alkylidyne, Organoimido, and Related NX Compounds, 123
     4.4 IR Data as a Measure of the Relative π-Donating Ability of Various Multiply Bonded Ligands, 125
     4.5 Hydrogen Stretching Modes, 126

4.6 ^{17}O NMR of Oxo Complexes, 127
4.7 Nitrogen NMR of Multiply Bonded Ligands, 129
4.8 ^{1}H NMR Spectroscopy, 132
4.9 ^{13}C NMR Spectroscopy, 133
4.10 NMR of the Metal Atom, 136

5 STRUCTURAL STUDIES 145
5.1 Distribution of Compounds with Metal–Ligand Multiple Bonds, 145
5.2 Coordination Geometries, 147
5.3 Metrical Data, 148
5.4 Structural Tables, 158

6 REACTIONS OF MULTIPLY BONDED LIGANDS 220
6.1 A Simple Conceptual Model, 221
6.2 Reactions with Electrophiles, 223
6.3 Reactions with Nucleophiles, 240
6.4 "α Cleavage" Reactions, 270
6.5 Modification of Organic Moiety in Organoimido and Alkylidene Ligands, 273
6.6 Coupling of Two Multiply Bonded Ligands, 273
6.7 Connections, 275

7 ROLE OF METAL–LIGAND MULTIPLE BONDS IN CATALYSIS 288
7.1 Oxometal Species in Catalysis, 288
7.2 Imidometal Species in Catalysis, 299
7.3 Hydrazido Intermediates in Enzymatic Nitrogen Fixation, 302
7.4 Alkylidene Intermediates in Catalysis, 304
7.5 Alkylidyne Complexes in Acetylene Metathesis, 310
7.6 Concluding Remarks, 312

INDEX 321

LIGAND ABBREVIATIONS

acac	2,4-pentanedionate anion (acetylacetonate)
bipy	2,2'-bipyridine
Bu	butyl (superscript n or t for normal or tertiary)
Cp	cyclopentadienyl
Cp*	pentamethylcyclopentadienyl
depe	1.2-bis(diethlylphosphino)ethane
DME	dimethoxyethane
dmpe	1,2-bis(dimethylphosphino)ethane
DMSO	dimethyl sulfoxide (Me_2SO)
dppe	1,2-bis(diphenylphosphino)ethane
dtc	dialkyldithiocarbamate ($R_2NCS_2^-$)
Et	ethyl
L	neutral ligand
Me	methyl
Np	neopentyl (CH_2CMe_3)
Ph	phenyl
Pr	propyl (subscripts n or i for normal or iso)
py	pyridine
salen	bis(salicylaldehyde)ethylenediimine
THF	tetrahydrofuran
TMEDA	tetramethyl ethylenediamine
TMP	tetramesityl porphyrin
TPP	tetraphenyl porphyrin
TTP	tetra(p-tolyl) porphyrin
trpy	terpyridine
X	anionic ligand
Y	anionic ligand

METAL-LIGAND MULTIPLE BONDS

CHAPTER 1

LIGAND TYPES AND OVERVIEW

This book is concerned with transition metal complexes containing oxo, nitrido, imido, alkylidene, and alkylidyne ligands. Throughout our study, we will discover striking similarities in the synthesis, molecular structure, and reactivity of complexes containing these various ligand types. These parallel properties are, of course, no accident. They result from a common feature in the electronic structure of these compounds, the presence of one or more π bonds involving ligand p and metal d orbitals. By understanding the nature of this d-π–p-π interaction we can understand not only the similarities but also the differences among these ligand types.

This field has grown dramatically in the last dozen years. Some notion of the extent of this growth can be gleaned from Figure 1.1. (Moreover, the graph excludes research on oxo–metal complexes. Broadly viewed, publication activity on oxo species is currently an order of magnitude greater than for the other ligands combined!) This proliferation has resulted in a wealth of new mechanistic ideas, an infusion of new compounds, and theoretical advances. It has become evident that even from the standpoint of "pure science" this would remain an active area for many years to come. Just as carbon π systems have become an important testing ground for the physical organic chemist, the nature of structure and bonding in unsaturated transition metal species are of fundamental interest.

There is another significant factor firing the growth of this field, namely, the range of important chemical processes in which multiply bonded ligands play a role. Such species are present on the catalyst surface in a variety of crucial industrial processes. They constitute the "business end" of some of the most useful reagents for laboratory-scale synthesis. They are involved in a fascinating array of enzymatic transformations. The oxo and nitrido moieties in particular are essential building blocks for a new generation of electronic materials. After providing some additional details on the coverage in this book, we will devote the remainder of this chapter to a brief historical overview of each ligand type, with emphasis on events of the last decade.

1.1 COVERAGE IN THIS BOOK

This volume is concerned primarily with mononuclear complexes in which a first row element—carbon, nitrogen, or oxygen—is multiply bonded to a transition metal. Related types of complexes are discussed where appropriate as comparison examples. The nature of the ligands under consideration leads to an emphasis on early transition metals and high oxidation states.

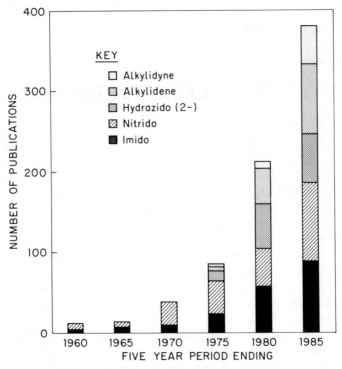

Figure 1.1 Publications on the various types of multiply bonded ligands by 5-year periods. (Excludes oxo and Fischer-type carbene ligands.)

Terminal oxo complexes **1** of the titanium through iron triads are included. Bridging dinuclear oxo derivatives **2** are not. Homo- and heteropolyanions receive only brief mention since these topics have recently been thoroughly reviewed [1].

$$M\equiv O \qquad \underset{M \quad M}{\overset{O}{\diagup\diagdown}}$$
$$\textbf{1} \qquad\qquad \textbf{2}$$

Terminal nitrido (**3**), imido (**4**), and hydrazido (**5**) ligands are covered, as are related "NX"-type ligands where X is a halogen or other electronegative group. Doubly bridging imido (**6**) and hydrazido (**7**) ligands and triply bridging nitrido species (**8**) are not covered.

3   **4**   **5**

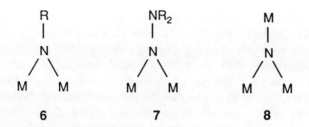

"Schrock-type" alkylidene (**9**) and alkylidyne (**10**) complexes are discussed. "Fischer-type" carbene and carbyne derivatives (see section 1.6) are included only in cases where comparison with Schrock-type systems is deemed instructive. The chemistry of both carbene [2] and carbyne [3] complexes has been reviewed recently with emphasis on the Fischer-type systems.

$$M=CRR' \qquad\qquad M\equiv C-R$$

$$\textbf{9} \qquad\qquad\qquad \textbf{10}$$

We have included in our coverage binuclear complexes containing bridging nitrido ligands **11** or bridging alkylidyne ligands **12** on the grounds that these derivatives still contain a ligand-to-metal double bond, at least as a resonance contributor. Also included are certain bridging N$_2$ species (**13**) since there is considerable evidence that these are best regarded as hydrazido(4−) complexes.

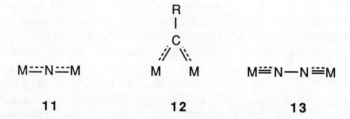

A case can also be made (and correctly so) for multiple bonding in other ligand types, including alkoxide [4], amido [5], sulfido [6], and even fluoride [7] ligands. The consequences of p-π-d-π bonding are especially evident in dialkylamido species (which are formally isoelectronic with alkylidene) in that their structure is universally observed to be planar. (However, thermodynamic studies suggest the bonding in early transition metal dialkylamides is dominated by the σ component [8].) Nevertheless, OR$^-$, NR$_2^-$, S^{2-}, and F$^-$ ligands will, in general, be excluded from coverage.

1.2 OXO LIGANDS

Among the types of multiply bonded ligands, oxo compounds have been known the longest and their chemistry has been most extensively developed. The oxometal compound sodium ferrate, Na_2FeO_4, was synthesized as early as 1702 [9] and others, including OsO_4, $KMnO_4$, and K_2CrO_4, were well known to chemists in the early nineteenth century. By the time that Pauling's classic "The Nature of the Chemical Bond" was published in 1938, the multiply bonded character of metal–oxo complexes was widely accepted [10]. The concepts were confirmed by the 1938 electron diffraction study of CrO_2Cl_2 and $VOCl_3$ by Palmer, who concluded that "structures involving multiple Cr—O and V—O bonds make important contributions to the normal state" [11]. The development of metal–oxo chemistry through about 1969 is summarized in a review by Griffith [12].

Today, complexes containing terminal oxo substituents are known for all of the transition metals of the vanadium through iron triads. For reasons we shall discuss below, the metals of the titanium triad tend to form bridged rather than terminal oxo structures. Nevertheless, the terminal oxo structure can be enforced by coordinative saturation of the metal and several terminal oxotitanium complexes have been structurally characterized [13].

One important and long-standing application of metal oxo derivatives is their use as oxidants in organic synthesis. The use of oxo derivatives of chromium(VI) [14], $KMnO_4$ [15], RuO_4 [16], and OsO_4 [17] in this regard have been reviewed. Moreover, the need for more efficient and selective oxidizing agents continues to provide an impetus for research in this area. Pyridinium chlorochromate (PCC) and pyridinium dichromate (PDC), introduced in 1975 and 1979, respectively, represent examples of novel oxidants that have gained widespread acceptance among organic chemists [18].

Our understanding of the mechanisms by which oxo derivatives oxidize olefins has increased considerably in the last decade. Historically, such reactions had been viewed as an electrophilic attack of the oxo–metal moiety on an electron-rich olefinic π system. In 1977 Sharpless et al. published a thought-provoking paper in which they pointed out that such a mechanism was incompatible with the then-prevalent notion of charge control [19]. Instead they proposed that these reactions proceed through prior coordination of the olefin and collapse of the resultant complex to an oxametallacyclic intermediate. In 1986 Jorgenson and Hoffmann, building on new experimental insight from the Sharpless group, invoked the notion of frontier orbital control and suggested that olefin oxidation by osmium tetroxide and related reagents is a concerted 3 + 2 cycloaddition reaction [20]. In their model, a key role is played by the coordination of amine ligands that convert the intrinsically unreactive tetrahedral osmium tetroxide to a reactive complex with C_{2v} symmetry. Other significant mechanistic insights emerged during this period. As one example, Kochi demonstrated that epoxidation of olefins by chromium reagents, long a mechanistic puzzle, was promoted by pre–reduction of the chromium to the +5 oxidation state [21].

These advances in mechanistic understanding have been accompanied by exciting developments in oxidation chemistry itself. Thus Meyer has succeeded in rationally designing ruthenium–oxo complexes that will catalytically oxidize a variety of organic substrates when continually reoxidized at the anode of an electrochemical cell [22]. Another remarkable advance is work by Groves in which an oxo–ruthenium porphyrin system is employed to epoxidize olefins catalytically—albeit slowly— with air as the stoichiometric oxidant [23]. Osmium-based systems have been developed for the asymmetric oxidation of olefins to diols; excellent chemical and optical yields have been achieved [24].

Nature also utilizes metal–oxo complexes in a ubiquitous series of important enzymes. Enzymes of the cytochrome P-450 family contain an oxo–iron porphyrin system and are involved in a wide range of biological oxidation processes including drug detoxification, activation of carcinogens, and steroid metabolism. Moreover, the unique capability of such enzymes to hydroxylate hydrocarbons selectively has long been a subject of fascination. An unprecedented amount of research activity has focused both on the P-450 enzymes themselves and on a variety of elegant model systems. It has been estimated that some 10,000 workers worldwide are involved in this effort. A noteworthy recent achievement is the X-ray crystal structure of one such enzyme, camphor 5-oxidase, in which a substrate molecule is actually in place at the active site [25]. Recent research in the P-450 area is the subject of an excellent review [26].

A second family of oxo–metal based enzymes are the molybenum- or tungsten-containing "oxo-transferases," Like P-450, these compounds play a variety of roles in living systems and. unlike P-450, are involved in both oxidative and reductive processes. Recently, significant advances have been made in developing model systems that mimic the enzymatic transformations [27].

Metal–oxo species are also present on the surface of industrially important heterogeneous catalysts (see Section 7.1.1). For example bismuth molybdate catalysts are used for the oxidation of C4's to butadiene and for the oxidation of propylene to acrolein. Similarly, iron molybdate catalysts are utilized for the oxidation of methanol to formaldehyde. A variety of evidence suggests that it is the terminal oxo groups on the catalyst surface that are directly involved in the catalytic chemistry [28]. Supported metal–oxo compounds also form the basis of commercial catalysts for olefin metathesis. In this case the oxo ligand remains bound to the metal but nevertheless seems to play an active role in promoting the reaction. Goddard has applied *ab initio* molecular orbital calculations to probe the origin of this "spectator oxo effect" [29]. Oxo ligands are also utilized in industrial processes involving homogeneous catalysis. For example, vanadium–oxo complexes catalyze the rearrangement of allylic or propargylic alcohols in the manufacture of terpene alcohols and of vitamin A (Section 7.1.2).

Transition metal oxides such as lithium niobate and potassium titanyl phosphate are important new electro-optic materials [30]. Although these materials are themselves network solids containing bridging oxo groups, discrete oxo transition metal compounds are of potential interest as precursors owing to the need to grow large, perfect crystals. In the future, oxo coordination chemistry may play a role both in the search for new optical materials and in their rational synthesis from solution under mild conditions.

During the past decade, organometallic chemists have begun to appreciate the ability of the oxo ligand to stabilize high oxidation states. This stabilization has been applied to the synthesis of both σ and π organotransition metal derivatives, for example, of vanadium(V), molybdenum(VI), tungsten(VI), and rhenium(VII). Several of these are shown as structures **14–16** [31–33]. Also remarkable for somewhat different reasons is the rhenium(III) complex **17,** an unusual example of a d^4 complex with a terminal oxo substituent [34].

14, **15**, **16**, **17**

N⌒N = bipyridyl

The group 5 and 6 transition metals, and molybdenum and tungsten in particular, are characterized by a tendency to form complex polymetallate acids and salts. The polyoxoanions of molybdenum and tungsten are of two types: (1) the isopoly anions such as $[Mo_8O_{26}]^{4-}$ containing only tungsten or molybdenum and oxygen and (2) the heteropolyanions, which additionally contain 1 or 2 atoms of another element as exemplified by $[P_2W_{18}O_{62}]^{6-}$. Recently there have been a number of important developments in the characterization of these compounds [35], their use in catalysis [36], and even in medicine as a treatment for AIDS [37]. Although these compounds typically contain both bridging and terminal oxo groups, they are not included in the coverage of this book. The reader is referred to the review of Pope [1].

1.3 NITRIDO COMPLEXES

The first nitrido complex to be synthesized was **18a,** the so-called "potassium osmiamate" reported by Fritsche and Struve in 1847 [38]. However, the structure was initially assigned as **18b** and it was not until 1901 that the presence of a nitrido ligand in this complex was recognized [39]. Similarly, the first bridging nitride to be prepared was the potassium salt of $[(H_2O)Cl_4Ru=N=RuCl_4(OH_2)]^{3-}$ in 1921 [40], but the nature of the product was not appreciated until much later [41]. The first neutral mononuclear nitrido complex did not appear until Chatt and coworkers' report of **19** in 1963 [42]. The development of nitrido chemistry through 1971 was reviewed by Griffith [43].

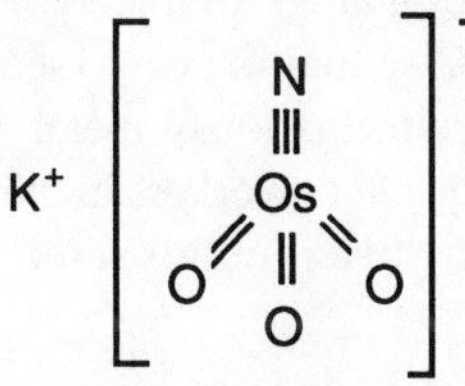

18a 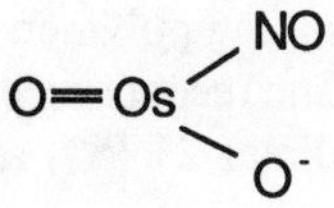   **18b** 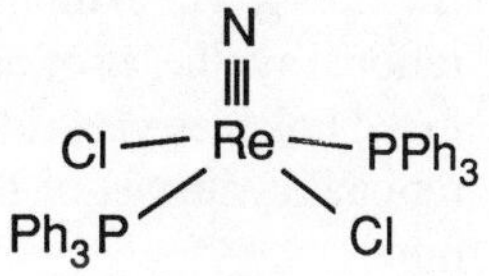   **19**

Multiply bonded nitrido complexes are now known for all of the metals of the vanadium through iron triads. However, the known examples for iron, niobium, and tantalum are of the bridged binuclear variety. Only recently have nitrido complexes of the first-row metals vanadium, chromium, manganese, and iron been prepared [44]. Their synthesis in some cases seems to require the use of strongly chelating or macrocylic ancillary ligands.

Applications of nitrido ligands typically reflect the great stability of the metal nitrogen triple bond. The ability of the nitrido group to stabilize organotransition metal species was recognized quite early by Chatt who in 1966 reported a series of arylrhenium nitride complexes, $Re(N)Ar_2(PR_3)_2$ [45]. Recently, Belmonte-Shapley has harnessed this stabilization to prepare a series of remarkable ruthenium(VI) and osmium(VI) alkyl derivatives [46]. An area where the energetics of ligation are critically important is that of technetium radiopharmaceuticals. The requisite "shake and shoot" hospital regimen demands that complexes form readily from very dilute aqueous solutions of pertechnetate(VII). A pretty solution is the direct conversion of TcO_4^- to lipophilic nitrido complexes such as **20**. This approach was first reported by Baldas et al. and has been developed extensively by others [47,48].

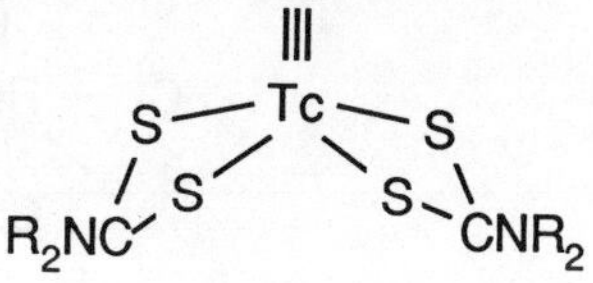

20

An emerging application of nitrido chemistry is the field of electronic materials. Niobium(III) nitride has been of interest for some time because of its ability to form films that are strong, stable, and, most importantly, superconducting [49]. Niobium nitride is itself a lattice solid. However, discrete (amido, imido, nitrido) species are presumably involved in the synthesis of this and other early transition metal nitrides from the metal halides and liquid ammonia. Our knowledge of such reactions rests on a body of speculative and very dated chemistry; little additional work has been reported since the field was reviewed in 1966 [50]. This area is ripe for new studies using ^{15}N

NMR. Further fueling current interest in the electronic properties of transition metal nitrides is a recent theoretical study by Hoffmann et al. pointing out a number of parallels between transition metal nitrides and such organic conductors as polyenes and phosphazenes [51]. (This includes the prediction that benzene analogs of the type [L_nMoN]$_3$ will exhibit a stable, delocalized π system.) Moreover, we may expect to see rational syntheses of new metal-containing polymers by controlled formation of metal–metal linkages via bridging ligands. The recent success of Doherty and coworkers in the rational synthesis of the bimetallic dimer **21** [52] represents a first step in this direction.

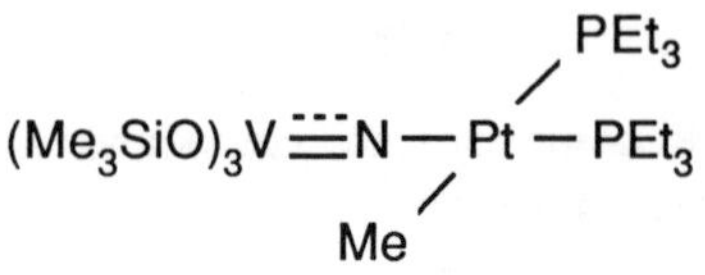

$$(Me_3SiO)_3V\equiv N-Pt(PEt_3)(PEt_3)(Me)$$

21

1.4 IMIDO COMPLEXES

The first organoimido transition metal complex to be prepared was *t*-butylimidotrioxo osmium(VIII), **22**, reported by Clifford and Kobayashi in 1956 [53]. By 1962 an extensive series of arylimido rhenium complexes of structure **23** was known [54]. Even today only a few examples of the parent imido (M=NH) derivatives have appeared. The first were the complexes $MoX_2(NH)(dppe)_2$ reported in 1975 [55]. A review surveying organoimido chemistry through the end of 1978 has been published [56].

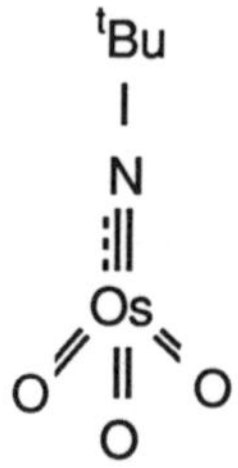

22 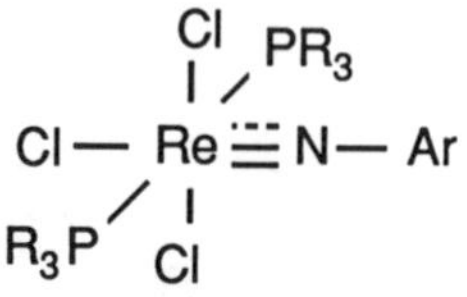   **23**

In addition to osmium and rhenium, terminal organoimido complexes are currently known for all the metals of vanadium and chromium triads. In an intriguing series of papers, Stone has suggested that the products obtained from treatment of a variety of low-valent group 8 (Ru, Rh, Ir, Pd, Pt) compounds with 2-*H*-hexafluoropropyl azide should also be formulated as mononuclear perfluoroalkylimido complexes [57]. Clearly, further studies on these puzzling compounds would be of interest.

The observation that **22** reacts with olefins was recorded by Milas and Iliopulos in

1959 [58]. However, it remained for Sharpless to recognize the importance of that observation which now serves as the basis for a useful set of transformations for organic synthesis. Compound **22** and its analogs effect the cis vicinal oxyamination (Eq. 1) of a variety of alkenes in either a stoichiometric or catalytic manner [59,60]. Subsequently the Sharpless group was able to synthesize di-, tri-, and even tetraimido analogs of osmium tetroxide that promote the corresponding vicinal diamination reaction [61].

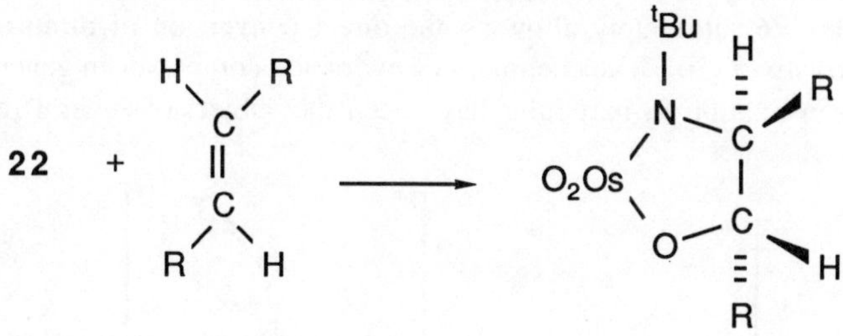

$$\mathbf{22} + \underset{RH}{\overset{HR}{\underset{\|}{\overset{C}{\underset{C}{\|}}}}} \longrightarrow \text{(structure)} \tag{1}$$

Imido species have been postulated as intermediates in catalytic processes. For example, surface molybdenum imido species have been suggested by workers at SOHIO to be the key intermediates in the industrial "ammoxidation" of propylene to acrylonitrile [62]. To date no direct evidence for the existence of such species or their involvement in ammoxidation has been forthcoming. Nevertheless, studies using organoimido derivatives have succeeded in providing structural models for these putative intermediates [63]. The key (C—N bond forming) steps from the proposed reaction mechanism have been reproduced using discrete, soluble, imido species under mild conditions [64]. Imido species are also believed to be involved in such enzymatic pathways as nitrogen fixation and the metabolism of certain hydrazines [65,66].

Organoimido ligands recently have come into their own as ancillary ligands in organometallic chemistry and homogeneous catalysis. Of particular interest is Schrock's use of the arylimido ligand in compounds **24a**, an important new class of homogeneous olefin metathesis catalysts [67]. Moreover **24b** surely represents one of the most remarkable organotransition metal derivatives reported to date [68].

Ar = 2,6-isopropyl-phenyl
R_f = OCMe(CF$_3$)$_2$ 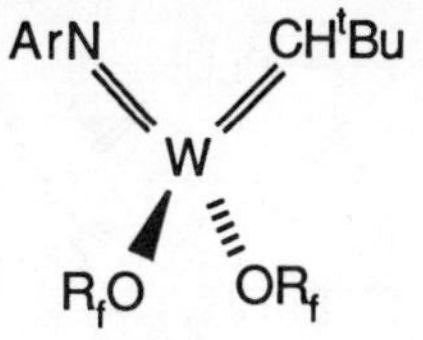
24a

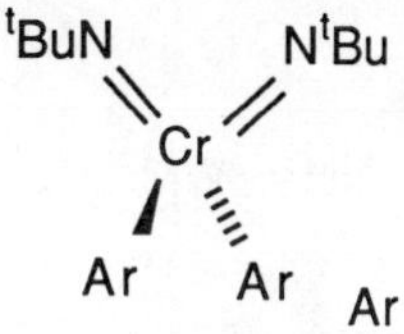

Ar = mesityl
24b

1.5 HYDRAZIDO(2-) AND RELATED LIGANDS

The chemistry of hydrazido(2-) complexes has been tied intimately to studies on dinitrogen fixation—both to efforts to develop models for the enzymatic process and the search for systems that would allow abiotic nitrogen fixation at low temperature and pressure. The focus of much of this activity has been the Institute for Nitrogen Fixation at Sussex University, initially under the direction of Joseph Chatt and subsequently under G.J. Leigh. In 1972, this organization reported the first synthesis of hydrazido(2-) complexes such as **25**; this was accomplished by protonation of tungsten and molybdenum dinitrogen complexes [69]. Such complexes, it was shown, undergo a number of C—N bond-forming reactions, a simple example being formation of the diazoalkane complex **26**, ultimately allowing the direct conversion of dinitrogen to organonitrogen derivatives [70]. The chemistry of hydrazido complexes in general and their role in nitrogen fixation in particular have been the subject of several reviews [71].

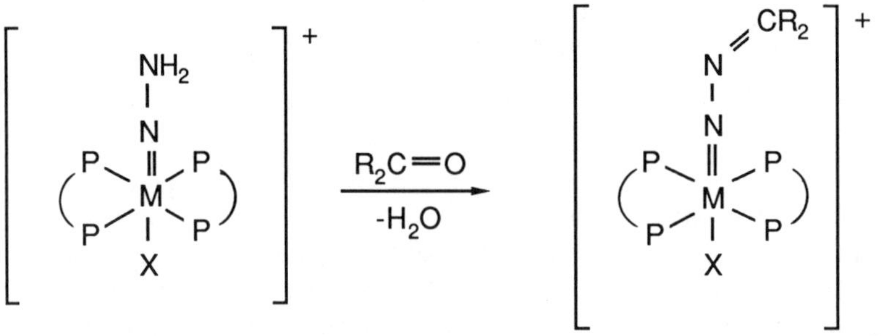

25 → (R₂C=O, −H₂O) → **26**   P⌒P = DPPE

Terminal hydrazido(2-) complexes are now known for the metals of the vanadium triad and for titanium, molybdenum, tungsten, and rhenium. In addition, it has recently been recognized that bridging "dinitrogen" complexes such as **27** and **28** are better regarded as doubly metallated hyrazido(4-) derivatives [72, 73]. This realization has breathed new life into the point of view, long advocated by Soviet scientists, that bridging rather than terminal nitrogen species may represent the key intermediates in enzymatic nitrogen fixation. The experimental work was followed by new theoretical studies in support of this view [74].

Cp*Me₃W=NN=WMe₃Cp*

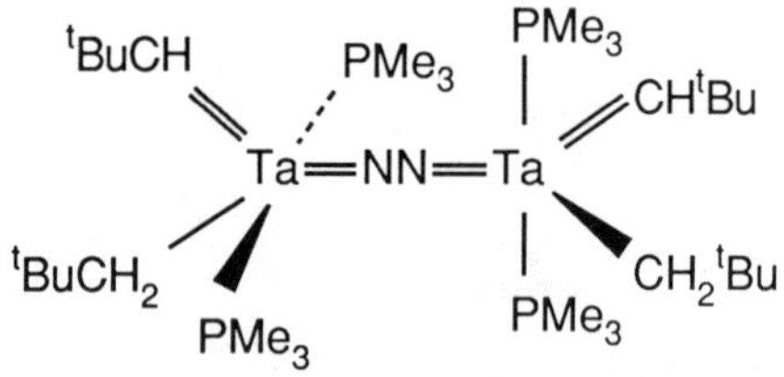

27   **28**

Diazoalkane complexes related to **26** have been used by Schwartz in an economically attractive variation of the Wittig reaction that does not stoichiometrically convert phosphine to phosphine oxide [75]. Another somewhat related set of M≡NX-type complexes are those in which X is a halogen. Their synthesis and chemistry have been explored extensively by the groups of Dehnicke and of Strähle, who recently have reviewed this area [76]. A fascinating recent development has been the synthesis of cyclic derivatives such as **29** and **30**. For each of these compounds there is some evidence that the resultant metallacycle has aromatic character [77,78].

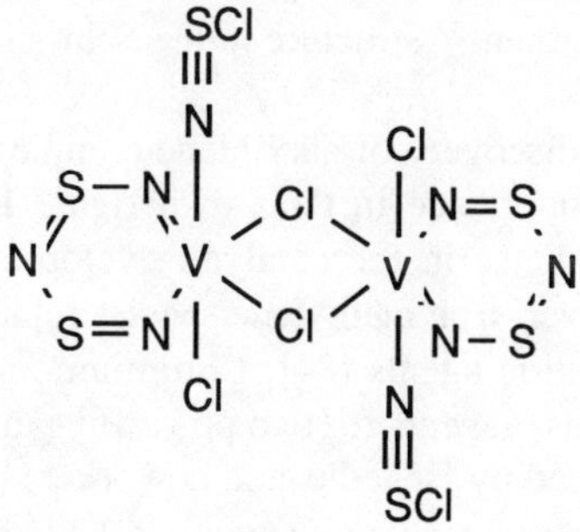

29 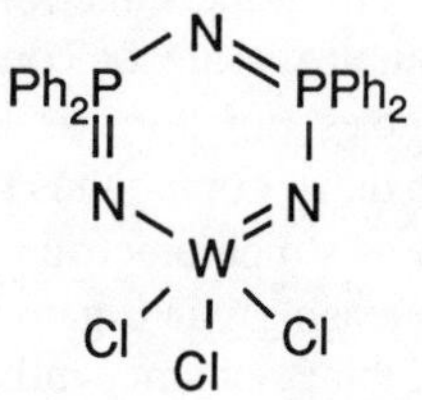 **30**

1.6 "FISCHER-TYPE" CARBENES

The first carbene complex was **31** reported by Fischer and Maasboel in 1964 [79]. Subsequently, dozens of tungsten, molybdenum, and chromium analogs of **31** have been prepared that incorporate a variety of heteroatom-containing substituents in place of the methoxy group. The carbene moiety in these compounds has a pronounced tendency to behave as a carbon electrophile—to the extent that some have referred to such species as "metal-stabilized carbonium ions." It has become evident that such electrophilic properties are shared by other carbene complexes, some of which do not contain an heteroatom substituent. Examples include **32** [80] and many other complexes that typically contain the metal in a low-valent ($>d^2$) oxidation state.

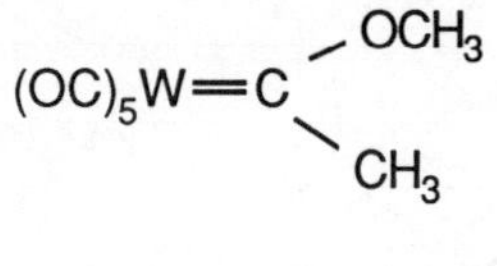

31 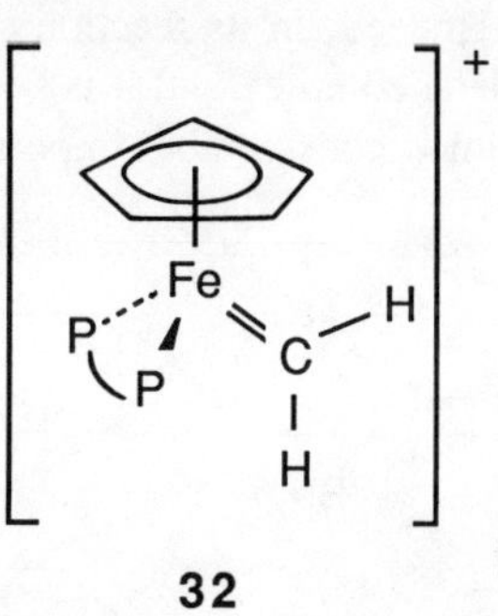 **32** P⌒P = DPPE

The bonding in Fischer-type carbenes is qualitatively different from that in the high-valent "Schrock" variety, which constitute one of the topics of this book. In the conventional model (Section 2.1.2), the bonding of Fischer carbenes is dominated by a single dative bond from a doubly occupied orbital on the carbene ligand. The system is stabilized by π back-bonding from filled metal d orbitals. (Some authors, including

Roper [81], have downplayed the differences between these two types of complexes. However, recent *ab initio* calculations (Section 2.3) indicate that the interplay of metal, substitutents, and ancillary ligands in these systems leads to significantly different bonding schemes: a coordinated "singlet carbene" fragment in the Fischer complexes in contrast to a covalently bound "triplet carbene" structure in the Schrock complexes [82].)

Besides setting the stage for Schrock's discovery of alkylidene complexes, Fischer carbene complexes are of considerable importance in their own right. This type of carbene species functions as key intermediates in the catalytic cyclopropanation of olefins with diazo compounds [83]. Moreover, iron methylene species related to **32** are finding use as stoichiometric cyclopropanation agents [84]. Chromium analogs of **31** are used increasingly in organic chemistry as reagents for two powerful synthetic transformations: the β-lactam synthesis developed by Hegedus and coworkers [85] and the Dötz naphthoquinone synthesis [86]. A comprehensive review is available for the interested reader [87].

1.7 "SCHROCK-TYPE" ALKYLIDENE LIGANDS

In 1973, in a single laboratory of DuPont's Central Research Department, the field of transition metal alkylidene chemistry was born. On the left side of the partition, Dick Schrock, in the course of an attempted synthesis of pentakisneopentyl tantalum (V), instead isolated complex **33**, a tantalum neopentylidene derivative. It is evident from Figure 1.2 that Schrock quickly recognized the true identity of **33**. Schrock has written an entertaining account of this discovery and the truly prolific research effort that ensued at DuPont and subsequently at Massachusetts Institute of Technology [88]. To the right of the partition, Fred Tebbe was conducting fundamental studies on Ziegler–Natta olefin polymerization. He found that treatment of Cp_2TiCl_2 with trimethylaluminum afforded **34**, a "masked" form of the titanium methylene complex $Cp_2Ti=CH_2$ [89]. Indeed, in its reaction chemistry this adduct behaves as though it were a free methylene complex and it is because of this that **34**, "the Tebbe–Grubbs reagent," has assumed considerable importance in synthetic organic chemistry [90].

33

34

To date, free (uncomplexed) high-valent alkylidene compounds have been isolated and characterized for six of the early transition metals: zirconium, niobium, tantalum, molybdenum, tungsten, and rhenium. In contrast to the Fischer carbenes, these alkylidene complexes behave chemically as carbon nucleophiles. The aspects of electronic structure underlying this dichotomy will be discussed in Sections 2.1.2 and 2.3. One

E. I. DU PONT DE NEMOURS & CO. (INC.)

106 Page No. TITLE LiP + TaP₃Cl₂ DATE 7-27-73

E 1338 Book No. PURPOSE Prep of TaP₅ ?

P = neopentyl

(a) .5g TaP₃Cl₂ and .17g LiP in 4ml pentane stood wrapped in foil for 24 hr. Filtered to give a small amount of white solid which gave a very strong Cl⁻ test @ AgNO₃. Orange filtrate stood again at room temp. wrapped in foil for 24 hr. but no additional solid formed. Stripped to yield an orange solid.

Nmr series 238 shows CH₃'s close to those in Li neopentyl (238(c)). Addition of Li neopentyl gave similar spectrum implying exchange. Only one different peak found in second case, others relatively unaltered.

* Note — Nmr #238(c)' shows exchange prob. not likely

(b) Repeat @ 1.0g TaP₃Cl₂ & .34g LiP standing 24 hr.
  → 2.15 mmoles

Filtration gave .19g LiCl (4.48 mmoles).

Stripping gave .85g orange solid (106(b)-1). Combined @ above product and sublimation attempted at 80°C, .5μ.

Solid melted + sublimed to yield .3g orange sublimate @ nmr #238(d) - unchanged.

Sent down for mass spec. A sample sent to catalyst screen.

Mass spec shows 464 in the main. Could be

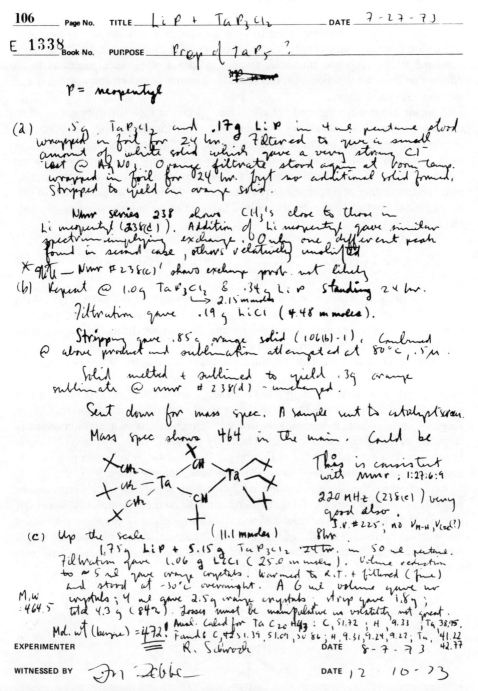

This is consistent with nmr; 1:27:6:9

220 MHz (238(c)) very good also.
J.V. #225; no ν_{M-H}, ν_{C=d}?

(c) Up the scale (11.1 mmoles) 8 hr

1.75g LiP + 5.15g TaP₃Cl₂ 24 hr. in 50 ml pentane. Filtration gave 1.06g LiCl (25.0 mmoles). Volume reduction to ~5 ml gave orange crystals. Warmed to R.T. + filtered (fine) and stood at -30°C overnight. A 6 ml volume gave no crystals; 4 ml gave 2.5g orange crystals; strip gave 1.8g; total 4.3g (84%). Losses must be manipulative - volatility not great.

M.W. :464.5

Mol. wt (benzene) = 472! Anal. Calcd for TaC₂₀H₄₃: C, 51.72; H, 9.33; Ta, 38.95.
Found: C, 51.39, 51.09, 50.86; H, 9.31, 9.24, 9.22; Ta, 41.22, 42.77

EXPERIMENTER R. Schrock DATE 8-7-73

WITNESSED BY DATE 12 10-73

Figure 1.2 A page from the laboratory notebook of R.R. Schrock. His single exclamation point marks the birth of transition metal alkylidene chemistry. (Courtesy of E.I. du Pont de Nemours and Company. Used by permission.)

13

noteworthy feature of alkylidene reactivity is the tendency to undergo "Wittig-like" olefination reactions with organic carbonyl compounds (Section 6.2.4). Unlike conventional Wittig reagents, transition metal alkylidenes allow such reactions to be effected on the ester functionality; this provides a unique transformation of esters and lactones into vinyl ethers.

A major impetus for the study of alkylidene species is their involvement as intermediates in the olefin metathesis reaction [91]. This transformation, schematically exemplified by Eq.2, is practiced industrially using highly active heterogeneous catalysts. A variant of considerable commercial importance, ring-opening polymerization (Section 7.4.1), utilizes a homogeneous catalyst. Remarkable advances in the mechanistic understanding of ring-opening polymerization and in the rational design of catalysts for this reaction have recently evolved from the growing understanding of alkylidene chemistry.

$$2 \; RCH=CH_2 \rightleftharpoons CH_2=CH_2 + RCH=CHR \qquad (2)$$

An additional spur to research in this area was provided by the thought-provoking proposal of Green, Rooney, and coworkers [92] that Ziegler–Natta olefin polymerization may in some cases proceed via the intermediacy of alkylidene species (Section 7.4.3). Several subsequent mechanistic studies appear to favor the conventional "Cossee" mechanism for cobalt, lutetium, and titanium catalysts, although it might be argued that the metals involved are not those with the highest proclivity toward alkylidene formation. A tantalum catalyst has been suggested to effect olefin polymerization via an alkylidene complex [93]. Given the tremendous commercial importance of the Ziegler–Natta process, it seems likely that the "Green mechanism" will continue to stimulate added interest in alkylidene chemistry for some time. Moreover, there is growing evidence supporting the 1975 proposal [99] that the coordination polymerization of acetylenes may proceed by way of alkylidene intermediates.

Recently it has become evident that oxidation state is not a foolproof criterion for predicting carbene versus alkylidene bonding in transition metal complexes. Roper has shown that the low-valent compound **35** reacts with a variety of electrophilic reagents at the methylene carbon [94]. This was followed by theoretical studies that led Goddard to conclude that the electronic configuration of the metal center—not merely oxidation state—must be taken into consideration [82].

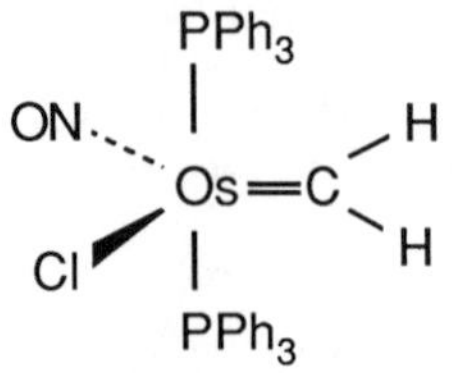

35

1.8 ALKYLIDYNE LIGANDS

The first carbyne complexes were compounds **36**, where M is chromium, molybdenum, or tungsten, reported by Fischer and coworkers in 1973 [95]. Since then, other examples of what can be regarded as low-valent carbynes have been prepared for manganese, rhenium, ruthenium, and osmium. In 1978 Schrock reported [96] a neutral d^0 alkylidyne complex, **37**, and examples of d^0 alkylidynes of molybdenum and tungsten have been reported subsequently. In addition, bridging alkylidyne complexes are known for the early transition metals niobium, tantalum, and tungsten.

36

37

Although there again seems to be a chemical basis for distinguishing between the low-valent carbynes on the one hand and the d^0 alkylidynes on the other, the distinction is less clear-cut than in the case of the carbene/alkylidene dichotomy. One complication arises in assigning the oxidation state of these species: The apparent oxidation state is altered by three units, depending on whether the alkylidyne moiety is regarded as a neutral ligand or a trianion. For example, **36** could in principle be regarded as a complex of either tungsten(I) or tungsten(IV). The difference becomes even hazier given the recent discovery of Mayr that Fischer-type carbynes can be converted into the Schrock variety by a simple bromine oxidation, thus providing a sort of "Northwest Passage" between these areas of research [97].

Studies on alkylidyne complexes have afforded the first well-defined catalysts for acetylene metathesis and suggest that known catalysts for this reaction may be converted to alkylidyne species under the reaction conditions [98].

A discovery of considerable potential importance is the reaction of tungsten–tungsten triple bonds such as that in **38** with acetylenes or nitriles, to afford alkylidyne complexes [100]. This chemistry, particularly if **38** becomes commercially available, promises to make tungsten alkylidynes among the most readily accessible of organometallics, surpassing even Grignard reagents in their ease of preparation. This bodes well for the use of such compounds in synthetic organic chemistry. In fact, Wittig-like reactions of compounds **39** with nitriles and with carbonyl derivatives have already been reported [101]. This synthetic approach has also allowed the preparation of novel heteronuclear bridging carbido complexes such as $(^tBuO)_3W{\equiv}C{-}Ru(CO)_2Cp$ [102].

$$(^tBuO)_3W\equiv W(O^tBu)_3 \quad \xrightarrow{RC\equiv CR} \quad (^tBuO)_3W\equiv CR$$

38   **39**

1.9 SUMMARY

Throughout the preceding discussion several recurrent themes can be discerned in the chemistry and applications of multiply bonded systems:

1. Multiply bonded ligands have a unique ability to stabilize early transition metals in their highest oxidation states.
2. The formation of a strong metal–ligand multiple bond can provide an important stabilizing effect on the intermediates of a catalytic cycle (or provide a powerful driving force for a stoichiometric reaction.)
3. The metal–ligand multiple bond nevertheless represents an unsaturation and is therefore a potential site for further chemical reactions.

All of these characteristics are a natural consequence of the presence of the d_π–p_π bond. Indeed, the molecular structure, the spectroscopic properties, and even the chemistry of these compounds are dominated by the effects of this bonding. Therefore in Chapter 2 we turn our attention to understanding the electronic structure of this class of compounds.

REFERENCES

1. Pope, M.T. "Heteropoly and Isopoly Oxometallates" Springer-Verlag, New York, 1983.
2. Seyferth, D., ed. "Transition Metal Carbene Complexes" Verlag Chemie, Weinheim, 1983.
3. Kim, H.P.; Angelici, R.J. *Adv. Organomet. Chem.* **1987**, *27*, 51.
4. Bradley, D.C.; Mehrotra, R.C.; Gaur, D.P. "Metal Alkoxides" Academic Press, London, 1978.
5. Lappert, M.F.; Power, P.P.; Sanger, R.R.; Srivastava, R.C. "Metal and Metalloid Amides" Wiley, New York, 1980.
6. Rice, D.R. *Coord. Chem. Rev.* **1978**, *25*, 199–227.
7. Sharpe, A.G. *Adv. Fluorine Chem.* **1960**, *1*, 29–67.
8. Lappert, M.F.; Patil, D.S.; Pedley, J.B. *J. Chem. Soc., Chem. Comm.* **1975**, 830–831, and especially the discussion thereof in reference 5.
9. Sharpless, K.B.; Flood, T.C. *J. Am. Chem. Soc.* **1971**, *93*, 2316–2318.
10. Pauling L. "The Nature of the Chemical Bond" 3rd edition, Cornell University Press, Ithaca, New York, 1960.
11. Palmer, K.J. *J. Am. Chem. Soc.* **1938**, *60*, 2360–2369.
12. Griffith, W.P. *Coord. Chem. Rev.* **1970**, *5*, 459–517.
13. Haase, W.; Hoppe, H. *Acta Cryst.* **1968**, *B24*, 282–283. Dwyer, P.N.; Puppe, L.; Buchler, J.W.; Scheidt, W.R. *Inorg. Chem.* **1975**, *14*, 1782–1785. Hiller, W.; Strähle, J.; Kobel, W.; Hanack, M. *Z. Kristallogr.* **1982**, *159*, 173.
14. Cainelli, G. "Chromium Oxidations in Organic Chemistry" Springer-Verlag, New York, 1984.
15. Fatiadi, A.J. *Synthesis* **1987**, 85–127.
16. Lee, D.G.; Van den Engh, M. in "Oxidation in Organic Chemistry, Part B" Trahanovsky, W.S., ed. Academic, New York, 1973.
17. Schröder, M. *Chem. Rev.* **1980**, *80*, 187–213.
18. Corey, E.J.; Suggs, J.W. *Tetrahedron Lett.* **1975**, 2647–2650. Dirand, J.; Ricard, L.; Weiss, R. *J. Chem. Soc., Dalton Trans.* **1976**, 278–282.
19. Sharpless, K.B.; Teranishi, A.Y.; Bäckvall, J.-E. *J. Am. Chem. Soc.* **1977**, *99*, 3120–3128.
20. Jørgensen, K.A.; Hoffmann, R. *J. Am. Chem. Soc.* **1986**, *108*, 1867–1876.
21. Miyaura, N.; Kochi, J.K. *J. Am. Chem. Soc.* **1983**, *105*, 2368–2378.
22. Moyer, B.A.; Thompson, M.S.; Meyer, T.J. *J. Am. Chem. Soc.* **1980**, *102*, 2310–2312.
23. Groves, J.T.; Quinn, R. *J. Am. Chem. Soc.* **1985**, *107*, 5790–5792.
24. Tokles, M.; Snyder, J.K. *Tetrahedron Lett.* **1986**, 3951–3954. Yamada, T.; Narasaka, K. *Chem. Lett.* **1986**, 131–134.
25. Poulos, T.L.; Finzel, B.C.; Gunsalus, I.C.; Wagner, G.C.; Krautt, J. *J. Biol. Chem.* **1985**, *260*, 16122–16130.
26. Ortiz de Montellano, P.R., ed. "Cytochrome P-450: Structure, Mechanism, and Biochemistry" Plenum, New York, 1986.
27. Holm, R.H.; Berg, J.M. *Acc. Chem. Res.* **1986**, *19*, 363–370.

28. Iwasawa, Y.; Nakamura, T.; Takamatsu, K.; Ogasawara, S. *J. Chem. Soc., Faraday Trans. 1* **1980**, *76*, 939–951. Trifiro, F.; Pasquon, I. *J. Catal.* **1968**, *12*, 412–416.
29. Rappé, A.K.; Goddard, III, W.A. *J. Am. Chem. Soc.* **1982**, *104*, 448–456.
30. Kaminow, I.V. "An Introduction to Electro-optic Devices" Academic Press, London, 1974.
31. Herrmann, W.A. *J. Organomet. Chem.* **1986**, *300*, 111–137.
32. Schrauzer, G.N.; Schlemper, E.O.; Liu, N.H.; Wang, Q.; Rubin, K.; Zhang, X.; Long, X.; Chin, C.S. *Organometallics* **1986**, *5*, 2452–2456.
33. Bokiy, N.G.; Gatilov, Yu.V.; Struchov, Yu.T.; Ustynyuk, N.A. *J. Organomet. Chem.* **1973**, *54*, 213–219.
34. Mayer, J.M.; Thorn, D.L.; Tulip, T.H. *J. Am. Chem. Soc.* **1985**, *107*, 7454–7462.
35. Klemperer, W.G. *Angew. Chem., Int. Ed. Engl.* **1978**, *17*, 246–254. Domaille, P.J. *J. Am. Chem. Soc.* **1984**, *106*, 7677–7687.
36. Matoba, Y.; Inoue, H.; Agaki, J.-I.; Okabayshi, T.; Ishii, Y.; Ogawa, M. *Synth. Comm.* **1984**, *14*, 865–873. Konishi, Y.; Sakata, K.; Misono, M.; Yoneda, Y. *J. Catal.* **1982**, *77*, 169–179. Zhizhina, E.G.; Kuznetsova, L.I.; Maksimovskaya, R.I.; Pavlova, S.N.; Matveev, K.I. *J. Mol. Catal.* **1986**, *38*, 345–353.
37. Rozenbaum, W.; Dormont, D.; Spire, B.; Vilmer, E.; Gentilini, M.; Griscelli, C.; Montagnier, L.; Barré-Sinoussi, F.; Chermann, J.C. *Lancet* **1985**, 450–451.
38. Fritzsche, J.; Struve, H. *J. Prakt. Chem.* **1847**, *41*, 97–113.
39. Werner, A.; Dinklage, K. *Chem. Ber.* **1901**, *34*, 2698–2703.
40. Krauss, F. *Z. Anorg. Chem.* **1921**, *119*, 217–220.
41. Cleare, M.J.; Griffith, W.P. *J. Chem. Soc., Chem. Comm.* **1968**, 1302.
42. Chatt, J.; Garforth, J.D.; Rowe, G.A. *Chem. Ind.* **1963**, 332.
43. Griffith, W.P. *Coord. Chem. Rev.* **1972**, *8*, 369–396.
44. Scherfise, K.D.; Dehnicke, K. *Z. Anorg. Allg. Chem.* **1986**, *538*, 119–122. Buchler, J.W.; Dreher, C. *Z. Naturforsch.* **1984**, *39B*, 222–230. Scheidt, W.R.; Summerville, D.A.; Cohen, I.A. *J. Am. Chem. Soc.* **1976**, *98*, 6623–6628.
45. Chatt, J.; Garforth, J.D.; Rowe, G.A. *J. Chem. Soc. A* **1966**, 1834–1836.
46. Belmonte, P.A.; Own, Z.-Y. *J. Am. Chem. Soc.* **1984**, *106*, 7493–7496.
47. Baldas, J.; Bonnyman, J.; Pojer, P.M.; Williams, G.A.; Mackay, M.F. *J. Chem. Soc., Dalton Trans.* **1981**, 1798–1801.
48. Abram, U.; Spies, H.; Goerner, W.; Kirmse, R.; Stach, J. *Inorg. Chim. Acta* **1985**, *109*, L9–L11.
49. Toth, L.E. "Transition Metal Carbides and Nitrides" Academic, New York, 1971.
50. Fowles, G.W.A. in "Developments in Inorganic Nitrogen Chemistry" Colburn, C.B., ed. Elsevier, Amsterdam, **1966**, pp. 522–576.
51. Wheeler, R.A.; Hoffmann, R.; Strähle, J. *J. Am. Chem. Soc.* **1986**, *108*, 5381–5387.
52. Doherty, N.M.; Critchlow, S.C. *J. Am. Chem. Soc.* **1987**, *109*, 7906–7908.

53. Clifford, A.F.; Kobayashi, C.S. Abstracts, 130th National Meeting of the American Chemical Society, Atlantic City, NJ, Sept. 1956, p.50R.
54. Chatt, J.; Rowe, G.A. *J. Chem. Soc.* **1962**, 4019-4033.
55. Chatt, J.; Dilworth, J.R. *J. Chem. Soc., Chem. Comm.* **1975**, 983-984.
56. Nugent, W.A.; Haymore, B.L. *Coord. Chem. Rev.* **1980**, *31*, 123-175.
57. McGlinchey, M.J.; Stone, F.G.A. *J. Chem. Soc., Chem. Comm.* **1970**, 1265. Ashley-Smith, J.; Green, M.; Stone, F.G.A. *J. Chem. Soc., Dalton Trans.* **1972**, 1805-1809.
58. Milas, N.A.; Iliopulos, M.I. *J. Am. Chem. Soc.* **1959**, *81*, 6089.
59. Patrick, D.W.; Truesdale, L.K.; Biller, S.A.; Sharpless, K.B. *J. Org. Chem.* **1978**, *43*, 2628-2638.
60. Herranz, E.; Sharpless, K.B. *J. Org. Chem.* **1978**, *43*, 2544-2548.
61. Chong, A.O.; Oshima, K.; Sharpless, K.B. *J. Am. Chem. Soc.* **1977**, *99*, 3420-3426. Hentges, S.G.; Sharpless, K.B.; Tulip, T.H. unpublished results.
62. Burrington, J.D.; Kartisek, C.T.; Grasselli, R.K. *J. Catal.* **1984**, *87*, 363-380.
63. Chan, D.M.-T.; Fultz, W.C.; Nugent, W.A.; Roe, D.C.; Tulip, T.H. *J. Am. Chem. Soc.* **1985**, *107*, 251-253.
64. Chan, D.M.-T.; Nugent, W.A. *Inorg. Chem.* **1985**, *24*, 1422-1424.
65. Chatt, J.; da Camera Pina, L.M.; Richards, R.L. eds. "New Trends in the Chemistry of Nitrogen Fixation" Academic Press, London, 1980.
66. Hines, R.N.; Prough, R.A. *J. Pharm. Exp. Ther.* **1980**, *214*, 80-86.
67. Schaverien, C.J.; Dewan, J.C.; Schrock, R.R. *J. Am. Chem. Soc.* **1986**, *108*, 2771-2773.
68. Hursthouse, M.B.; Motevalli, M.; Sullivan, A.C.; Wilkinson, G. *J. Chem. Soc., Chem. Comm.* **1986**, 1398-1399.
69. Chatt, J.; Heath, G.A.; Richards, R.L. *J. Chem. Soc., Chem. Comm.* **1972**, 1010-1011.
70. Hidai, M.; Mizobe, Y.; Sato, M.; Kodama, T.; Uchida, Y. *J. Am. Chem. Soc.* **1978**, *100*, 5740-5748.
71. Chatt, J.; Dilworth, J.R.; Richards, R.L. *Chem. Rev.* **1978**, *78*, 589-625. Hidai, M. in "Molybdenum Enzymes"; Spiro, T.G., ed. Wiley, New York, 1985. See also reference 65.
72. Rocklage, S.M.; Schrock, R.R. *J. Am. Chem. Soc.* **1982**, *104*, 3077-3081.
73. Murray, R.C.; Schrock, R.R. *J. Am. Chem. Soc.* **1985**, *107*, 4557-4558.
74. Powell, C.B.; Hall, M.B. *Inorg. Chem.* **1984**, *23*, 4619-4627. Rappé, A.K. *Inorg. Chem.* **1984**, *23*, 995-996.
75. Smegal, J.A.; Meier, I.K.; Schwartz, J. *J. Am. Chem. Soc.* **1986**, *108*, 1322-1323.
76. Dehnicke, K.; Strähle, J. *Angew. Chem., Int. Ed. Engl.* **1981**, *20*, 413-426.
77. Hanich, J.; Krestel, M.; Müller, U.; Dehnicke, K.; Rehder, D. *Z. Naturforsch.* **1984**, *39B*, 1686-1695.
78. Roesky, H.W.; Katti, K.V.; Seseke, U.; Witt, M.; Egert, E.; Herbst, R.; Sheldrick, G.M. *Angew. Chem., Int. Ed. Engl.* **1986**, *25*, 477-478.

79. Fischer, E.O.; Maasböl, A. *Angew. Chem., Int. Ed. Engl.* **1964**, *3*, 580-581.
80. Brookhart, M.; Tucker, J.R.; Flood, T.C.; Jensen, J. *J. Am. Chem. Soc.* **1980**, *102*, 1203-1205.
81. Gallop, M.A.; Roper, W.R. *Adv. Organometal. Chem.* **1986**, *25*, 121-198.
82. Carter, E.A.; Goddard, III, W.A. *J. Am. Chem. Soc.* **1986**, *108*, 4746-4754.
83. Doyle, M.P. *Chem. Rev.* **1986**, *86*, 919-939.
84. Brookhart, M.; Studabaker, W.B. *Chem. Rev.* **1987**, *87*, 411-432.
85. Hegedus, L.S.; McGuire, M.A.; Schultze, L.M.; Yijun, C.; Anderson, O.P. *J. Am. Chem. Soc.* **1984**, *106*, 2680-2687.
86. Dötz, K.H. *Pure Appl. Chem.* **1983**, *55*, 1689-1706.
87. Seyferth, D., ed. "Transition Metal Carbene Complexes" Verlag Chemie, Weinheim, 1983.
88. Schrock, R.R. *J. Organomet. Chem.* **1986**, *300*, 249-262.
89. Tebbe, F.N.; Parshall, G.W.; Reddy, G.S. *J. Am. Chem. Soc.* **1978**, *100*, 3611-3613.
90. Brown-Wensley, K.A.; Buchwald, S.L.; Cannizzo, L.; Clawson, L.; Ho, S.; Meinhardt, D.; Stille, J.R.; Straus, D.; Grubbs, R.H. *Pure Appl. Chem.* **1983**, *55*, 1733-1744.
91. Ivin, K.J. "Olefin Metathesis" Academic Press, London, 1983.
92. Ivin, K.J.; Rooney, J.J.; Stewart, C.D.; Green, M.L.H.; Mahtab, R. *J. Chem. Soc., Chem. Comm.* **1978**, 604-606.
93. Turner, H.W.; Schrock, R.R.; Fellman, J.D.; Holmes, S.J. *J. Am. Chem. Soc.* **1983**, *105*, 4942-4950.
94. Hill, A.F.; Roper, W.R.; Waters, J.M.; Wright, A.H. *J. Am. Chem. Soc.* **1983**, *105*, 5939-5940.
95. Fischer, E.O.; Kreis, G.; Kreiter, C.G.; Müller, J.; Huttner, G.; Lorenz, H. *Angew. Chem., Int. Ed. Engl.* **1973**, *12*, 564-565.
96. McLain, S.J.; Wood, C.D.; Messerle, L.W.; Schrock, R.R.; Hollander, F.J.; Youngs, W.J.; Churchill, M.R. *J. Am. Chem. Soc.* **1978**, *100*, 5962-5964.
97. Mayr, A.; McDermott, G.A. *J. Am. Chem. Soc.* **1986**, *108*, 548-549.
98. Sancho, J.; Schrock, R.R. *J. Mol. Catal.* **1982**, *15*, 75-79. Freudenberger, J.H.; Schrock, R.R.; Churchill, M.R.; Rheingold, A.L.; Ziller, J.W. *Organometallics* **1984**, *3*, 1563-1573.
99. Masuda, T.; Susaki, N.; Higashimura, T. *Macromolecules* **1975**, *8*, 717-721.
100. Schrock, R.R.; Listemann, M.L.; Sturgeoff, L.G. *J. Am. Chem. Soc.* **1982**, *104*, 4291-4293.
101. Wallace, K.C.; Dewan, J.C.; Schrock, R.R. *Organometallics* **1986**, *5*, 2162-2164.
102. Latesky, S.L.; Selegue, J.P. *J. Am. Chem. Soc.* **1987**, *109*, 4731-4733.

CHAPTER 2

ELECTRONIC STRUCTURE

Oxo, nitrido, imido, alkylidyne, and alkylidene ligands all form multiple bonds with metal centers, bonds involving interactions of both σ and π symmetry. The requirement of strong π-bonding distinguishes these from almost all other ligands and provides a focus for this chapter and a theme for much of the book. A basic knowledge of the electronic structure of metal–ligand multiple bonds and their complexes is critical to an understanding of the chemistry of these compounds. The structure and reactivity of a multiple bond are directly related to its electronic structure; furthermore, a multiple bond often dominates the electronic structure of a metal complex as a whole.

This chapter begins with a very qualitative discussion of the nature of metal–ligand multiple bonds and an overview of their periodic trends. Molecular orbital (ligand field) analyses are presented for complexes of these ligands in their most common coordination geometries. Recent quantitative studies employing higher levels of theory are then described, concluding with calculations dealing with particular compounds. Theoretical studies that focus on specific reactions of metal–ligand multiple bonds are deferred to Chapter 6.

2.1 THE NATURE OF THE METAL–LIGAND MULTIPLE BOND

Metal–ligand multiple bonds are usually considered to consist of a σ bond plus one or two π bonds. The π interactions involve overlap of metal d orbitals with p orbitals on the ligand: If the z axis is taken as coincident with the metal–ligand multiple bond, overlap occurs between the d_{xz} and p_x orbitals and/or between d_{yz} and p_y (Fig. 2.1). This simple picture will suffice for most of the following discussion, but it should be noted that more sophisticated views of these interactions have been developed over the last few years using *ab initio* calculations (Section 2.4).

The p orbitals of oxo, nitrido, and imido ligands are lower in energy than the metal d orbitals, due to the high electronegativity of oxygen and nitrogen. In an oxidation state formalism, these ligands are best described as closed-shell anions, N^{3-}, NR^{2-}, and O^{2-}. Alkylidene and alkylidyne ligands can also be considered as CR_2^{2-} and CR^{3-}, but this is a good approximation only for complexes of the more electropositive transition metals, as discussed below. Therefore the simpler case of metal–oxygen and metal–nitrogen multiple bonding is described first, followed by a comparison with metal–carbon multiple bonds. The section concludes with a discussion of the effect of substituents on the nature of the multiple bond.

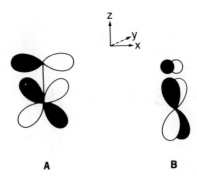

Figure 2.1 Drawings of the metal d_π–p_π interactions. (**A**) overlap of d_{xz} with p_x. (**B**) overlap of d_{yz} with p_y.

2.1.1 Metal–Oxygen and Metal–Nitrogen Multiple Bonds

The description of oxo, nitrido, and imido ligands as closed-shell anions implies that their p orbitals are filled. Productive π-bonding with a metal center therefore requires that the metal d orbitals be empty, in other words that the metal center be in a high oxidation state with a low d electron count. In fact the vast majority of complexes of these ligands have d^0, d^1, or d^2 electronic configurations [1–4] (Section 5.1). Only in the last few years have a few d^4 oxo compounds [5–12] and two d^5 species been prepared [12,13]. There are as yet no fully characterized transition metal imido or nitrido compounds with a greater than d^2 configuration, although d^6 and d^8 fluoroalkyl–imido complexes have been suggested [14].

In simple valence bond pictures, nitrido (and alkylidyne) ligands form triple bonds with a metal center while oxo and imido (and alkylidene) ligands form double bonds. However, examination of the orbitals involved shows that the N^{3-}, NR^{2-}, O^{2-}, and CR^{3-} ligands are isoelectronic, with one filled orbital of σ symmetry and two filled p orbitals of π symmetry perpendicular to the metal ligand axis (Fig. 2.1). Thus all four ligands have the capability to form triple bonds [1–4].

In the case of imido ligands, the valence bond description suggests that the metal–nitrogen bond order can be inferred from the position of the substituent, at least to a first approximation. A linear M—N—R unit implies that the nitrogen is sp hybridized and that there is a metal–nitrogen triple bond (**A**, Eq. 1), while substantial bending of the M—N—R linkage (**B**) indicates the presence of a lone pair on the nitrogen and is usually taken as evidence for a reduced bond order [4]. (The Lewis-dot structures in this chapter, such as **A** and **B**, are drawn without formal charges or dative bonds. More detailed and more accurate descriptions of metal–ligand multiple bonding are presented in Sections 2.3.1 and 2.4.)

$$M\equiv N\text{—}R \qquad M=\!\!N\!\!:\!\!\diagup\!\!R \tag{1}$$
$$\quad\ \ \textbf{A} \qquad\qquad\ \textbf{B}$$

In these simple terms, a bent structure (**B**) is expected only when the metal center cannot form a bond with the nitrogen lone pair. This occurs when a linear, triply bonded

NR ligand would cause the electron count of the complex to exceed 18 electrons (the effective atomic number rule) [4]. [A triply bonded imido or oxo ligand is a six-electron donor when counted as a dianion (NR^{2-}, O^{2-}) or a four-electron donor when counted as a neutral ligand (NR, O).] In an orbital description, structure **B** should be observed when there is only one metal orbital of π symmetry available for bonding to the nitrogen, either because of competition with another π-bonding ligand or because the metal d orbitals are filled.

There is only one clear example of a bent imido ligand, **B**: The structure of $Mo(NPh)_2(S_2CNEt_2)_2$, **1**, shows both bent and roughly linear phenyl imido ligands. The bent imido group has a small Mo—N—C angle of 139.4(4)° and a long molybdenum–nitrogen bond of 1.789(4) Å, compared to the other imido ligand, <Mo—N—C = 169.4(4)°, Mo—N = 1.754(4) Å [15]. This molecule is described as having one metal–nitrogen triple bond and one M—N double bond, using the three d orbitals of π symmetry on the molybdenum center.

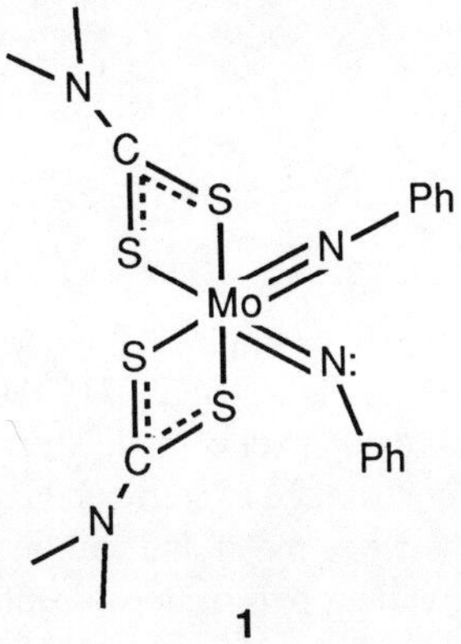

1

Other examples of bisimido and imido–oxo complexes, including three species isoelectronic with **1** [16], have M—N—R angles > 153° (Section 5.3.2). These complexes have more delocalized bonding with M—N bond orders between two and three, which apparently favors the linear structure [17]. Thus bent imido ligands are less common than the simple bonding picture above would predict. In fact all monoimido complexes exhibit M—N—R angles > 155° (Section 5.3.2).

Complexes with terminal nitrido ligands are believed to contain metal–nitrogen triple bonds, or at least to be best described by that resonance structure, on the basis of crystallographic and spectroscopic data [2,3,18]. Metal–oxo compounds, however, appear to have bond orders from three to possibly as low as one (Eq.2).

$$M\equiv\ddot{O}: \longleftrightarrow M=\ddot{\ddot{O}}: \longleftrightarrow M-\ddot{\ddot{O}}: \qquad (2)$$

Thus, for instance, studies of OsO_3N^- indicate that the osmium–oxygen bond order is low and that the nitrido ligand "dominates" the π-bonding [18]. The ability of oxygen to tolerate three lone pairs is due to its high electronegativity. Monooxo complexes usually contain metal–oxygen triple bonds, but di-, tri-, and tetraoxo compounds have lower bond orders due to competition for the metal π orbitals (Section 2.3).

2.1.2 Metal–Carbon Multiple Bonds

The bonding of an alkylidyne ligand to a metal center is very similar to the bonding of the oxygen and nitrogen ligands discussed above. The principal differences are due to the much lower electronegativity of carbon compared to oxygen and nitrogen. Alkylidyne ligands have been described as CR^{3-}, CR neutral, and CR^+, so there is often an ambiguity in assigning oxidation states in alkylidyne compounds. In this volume alkylidyne ligands are taken as closed-shell anions, CR^{3-}, because this emphasizes the similarities with imido, nitrido, and oxo ligands; this description is most accurate in high-oxidation-state complexes of electropositive metals such as tantalum or tungsten where our attention will be concentrated. As CR^{3-}, the alkylidyne ligand is a better π-donor than N^{3-}, NR^{2-}, and O^{2-} (see below); when taken as CR^+ it is a better π-acceptor than CO [19].

The low electronegativity of carbon destabilizes structures with a lone pair on the carbon, such as **D** (Eq. 3), so that alkylidyne ligands always form metal–carbon triple bonds.

$$M\equiv C-R \quad \longleftrightarrow\!\!\!\!\!\times\!\!\!\!\!\longrightarrow \quad M=C\!\!\overset{R}{\underset{..}{\diagup}} \tag{3}$$
$$\quad\textbf{C} \qquad\qquad\qquad\qquad \textbf{D}$$

Unlike imido ligands, there are no examples of even moderately bent alkylidyne groups: M—C—R angles are found in the range 171–180° (Section 5.3.2). (Structure **D** has, however, been suggested as an excited-state structure [20].) Slight bending of the M—C—R linkage has been attributed to steric and crystal packing influences [21] and to asymmetric π-bonding between metal and carbon [22]. Only one compound is known with both an alkylidyne and an oxo, imido, or nitrido ligand, complex **2** [23]. The imido ligand in **2** was proposed to be bent to avoid competition between the two multiply bonded ligands for a metal π-symmetry orbital; a crystal structure would be of interest.

2

The alkylidene ligand is different from the ligands discussed above because it is a single-faced π-donor, and can form at most a double bond. There are many similarities with alkylidyne ligands, however, beginning with the oxidation-state ambiguity: Alkylidene ligands are best described as CR_2^{2-} only when they lack heteroatom substituents and are bound to high-oxidation-state metal centers ("Schrock-type" alkylidene ligands) [24]. CR_2^{2-} is isoelectronic with amide (NR_2^-) [25], phosphide (PR_2^-) [26], and bent alkoxide (OR^-) groups [27]. The CR_2 group is also often treated as a neutral ligand (a carbene), in particular when the carbon bears a substituent with lone pairs such as an alkoxy or an amino group ("Fischer carbene" complexes) [28]. Using simple pictures, the bonding in Schrock and Fischer alkylidene ligands is similar, except that Schrock alkylidenes tend to be more electron rich and nucleophilic due to the diffuse, high-energy orbitals of the early transition metal [28,29]. However recent calculations using higher levels of theory (discussed in Section 2.4.1) suggest that Schrock alkylidenes form ethylene-like covalent double bonds, but Fischer carbenes bond to a metal center via donor–acceptor interactions.

NMR and structural studies (Chapters 4 and 5), as well as calculations, indicate that alkylidene ligands without heteroatom substituents form metal–carbon double bonds with substantial metal–carbon π-bonding. For instance, M=C bond lengths are shorter than single-bond distances and high barriers to rotation about the M—C bond axis are observed frequently. The importance of these metal–carbon π interactions is also evident in compounds with a second π-donor ligand: alkylidene–oxo,–imido, and –alkylidyne complexes adopt geometries in which the two ligands do not interact with the same metal π-symmetry orbital [30].

2.1.3 The Importance of Substituents

The substituent(s) in alkylidene, alkylidyne, or imido ligands can be very important to the chemistry of the metal–ligand multiple bond. This is particularly the case when the substituent has an accessible π-symmetry orbital, since this orbital is usually in conjugation with the metal–ligand π orbitals. The effects of substituents have been studied in most detail for Fischer carbene complexes, in which bonding between a substituent lone pair and the carbene carbon atom plays a major role (Eq. 4) [28]. As indicated by the resonance structures below, this bonding significantly reduces the metal–carbon bond order and the electrophilic character of the alkylidene carbon (see also Section 2.4).

$$M{=}C\diagup^{R}_{\diagdown OR} \longleftrightarrow M{-}C\diagup^{R}_{\diagdown OR} \qquad (4)$$

Large substituent effects are apparent in the electronic spectra of metal alkylidene complexes, particularly in the energy of the metal-to-alkylidene charge transfer transition (essentially $n \rightarrow \pi\ ^*$) [31]. Alkylidene (carbene) complexes have been prepared with a wide variety of π-donating substituents, including alkoxide, amide, and thiolate groups, but few alkylidene ligands are known with substituents more π-accepting than a phenyl group [32].

The presence of a substituent lone pair also profoundly affects the properties of imido and alkylidyne ligands, to the extent that these are essentially different types of ligands. Amino– and alkoxy–alkylidyne ligands (CNR_2, COR), for instance, are often thought of as alkylated isonitrile or carbonyl ligands. A variety of structural types and resonance forms are possible for these ligands, as illustrated in Eq. 5 for hydrazido (NNR_2) ligands. Hydrazido and diazenido (NNR) ligands are of particular importance because of their suggested intermediacy in the reduction of dinitrogen by metal complexes and metalloenzymes (Sections 3.4 and 7.3).

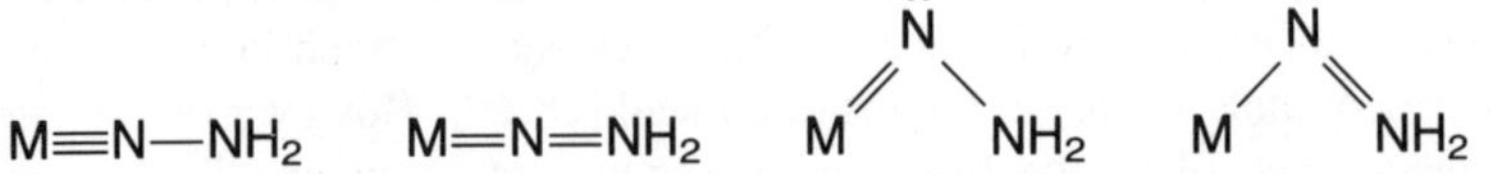

$$M\equiv N-NH_2 \quad M=N=NH_2 \quad M\overset{\ddot{N}}{\diagup}\diagdown NH_2 \quad M\diagup\overset{\ddot{N}}{\diagdown}NH_2 \quad (5)$$

Conjugation of a substituent lone pair with a metal–ligand triple bond splits the degeneracy of the π orbitals of the triple bond. Calculations on hydrazido [33] and amino–alkylidyne complexes [19,22,34] such as **3** and **4** indicate that the amino lone pair is very involved in the bonding and that there is significant multiple bond character to the N—N and C—N bonds. Therefore, these ligands are similar to vinylidene ligands (C=CR$_2$). The interaction with an empty substituent p orbital has been noted in studies of both phenylimido (Cp$_2$VNPh) [17] and benzylidyne complexes [Cr(CPh)(CO)$_4$Br] [19,22].

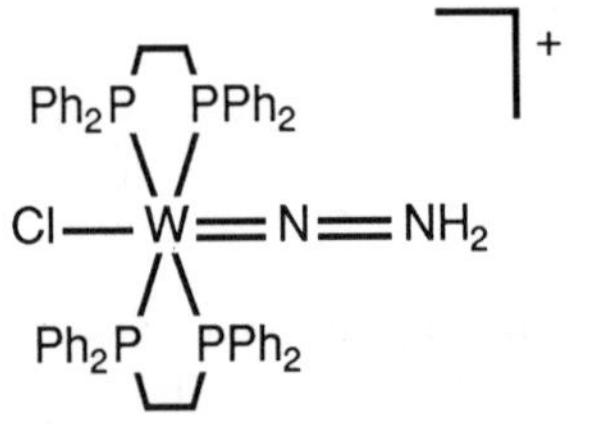

3 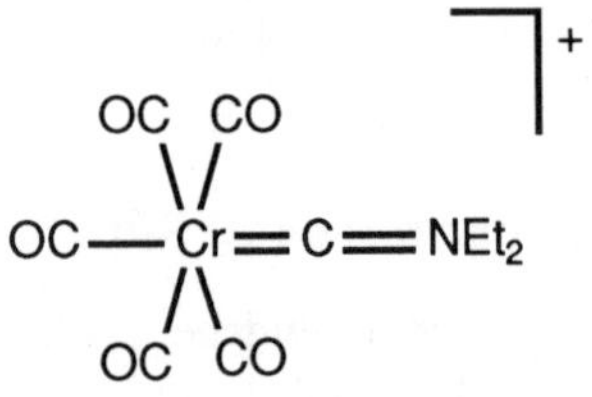   **4**

There is an interesting parallel between the π-donor ligands such as NR^{2-} and CR^{3-} that are the focus of this volume and π-acceptor ligands such as NO$^+$ or CO. The symmetry of the metal–ligand interactions (Fig. 2.1) is very similar in the two cases, the principal difference being whether the accessible ligand π orbitals are empty or filled. There is a clear progression from one extreme to the other on changing the nature of the substituent, for instance from CR to COR to CO. The π-bonding substituent changes both the energy and the shape of the ligand π orbitals, so that in N$_2$ and CO only the ligand π^* orbitals interact strongly with a metal center.

2.2 PERIODIC TRENDS

The properties and reactivity of the metal–ligand multiple bond vary over a wide range, depending on the nature of both the metal and the ligand. This section is a description of these periodic trends, with attempts to rationalize them on the basis of the bonding pictures presented above. Our discussion starts with the patterns among oxygen, nitrogen, and carbon ligands, followed by the trends across the transition series. The last section presents the limited data available on metal–ligand bond strengths.

2.2.1 Ligand Comparisons

Alkylidyne, nitrido, imido, and oxo groups, considered as closed-shell anions, are isoelectronic ligands and frequently form closely related compounds. As a general rule,

TABLE 2.1 Carbonyl Stretching Frequencies in Tungsten(IV) Compounds

Compound	$\nu(CO)$ (cm^{-1})	Reference
W(O)Cl$_2$(CO)(PMePh$_2$)$_2$	2006	35
W(S)Cl$_2$(CO)(PMePh$_2$)$_2$	1986	36
W(NCMe$_3$)Cl$_2$(CO)(PMePh$_2$)$_2$	1965	36
W(NTol)Cl$_2$(CO)(PMePh$_2$)$_2$	1966	36
W(CHTol)Cl$_2$(CO)(PMe$_3$)$_2$[a]	1938	37
W(CPh)Cl(CO)(py)(PMe$_3$)$_2$[a,b]	1870	37
W(CCMe$_3$)Cl(CO)$_2$(dppe)[c]	2000, 1926	38
W(CPh)Cl(CO)$_2$(dppe)[c]	2003, 1937	38

The first four compounds are identical except for the multiply bonded ligand; the last four have additional differences, as indicated:
[a] Complex contains PMe$_3$ instead of PMePh$_2$.
[b] Complex contains a pyridine in place of one of the chloride ligands.
[c] Complex contains *two* CO ligands trans to phosphorus, not chlorine.

oxo groups seem to form the most polar bonds and seem to be the least electron-donating of the ligands, while metal–carbon bonds are the most covalent and the most electron-donating (CR^{3-} is a very strong donor). This is consistent with the electronegativity of the elements.

The trend is illustrated by the CO-stretching frequencies in carbonyl complexes containing these ligands, because the electron density at the metal will be reflected in the amount of backbonding to the CO ligand(s). The CO-stretching frequencies of a number of tungsten carbonyl complexes are given in Table 2.1. The oxo, imido, and sulfido complexes listed are identical except for the multiply bonded ligand, but the alkylidene and alkylidyne complexes have different structures and/or supporting ligands (as indicated) so that the comparison is not exact. The trend, however, is clear: the CO-stretching frequencies fall in the order oxo > sulfido > imido > alkylidene > alkylidyne, indicating that the electron density at the tungsten center varies in the inverse order. This trend is also illustrated by the variety of carbonyl complexes of these ligands since the π-acidic CO ligand binds best to electron-rich metal centers. Only a few oxo, imido, or nitrido complexes with carbonyl ligands are known and all are monocarbonyl adducts [35,36,39,40], while alkylidene and alkylidyne ligands are found in a variety of complexes containing carbonyl ligands [28,41], including tetra- and pentacarbonyl compounds such as W(CR)(CO)$_4$X and W(CPh$_2$)(CO)$_5$ [21,42].

The relative strength of metal–ligand π-bonding can be probed using complexes with more than one π-bonding ligand, because there is competition for a limited number of π-symmetry metal orbitals. Studies of OsO$_3$N$^-$ and ReO$_3$N^{2-} indicate that nitrido ligands are more effective π donors than oxo groups [2], on the basis of optical spectra [18] and comparisons of stretching force constants with the tetraoxo derivatives (Section 4.4 and [43]). The strong preference of nitrido and alkylidyne ligands for the formation of triple bonds is illustrated by the lack of dinitrido or bisalkylidyne compounds, in which the ligands would have to share a metal π orbital. (There are examples, however, of bis(amino–alkylidyne) complexes [44].) A recent attempted synthesis of a bisalkylidyne derivative yielded instead an acetylene complex, presumably formed by coupling of the two alkylidyne ligands [45].

Imido ligands appear to form as strong or stronger π bonds than oxo groups, based on the relative bond distances in a few structures. In $MoO(NH)Cl_2(OPPh_2Et)_2$ the molybdenum–nitrogen distance is 0.04(1) Å longer than the Mo—O [46], consistent with equal π-bonding given the 0.03 Å larger radius of multiply bonded nitrogen [47] (see Fig. 5.2). However, shorter metal–nitrogen than metal–oxygen bonds are observed in two of the three osmium oxo–imido structures (OsO_3(N-adamantyl), $OsO_2(N^tBu)_2$ [48], $\{OsO_3(NCMe_2CH_2CMe_3)\}_2$(DABCO) [49]). It should be noted that the *strength* of multiple bonding to oxygen versus nitrogen versus carbon ligands is also a function of the metal, as discussed in Section 2.2.3.

2.2.2 Ligand Reactivity

The reactivity of multiply bonded ligands ranges from electrophilic to nucleophilic, depending on the ligand, the metal, and the metal oxidation state and ancillary ligands. The most electrophilic complexes are observed for transition metals on the left side of the periodic table in their highest oxidation states; low-valent complexes of later transition metals with π-acid ancillary ligands usually exhibit nucleophilic reactivity. Carbon ligands seem to exhibit the widest range of reactivity, while oxo ligands are not usually very reactive.

The reactivity of multiply bonded ligands has been discussed in terms of both frontier orbital control [19,22,50–53] and charge control [29,53], with the consensus moving toward the former in recent years (see Chapter 6). Studies by Fenske and others have shown that nucleophilic attack at alkylidene and alkylidyne ligands is frontier orbital controlled, occurring at the M—C π^* LUMO [19,22,51,52]. In $(CO)_5Cr{=}C(OR)R'$, for instance, the alkylidene carbon is not the most positively charged atom in the molecule (in fact *ab initio* calculations suggest it carries a net negative charge [51]), but the LUMO is to a great extent localized on this carbon [51,53]. On the other hand the protonation of $CpMo({\equiv}CCH_2{}^tBu)[P(OMe)_3]_2$ apparently at the alkylidyne carbon has been described as a charge-controlled reaction [54].

The nature of the frontier orbitals (and the ligand charge) vary significantly with the metal and the ligand [29,53]. At one extreme, the ligand π orbitals are lower in energy than the metal d orbitals and the ligand acts as a closed-shell anion and as a π donor to the metal center (**A** in Fig. 2.2). The π-bonding molecular orbitals resemble ligand orbitals while the π^* orbitals are principally metal d; the ligand reacts as a nucleophile. The other extreme case (**C**) considers the ligand orbitals empty and higher in energy than the metal d orbitals, for instance, considering an alkylidyne as CR^+. The bonding MOs are principally metal in character and the empty π^* levels are based largely on the

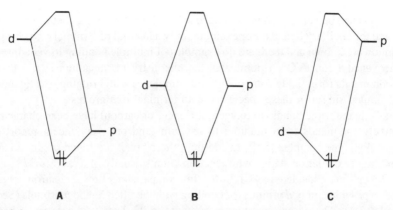

Figure 2.2 Changes in metal d–ligand p π interactions as the relative energies of the atomic orbitals are changed. (Adapted from ref. 53 with permission of Elsevier Sequoia.)

ligand, giving rise to electrophilic reactivity. There is a continuum between the two extremes, which is partially traversed on moving from oxygen to nitrogen to carbon ligands, and on moving from early to later transition metals. The compounds of multiply bonded ligands that are the focus of this volume generally fall in the range between **A** and **B**; **C** is a good description of bonding involving π-acid ligands. It should be noted that in addition to these arguments based on orbital energies, metal–ligand bonding and reactivity are also significantly influenced by the size of the metal and ligand orbitals and the magnitude of the metal–ligand overlap [29,55].

A few examples serve to illustrate these trends. d^0 Compounds of tantalum such as $Cp_2Ta(=CHR)R'$ react with electrophiles at the alkylidene ligand, both because the ligand carries a significant negative charge and because the HOMO is a π-bonding orbital primarily on the ligand [29]. On the other hand alkylidene and alkylidyne complexes of low-valent later metals, such as $(CO)_5Cr=CXY$, $X(CO)_4Cr\equiv CR$, and $[Cp(CO)_2Mn\equiv CR]^+$, usually are reactive with nucleophiles at the carbon due to the significant ligand character of the π^* LUMO [19,22,28]. The possibility of nucleophilic attack at nitrogen and oxygen ligands is a subject of current interest [3,50,56] (Chapter 6).

2.2.3 The Importance of the Metal

The nature of the metal center has a substantial influence not only on the reactivity of the ligand, but also on the preference for or against metal–ligand multiple bonding. The best way to quantify this preference would be a comparison of multiple bond strengths, but only a very limited amount of thermodynamic data is available (Section 2.2.4). The periodic trends are most striking in the preference for terminal, multiply bonded ligands versus bridged structures with single bonds. In carbon chemistry two single bonds are almost always favored over a double bond in the absence of steric effects: For instance, olefin polymerization is a very exothermic process and ketones are only a few kilocalories per mole more stable than geminal diols. For metals this preferences is much more variable and can be very substantial; it depends on the nature of the metal, its oxidation state, and the ancillary ligands. For example, only a handful of terminal-oxo complexes of titanium are known and there are no examples of multiple bonding to vanadium(III), but the vanadyl ion (VO^{2+}) dominates the chemistry of vanadium(IV). All of the transition metals form oxide compounds and complexes with bridging oxo ligands, but only a limited subset of these species contain terminal oxo groups.

Metal–ligand triple bonds (to oxygen, nitrogen, or carbon) have been characterized definitively for metals from titanium to osmium (and one very recent report of an iridium alkylidyne complex [57]). In addition, these bonds are found predominantly in +4 and higher oxidation states, with electron counts usually of d^0, d^1, or d^2. (Alkylidyne ligands are considered as CR^{3-} for this comparison.) Low d electron counts are needed to provide empty d orbitals for bonding with the filled ligand p orbitals (Sections 2.1 and 2.3.1). The d^4 oxo complexes prepared in the last few years are either very reactive (Section 2.5.1) or adopt an usual structure to promote multiple bonding (Section 2.3.3). A few alkylidyne compounds such as **5** can be described as d^4 species with a CR^{3-} ligand or (perhaps better) as d^8 species with a CR^+ group [57,58]. In the latter view the alkylidyne ligand is a π acid that requires filled metal d orbitals so that the preference for high-oxidation states described above need not be followed. It is interesting that the alkylidyne ligand in **5** reacts with electrophilic reagents in contrast to most group VII and VIII alkylidyne complexes that are very reactive with nucleophiles [58]; a theoretical study of these molecules would certainly be of interest.

5

The formation of metal–ligand multiple bonds seems to be most favorable along a diagonal from vanadium to osmium (see Section 5.1 and Fig. 5.1). This is illustrated in the following brief survey, which emphasizes oxo complexes because they are the most common. Titanium, zirconium, and hafnium prefer oligomeric structures with bridging ligands to complexes with multiple bonds [59–62]: the solid dioxides MO_2, for instance, have lattice structures without short metal–oxygen distances [63]. Group IV complexes with multiple bonds are known only in the +4 (d^0) oxidation state, and have been fully characterized only for dicyclopentadienyl alkylidene compounds [64,65] and a few titanium–oxo complexes [66]. In group V, however, multiple bonds are ubiquitous in the higher oxidation states of vanadium (+4 and +5, d^0 and d^1), for example, the vanadyl (VO^{2+}) ion [62] and the short V—O multiple bond distance in V_2O_5 [63]. Niobium and tantalum seem to have a high affinity for multiple bonds to carbon [24], and they form both bridging and terminal oxo compounds [1,62,63,67,68]. $VOCl_3$ is a monomeric liquid with a terminal V—O bond while $NbOCl_3$ is a solid with bridging oxygens. In group VI, $MoOCl_3$ and $MoOCl_4$ are solids with bridging chlorides [62,63] and multiple bonds are favored in d^0, d^1, and d^2 electronic configurations. This progression can be continued to the chemistry of osmium(VIII), in which multiple bonding seems to be strongly preferred to bridge structures in the higher oxidation states [62, 68, 69]. OsO_4 is a monomeric, volatile material, and there are no osmium or rhenium analogs of the condensed bridged structures (isopolyanions) so common for tungsten and tantalum [68]. To the right of the osmium, bridged structures are again favored and there are no well characterized metal–ligand triple bonds.

While this survey focused on oxo compounds, other multiply bonded ligands follow a similar periodic trend. Alkylidene ligands deviate the most from the pattern because they bind to a metal center through only a single π interaction and therefore have less stringent requirements for the electronic structure of a complex than a triply bonded ligand. Still, alkylidene ligands without heteroatom substituents are frequently observed as terminal ligands for groups V–VII [24], while for other metals they are more commonly bridging ligands [70] (for group VIII terminal alkylidene complexes, see [58,71,72]). Bisneopentyl complexes of tantalum decompose to form an alkylidene species [24] but platinum and iridium analogs yield metallacycle complexes [73] (Eqs. 6 and 7).

$$2\ Cl_3Ta(CH_2CMe_3)_2 + 2\ PMe_3 \rightarrow [(PMe_3)Cl_3Ta(=CHCMe_3)]_2 + CMe_4 \quad (6)$$

$$(Et_3P)_2Pt(CH_2CMe_3)_2 \xrightarrow{\Delta} (Et_3P)_2Pt\begin{array}{c} CH_2 \\ \diagup \quad \diagdown \\ \diagdown \quad \diagup \\ CH_2 \end{array}CMe_2 + CMe_4 \quad (7)$$

The origin of the periodic trends is not well understood, although they are undoubtedly related to the changes in energy and extension of the metal d orbitals across the pericdic table. The very early transition metals have higher energy, diffuse d orbitals, and therefore form more ionic, less covalent bonds than the later metals. The more polar the bond, it could be argued, the greater the basicity of the ligand and the greater the tendency to bridge. To the right of the iron triad, the metal d orbitals become too contracted for good π-bonding and bridged structures are again favored [55]. Ab initio calculations suggest that exchange and promotion energies also play an important role [55].

It is interesting that the metals favoring metal–ligand multiple bonding also form strong metal–metal multiple bonds [74], suggesting that π-bonding is in general favored in these systems. The preference for or against multiple bonding in the transition series also parallels the chemistry of main group compounds: silicon and titanium prefer not to form multiple bonds, while pentavalent phosphorus and tetra- and hexavalent sulfur, like vanadium(V) and molybdenum(IV) and (VI), form strong bonds that can be described as multiple [68].

The nature of the metal influences not only the overall preference for multiple bonding, but also the preference for a particular type of ligand. For instance, while vanadium and tantalum imido complexes hydrolyze in the presence of even trace amounts of water to give oxo species, osmium imido complexes can be prepared by adding *t*-butylamine to an *aqueous solution* of OsO$_4$. While this appears to be a preference of the more electropositive, "harder" metal centers for more electronegative, "harder" ligands, more data are clearly needed.

2.2.4 Metal–Ligand Multiple Bond Strengths

The little thermodynamic data available concerning metal–ligand multiple bonds is almost all for simple gas-phase metal-oxide species, calculated from heats of formation [75–77]. RuO$_4$ and OsO$_4$ are the only species that are molecular in condensed phase whose average bond strength has been determined: 109 and 127 kcal/mole, respectively [75]. Stronger bonds are found in MoO$_3$ and WO$_3$ (141 and 150 kcal/mole [75]), which although not molecular in solution do form simple complexes with neutral ligands. Calorimetric methods have been used to determine the energy required to cleave the first molybdenum–oxo bond in Mo(O)$_2$(S$_2$CNEt$_2$)$_2$ (Eq. 8), which is a reactive metal–oxo group [78].

$$\text{Mo(O)}_2(\text{S}_2\text{CNR}_2)_2 \rightarrow \text{Mo(O)}(\text{S}_2\text{CNR}_2)_2 + [\text{O}] \quad \Delta H = 98 \text{ kcal/mole} \quad (8)$$

The average Mn—O bond strength in the permanganate ion has been estimated from Raman spectra to be 126 ± 25 kcal/mole [79].

Sanderson has recently calculated bond energies in more complex molecules using experimental heats of formation and an empirical procedure based on electronegativities [77]. The values obtained for gas-phase WOCl$_4$ and WO$_2$Cl$_2$ (which reproduce H_f within 2%) include tungsten–oxygen bond strengths of 195 kcal/mole. *Ab initio* GVB calculations (not compared to H_f) yield much lower bond energies for the chromium and molybdenum analogs: 82 (CrOCl$_4$), 102 (MoOCl$_4$), 51 (CrO$_2$Cl$_2$), and 79 kcal/mole (MoO$_2$Cl$_2$) [80]. These calculated values are not, however, comparable to experimental values: While the bond strengths are taken as the difference in energy between the molecule and the fragments formed on bond breaking (e.g., CrOCl$_4$ and O + CrCl$_4$), the geometry and therefore the energy of the metal-containing fragment (e.g., CrCl$_4$) are not optimized.

The bond strengths in most metal–oxo diatomic and metal–dioxo triatomic species have been determined [75], but these are not in general directly related to the chemistry of complexes in condensed phase because of differences in structure and electronic configuration. The bond strengths (in kcal/mole) generally decrease on moving to the right in the periodic table: from 192 (ZrO), 168 (TiO), and 167 (ZrO$_2$) to 124 (CrO) and 161 (WO), to 98 (FeO) and 127 (RuO), to 92 (MnO) and lower. The bond strengths given above indicate that M—O bonds can be very strong, often stronger than the double bonds in CO$_2$ (127) and O$_2$ (119) but weaker than the triple bond in CO (257) [81].

Very little information is available concerning the strength of metal–nitrogen or metal–carbon multiple bonds. Gas-phase ion–molecule reactions indicate that the M$^+$—CH$_2$ bond energies lie in the range 65–96 kcal/mole for first-row transition metals [82]. A few metal–alkylidene and –carbene bond strengths have been calculated by *ab initio* methods giving values in the range of 37–48 kcal/mole for first-row metals and 71–74 kcal/mole for second-row species [51, 55, 80]. Seyferth has estimated that the metal–methylene bond strengths in molybdenum and rhenium metathesis catalysts must be >90 kcal/mole based on the lack of formation of cyclopropanes [83].

2.3 LIGAND FIELD DESCRIPTIONS

The electronic structure of a metal–ligand multiple bond not only determines the reactivity of that bond, but also has a profound influence on the structure and reactivity of the complex as a whole. In this section we present simple molecular orbital diagrams—ligand field theory pictures—for octahedral and tetrahedral complexes with multiply bonded ligands. These descriptions are by no means the last word in the theoretical analysis of the compounds, but they are a necessary prerequisite to even a basic understanding of the chemistry of multiply bonded ligands.

2.3.1 Octahedral Complexes

The majority of complexes containing multiply bonded ligands are six-coordinate and adopt a geometry best described as octahedral. Octahedral complexes are also the easiest to analyze in molecular orbital terms, because the σ and π orbitals are separate due to the high symmetry. The discussion below can be applied to all compounds of octahedral geometry, although the arguments will be less directly applicable the larger

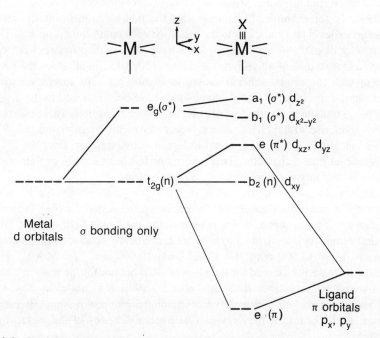

Figure 2.3 Partial molecular orbital diagram for an octahedral complex with one oxo, nitrido, (linear) imido, or alkylidene ligand.

the deviations from the high-symmetry structure, whether the deviations are due to distortions in bond angles and distances or to different bonding capabilities of the ligands. The basic conclusions seem also to apply well to square pyramidal structures (Section 2.3.2).

All octahedral complexes have essentially the same σ-bonding framework, regardless of π interactions. In a molecule with full O_h symmetry, the five metal d orbitals split into a degenerate e_g set ($d_{x^2-y^2}$, d_{z^2}) of σ^* character and a nonbonding t_{2g} set (d_{xy}, d_{xz}, d_{yz}). Introduction of a cylindrically symmetric π-bonding ligand (a ligand that can form a triple bond, O^{2-}, N^{3-}, CR^{3-}, or linear NR^{2-}) lowers the symmetry of the complex to C_{4v} and splits the degeneracies of both the e_g and t_{2g} orbitals (Fig. 2.3). Qualitatively the e_g set changes little since both orbitals remain σ^*, but the t_{2g} orbitals are substantially split because two of them are involved in π-bonding (d_{xz} and d_{yz} if the z axis is taken as coincident with the metal–ligand bond axis). Thus the ligand field portion of the MO diagram contains a nonbonding d_{xy} orbital, a π^* e set, and two σ^* levels. It should be noted that in this symmetry the two π-bonding orbitals are degenerate and the multiple bond must be considered a triple bond, unless the π^* orbitals are occupied. (This case is similar to the bonding in phosphine oxides, where the P—O bond should be described as a hybrid of single- and triple-bond resonance structures.)

Support for this ligand field description includes spectroscopic data, chemical reactivity, structural data, and theoretical calculations [33, 80, 84–99]. The spectroscopy of d^1 complexes provides the most readily interpretable data about the one-electron energy levels. Electron spin resonance studies of vanadyl and molybdenyl complexes (VOX_5^{3-} and $MoOX_5^{2-}$, for example, [84]) show that the odd electron primarily occupies a nondegenerate level perpendicular to the metal–oxygen bond, the d_{xy} orbital. The electronic spectra of oxo and nitrido complexes have been studied in detail [85–96] and there are a few studies of alkylidyne complexes [20, 100]; in all cases the results are consistent with the orbital scheme shown in Figure 2.3. The lowest energy bands observed in the spectra of d^1 and d^2 compounds have been assigned to the $d_{xy} \rightarrow d_{xz}$, d_{yz} ($b_2 \rightarrow e$) transition, in other words to an $n \rightarrow \pi^*$ transition. This absorption, for instance, gives rise to the characteristic blue color of the vanadyl ion [85–87]. The transition can also be described as metal-to-ligand charge transfer, since the π^* orbitals have significant ligand character. This description has been used to explain the photochemistry of alkylidyne complexes [20, 31, 100–102].

The energy of the $n \rightarrow \pi^*$ transition ($d_{xy} \rightarrow d_{xz}$, d_{yz}; $^2B_2 \rightarrow {}^2E$) for d^1 oxo–halide complexes is remarkably constant across and down the periodic table (Table 2.2). In contrast, $n \rightarrow \sigma^*$ transitions, which correspond to the ligand field splitting (10 Dq in octahedral symmetry), are known to increase dramatically on descending a group. For instance 10 Dq = 13,000 cm^{-1} for $CrCl_6^{3-}$ and 19,000 cm^{-1} for $MoCl_6^{3-}$ [103]. A substantial increase is observed for the $n \rightarrow \sigma^*$ (Cl) but not for the $n \rightarrow \pi^*$ transition in the oxo–halide complexes listed in Table 2.2 [90]. It is not clear why $n \rightarrow \pi^*$ transitions do not markedly increase on descending a group; this is an interesting area for future work because it may reveal fundamental features of the metal–ligand π interactions.

In octahedral oxo and nitrido complexes the $n \rightarrow \pi^*$ absorption band (and its coun-

TABLE 2.2 $^2B_2 \rightarrow \ ^2E$ Transition Energies in d^1 MOCl$_4$(L)$^{n-}$ Complexes

Complex	Energy (cm^{-1})	Reference
[V(O)Cl$_5$]$^{3-}$	15,500	87
[Cr(O)Cl$_4$]$^-$	13,400	88
[Cr(O)Cl$_5$]$^{2-}$	12,900	89
[Mo(O)Cl$_4$]$^-$	15,400	90, 91
[Mo(O)Cl$_5$]$^{2-}$	14,400	87
[Mo(O)Cl$_4$(H$_2$O)]$^-$	13,400	92
[W(O)Cl$_4$(H$_2$O)]$^-$	~14,700	93

terpart in emission) often exhibit a vibrational progression that is due to the metal–ligand multiple bond stretching mode. The energy of this stretch is usually substantially reduced in the excited state (from 1008 to 900 cm^{-1} for Mo(O)Cl$_4^-$ [90]) and the metal–ligand bond is significantly longer [90, 104–106], both of which indicate the strongly antibonding character of the π^* (d_{xz}, d_{yz}) orbitals. This provides a rationale for the observation that metal–ligand triple bonds are found predominantly in d^0, d^1, or d^2 electronic configurations: in d^3 or d^4 complexes electrons must occupy high-energy metal–ligand π^* orbitals. The excited states of oxo and alkylidyne complexes, in which these orbitals are probably populated, seem to be much more reactive than the ground states [20, 100–102, 104, 105, 107, 108]; the photochemistry of these compounds promises to be an exciting area [31]. The few compounds in which metal–ligand π^* levels are apparently populated in the ground state are very reactive compounds, as described in Section 2.5.1. Population of metal–ligand π^* orbitals thus seems to have a profound effect on the metal–ligand multiple bond. Calculations on porphyrin–metal–methylene complexes also indicate that population of metal–ligand π^* orbitals is strongly destabilizing [109].

The spectra and calculations indicate that the gap between the nonbonding d_{xy} and the π^* d_{xz}, d_{yz} orbitals is sizable. This is consistent with the observed diamagnetism of all octahedral d^2 complexes with multiply bonded ligands: The two electrons are paired in the d_{xy} orbital. d^2 Complexes with π-acid ligands, such as the carbonyl complexes given in Table 2.1, have a strong preference for the π acid to lie cis to the multiply bonded ligand, so that the filled d_{xy} is used for backbonding. For example, the carbonyl trans to the triple bond in (CO)$_5$W≡CR, which cannot backbond with this orbital, is readily replaced by a σ-donor ligand [110]. (Equivalently, if the alkylidyne ligand is considered to be CR$^+$, its very large π-acidity reduces the backbonding to the trans CO ligand.) Single-faced π-acid ligands such as olefins or acetylenes prefer to lie perpendicular to a metal–ligand triple bond to overlap with the d_{xy} orbital [35, 111–115].

Octahedral compounds with more than one multiply bonded ligand adopt geometries that maximize π-bonding and minimize the population of π-antibonding orbitals. Thus d^0 dioxo and diimido prefer cis geometries so that all three d_π orbitals can be used in bonding, and the net bond order to each ligand is 2.5 [1, 116]. Octahedral trioxo complexes are facial for the same reason, with double bonds between metal and oxygen to a first approximation. Compounds with both an alkylidene ligand and an alkylidyne, imido, or oxo group are all d^0 and have the alkylidene substituents aligned coplanar with the CR, NR, or O ligand so that the alkylidene ligand π bonds with the one metal d orbital not used in multiple bonding to the other ligand [30]. While d^0 dioxo complexes have a cis configuration, d^2 dioxo species adopt a trans structure so that the two d electrons can occupy a nonbonding orbital (d_{xy}) [1, 104, 105, 116]. The ligand field description of this case is very similar to the monooxo example given above, except that the d_{xz} and d_{yz} orbitals form π bonds to two ligands and the net bond order to each ligand is two.

Although these rules hold for the vast majority of compounds, there are a few exceptions. The uranyl ion UO_2^{2+}, despite its $d^0 f^0$ configuration, always adopts a trans geometry [62]. Theoretical studies indicate that the involvement of f orbitals in the bonding favors this stereochemistry [117], although one report emphasizes the role of nonvalence $6p$ orbitals [116]. While almost all d^2 dioxo compounds are trans, there are two well-characterized cis complexes: $[OsO_2(O_2CMe)_3]^-$ [118] and $[OsO_2bipy_2]^{2+}$ [119]. The bipyridine derivative has been shown to be thermodynamically unstable with respect to the trans isomer [119]. In both compounds the O—Os—O angle is substantially larger than 90° (125.2(3)° in $[OsO_2(O_2CMe)_3]^-$), which may be important in maximizing π-bonding and splitting the HOMO and LUMO to give the observed diamagnetic ground states (cf. theoretical studies of d^4 dicarbonyl complexes [120]).

2.3.2 Square Pyramidal Complexes

The molecular orbital scheme for octahedral complexes presented above is also applicable to square pyramidal species. The optical spectra of oxo and alkylidyne complexes are only slightly affected by the nature or even the presence of a ligand trans to the multiply bonded ligand (see Table 2.2, and [84,100]), indicating that the sixth ligand does not significantly alter the relative d_{xy} and π^* orbital energies. This is an indication that the trans ligand does not bind very strongly to the metal and is related to the trans effect and trans influence of multiply bonded ligands (Section 5.3.3.1).

The absence of a sixth ligand in square pyramidal complexes has been suggested to have one significant effect, that of lowering the energy of a σ^* component of the metal–ligand multiple bond (a_1 in Fig. 2.3). Calculations suggest that in Os(N)Cl$_4^-$ a σ^* orbital, composed of roughly equal amounts of osmium d_{z^2} and nitrogen p_z, is close in energy to the π^* orbitals [33] (although this is not the conclusion of spectroscopic studies [96]). This orbital was suggested to be involved in nucleophilic attack of triphenylphosphine at the nitrido ligand (Section 6.3.2, [121]).

2.3.3 Tetrahedral Complexes

2.3.3.1 Tetraoxo and Related Complexes. Theoretical studies of metal–ligand multiple bonding begin with the work of Wolfsberg and Helmholz in 1952 on the permanganate and chromate ions [122]. This classic paper also describes the invention

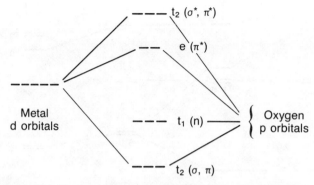

Figure 2.4 Partial molecular orbital diagram for tetrahedral MO_4^{n-}.

of a semiempirical molecular orbital treatment now known as the extended Hückel method. Although the derived orbital ordering was not correct, the authors conclude that "bonding between [the metal] atom orbitals and π-orbitals on [the ligand] is of no small importance. Thus in predicting configurations of complex molecules one probably has to consider such bonding in addition to the usual sigma bonding." A history of the study of the electronic structure of permanganate, starting from spectroscopic studies by Teltow in 1938 [123], has been given by Ballhausen and Gray [124]. Additional theoretical and spectroscopic studies of tetrahedral oxo, oxo–nitrido, and oxo–halo complexes can be found in [18, 125–130] and in an excellent review [131].

The d orbitals in tetrahedral complexes split into a "three above two" pattern (t_2 above e, Fig. 2.4). σ- and π-bonding are not distinct in this symmetry: The upper t_2 set forms bonds of both σ and π symmetry. This complicates the assignment of bond orders in tetrahedral complexes, since from two to five π bonds can be formed. Two strong π bonds are made by the lower e set of orbitals, since they are not involved in σ-bonding. In d^0 complexes such as the permanganate (MnO_4^-) and osmiamate (OsO_3N^-) ions, this π* e set is the LUMO; the intense color of permanganate is due to a charge-transfer transition from filled oxygen p orbitals to the e and t_2 orbitals.

The chemistry of d^0 tetrahedral complexes illustrates the preference described above to avoid populating π* orbitals. These compounds are not exceptional one-electron (outer sphere) oxidants, but they can be very reactive to inner-sphere processes accompanied by a change in structure and π-bonding. Reactions of osmium tetroxide with olefins, for instance, typically lead to octahedral *trans* dioxo-osmium(VI) compounds [132]. Only manganese, iron, and ruthenium form reduced tetraoxo compounds that are isolable, and these species are very reactive; the manganese–oxygen distances in d^1 MnO_4^{2-} are 0.04 Å longer than those in d^0 MnO_4^- [133].

2.3.3.2 d^4 Rhenium–Oxo Complexes.

A number of oxo complexes of rhenium(III), d^4 have been prepared recently [5, 6] that have an unusual tetrahedral structure (for instance, **6**). Tetrahedral complexes of multiply bonded ligands are commonly found only for d^0 species such as the tetraoxo compounds discussed above. However, a tetrahedral structure with one oxo or similar ligand can have two low-lying d orbitals (Fig. 2.5), and thus stable, diamagnetic d^4 complexes can be formed. In contrast, two

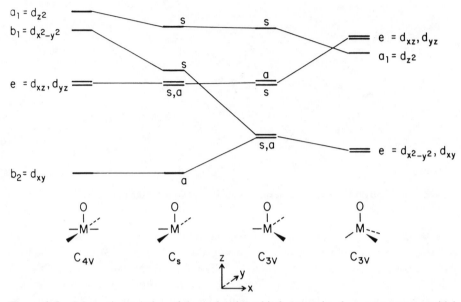

Figure 2.5 Schematic evolution of the molecular orbital energy levels as a square-pyramidal MOL$_4$ compound is converted to a tetrahedral MOL$_3$ compound. Energy levels are labeled s,a for orbitals symmetric or antisymmetric to the xz plane. (Reprinted with permission from J.M. Mayer et al., *J. Am. Chem. Soc.* **1985**, *107*, 7457. Copyright 1985 American Chemical Society.)

of the electrons in octahedral d^4 compounds must occupy high-energy metal–ligand π^* orbitals (Fig. 2.3, Sections 2.3.1 and 2.5.1). The rhenium(III) oxo-acetylene complexes appear to adopt tetrahedral structures to avoid placing electrons in metal–oxygen antibonding orbitals [5]. This preference not only determines the solid-state structures of these compounds but has also been used to explain their lack of fluxionality [5] and the stereochemistry of their ligand exchange reactions [6].

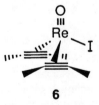

6

The critical difference between the two coordination geometries is the energy of the $d_{x^2-y^2}$ orbital, which in octahedral and square pyramidal complexes (C_{4v} symmetry) is σ-antibonding with respect to the ligands in the equatorial plane. This antibonding interaction is greatly reduced in a tetrahedral structure because there is effectively a threefold axis about the metal–oxygen bond and the overlap between $d_{x^2-y^2}$ and the equatorial ligands is therefore lower (Fig. 2.5). Similar *d* orbital splitting patterns have been calculated for tetrahedral nitrido complexes [99]. Thus complexes with d^4 electron counts seem to have a substantial preference for structures with pseudothreefold symmetry about the metal–oxygen axis, while d^2 compounds prefer octahedral or square pyramidal structures.

2.3.4 Other Coordination Geometries

The electronic structure of complexes of multiply bonded ligands in other coordination geometries have not been studied extensively (see, however, Section 2.5.2). A rough idea of the ligand field splitting can sometimes be constructed by adding ligand π orbitals as a perturbation to the σ level ordering that is well known for most regular geometries. Studies of π-backbonding in a given geometry indicate which metal orbitals are most likely to interact in a π fashion (see, for example, [20] and [134]).

2.4 AB INITIO CALCULATIONS

The ligand field analyses above have been presented above in terms of simple one-electron theory. In the last few years much more sophisticated *ab initio* calculations have been used to probe metal–ligand multiple bonds, a very difficult task because of the large number of electrons involved. The calculations provide important information on electron correlation energies, and hold the promise of accurately determining bond lengths and bond strengths. While simple models are more than adequate for many questions, it is clear that higher levels of theory will play an increasing role in studies of metal–ligand multiple bonding.

2.4.1 Alkylidene Complexes

Metal–alkylidene bonding has been the subject of a number of *ab initio* studies [51,55,80,135–137]. The calculations support earlier results [28,29] that the metal–carbon interaction consists of a σ and a π bond, but they can go a step further because they include electron correlation. Both Goddard [55,80] and Hall [135] have suggested that the M=CR$_2$ double bond can be either a covalent bond resembling ethylene, or a donor–acceptor bond (Fig. 2.6). The former is described as the interaction of a triplet metal center bonding to a triplet carbene fragment (**A** in Fig. 2.6), while the latter involves donation from a singlet carbene lone pair into an empty metal orbital and backbonding from a filled metal orbital into the vacant carbene p orbital (**B**). The donor–acceptor bonding of a singlet carbene to a metal center (**B**) is essentially the same as the well known Dewar–Chatt–Duncanson model for olefin binding to a metal center [138].

"Schrock-type" alkylidene compounds appear to have ethylene-like covalent bonds as in **A**, while "Fischer-type" carbene complexes resemble **B**. This pattern results from the preferences of both the metal center and the carbene fragment. Free carbenes with only hydrogen or alkyl substituents (CH$_2$, CHR, and CR$_2$), commonly found in Schrock alkylidenes, have triplet ground states and thus prefer binding as in **A**. On the other hand the carbene fragments commonly found in Fischer complexes contain sub-

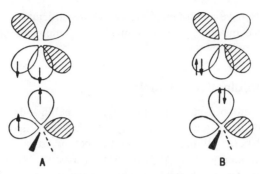

Figure 2.6 Schematic picture of two types of metal–alkylidene bonding: ethylene-like covalent bonding (**A**) and donor–acceptor bonding (**B**). (Adapted with permission from T.E. Taylor and M.B. Hall, *J. Am. Chem. Soc.* **1984**, *106*, 1576–1584. Copyright 1984 American Chemical Society.)

stituents with $p\pi$ lone pairs (X, OR, SR, NR$_2$) and have singlet ground states. Similarly, donor–acceptor bonding is favored by metal fragments with doubly occupied d orbitals, commonly low-valent, late-metal species. Compounds of the early transition metals in high oxidation states, with few d electrons, prefer ethylene-like covalent bonding as in **A** [55,135].

This description not only explains the dichotomy between Fischer and Schrock alkylidene complexes but also rationalizes their different reactivity. Complexes of singlet carbenes (**B**) are electrophilic because they have an empty orbital on the carbon, albeit stabilized by backbonding from the metal. Triplet carbene compounds (**A**) have two covalent bonds and the carbon center is nucleophilic, as in ethylene.

2.4.2 Oxo and Nitrido Complexes

Rappé and Goddard have also applied their GVB method to the study of oxo, nitrido, and hydrazido [80,97,139] complexes. Their conclusions basically agree with the ligand field theory arguments of Section 2.3, that in complexes with one oxo or nitrido group the metal–ligand bond is a triple bond while in dioxo compounds the bond order is significantly lower. The triple bonds, however, were suggested to be composed of two covalent π bonds and a donor–acceptor σ bond. The σ interaction involves overlap of an empty metal d orbital with a lone pair on the oxygen pointing directly at the metal. The symbol M≡O was coined to denote the triple bond. The metal–oxygen and metal–carbon bonds in M(O)$_2$Cl$_2$ and M(O)(CH$_2$)Cl$_2$ (M = Cr, Mo) were found to be covalent double bonds, the latter a typical Schrock alkylidene ligand.

The most important conclusion of these studies was that the conversion of a metal–oxygen double bond to a triple bond can provide a driving force for reactions occurring at a metal center. Thus reaction of an olefin with an oxo or alkylidene group in M(O)$_2$Cl$_2$ or M(O)(CH$_2$)Cl$_2$ to form a metallacycle (e.g., Eq. 9) was calculated to be thermodynamically favorable only because the bond to the "spectator" oxo group becomes much stronger as the bond order increases from two to three. This is not a subtle effect: The difference in bond strength between the double- and triple-bonded forms was calculated to be 20–30 kcal/mole and the reaction of $M(O)_2Cl_2$ with olefins was roughly 70 kcal/mole more favorable than the analogous reaction of $M(O)Cl_4$ (Eq. 9).

$$\begin{array}{c} Cl_{\prime\prime\prime\prime}\!\!\!\underset{Cl}{\overset{O}{\underset{\|}{M}}}\!\!=\!\!O + H_2C\!\!=\!\!CH_2 \longrightarrow \underset{Cl}{\overset{O}{\underset{\|}{\overset{Cl}{M}}}}\!\!\underset{H_2C-CH_2}{\overset{O}{\underset{|}{-O}}} \end{array} \qquad (9)$$

$$\underset{Cl}{\overset{Cl}{\underset{|}{Cl}}}\!\!\underset{Cl}{\overset{|}{M}}\!\!=\!\!O + H_2C\!\!=\!\!CH_2 \longrightarrow Cl_4M\!\!\underset{H_2C-CH_2}{\overset{O}{\underset{|}{-O}}}$$

The lower bond order and higher reactivity of dioxo versus monooxo complexes had been known for many years [1], but this was the first suggestion that a multiply bonded ligand acting only as a "spectator" could play a critical role in reaction chemistry.

2.5 THEORETICAL STUDIES OF SPECIFIC SYSTEMS

2.5.1 Metal–Oxo Complexes

An oxo–iron–porphyrin group is believed to be at the active site of cytochromes P-450, peroxidases, and other enzymes [140–144]. These enzymes exhibit remarkable reactivity, including the hydroxylation of alkanes and the epoxidation of olefins. This has prompted a great deal of study of the nature of the iron–oxo unit, especially the oxidation state of the iron and the electronic structure of the Fe—O bond. Two intermediates have been observed for horseradish peroxidase (HRP), one (HRP-I) apparently two oxidation states above the resting Fe(III) enzyme, the other (HRP-II) only one electron oxidized. HRP-II appears to be a porphyrin–iron(IV)–oxo species, while HRP-I is best described as an iron(IV)–oxo unit bound to a porphyrin radical cation. The as yet unobserved reactive intermediate in cytochrome P-450 is thought to resemble HRP-I, based at least in part on model systems that mimic much of the reactivity of the enzyme. (See Section 7.1.3)

The iron(IV)–oxo complex in the enzymes and models is a rare example of a d^4 oxo complex. Spectroscopic [141] and theoretical [93,104,144] studies indicate that it has a triplet ground state, with the unpaired electrons primarily occupying metal–oxygen π-antibonding orbitals, consistent with the molecular orbital scheme described above (Fig. 2.3). The presence of unpaired electrons in π^* levels gives the oxo ligand significant radical character and may be in part responsible for the high reactivity. It should be noted, however, that much of the same chemistry has been observed for oxo–manganese(V) species [145], which are d^2 and should therefore be diamagnetic (like nitrido–manganese(V) [146] and oxo–chromium(IV) compounds [147]).

Related d^4 ruthenium(IV)–oxo complexes, such as **7**, are also reactive oxidants; a number recently have been isolated and fully characterized [7–12]. The same bonding scheme, with two electrons in M—O π^* levels, has been proposed for these compounds since their magnetic moments ($\mu_{eff} = 2.9$ μ_B) are close to the spin-only value for two unpaired electrons (2.83 μ_B) [7,8,10]. Detailed magnetic studies on (bipy)$_2$(py)Ru(O)$^{2+}$ indicate that it has a singlet ground state with a low-lying triplet state [8].

7

2.5.2 Metal–Imido and Metal–Nitrido Complexes

The electronic structures of metal–imido and metal–nitrido compounds have not been extensively studied theoretically, in part because they are well described by the results obtained for oxo complexes. Studies concerning the conversion of N$_2$ to nitrido complexes [33,139], the electronic properties of nitrido-bridged polymers [99], electrophilic versus nucleophilic reactivity of imido ligands [53], and the attack of olefins at imido ligands [50] have been discussed above and/or are incorporated into the discussion of reactivity in Chapter 6.

The X-ray structures and the electronic structures of vanadocene imido complexes, Cp$_2$V(NR) (**8**), have been studied by Trogler and others [17,148,149]. This work is of interest not only for the details of the metal–nitrogen bond, but also because the bond is incorporated into an organometallic coordination environment that is not easily related to the octahedral or tetrahedral structures described above. (This situation is increasingly common since the organometallic chemistry of metal–ligand multiple bonds is growing very rapidly.) The frontier orbitals of a "bent-sandwich" fragment such as Cp$_2$V are well known [150], consisting of two orbitals of σ symmetry and one of π symmetry. In this simple picture only one vanadium–nitrogen π bond can be formed and Cp$_2$V(NR) should have a doubly bonded, bent imido ligand. However, a linear structure is observed [17,149]. Calculations [17] indicate that the imido ligand does form a triple bond to the vanadium, utilizing an orbital normally taken as bonding to the cyclopentadienyl ligands. The fragment approach is not successful because the interactions of the vandium with the imido ligand are at least as important as those with the cyclopentadienyl groups.

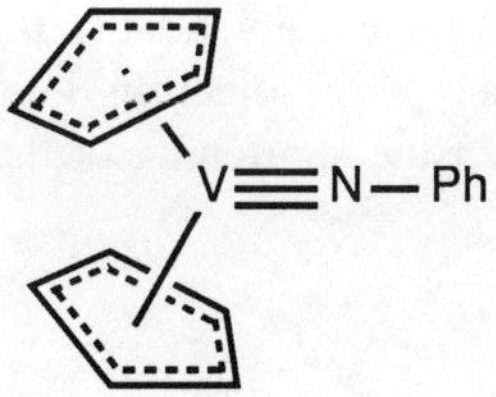

8

2.5.3 Alkylidyne and Alkylidene Complexes

The electronic structure of alkylidene complexes has been the subject of considerable study, in particular low-valent "Fischer-type" carbene complexes such as $(CO)_5Cr=CR(OR)$. An excellent summary of these results and a survey of metal–alkylidene binding in various coordination geometries have been recently given by P. Hoffmann [28] and will not be repeated here. Detailed analyses based on *ab initio* calculations have been presented in Section 2.4. The important studies of Fenske and others on the reactivity of alkylidene and alkylidyne ligands are described in Section 2.2.2.

"Schrock-type" alkylidene complexes of the form $L_nM=CHR$ often display a remarkable distortion about the alkylidene carbon, with large M—C—R angles, short M···H distances, and low C—H stretching frequencies and coupling constants (Sections 4.5, 4.9, and 5.3.2). The distortion has been explained theoretically in at least two ways [29,151], but there is a consensus that the electronic driving force derives from coordinative unsaturation at the metal. Hoffmann has modeled the distortion as a rotation of the alkylidene ligand (**A**, Eq. 10), which allows the alkylidene σ orbital to overlap with an additional empty metal *d* orbital [29]. Metal–hydrogen overlap and weakening of the C—H bond occur as a secondary interaction. In contrast, Hehre and coworkers view the geometrical distortions in terms of hyperconjugation (**B**), by comparison with carbocation systems (**C**) [151].

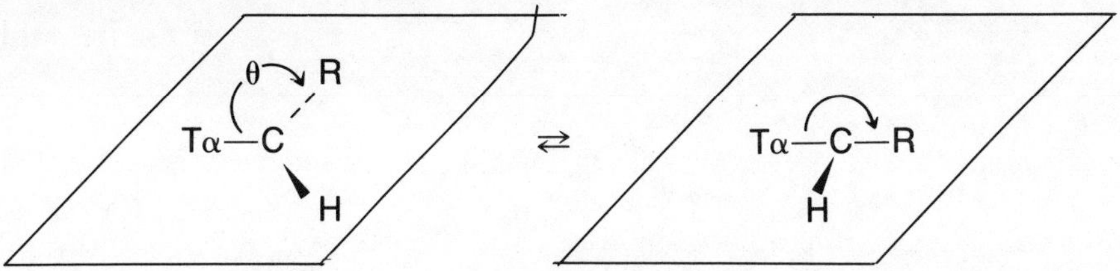

A (10)

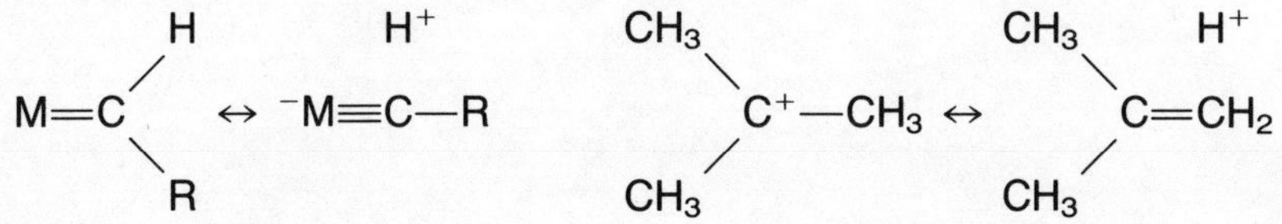

B

C

The hyperconjugation approach is closely related to Schrock's suggestion that the C—H bond can act as a donor to the electron-deficient metal center [24]; it also provides a simple explanation of the shortening of the metal–carbon bond on distortion.

REFERENCES

1. Griffith, W.P. *Coord. Chem. Rev.* **1970**, *5*, 459-517.
2. Griffith, W.P. *Coord. Chem. Rev.* **1972**, *8*, 369-396.
3. Dehnicke, K.; Strähle, J. *Angew. Chem., Int. Ed. Engl.* **1981**, *20*, 413-426.
4. Nugent, W.A.; Haymore, B.L. *Coord. Chem. Rev.* **1980**, *31*, 123-175.
5. Mayer, J.M.; Thorn, D.L.; Tulip, T.H. *J. Am. Chem. Soc.* **1985**, *107*, 7454-7462.
6. Mayer, J.M.; Tulip, T.H.; Calabrese, J.C.; Valencia, E. *J. Am. Chem. Soc.* **1987**, *109*, 157-163.
7. Moyer, B.A.; Meyer, T.J. *Inorg. Chem.* **1981**, *20*, 436-444. Takeuchi, K.J.; Samuels, G.J.; Gersten, S.W.; Gilbert, J.A.; Meyer, T.J. *Inorg. Chem.* **1983**, *22*, 1047-9. Takeuchi, K.J.; Thompson, M.S.; Pipes, D.W.; Meyer, T.J. *Inorg. Chem.* **1984**, *23*, 1845-1851.
8. Roeker, L.; Meyer, T.J. *J. Am. Chem. Soc.* **1986**, *108*, 4066-4073.
9. Marmion, M.E.; Takeuchi, K.J. *J. Am. Chem. Soc.* **1986**, *108*, 510-511.
10. Che, C.-M.; Tang, T.-W.; Poon, C.-K. *J. Chem. Soc., Chem. Comm.* **1984**, 641-642. Che, C.-M.; Wong, K.-Y.; Mak, T.C.W. *ibid.* **1985**, 546-548, 988. Che, C.-M.; Lai, T.-F.; Wong, K.-Y. *Inorg. Chem.* **1987**, *26*, 2289-2299.
11. Aoyagi, K.; Yukawa, Y.; Shimuzu, K.; Mukaida, M.; Takeuchi, T.; Kakihana, H. *Bull. Chem. Soc. Jpn.* **1986**, *59*, 1493-1499.
12. Pipes, D.W.; Meyer, T.J. *Inorg. Chem.* **1986**, *25*, 3256-3262.
13. Valencia, E.; Santarsiero, B.D.; Geib, S.J.; Rheingold, A.L.; Mayer, J.M. *J. Am. Chem. Soc.* **1987**, *109* 6896-6898.
14. McGlinchey, M.J.; Stone, F.G.A. *J. Chem. Soc., Chem. Comm.* **1970**, 1265. Ashley-Smith, J.; Green, M.; Mayne, N.; Stone, F.G.A. *ibid.* **1969**, 409. Ashley-Smith, J.; Green, M.; Stone, F.G.A. *J. Chem. Soc., Dalton Trans.* **1972**, 1805-1809.
15. Haymore, B.L.; Maatta, E.A.; Wentworth, R.A.D. *J. Am. Chem. Soc.* **1979**, *101*, 2063-2068.
16. Chan, D.M.-T.; Fultz, W.C.; Nugent, W.A.; Roe, D.C.; Tulip, T.H. *J. Am. Chem. Soc.* **1985**, *107*, 251-253. Ashcroft, B.R.; Bradley, D.C.; Clark, G.R.; Errington, R.J.; Nielson, A.J.; Rickard, C.E.F. *J. Chem. Soc., Chem. Comm.* **1987**, 170-171. Nugent, W.A.; Harlow, R.L. *J. Chem. Soc., Chem. Comm.* **1979**, 1105-1106.
17. Osborne, J.H.; Rheingold, A.L.; Trogler, W.C. *J. Am. Chem. Soc.* **1985**, *107*, 7945-7952.
18. Miskowski, V.; Gray, H.B.; Poon, C.K.; Ballhausen, C.J. *Mol. Phys.* **1974**, *28*, 747-757.
19. Kostic, N.M.; Fenske, R.F. *J. Am. Chem. Soc.* **1981**, *103*, 4677-4685.
20. Vogler, A.; Kisslinger, J.; Roper, W.R. *Z. Naturforsch.* **1983**, *38B*, 1506-1509.
21. Huttner, G.; Frank, A.; Fischer, E.O. *Isr. J. Chem.* **1977**, *15*, 133-142.
22. Connor, J.A.; Ebsworth, E.A.V. *Adv. Inorg. Chem. Radiochem.* **1964**, *6*, 279-381.

23. Rocklage, S.M.; Schrock, R.R.; Churchill, M.R.; Wasserman, H.J. *Organometallics* **1982**, *1*, 1332–1338.
24. Schrock, R.R. *Acc. Chem. Res.* **1979**, *12*, 98–104.
25. Lappert, M.F.; Power, P.P.; Sanger, R.R.; Srivastava, R.C. "Metal and Metalloid Amides" Wiley, New York, 1980.
26. Baker, R.T.; Whitney, J.F.; Wreford, S.S. *Organometallics* **1983**, *2*, 1049–1051 and references therein. Baker, R.T.; Krusic, P.J.; Tulip, T.H.; Calabrese, J.C.; Wreford, S.S. *J. Am. Chem. Soc.* **1983**, *105*, 6763–6765.
27. Bradley, D.C.; Mehrotra, R.C.; Gaur, D.P. "Metal Alkoxides" Academic Press, London, 1978. Mehrotra, R.C. *Adv. Inorg. Chem. Radiochem.* **1983**, *26*, 269–335.
28. Seyferth, D., ed. "Transition Metal Carbene Complexes" Verlag Chemie, Weinheim, 1983.
29. Goddard, R.J.; Hoffmann, R.; Jemmis, E.D. *J. Am. Chem. Soc.* **1980**, *102*, 7667–7676.
30. Wengrovius, J.H.; Schrock, R.R.; Churchill, M.R.; Missert, J.R.; Youngs, W.J. *J. Am. Chem. Soc.* **1980**, *102*, 4515–4516. Wengrovius, J.H.; Schrock, R.R. *Organometallics* **1982**, *1*, 148–155. Churchill, M.R.; Rheingold, A.L.; Youngs, W.J.; Schrock, R.R.; Wengrovius, J.H. *J. Organomet. Chem.* **1981**, *204*, C17–C20. Churchill, M.R.; Missert, J.R.; Youngs, W.J. *Inorg. Chem.* **1981**, *20*, 3388–3391. Churchill, M.R.; Rheingold, A.L. *Inorg. Chem.* **1982**, *21*, 1357–1359. Pedersen, S.F.; Schrock, R.R. *J. Am. Chem. Soc.* **1982**, *104*, 7483–7491. Schaverien, C.J.; Dewan, J.C.; Schrock, R.R. *J. Am. Chem. Soc.* **1986**, *108*, 2771–2773. Holmes, S.J.; Clark, D.N.; Turner, H.W.; Schrock, R.R. *J. Am. Chem. Soc.* **1982**, *104*, 6322–6329. Edwards, D.S.; Biondi, L.V.; Ziller, J.W.; Churchill, M.R.; Schrock, R.R. *Organometallics* **1983**, *2*, 1505–1513.
31. Pourreau, D.B.; Geoffroy, G.L. *Adv. Organomet. Chem.* **1985**, *24*, 249–352.
32. Wanat, R.A.; Collum, D.B. *Organometallics* **1986**, *5*, 120–127.
33. DuBois, D.L.; Hoffmann, R. *Nouv. J. Chim.* **1977**, *1*, 479–492.
34. Schubert, U.; Neugebauer, D.; Hofmann, P.; Schilling, B.E.R.; Fischer, H.; Motsch, A. *Chem. Ber.* **1981**, *114*, 3349–3365.
35. Su, F.-M.; Cooper, C.; Geib, S.J.; Rheingold, A.L.; Mayer, J.M. *J. Am. Chem. Soc.* **1986**, *108*, 3545–3547.
36. Bryan, J.C.; Geib, S.J.; Rheingold, A.L.; Mayer, J.M. *J. Am. Chem. Soc.* **1987**, *109*, 2826–2828.
37. Mayr, A.; Asaro, M.F.; Kjelsberg, M.A.; Lee, S.L.; Van Engon, D. *Organometallics* **1987**, *6*, 432–434.
38. Mayr, A.; McDermott, G.A.; Dorries, A.M. *Organometallics* **1985**, *4*, 608–610.
39. Hillhouse, G.L.; Haymore, B.L. *J. Organomet. Chem.* **1978**, *162*, C23–C26; *J. Am. Chem. Soc.* **1982**, *104*, 1537–1548.
40. LaMonica, G.; Cenini, S. *J. Chem. Soc., Dalton Trans.* **1980**, 1145–1149. LaMonica, G.; Cenini, S. *Inorg. Chim. Acta* **1973**, *29*, 183–187.
41. Kim, H.P.; Angelici, R.J. *Adv. Organomet. Chem.* **1987**, *27*, 51.
42. Casey, C.P.; Burkhardt, T.J.; Bunnell, C.A.; Calabrese, J.C. *J. Am. Chem. Soc.* **1977**, *99*, 2127.

43. Woodward, L.A.; Creighton, J.A.; Taylor, K.A. *Trans. Faraday Soc.* **1960**, *56*, 1267-1272. Krebs, B.; Müller, A. *J. Inorg. Nucl. Chem.* **1968**, *30*, 463-466. Müller, A.; Krebs, B.; Höltje, W. *Spectrochim. Acta, Part A* **1967**, *23*, 2753-2760.
44. Chatt, J.; Pombeiro, A.J.L.; Richards, R.L. *J. Chem. Soc. Dalton Trans* **1980**, 492-498.
45. McDermott, G.A.; Mayr, A. *J. Am. Chem. Soc.* **1987**, *109*, 580-582.
46. Chatt, J.; Choukroun, R.; Dilworth, J.R.; Hyde, J.; Vella, P.; Zubieta, J. *Trans. Met. Chem.* **1979**, *4*, 59-63.
47. Pauling, L. "The Nature of the Chemical Bond" 3rd. edition, Cornell University Press, Ithaca, New York, 1960, pp. 224ff.
48. Nugent, W.A.; Harlow, R.L.; McKinney, R.J. *J. Am. Chem. Soc.* **1979**, *101*, 7265-7268.
49. Griffith, W.P.; McManus, N.T.; Skapski, A.C.; White, A.D. *Inorg. Chim. Acta* **1985**, *105*, L11.
50. Jørgensen, K.A.; Hoffmann, R. *J. Am. Chem. Soc.* **1986**, *108*, 1867-1876.
51. Nakatsuji, H.; Ushio, J.; Han, S.; Yonezawa, T. *J. Am. Chem. Soc.* **1983**, *105*, 426-434. Ushio, J.; Nakatsuji, H.; Yonezawa, T. *J. Am. Chem. Soc.* **1984**, *106*, 5892-5901.
52. Block, T.F.; Fenske, R.F.; Casey, C.P. *J. Am. Chem. Soc.* **1976**, *98*, 441-443.
53. Nugent, W.A.; McKinney, R.J.; Kasowski, R.V.; Van-Catledge, F.A. *Inorg. Chim. Acta* **1982**, *65*, L91-L93.
54. Green, M.; Orpen, A.G.; Williams, I.D. *J. Chem. Soc., Chem. Comm.* **1982**, 493-495.
55. Carter, E.A.; Goddard, III, W.A. *J. Am. Chem. Soc.* **1986**, *108*, 2180-2191; 4746-4754.
56. Jacobsen, E.N.; Sharpless, K.B. Abstract 154a, 42nd Northwest Regional Meeting, American Chemical Society, June, 1987.
57. Höhn, A.; Werner, H. *Angew. Chem., Int. Ed. Eng.* **1986**, *25*, 737-738.
58. Gallop, M.A.; Roper, W.R. *Adv. Organomet. Chem.* **1986**, *25*, 121-198.
    Roper, W.R. *J. Organomet. Chem.* **1986**, *300*, 167-190.
59. Fachinetti, G.; Floriani, C.; Chiesi-Villa, A.; Guastini, C. *J. Am. Chem. Soc.* **1979**, *101*, 1767-1775.
60. Nugent, W.A.; Harlow, R.L. *Inorg. Chem.* **1979**, *18*, 2030.
61. Ott, K.C.; Grubbs, R.H. *J. Am. Chem. Soc.* **1981**, *103*, 5922-5923.
62. Cotton, F.A.; Wilkinson, G. "Advanced Inorganic Chemistry" 4th edition, J. Wiley & Sons, New York, 1980.
63. Wells, A.F. "Structural Inorganic Chemistry" 5th edition, Oxford, New York, **1984**, pp. 327, 560ff.
64. Buchwald, S.L.; Grubbs, R.H. *J. Am. Chem. Soc.* **1983**, *105*, 5490-5491.
65. Hartner, Jr., F.W.; Schwartz, J.; Clift, S.M. *J. Am. Chem. Soc.* **1983**, *105*, 640-641. Schwartz, J.; Gell, K.I. *J. Organomet. Chem.* **1980**, *184*, C1-C2.
66. Haase, W.; Hoppe, H. *Acta Cryst.* **1968**, *B24*, 282-283. Dwyer, P.N.; Puppe, L.; Buchler, J.W.; Scheidt, W.R. *Inorg. Chem.* **1975**, *14*, 1782-1785. Guilard, R.; Latour, J.M.; Lecomte, C.; Marchon, J.-C.; Protas, J.; Ripoll, D. *Inorg. Chem.* **1978**, *17*, 1228-1237.

Hiller, W.; Strähle, J.; Kobel, W.; Hanack, M. *Z. Kristallogr.* **1982**, *159*, 173. Merbach, A.; Comba, P. *Inorg. Chem.* **1987**, *26*, 1315–1323 and references therein.
67. Fairbrother, F. "The Chemistry of Niobium and Tantalum" Elsevier, New York, 1967.
68. Greenwood, N.N.; Earnshaw, A. "Chemistry of the Elements" Pergamon Press, New York, 1984.
69. Gulliver, D.J.; Levason, W. *Coord. Chem. Rev.* **1982**, *46*, 1–127.
70. Herrmann, W.A. *Adv. Organomet. Chem.* **1982**, *20*, 159–263.
71. Empsall, H.D.; Hyde, E.M.; Markham, R.; McDonald, W.S.; Norton, M.C.; Shaw, B.L.; Weeks, B. *J. Chem. Soc. Chem. Comm.* **1977**, 589.
72. Fryzuk, M.D.; MacNeil, P.A.; Rettig, S.J. *J. Am. Chem. Soc.* **1985**, *107*, 6708–6710.
73. Tulip, T.H.; Thorn, D.L. *J. Am. Chem. Soc.* **1981**, *103*, 2448–2450.
74. Cotton, F.A.; Walton, R.A. "Multiple Bonds Between Metal Atoms" J. Wiley & Sons, New York, 1982.
75. Glidewell, C. *Inorg. Chim. Acta* **1977**, *24*, 149–157 and references therein.
76. Brewer, L.; Rosenblatt, G.M. *Chem. Rev.* **1961**, *61*, 257–263. Brewer, L. *Chem. Rev.* **1953**, *52*, 1–69.
77. Sanderson, R.T. *Inorg. Chem.* **1986**, *25*, 3518–3522.
78. Watt, G.D.; McDonald, J.W.; Newton, W.E. *J. Less-Common Met.* **1977**, *54*, 415–423.
79. Kiefer, W.; Bernstein, H.J. *Chem. Phys. Lett.* **1971**, *8*, 381–383.
80. Rappé, A.K.; Goddard, III, W.A. *J. Am. Chem. Soc.* **1982**, *104*, 448–456.
81. "Handbook of Chemistry and Physics" R.C. Weast. ed. CRC Press, 1973, 54th edition, pp. F200–207.
82. Armentrout, P.B.; Halle, L.F.; Beaushamp, J.L. *J. Am. Chem. Soc.* **1981**, *103*, 6501–6502. Aristov, N.; Armentrout, P.B. *ibid* **1984**, *106*, 4065–4066.
83. Taube, R.; Seyferth, K. *J. Organomet. Chem.* **1983**, *249*, 365–369.
84. Sunil, K.K.; Rogers, M.T. *Inorg. Chem.* **1981**, *20*, 3283–3287 and references therein.
85. Ballhausen, C.J.; Gray, H.B. *Inorg. Chem.* **1962**, *1*, 111–122.
86. Ballhausen, C.J.; Djurinskig, B.F.; Watson, K.J. *J. Am. Chem. Soc.* **1968**, *90*, 3305–3309.
87. Wentworth, R.A.D.; Piper, T.S. *J. Chem. Phys.* **1964**, *41*, 3884–3889.
88. Garner, C.D.; Kendrick, J.; Lambert, P.; Mabbs, F.E.; Hillier, I.H. *Inorg. Chem.* **1976**, *15*, 1287–1291.
89. Gray, H.B.; Hare, C.R. *Inorg. Chem.* **1962**, *1*, 363–368.
90. Winkler, J.R.; Gray, H.B. *Comm. Inorg. Chem.* **1981**, *1*, 257–263.
91. Garner, C.D.; Hill, L.H.; Mabbs, F.E.; McFadden, D.L.; McPhail, A.T. *J. Chem. Soc., Dalton Trans.* **1977**, 853; 1202.
92. Garner, C.D.; Hill, L.H.; Mabbs, F.E.; McFadden, D.L.; McPhail, A.T. *J. Chem. Soc., Dalton Trans.* **1977**, 1202.
93. Hill, L.H.; Howlader, N.C.; Mabbs, F.E.; Hursthouse, M.B.; Malik, K.M.A. *J. Chem. Soc., Dalton Trans.* **1980**, 1475–1481.
94. Gray, H.B.; Hare, C.R.; Bernal, I. *Inorg. Chem.* **1962**, *1*, 831. Cowman, C.D.; Trogler, W.C.; Mann, K.R.; Poon, C.K.; Gray, H.B. *Inorg. Chem.* **1976**, *15*, 1747–1751.
95. Weber, J.; Garner, C.D. *Inorg. Chem.* **1980**, *19*, 2206–2209.

Collison, D.; Garner, C.D.; Mabbs, F.E.; King, T.J. *J. Chem. Soc., Dalton Trans.* **1981**, 1820–1824.
96. Collison, D.; Garner, C.D.; Mabbs, F.E.; Salthouse, J.A.; King, T.J. *J. Chem. Soc., Dalton Trans.* **1981**, 1812–1819.
97. Rappé, A.K.; Goddard, III, W.A. *J. Am. Chem. Soc.* **1982**, *104*, 3287–3294.
98. Antipas, A.; Buchler, J.W.; Gouterman, M.; Smith, P.D. *J. Am. Chem. Soc.* **1980**, *102*, 198–207.
99. Wheeler, R.A.; Whangbo, M.-H.; Hughbanks, T.; Hoffmann, R.; Burdett, J.K.; Albright, T.A. *J. Am. Chem. Soc.* **1986**, *108*, 2222–2236.
100. Bocarsly, A.B., Cameron, R.E.; Rubin, H.-D.; McDermott, G.A.; Wolff, C.R.; Mayr, A. *Inorg. Chem.* **1985**, *24*, 3976–3978.
101. Beevor, R.G.; Freeman, M.J.; Green, M.; Morton, C.E.; Orpen, A.G. *J. Chem. Soc., Chem. Comm.* **1985**, 65–70.
102. Sheridan, J.B.; Geoffroy, G.L.; Rheingold, A.L. *Organometallics* **1986**, *5*, 1514–1515.
103. Figgis, B.N. "Introduction to Ligand Fields" J. Wiley & Sons, New York, 1966.
104. Winkler, J.R.; Gray, H.B. *J. Am. Chem. Soc.* **1983**, *105*, 1373–1374; *Inorg. Chem.* **1985**, *24*, 346–355.
105. Preetz, W.; Scholz, H. *Z. Naturforsch.* **1983**, *38B*, 183–189.
106. Che, C.-M.; Cheng, W.-K. *J. Am. Chem. Soc.* **1986**, *108*, 4644–4645.
107. Nocera, D.G.; Maverick, A.W.; Winkler, J.R.; Che, C.-M.; Gray, H.B. *ACS Symposium Series* **1982**, *211*, 21–31.
108. Fischer, E.O.; Friedrich, P. *Angew. Chem., Int. Ed. Eng.* **1979**, *18*, 327–328.
109. Tatsumi, K.; Hoffmann, R. *Inorg. Chem.* **1981**, *20*, 3771–3784.
110. Fischer, E.O.; Schubert, U. *J. Organomet. Chem.* **1975**, *100*, 59–81.
111. Mayr, A.; Dorries, A.M.; McDermott, G.A.; Geib, S.J.; Rheingold, A.L. *J. Am. Chem. Soc.* **1985**, *107*, 7775–7776.
112. Newton, W.E.; McDonald, J.W.; Corbin, J.L.; Ricard, L.; Weiss, R. *Inorg. Chem.* **1980**, *19*, 1997–2006.
113. Templeton, J.L.; Winston, P.B.; Ward, B.C. *J. Am. Chem. Soc.* **1981**, *103*, 7713–7721.
114. Bennett, M.A.; Boyd, I.W. *J. Organomet. Chem.* **1985**, *290*, 165–180.
115. Chen, G.J.-J.; McDonald, J.W.; Newton, W.E. *Organometallics* **1985**, *4*, 422–423.
116. Tatsumi, K.; Hoffmann, R. *Inorg. Chem.* **1980**, *19*, 2656–2658 and references therein.
117. Wadt, W.R. *J. Am. Chem. Soc.* **1981**, *103*, 6053–6057, and references therein.
118. Behling, T.; Capparelli, M.V.; Skapski, A.C.; Wilkinson, G. *Polyhedron* **1982**, *1*, 840–841.
119. Dobson, J.C.; Takeuchi, K.J.; Pipes, D.W.; Geselowitz, D.A.; Meyer, T.J. *Inorg. Chem.* **1986**, *25*, 2357–2365.
120. Kubacek, P.; Hoffmann, R. *J. Am. Chem. Soc.* **1981**, *103*, 4320–4332.
121. Pawson, D.; Griffith, W.P. *J. Chem. Soc., Dalton Trans.* **1975**, 417–423.
122. Wolfsberg, M.; Helmholz, L. *J. Chem. Phys.* **1952**, *20*, 837–843.

123. Teltow, J. *Z. Phys. Chem.* **1938**, *B40*, 397; **1939**, *B43*, 198.
124. Ballhausen, C.J.; Gray, H.B. in "Coordination Chemistry" vol. I, A.E. Martell, ed. ACS Monograph #168, 1971, pp. 3-83.
125. Wells, E.J.; Jordan, A.D.; Alderdice, D.S.; Ross, I.G. *Aust. J. Chem.* **1967**, *20*, 2315-2322.
126. Güdel, H.U.; Ballhausen, C.J. *Theoret. Chim. Acta (Berlin)* **1972**, *25*, 331-337.
127. Miskowski, V.; Gray, H.B.; Poon, C.K.; Ballhausen, C.J. *Mol. Phys.* **1974**, *28*, 729-745; 747-757.
128. Clark, R.J.H.; Dines, T.J.; Doherty, J.M. *Inorg. Chem.* **1985**, *24*, 2088-2091 and references therein.
129. Foster, S.; Felps, S.; Cusachs, L.C.; McGlynn, S.P. *J. Am. Chem. Soc.* **1973**, *95*, 5521-5524.
130. Rauk, A.; Ziegler, T.; Ellis, D.E. *Theoret. Chim. Acta (Berlin)* **1974**, *34*, 49-59, and references therein.
131. Müller, A.; Diemann, E. in Sharp, D.W.A., ed. "MTP International Review of Science, Inorganic Chemistry, Series 2" Butterworths, London, **1974**, vol. 5, pp. 71-110.
132. Schröder, M. *Chem. Rev.* **1980**, *80*, 187-213.
133. Palenik, G.J. *Inorg. Chem.* **1967**, *6*, 503-507; 507-511.
134. Berg, D.M.; Sharp, P.R. *Inorg. Chem.* **1987**, *26*, 2959-2962.
135. Taylor, T.E.; Hall, M.B. *J. Am. Chem. Soc.* **1984**, *106*, 1576-1584.
136. Rappé, A.K.; Goddard, III, W.A. *J. Am. Chem. Soc.* **1977**, *99*, 3966.
137. Spangler, D.; Wendoloski, J.J.; Dupuis, M.; Chen, M.M.L.; Schaefer, III, H.F. *J. Am. Chem. Soc.* **1981**, *103*, 3985-3990.
138. Collman, J.P.; Hegedus, L.S.; Norton, J.R.; Finke, R.G. "Principles and Applications of Organometallic Chemistry" University Science Press, Mill Valley, CA, 1987.
139. Rappé, A.K. *Inorg. Chem.* **1984**, *23*, 995-996.
140. Ortiz de Montellano, P.R., ed. "Cytochrome P-450: Structure, Mechanism, and Biochemistry" Plenum, New York, 1986. Groves, J.T. *J. Chem. Ed.* **1982**, *59*, 179.
141. Groves, J.T.; Gilbert, J.A. *Inorg. Chem.* **1986**, *25*, 125-127. Boso, B.; Lang, G.; McMurray, T.J.; Groves, J.T. *J. Chem. Phys.* **1983**, *79*, 1122-1126. Penner-Hahn, J.E.; Eble, K.S.; McMurry, T.J.; Renner, M.; Balch, A.L.; Groves, J.T.; Dawson, J.H.; Hodgson, K.O. *J. Am. Chem. Soc.* **1986**, *108*, 7819-7825. Balch, A.L.; La Mar, G.N.; Latos-Grazynski, L.; Renner, R.W.; Thanabal, V. *J. Am. Chem. Soc.* **1985**, *107*, 3003-3007. Balch, A.L.; Latos-Grazynski, L.; Renner, R.W. *J. Am. Chem. Soc.* **1985**, *107*, 2983-2985. La Mar, G.N.; de Ropp, J.S.; Latos-Grazynski, L.; Balch, A.L.; Johnson, R.B.; Smith, K.M.; Parish, D.W.; Cheng, R.-J. *J. Am. Chem. Soc.* **1985**, *107*, 782-787. Simonneaux, G.; Scholz, W.F.; Reed, C.A.; Lang, G. *Biochim. Biophys. Acta* **1982**, *716*, 1-7. Schappacher, M.; Weiss, R.; Montiel-Montoya, R.; Trautwein, A.; Tabard, A. *J. Am. Chem. Soc.* **1985**, *107*, 3736-3738. Calderwood, T.S.; Lee, W.A.; Bruice, T.C. *J. Am. Chem. Soc.* **1985**, *107*, 8272-8273.
142. Collman, J.P.; Kodadek, T.; Brauman, J.I. *J. Am. Chem. Soc.* **1986**, *108*, 2588-2594.
143. Blair, D.F.; Witt, S.N.; Chan, S.I. *J. Am. Chem. Soc.* **1985**, *107*, 7389-7399.

144. Loew, G.H.; Kert, C.J.; Hjelmeland, L.M.; Kirchner, R.F. *J. Am. Chem. Soc.* **1977**, *99*, 3534–3536. Loew, G.H.; Herman, Z.S. *J. Am. Chem. Soc.* **1980**, *102*, 6173–6174. Hanson, L.K.; Chang, C.K.; Davis, M.S.; Fajer, J. *J. Am. Chem. Soc.* **1981**, *103*, 663–670. Sontum, S.F.; Case, D.A. *J. Am. Chem. Soc.* **1985**, *107*, 4013–4015. Sevin, A.; Fontecave, M. *J. Am. Chem. Soc.* **1986**, *108*, 3266–3272.
145. Hill, C.L.; Smart, B.C. *J. Am. Chem. Soc.* **1980**, *102*, 6374–6375. Groves, J.T.; Kruper, W.J.; Haushalter, R.C. *ibid.* 6375–6377. Collman, J.P.; Brauman, J.I.; Meunier, B.; Hayashi, T.; Kodadek, T.; Raybuck, S.A. *J. Am. Chem. Soc.* **1985**, *107*, 2000–2005. Srinivasan, K.; Michaud, P.; Kochi, J.K. *J. Am. Chem. Soc.* **1986**, *108*, 2309–2320.
146. Hill, C.L.; Hollander, F.J. *J. Am. Chem. Soc.* **1982**, *104*, 7318–7319.
Buchler, J.W.; Dreher, C.; Lay, K.-L.; Lee, Y.J.A.; Scheidt, W.R. *Inorg. Chem.* **1983**, *22*, 888–891.
147. Budge, J.R.; Gatehouse, B.M.K.; Nesbit, M.C.; West, B.O. *J. Chem. Soc., Chem. Comm.* **1981**, 370. Buchler, J.W.; Lay, K.L.; Castle, L.; Ullrich, V. *Inorg. Chem.* **1982**, *21*, 842–844. Groves, J.T.; Kruper, Jr., W.J.; Haushalter, R.C.; Butler, W.M. *Inorg. Chem.* **1982**, *21*, 1363.
148. Wiberg, N.; Häring, H.W.; Schubert, U. *Z. Naturforsch.* **1980**, *35B*, 599–603.
149. Gambarotta, S.; Chiesi-Villa, A.; Guastini, C. *J. Organomet. Chem.* **1984**, *270*, C49–C52.
150. Lauher, J.W.; Hoffman, R. *J. Am. Chem. Soc.* **1976**, *98*, 1729–1742.
151. Francl, M.M.; Pietro, W.J.; Hout, Jr., R.F.; Hehre, W.J. *Organometallics* **1983**, *2*, 281–286.

CHAPTER 3

REACTIONS RESULTING IN THE FORMATION OF MULTIPLY BONDED LIGANDS

3.1 GENERAL CONSIDERATIONS

This chapter provides a discussion of the ways in which multiply bonded ligands are introduced into a metal complex rather than a general survey of synthetic routes to the transition metal complexes that contain them. Regrettably, space does not permit us to cover the large body of chemistry involving the subsequent exchange or modification of ancillary ligands in complexes where a multiply bonded ligand is already present. The reader seeking such information is referred to the primary literature.

Reactions that afford multiply bonded ligands fall into three classes, depending on what bonds to the ligand precursor are being made or broken. By far the most common are those in which a bond to the eventual α atom is broken; these will be discussed in Section 3.2. Included in this category are reactions in which the α atom of the ligand precursor undergoes deprotonation, desilylation, dealkylation, or cleavage of some electronegative leaving group. Within this topic we have somewhat arbitrarily segregated the discussion of two subtopics of special interest: The Wittig-like (2 + 2) replacement of existing multiply bonded ligands is discussed in Section 3.2.6, while routes involving cleavage of symmetrical ligand precursors (dioxygen, diazobenzene, acetylene, etc.) are covered in Section 3.2.5. The two remaining broad classes of synthetic reactions are those involving formation of a bond to the α atom (Section 3.3) or formation of a bond to the β atom (Section 3.4).

3.2 BY CLEAVAGE OF α BOND OF PRECURSOR

3.2.1 Cleavage of α Hydrogen

Routes involving cleaving of hydrogen from the α atom of the precursor are an important entry into metal–ligand multiple bonds. The majority of such reactions are most easily understood as deprotonation reactions; however, cases that are represented better as a transfer of a hydrogen atom (e.g., Eq. 5) are also known. Most oxo complexes are prepared by procedures involving deprotonation of aquo intermediates. A review summarizing ^{18}O-labeling studies of oxo–aquo exchange has appeared [1]. Likewise, the majority of alkylidene and alkylidyne complexes have been prepared in this way, large-

TABLE 3.1 Formation of Alkylidene and Alkylidyne Ligands by Cleavage of an α C–H Bond[a]

Starting Complex	Additive	Product	Reference
$Ta(CH_2Ph)_3Cl_2$	CpTl	$Cp_2Ta(=CHPh)(CH_2Ph)$	2
$Ta(Np)Cl_4$	PMe_3, Na(Hg)	$Ta(=CH^tBu)H(PMe_3)Cl_2$	3
$Cp^*Ta(CH_2Ph)_2Cl_2$	$Ph_3P=CH_2$	$Cp^*Ta(=CHPh)(CH_2Ph)Cl$	4
$Ta[N(TMS)_2]_2Cl_3$	$LiCH_2TMS$	$Ta(=CHTMS)(CH_2TMS)[N(TMS)_2]_2$	5
$TaMe_3(OAr)_2$	$h\nu$	$Ta(=CH_2)Me(OAr)_2$	6
$Ta(Np)Cl_2(C_2H_4)(PMe_3)_2$	—	$Ta(=CH^tBu)EtCl_2(PMe_3)_2$	7
$Ta(=CH^tBu)(Np)_3$	PMe_3	$Ta(=CH^tBu)_2(Np)(PMe_3)_2$	8
$Mo(N^tBu)(Np)_3Cl$	LiNp	$Mo(N^tBu)(=CH^tBu)Np_2$	9
$MoCl_5$	$LiCH_2SiMe_3$	$Mo(=CHSiMe_3)(CH_2SiMe_3)_3$	10
$W(OAr)_2Cl_4$	$MgNp_2 \cdot dioxane$	$W(=CH^tBu)(OAr)_2Cl_2$	11
$W(NPh)(Np)_3Cl$	$Ph_3P=CH_2$	$W(NPh)(=CH^tBu)Np_2$	12
$Re(NAr)_2(Np)Cl_2$	DBU	$Re(NAr)_2(=CH^tBu)Cl$	13
$Re(N^tBu)_2Cl_3$	NpMgCl	$Re(N^tBu)_2(=CH^tBu)Np$	14
MCl_5 (M = Nb, Ta)	$LiCH_2SiMe_3$	$M_2(\mu\text{-}CSiMe_3)_2(CH_2SiMe_3)_4$	15
$CpTa(=CH^tBu)Cl_2$	PMe_3, $Ph_3P=CH_2$	$Cp(Ta\equiv C^tBu)Cl(PMe_3)_2$	16
$Ta(Np)Cl_4$	PMe_3, Na(Hg)	$Ta(\equiv C^tBu)(H)Cl_2(PMe_3)_3$	17
MoO_2Cl_2	NpMgCl	$Mo(\equiv C^tBu)(Np)_3$	18
$Mo_2Br_2(CH_2SiMe_3)_4$	PEt_3	$Mo(\equiv CSiMe_3)Br(PEt_3)_4$	19
WCl_4	Me_3SiCH_2MgCl	$W_2(\mu\text{-}CSiMe_3)_2(CH_2SiMe_3)_4$	20
WCl_6	$LiCH_2SiMe_3$	$W(\equiv CSiMe_3)(CH_2SiMe_3)_3$	10
$W(OMe)_3Cl_3$	NpMgCl	$W(\equiv C^tBu)(Np)_3$	21
$WCl_2(PMe_3)_4$	$AlMe_3$, tmeda	$W(\equiv CH)Cl(PMe_3)_4$	22
$W(=CHPh)Cl_2(CO)(PMe_3)_2$	BuLi, py	$W(\equiv CPh)Cl(CO)py(PMe_3)_2$	23
$Re(N^tBu)_2(=CH^tBu)Np$	Lut·HCl	$[Re(\equiv C^tBu)(CH^tBu)(NH_2{}^tBu)Cl_2]_2$	14

[a] Abbreviations: Np = neopentyl; TMS = trimethylsilyl;
BDU = 1,8-diazabicyclo[5.4.0]undec-7-ene;
tmeda = tetramethyl- ethylenediamine; Lut·HCl = lutidine hydrochloride.

ly by Schrock and his coworkers. A number of representative examples are tabulated in Table 3.1.

Consider the generalized equilibrium in Eq. 1. The position of this equilibrium will depend, to begin with, on the nature of metal. For example, replacing the tungsten atom in $[W(NH)X(dppe)_2]^+X^-$ by molybdenum increases the acidity of the imido proton

1000-fold [24] (see Section 6.1). Moreover, for a given metal four additional factors can be identified that can shift the equilibrium to the right.

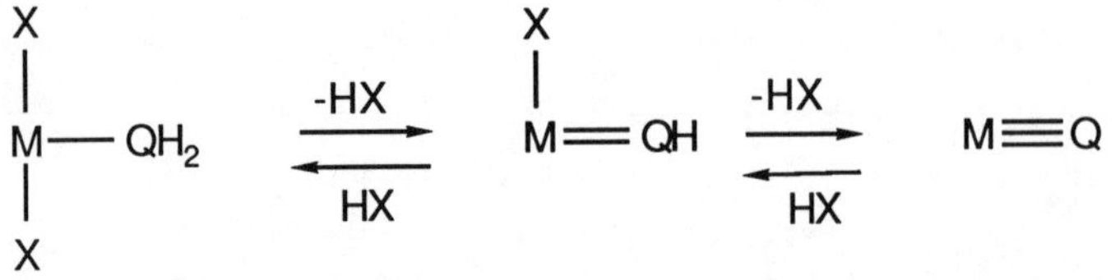

$$\underset{X}{\overset{X}{M}}-QH_2 \underset{HX}{\overset{-HX}{\rightleftarrows}} \underset{}{\overset{X}{M}}=QH \underset{HX}{\overset{-HX}{\rightleftarrows}} M\equiv Q \qquad (1)$$

These are (1) raising the oxidation state of the metal, (2) use of strongly basic leaving groups X, (3) addition of an external base to consume HX, or (4) increasing the steric congestion around the metal. Let us consider each of these effects.

Raising the oxidation state of the metal enhances the acidity of a coordinated group (NH_3, H_2O, etc.) which serves as precursor to the multiply bonded ligand. At the same time it empties a d orbital for π donation. Several examples of such oxidation-promoted hydrogen transfer processes are collected in Eqs. 2–5 [25–27]. Such reactions typically differ from the general example of Eq. 1 in that no anionic leaving group X is lost from the metal under these circumstances. Equation 5 represents an example in which the C—H bond is suggested to be cleaved via a hydrogen atom transfer process.

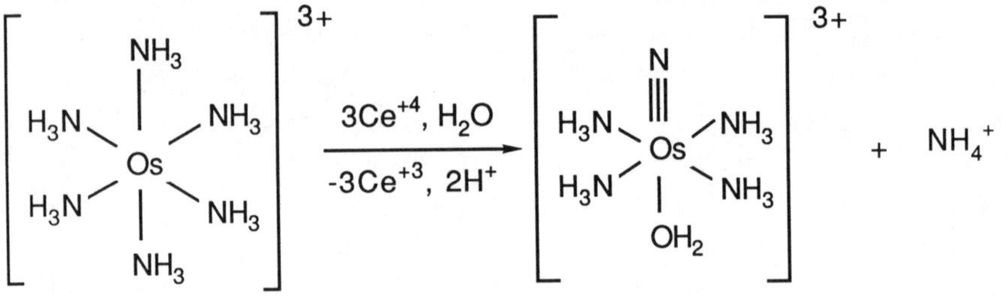

$$[Os(NH_3)_6]^{3+} \xrightarrow[-3Ce^{+3},\ 2H^+]{3Ce^{+4},\ H_2O} [Os(NH_3)_4(N)(OH_2)]^{3+} + NH_4^+ \qquad (2)$$

$$[(bipy)_2(py)Ru(OH_2)]^{2+} \xrightarrow{-e^-} [(bipy)_2(py)Ru(OH)]^{2+} + H^+ \qquad (3a)$$

$$2[(bipy)_2(py)Ru(OH)]^{2+} \rightleftarrows [(bipy)_2(py)Ru(OH_2)]^{2+} + [(bipy)_2(py)Ru(O)]^{2+} \qquad (3b)$$

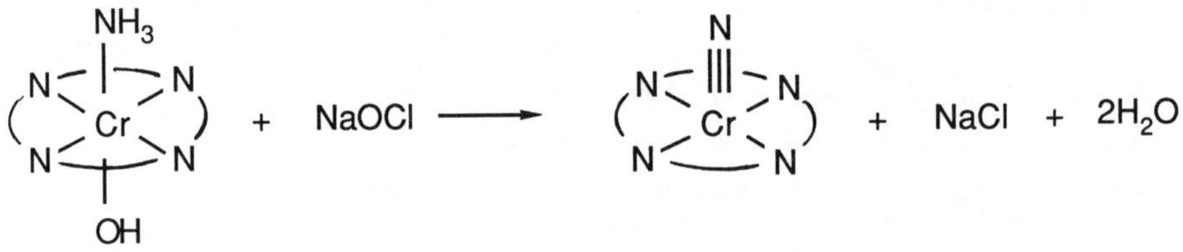

$$[Cr(N_4)(NH_3)(OH)] + NaOCl \longrightarrow [Cr(N_4)(N)] + NaCl + 2H_2O \qquad (4)$$

(N_4)  = 5,10,15,20-tetra-p-tolylporphyrin

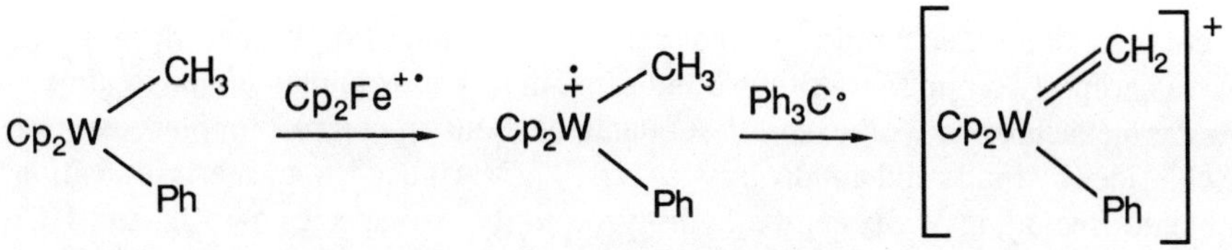

$$Cp_2W\begin{matrix}CH_3\\Ph\end{matrix} \xrightarrow{Cp_2Fe^{+\bullet}} Cp_2W^{+}\begin{matrix}CH_3\\Ph\end{matrix} \xrightarrow{Ph_3C^{\bullet}} \left[Cp_2W\begin{matrix}CH_2\\Ph\end{matrix}\right]^{+} \quad (5)$$

Of these, Eq. 3 has been studied in considerable detail by Meyer and coworkers [28] and serves as the basis for a number of interesting catalytic oxidation reactions (Section 7.1.5). It was shown that initial one-electron oxidation of the ruthenium(II) complex results in a ruthenium(III) hydroxo intermediate which disproportionates to the observed product and starting material. Because of the intrinsic stability of oxo ligands on d^0 metals, simply treating the metal with some appropriate oxidizing agent in an aqueous medium often cleanly affords an oxo derivative. A classic example is the conversion of osmium metal to osmium tetroxide. On the other hand, careful control of the oxidizing potential and judicious choice of coordinating ligands can provide a selective synthesis of interesting oxo derivatives. A pretty example is the electrochemical oxidation of a vanadium anode to oxo–vanadium(IV) dialkyldithiocarbamate complexes [29].

Typical strongly basic leaving groups used to promote Eq. 1 have been σ alkyl substituents or dialkylamido ligands as exemplified in Eqs. 6–8 [30,32]. The first two transformations are sufficiently facile that the intermediate (in brackets) is not observed. Many of the alkylidene syntheses in Table 3.1 likewise involve proton transfer to strongly basic alkyl anions.

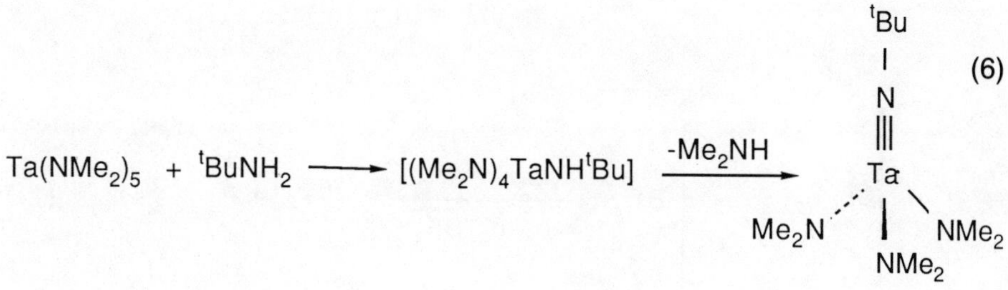

$$Ta(NMe_2)_5 + {}^tBuNH_2 \longrightarrow [(Me_2N)_4TaNH^tBu] \xrightarrow{-Me_2NH} \underset{\underset{NMe_2}{Me_2N}}{\overset{\overset{{}^tBu}{\underset{|}{N}}\equiv}{Ta}}\text{—}NMe_2 \quad (6)$$

$$Cp^*TaMe_3Cl \xrightarrow{LiNH^tBu} \left[Cp^*TaMe_3(NH^tBu)\right] \xrightarrow{-CH_4} Cp^*Ta(N^tBu)Me_2 \quad (7)$$

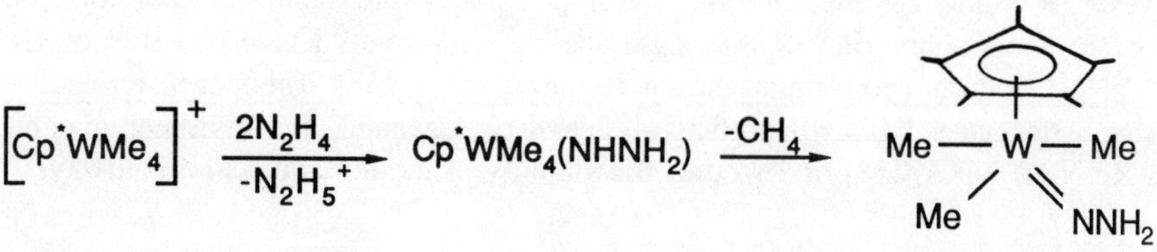

$$\left[Cp^*WMe_4\right]^+ \xrightarrow[-N_2H_5^+]{2N_2H_4} Cp^*WMe_4(NHNH_2) \xrightarrow{-CH_4} \underset{Me}{\overset{Cp^*}{\underset{|}{W}}}\!\!\begin{matrix}Me\\ \\NNH_2\end{matrix} \quad (8)$$

By contrast, some fairly mild external bases can promote Eq. 1 when the starting complex is properly set up for multiple bond formation. For example, simple treatment of tungsten hexachloride with excess *t*-butylamine results in complete replacement of all the chlorides by imido and amido groups (Eq. 9) [33]. (Indeed, some early-transition metal halides including WCl_6 can be hydrolyzed to the oxo species $WOCl_4$ and HCl with chloride ion itself acting as base.) Aqueous KOH converts the initially formed amine complex of OsO_4 to the "osmiamate anion" (Eq. 10) [34]. The phosphorane reagent $Me_3P=CH_2$, a very mild base, has proven useful for deprotonating a methyl-tantalum cation (Eq. 11) [35]. Even triethylamine suffices to deprotonate the tungsten analog $[Cp^*WMe_4]^+$ at $-78°C$ [36].

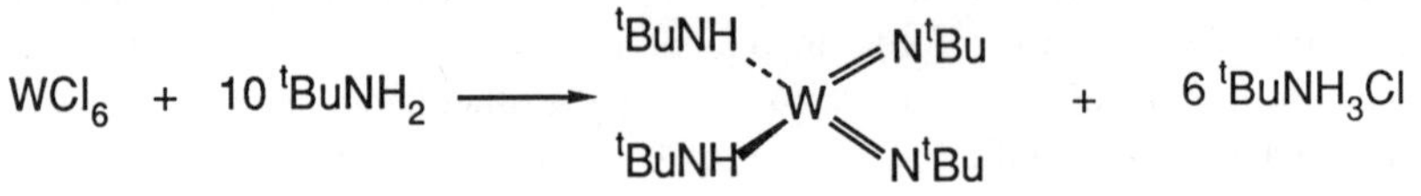

$$WCl_6 + 10\ ^tBuNH_2 \longrightarrow W(NH^tBu)_2(N^tBu)_2 + 6\ ^tBuNH_3Cl \qquad (9)$$

$$O=Os(O)_2(O)(NH_3) \xrightarrow[-2 H_2O]{KOH} K^+[Os(N)(O)_3]^- \qquad (10)$$

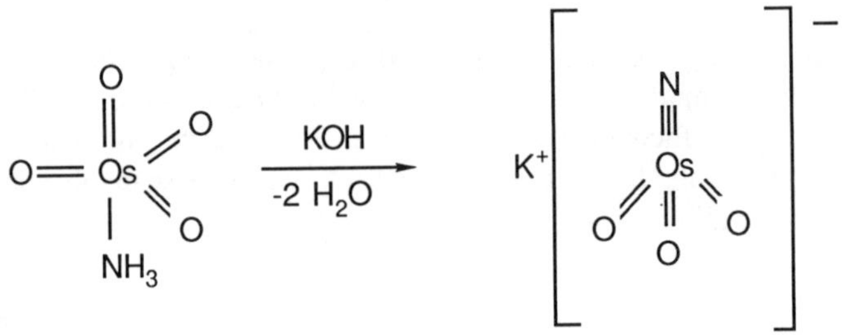

$$[Cp_2Ta(CH_3)_2]^+ BF_4^- \xrightarrow[-PMe_4^+BF_4^-]{Me_3P=CH_2} Cp_2Ta(=CH_2)(CH_3) \qquad (11)$$

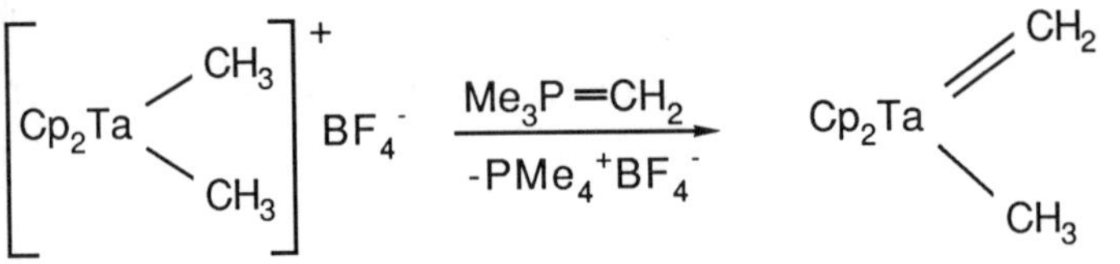

Often it is not easy to distinguish whether a strongly basic additive is promoting the formation of a multiply bonded ligand by direct deprotonation or by prior coordination to the metal. Indeed, it is still unknown which of these pathways [Eq. 12a versus 12b] is operant in Schrock's original synthetic route to an alkylidene tantalum compound [37]. However, it is now certain that either intramolecular or intermolecular deprotonation *can* occur under appropriate conditions. Labeling studies and detailed kinetics on the $Cp^*TaNp_2Cl_2$ system confirm an intramolecular pathway [38]. The fact that a neutral organic base such as DBU will suffice to "dehydrohalogenate" the alkylrhenium complex $Re(NPh)_2NpCl_2$ [13] underscores the viability of the intermolecular pathway.

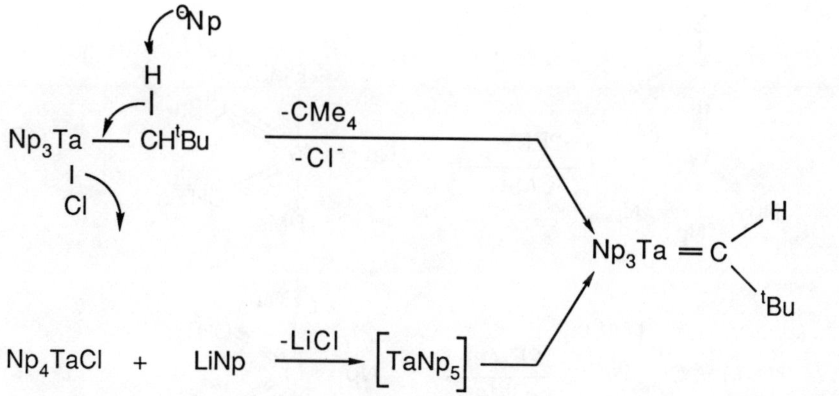

$$Np_3Ta\overset{Np}{\underset{Cl}{\overset{H}{\underset{}{-}}CH^tBu}} \xrightarrow[-Cl^-]{-CMe_4} \quad (12a)$$

$$Np_3Ta=C\overset{H}{\underset{{}^tBu}{\diagdown}}$$

$$Np_4TaCl + LiNp \xrightarrow{-LiCl} [TaNp_5] \quad (12b)$$

Steric congestion is frequently a critical factor in promoting formation of ligand–metal multiple bonds. For example, complex **1** is isolable and relatively stable when R = methyl or benzyl. With the bulkier R = neopentyl, elimination to complex **2** is instead observed [35,4,38]. Along similar lines, compound **3**, R = 1-adamantyl is a robust complex that has been characterized structurally [33]. On the other hand, efforts to prepare **3**, R = *t*-butyl, have instead afforded only **4**.

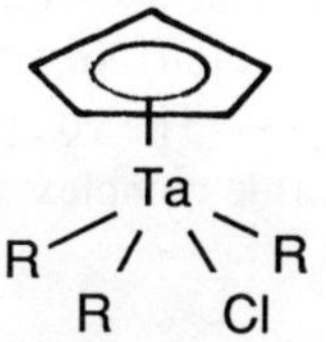

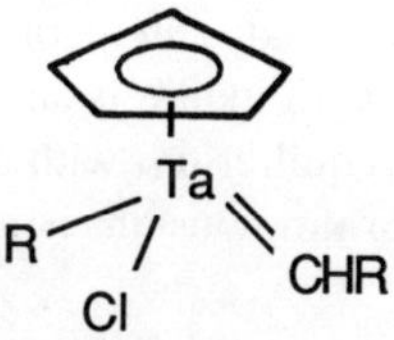

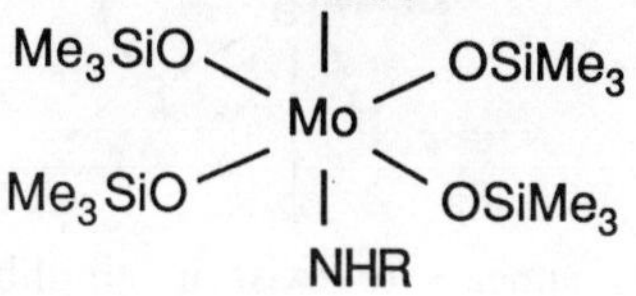

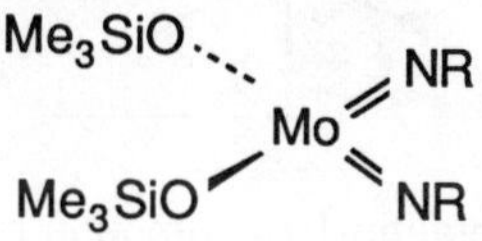

Schrock and coworkers have developed and utilized the technique of promoting α elimination by addition of strongly coordinating trialkylphosphine (especially PMe$_3$) to alkyltransition metal species. Pertinent examples include Eqs. 13 and 14 [39–41]. The function of the phosphine seems to be to increase the steric congestion at the metal.

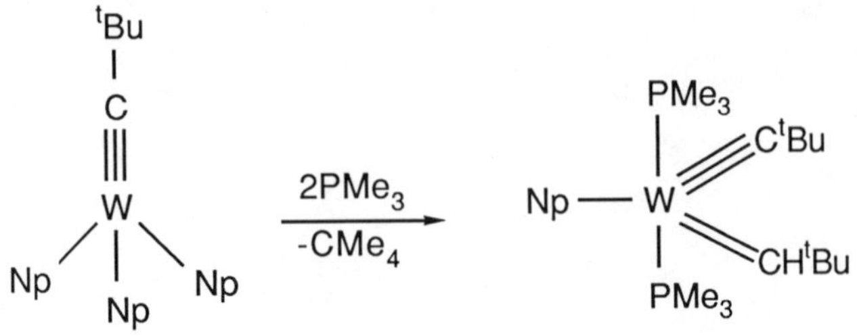

$$\text{(scheme 13)} \quad \text{Np}_3\text{W}\!\equiv\!\text{C}^t\text{Bu} \xrightarrow[-\text{CMe}_4]{2\text{PMe}_3} \text{Np}(\text{PMe}_3)_2\text{W}(\equiv\!\text{C}^t\text{Bu})(\!=\!\text{CH}^t\text{Bu}) \quad (13)$$

$$\text{Np}_3\text{M}\!=\!\text{CH}^t\text{Bu} \xrightarrow[-\text{CMe}_4]{2\text{PMe}_3} \text{Np}(\text{PMe}_3)_2\text{M}(\!=\!\text{CH}^t\text{Bu})_2 \quad (14)$$

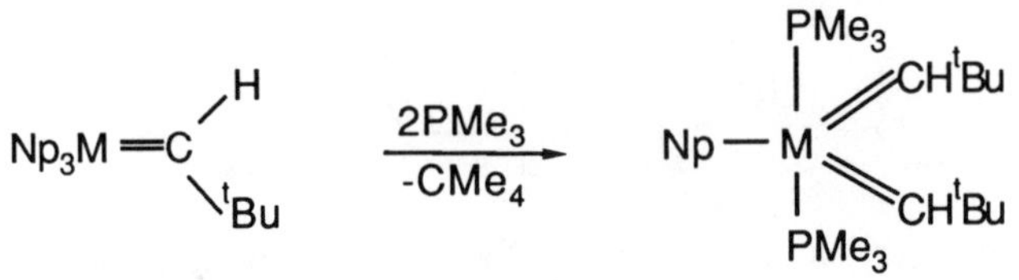

One special case for α deprotonation arises when a lower-valent metal is present. A "basic" d orbital on the metal can accept a proton from the α atom resulting in hydrido–metal complex. This is exemplified by the tantalum(III)–alkylidene complexes $(\text{dmpe})_2\text{Ta}(\text{CH}^t\text{Bu})\text{X}$ [42]. When X = chloride this complex exists as a (highly distorted) neopentylidene complex; its very small $J(H^{13}C)$ coupling constant of 57 Hz suggests a Ta—C—C bond angle of 177° (see Fig. 4.2). This distortion indicates that the α proton is already interacting with a filled metal d orbital. Simply replacing the chloride ligand with iodide is sufficient to complete the job. The complex where X is iodide exists in equilibrium with the neopentylidyne hydride complex as shown in Eq. 15. The equilibrium constant is about 1 at 323 K [42].

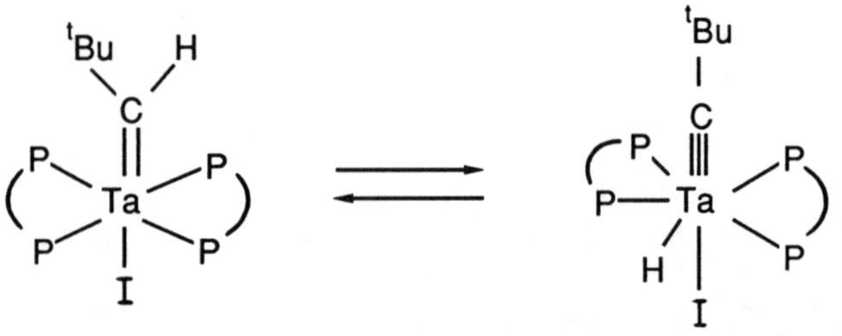

$$\text{(dmpe)}_2\text{Ta}(\text{CH}^t\text{Bu})\text{I} \rightleftharpoons \text{(dmpe)}_2\text{Ta}(\equiv\!\text{C}^t\text{Bu})(\text{H})(\text{I}) \quad (15)$$

The tantalum(III) compound $\text{Cp}^*_2\text{TaCH}_3$ appears to exist in equilibrium with $\text{Cp}^*_2\text{Ta}(\text{CH}_2)\text{H}$, with the methylidene hydride form predominating [43]. A similar equilibrium has been proposed for the isoelectronic tungsten cation $[\text{Cp}_2\text{W}(\text{CH}_3)]^+$ [44]. In both the tungsten and tantalum systems, the complex can be isolated in either of the equilibrating forms by addition of an appropriate ligand as trapping agent.

Another special case among α hydrogen transfer reactions is that in which the proton (or hydrogen) acceptor is itself another multiply bonded ligand. One might anticipate that such prototropy within the coordination sphere of a metal would always be rapid, at least for thermoneutral or exothermic processes. However, this is not the case. Hydroxo to oxo H transfer in Eq. 16, involving what might be viewed as a "metalla-carboxylic acid," proceeds with a half-life of about 11 hr at at ambient temperatures [45].

Although Eq. 17 is rapid when QH = hydroxo, the reaction exhibits a half-life of 12 min at 77°C for the phenylimido case QH = NHPh and does not proceed at all for the phosphido derivative, QH = PHPh [46]. It is noteworthy that both Eq. 16 and Eq. 17, QH = NHPh, are markedly accelerated by acidic additives.

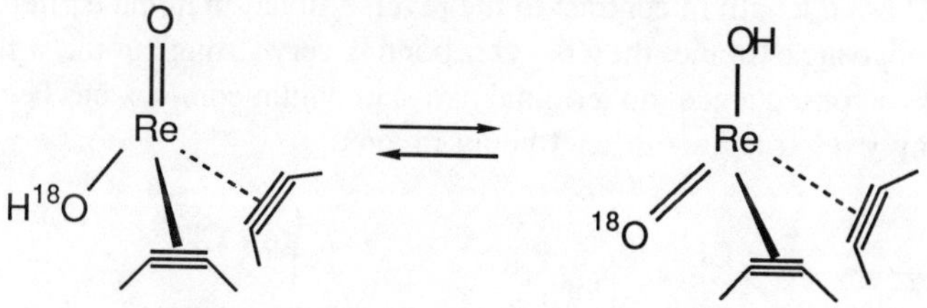

$$\begin{array}{c}\text{Re complex with } H^{18}O \text{ and oxo} \rightleftharpoons \text{Re complex with } ^{18}O \text{ and OH}\end{array} \quad (16)$$

$$\text{Cl}_2(\text{PEt}_3)_2\text{W}(\equiv\text{C}^t\text{Bu})(\text{QH}) \longrightarrow \text{Cl}_2(\text{PEt}_3)_2\text{W}(=\text{Q})(\text{CH}^t\text{Bu}) \quad (17)$$

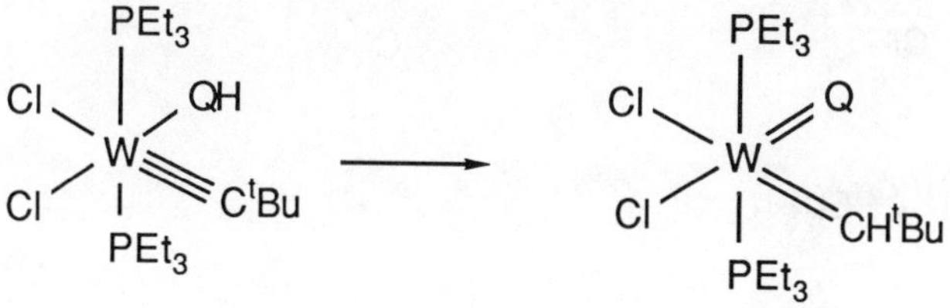

Interestingly, the existence of the equilibrium in Eq. 18 seems to be responsible for the instability of alkylidene complexes when β hydrogen is present, at least in the one system that has been thoroughly studied. Freudenberger and Schrock found that complexes of the type $W(CHEt)(OCMe_3)_2(O_2CR)_2$ rearrange to propylene complexes $W(propylene)(OCMe_3)_2(O_2CR)_2$ in a reaction that is second order in the tungsten complex [47]. The rearrangement is also markedly accelerated by addition of a small amount of carboxylic acid. In the proposed mechanism, the n-propyl tungsten derivative is generated as in Eq. 18 and then loses a β proton to afford the observed propylene complex.

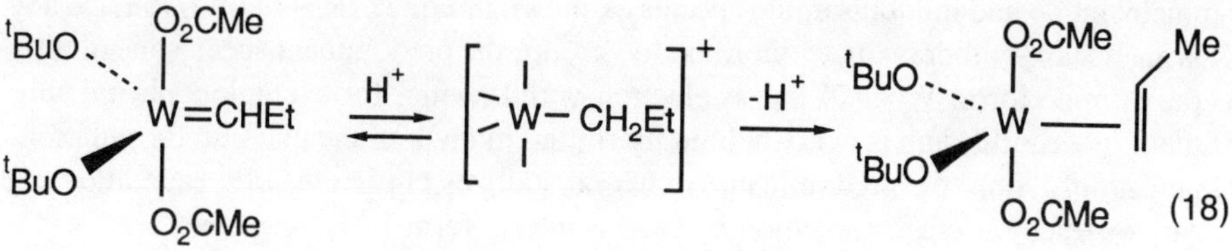

$$(^t\text{BuO})_2(O_2\text{CMe})_2\text{W}=\text{CHEt} \underset{-H^+}{\overset{H^+}{\rightleftharpoons}} [(^t\text{BuO})_2(O_2\text{CMe})_2\text{W}-\text{CH}_2\text{Et}]^+ \xrightarrow{-H^+} (^t\text{BuO})_2(O_2\text{CMe})_2\text{W}(\text{propylene}) \quad (18)$$

When such pathways for interconversion of multiply bonded ligands are present, it is not always clear *a priori* what combination of ligands will be preferred thermodynamically. In general, for early transition metals like tantalum the order of preference for multiple bond formation appears to fall off in the order oxo > imido > alkylidene. However, as one proceeds upward and/or to the right among the early transition metals of the periodic table, this preference is no longer clear cut. For example, the rhenium–oxo complex in Eq. 19 will undergo aminolysis with aniline to afford the corresponding phenylimido species [48]. Proceeding further to the right in the case of osmium, it is found that Eq. 20 proceeds even with alkylamines. To our knowledge, no explanation of this trend has been offered. It would appear, however, that the importance of the σ component of ligand bonding is decreasing relative to the importance of π-bonding as one, for example, proceeds along the third transition series from tantalum(V) to osmium(VIII). This point of view is supported by structural studies of osmium(VIII) oxo imido derivatives: It is found that the Os—N bond length is actually somewhat shorter than the Os—O bond length in contrast to the reverse situation in the earlier transition metals. At the opposite extreme, the Zr—O σ bond is very strong but the π interaction is weak [49]. As a consequence, no terminal oxo–zirconium complex has been isolated to date, bridging oxo structures instead being favored.

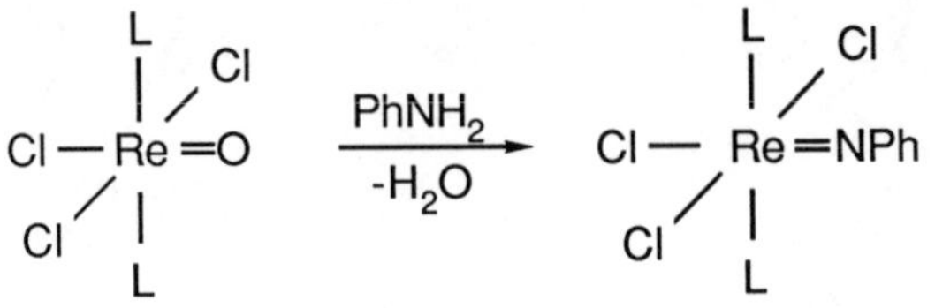

$$Cl_3(L)_2Re=O \xrightarrow[-H_2O]{PhNH_2} Cl_3(L)_2Re=NPh \qquad (19)$$

$L = PEt_2Ph$

$$OsO_4 + {}^tBuNH_2 \longrightarrow O_3Os(N^tBu) + H_2O \qquad (20)$$

Despite the usual preference for oxo over imido ligation in the early transition metals, it is worth noting that hydrazido(2−) ligands are preferred over oxo ligands. Indeed, the protonolysis of an oxo group by a hydrazine is a general synthetic entry into hydrazides [50,51]. Remarkably, even the robust oxo group of WOF_4 is reported to be replaced by 2,4-dinitrophenyhydrazine [52].

In some cases the ultimate disposition of the ligands in a given complex can hinge on fairly subtle electronic effects of the ancillary ligands. For example, treatment of $({}^tBuN)_2W(NH^tBu)_2$ with vicinal glycols gives rise to equilibrating mixtures of aminebisimido and imidobisamido species as shown in Eq. 21 (R is t-butyl) [53]. When X is an electon-withdrawing trifluoromethyl group, the predominant species in solution is the diimido form. When X is less electron-withdrawing, for example a phenyl substituent, the equilibrium is overwhelmingly shifted in favor of the diamido formulation. (Interestingly, only the predominant isomer crystallizes out in the solid state allowing separate structural characterization of each isomeric form.)

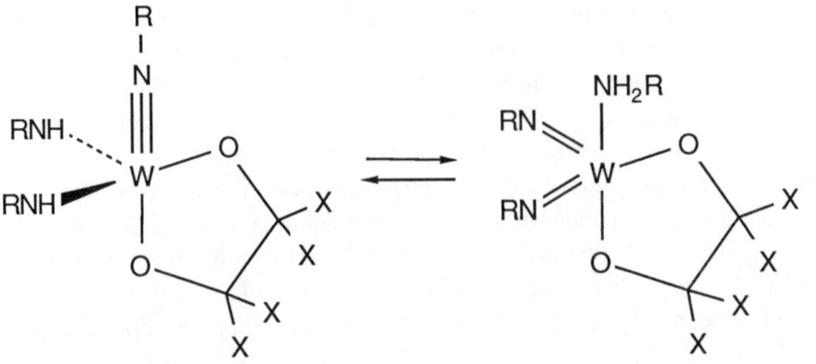

$$\text{(aminebisimido)} \rightleftharpoons \text{(imidobisamido)} \qquad (21)$$

As a final illustration, we consider systems where formal intramolecular hydrogen transfer leads to interconversion of alkyl, alkylidene, and alkylidyne ligands. Treatment of Cp*Ta(C^{t}Bu)Cl(PMe$_3$)$_2$ with one equivalent of neopentyllithium is suggested by Schrock [41] to result in a monomeric neopentyl–neopentylidyne complex, which then rearranges according to Eq. 22. Similarly, the complex Ta(CH^{t}Bu)$_2$–(CH$_2$^{t}Bu)(PMe$_3$)$_2$ exists as a bisalkylidene rather than rearranging to a alkylidyne structure [41]. Yet, just one position to the right on the periodic table, the complex W(C^{t}Bu)(CH$_2$^{t}Bu)$_3$ [54] possesses the alkyl–alkylidyne formulation. (The molybdenum analog is also known [18].) Increased steric congestion upon addition of trimethylphosphine results in the formation of an alkylidene ligand (Eq. 23). However, in this α-hydrogen cleavage it is an alkyl ligand rather than the alkylidyne which serves as ultimate base [40]. We suggest that these results could not have been predicted given the current level of knowledge. The different behavior of tantalum versus tungsten is again consistent with the apparent increasing importance of π- versus σ-bonded interactions in d^0 systems as one proceeds to the right on the periodic table.

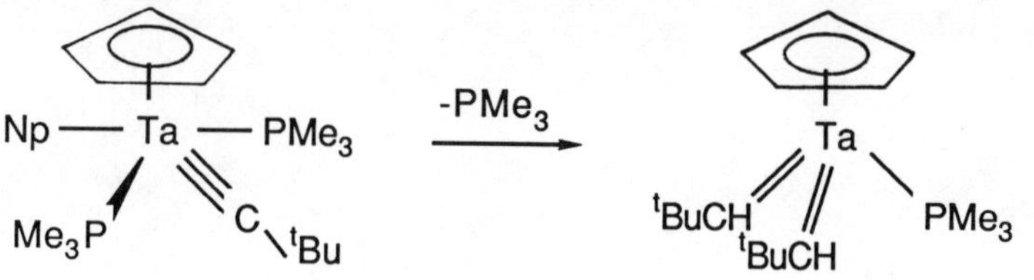

$$\text{Np—Ta(Cp*)(PMe}_3\text{)(PMe}_3\text{)(C}^t\text{Bu)} \xrightarrow{-\text{PMe}_3} \text{Cp*Ta(=CH}^t\text{Bu)}_2\text{(PMe}_3\text{)} \quad (22)$$

$$\text{W(≡C}^t\text{Bu)(Np)}_3 \xrightarrow[-\text{CMe}_4]{2\text{PMe}_3} \text{Np—W(≡C}^t\text{Bu)(=CH}^t\text{Bu)(PMe}_3\text{)}_2 \quad (23)$$

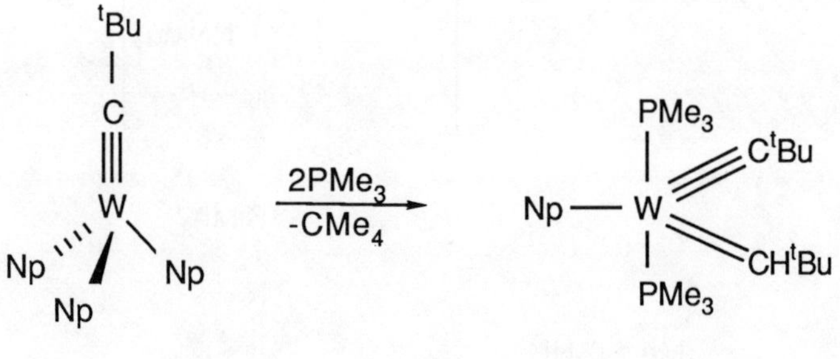

3.2.2 Cleavage of Silicon and Other Electropositive Leaving Groups

Routes to metal–ligand multiple bonds based on the cleavage of an α-silyl substituent offer a number of advantages. The formation of a strong covalent bond to silicon, particularly a Si—F or Si—O bond, can enhance the driving force for such reactions. Complications arising from hydrogen bonding in the product mixture are diminished or eliminated. Moreover, the typical side-products from such reactions are volatile species such as Me$_3$SiCl or Me$_3$SiOSiMe$_3$, which can simply be distilled away with the solvent. As a simple example, tungsten hexafluoride reacts with hexamethyldisiloxane in acetonitrile to afford a monooxo complex as shown in Eq. 24 [55]. In other media, introduction of two oxo groups can be achieved using the same reactants [56]. Other applications are shown in Eqs. 25–28 [57–60].

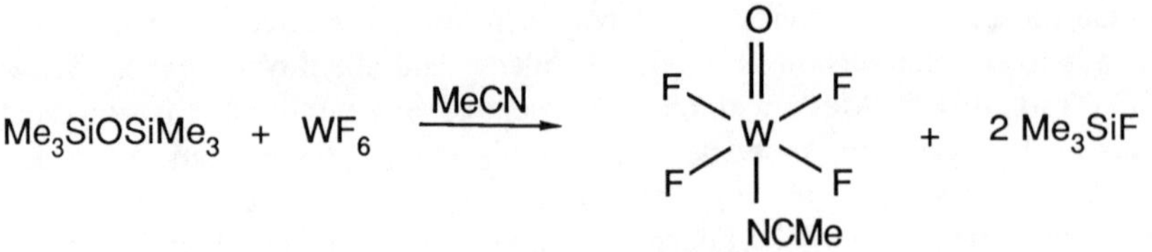

$$Me_3SiOSiMe_3 + WF_6 \xrightarrow{MeCN} \underset{NCMe}{F_4W(O)(NCMe)} + 2\,Me_3SiF \quad (24)$$

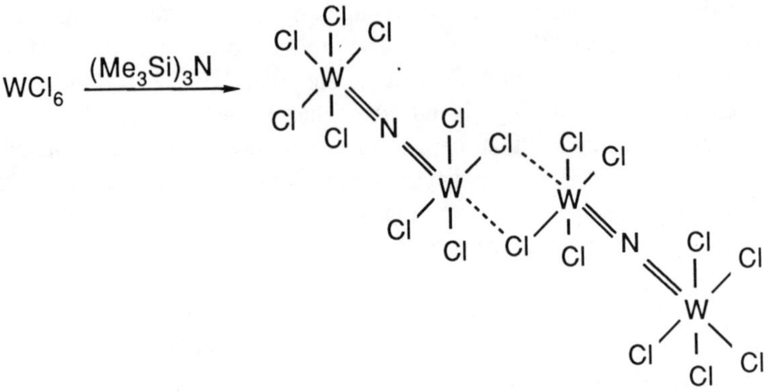

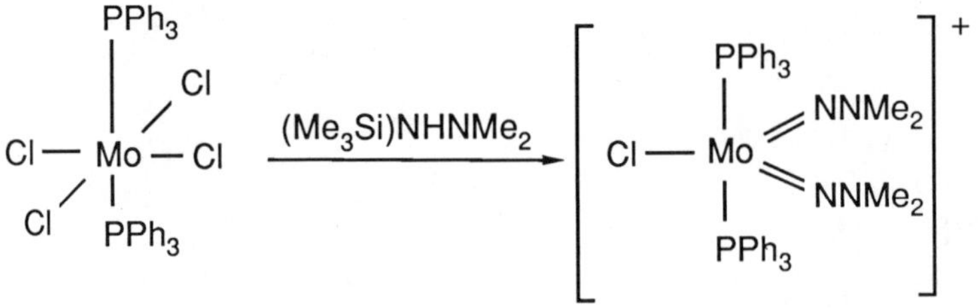

$$WCl_6 \xrightarrow{(Me_3Si)_3N} [\text{trinuclear W-N-W-N-W chloride imido cluster}] \quad (25)$$

$$\underset{PPh_3,\,Cl_3}{MoCl_3(PPh_3)_2} \xrightarrow{(Me_3Si)NHNMe_2} [Cl(PPh_3)_2Mo(=NNMe_2)_2]^+ Cl^- \quad (26)$$

$$NH_4VO_3 \xrightarrow[DMF]{(Me_3Si)_2NH} (Me_3SiO)_3V=N-SiMe_3 \quad (27)$$

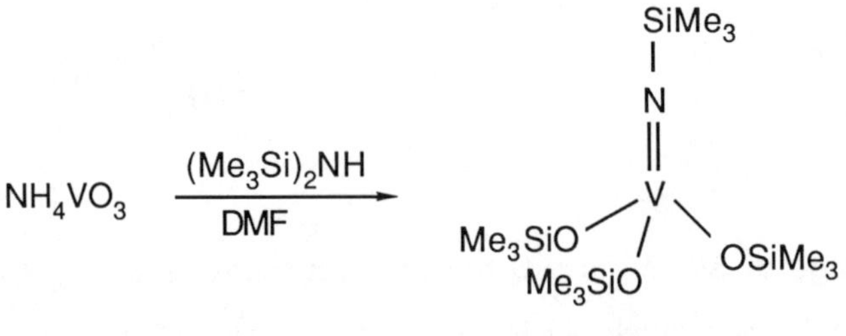

$$Cp_2V + Me_3SiN=NSiMe_3 \longrightarrow Cp_2V=N-N(SiMe_3)_2 \quad (28)$$

The silicon-based approach has been particularly productive for synthesis of imido compounds. An early example was Winfield's synthesis of a tungsten methylimido species shown in Eq. 29 [61,55]. Recently, many syntheses based on the use of mono-trimethylsilyl amine derivatives, $RNH(SiMe_3)$, have been reported. Selected examples are given in Table 3.2.

TABLE 3.2 Syntheses of *t*-Butylimido Complexes with ^{t}BuNH(SiMe$_3$)

Starting Complex	Additional Reactant	Product	Reference
NH$_4$VO$_3$	—	(Me$_3$SiO)$_3$V(N^{t}Bu)	62
V(NPh)Cl$_3$	PhNCO	V$_3$Cl$_2$(N^{t}Bu)$_3$(μ_2–NPh)$_2$(μ_3–PhNCONH^{t}Bu)	63
NbCl$_5$	PMe$_3$	Nb(N^{t}Bu)Cl$_3$(PMe$_3$)$_2$	64
TaCl$_5$	—	[TaCl(μ–Cl)(N^{t}Bu)(NH^{t}Bu)(NH$_2$^{t}Bu)]$_2$	65
TaCl$_5$	PMe$_3$	Ta(N^{t}Bu)Cl$_3$(PMe$_3$)$_2$	64
CrO$_2$Cl$_2$	—	(Me$_3$SiO)$_2$Cr(N^{t}Bu)$_2$	33
CrO$_3$	—	(Me$_3$SiO)$_2$Cr(N^{t}Bu)$_2$	66
MoO$_2$Cl$_2$	—	(Me$_3$SiO)$_2$Mo(N^{t}Bu)$_2$	33
WCl$_6$	bipy	(bipy)WCl$_2$(N^{t}Bu)$_2$	67
W(NPh)Cl$_4$	—	[WCl$_2$(μ–NPh)(N^{t}Bu)(NH$_2$^{t}Bu)]$_2$	68
(Me$_3$SiO)ReO$_3$	—	(Me$_3$SiO)Re(N^{t}Bu)$_3$	66
OsO$_4$	—	OsO$_2$(N^{t}Bu)$_2$	69

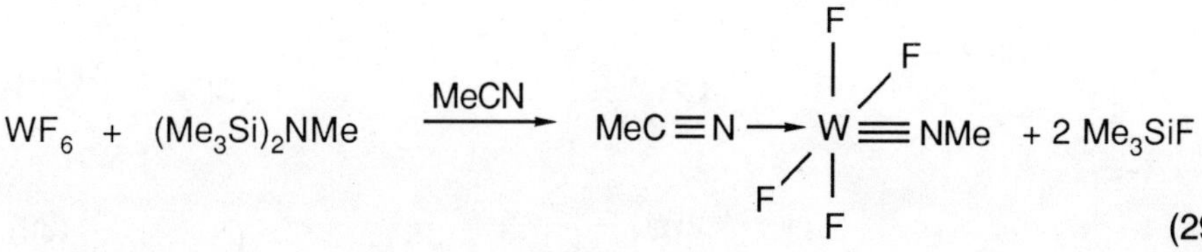

$$WF_6 + (Me_3Si)_2NMe \xrightarrow{MeCN} MeC\equiv N \rightarrow F_4W\equiv NMe + 2\,Me_3SiF \qquad (29)$$

Application of silicon-based methodology to the synthesis of alkylidene species has proven less useful. This is consistent with the lesser driving force for the formation of single-faced π-donor alkylidene when compared with the oxo and imido examples above. For example, (Me$_3$SiCH$_2$)$_3$V=O apparently shows no tendency toward C to O silicon migration [70]. But there appear to be additional complications: A number of early transition metal halides upon treatment with trimethylsilylmethyl lithium are found to undergo hydrogen transfer (to afford bridging alkylidynes) in preference to silicon transfer [15,71,20]. However, Soviet workers have suggested that treatment of tungsten hexachloride with substoichiometric amounts of trimethylsilylmethyl lithium proceeds via Eq. 30 to give a tungsten methylidene species [72]. Indirect evidence for Eq. 30 (ethylene formation, initiation of ring-opening polymerization) was given, but no discrete tungsten methylidene species were isolated. The use of α-silicon cleavage for the synthesis of main-group alkylidene species [73] provides some hope that this may yet prove an entry into transition metal analogs.

$$WCl_6 \xrightarrow[-LiCl,\,Me_3SiCl]{LiCH_2SiMe_3} \text{"W(CH}_2\text{)Cl}_4\text{"} \qquad (30)$$

The silicon strategy has analogies in synthetic routes based on other electropositive leaving groups. The Tebbe–Grubbs reagent [74,75] and its zirconium analogs [76] typically are generated in "masked" form with an aluminum atom bound to the α carbon. Treatment with nitrogen bases removes the aluminum (Eq. 31); in the case of zirconium, the free alkylidene species can be observed and characterized. The use of a more electropositive transition metal as the leaving group is a largely unexplored area. One fascinating example exists in the transfer of an existing alkylidene ligand from tantalum to tungsten in Eq. 32 [77].

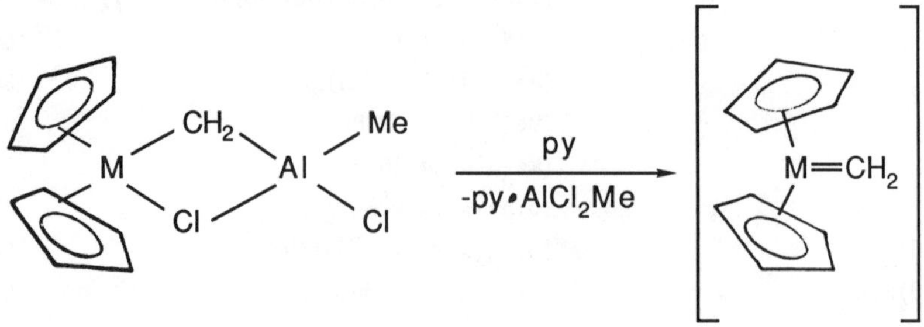

$$\text{Cp}_2\text{M}(\mu\text{-CH}_2)(\mu\text{-Cl})\text{AlMeCl} \xrightarrow[-\text{py}\cdot\text{AlCl}_2\text{Me}]{\text{py}} [\text{Cp}_2\text{M}=\text{CH}_2] \quad (31)$$

$$\text{Cl}_2(\text{PMe}_3)_2\text{Ta}=\text{CHR} + \text{W(OR)}_4\text{O} \longrightarrow 1/2[\text{Ta(OR)}_4\text{Cl}]_2 + \text{Cl}_2(\text{PMe}_3)_2\text{W(O)}=\text{CHR}$$

R = ᵗButyl (32)

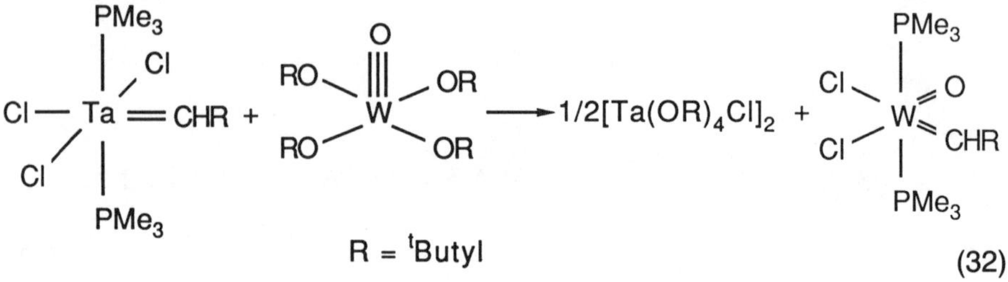

3.2.3 Routes Based on Dealkylation

In favorable cases, the driving force of multiple bond formation promotes synthetic routes based on the cleavage of carbon from the α atom of the ligand precursor. Such reactions fall into two categories. The first category includes those in which the oxidation state of the transition metal does not change, a class of reactions that generally requires significant Lewis acid character in the transition metal reactant. The other type of dealkylation reaction utilizes a low-valent transition metal which is oxidized during the course of the reaction, providing additional driving force for the transformation.

As an example of the first type of reaction, tungsten hexafluoride is reported to dealkylate dimethyl ether (Eq. 33) and a variety of other oxygen-containing organic molecules [78,79]. (Complete dealkylation of dialkylamines to the nitride has similarly been reported [80].) Reaction of CpMoCl$_4$ with propylene oxide is a uniquely effective route to CpMoOCl$_2$ [81]. Deoxygenation of 2,4-pentanedione by WOCl$_4$ affords, in addition to a dioxotungstate anion, an unusual organic product, the 2,4,6-trimethyl-3-acetyl-pyrilium cation (Eq. 34) [82]. Cleavage of tertiary butyl groups (Eqs. 35 and 36) appears especially facile, suggesting that at least in some cases dealkylation may proceed via an S_N1-type mechanism [83,34]. However in other cases the mechanism can be quite obscure; see, for example, the *N*-demethylation reaction in ref. 83a. $MoCl_5$ and $TaCl_5$ will also deoxygenate ethers but the products are usually oxo-bridged polymers; a possible exception is the complex $(dioxane)_2TaCl_3O$ prepared from $TaCl_5$ and *p*-dioxane [84].

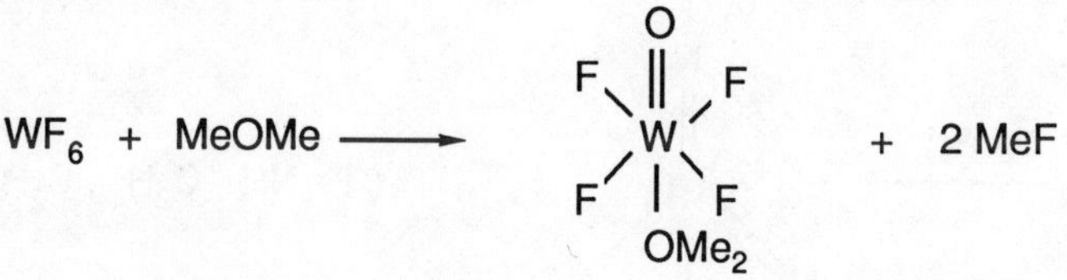

$$WF_6 + MeOMe \longrightarrow WF_5(OMe_2)(O) + 2\,MeF \qquad (33)$$

$$(WOCl_4)_n + 3\,MeCCH_2CMe \xrightarrow[-2HCl]{-H_2O} [\text{trimethylpyrylium}]^+ [WO_2Cl_2(\text{acac})]^- \qquad (34)$$

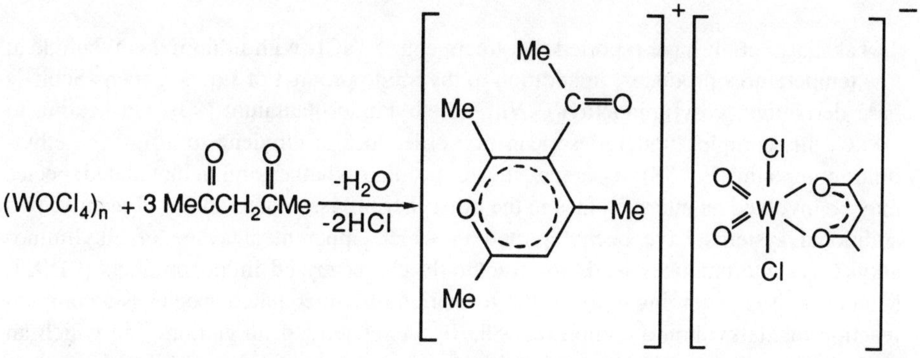

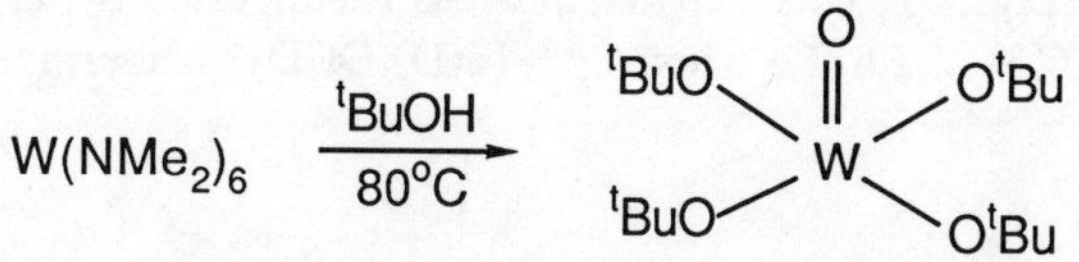

$$W(NMe_2)_6 \xrightarrow[80°C]{^tBuOH} W(O^tBu)_4(O) \qquad (35)$$

$$\text{}^tBuN{=}Os(O)_3 \xrightarrow[-H_2O,\,Cl_2]{HCl} [Cl_4Os{\equiv}N]^- \qquad (36)$$

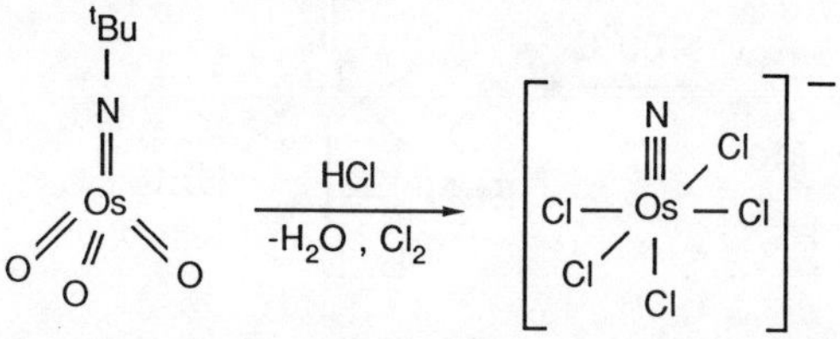

Under controlled conditions it is possible to isolate the species $Cl_5W(OMe)$ [85]. Consistent with the above discussion, this material is found to be a powerful methylating agent roughly comparable to methyl triflate. Thermolysis of $Cl_5W(OMe)$ has been used as a route to very pure $WOCl_4$ [12].

One of the first reported routes to alkylimido derivatives involved the *in situ* thermolysis of pentakis(dialkylamido) tantalum derivatives (Eq. 37) reported by Bradley

[86,87]. An interesting feature of this reaction is that for unsymmetrical amides such as the *N*-methyl-*N*-butyl analog, it is the smaller methyl group that is cleaved. Related dealkylations of dialkylamido tungsten and niobium compounds subsequently have been reported [88,89].

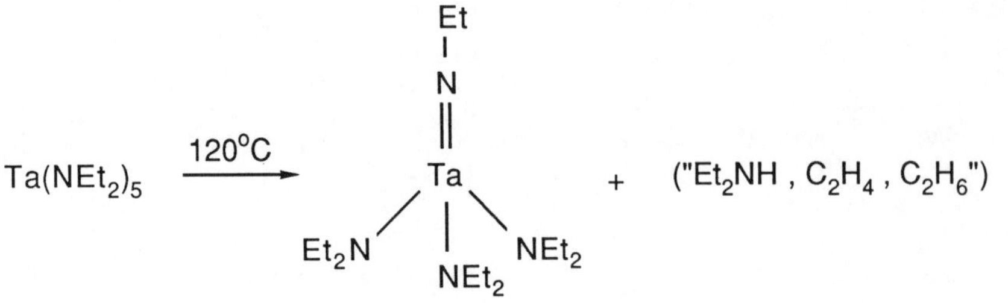

$$Ta(NEt_2)_5 \xrightarrow{120°C} (Et_2N)_3Ta(=NEt) + (\text{"Et}_2NH, C_2H_4, C_2H_6\text{"}) \quad (37)$$

Takahashi et al. have reported that treatment of TaCl$_5$ with lithium diethylamide at low temperatures produces, in addition to the imido product of Eq. 37, an η^2-Schiff's base derivative, ethyliminoethyl*(C,N)*tris(diethylamido)tantalum [90]. On heating to 100°C, this complex undergoes clean first-order loss of ethylene to afford the ethylimido compound (Eq. 38). It is interesting to speculate whether similar metallated species may be involved as intermediates in the other thermolyses noted above. Recently, two additional systems have been reported in which apparent cleavage of alkyliminoalkyl*(C,N)* intermediates leads to structurally characterized imido products [91,92]. Moreover, it is becoming evident that formation of β-metallated species is a common reaction of dialkylamido complexes [88,30]. A related "dealkylation," in which an η^2-formaldehyde complex is cleaved to a terminal oxo ligand, has also been reported. Interestingly, the proposed mechanism involves prior reversion to a tantalum(III) methoxide species (Eq. 39). This stepwise process is supported by an inverse isotope effect (k_H/k_D = 0.46(3) at 140°C) when Cp*$_2$Ta(D)(OCD$_2$) is thermolyzed [43].

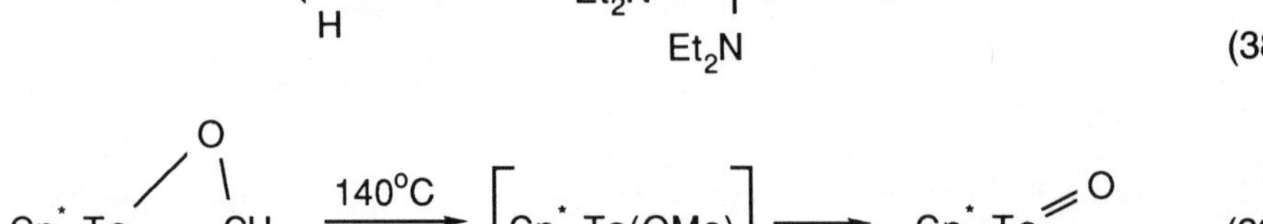

$$(Et_2N)_3Ta\begin{pmatrix}N-Et\\|\\C-Me\\|\\H\end{pmatrix} \xrightarrow{100°C} (Et_2N)_3Ta(=NEt) + CH_2=CH_2 \quad (38)$$

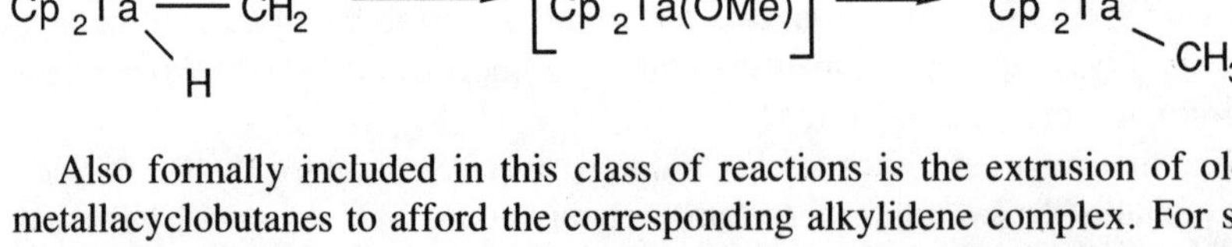

$$Cp^*_2Ta\begin{pmatrix}O\\|\\CH_2\\|\\H\end{pmatrix} \xrightarrow{140°C} [Cp^*_2Ta(OMe)] \longrightarrow Cp^*_2Ta\begin{pmatrix}O\\\\CH_3\end{pmatrix} \quad (39)$$

Also formally included in this class of reactions is the extrusion of olefins from metallacyclobutanes to afford the corresponding alkylidene complex. For such applications it is desirable to use metallacyclobutanes in which β hydrogen atoms are absent so that competing decomposition pathways are precluded [93]. The preequilibrium in Eq. 40 lies heavily to the left; nevertheless, such metallacyles provide a convenient source of alkylidene species free of Lewis acid/base contaminants [94,95]. Eq. 40 and some related reactions show a first-order dependence on the titanacyclobutane concentration and zeroth-order dependence on substrate [96]. This observation, as well as labeling studies on the rearrangement of titanacyclobutanes [97], suggests that such reactions involve a rate-limiting extrusion of the olefin to give the titanium methylidene complex as reactive intermediate.

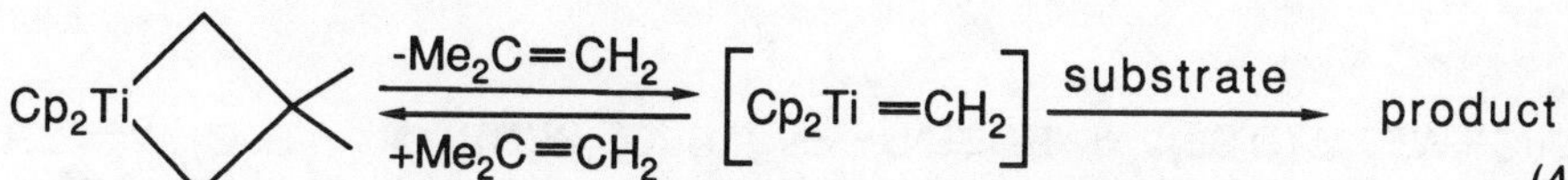

$$Cp_2Ti\diagup\diagdown \underset{+Me_2C=CH_2}{\overset{-Me_2C=CH_2}{\rightleftarrows}} [Cp_2Ti=CH_2] \xrightarrow{substrate} product \qquad (40)$$

A particularly interesting example of such a process is the intramolecular case shown in Eq. 41 [98]. The resultant substituted alkylidene is free of β hydrogen; it was shown to be a uniquely effective catalyst for ring-opening polymerization of norbornylene [99]. In contrast to the titanacyclobutanes, the corresponding titanacyclobutenes will not in general revert thermally to the alkylidene complex and acetylene. An exception is the adduct of bis(trimethylsilyl)acetylene in Eq. 42 which reversibly extrudes the acetylene at 85°C [100,101].

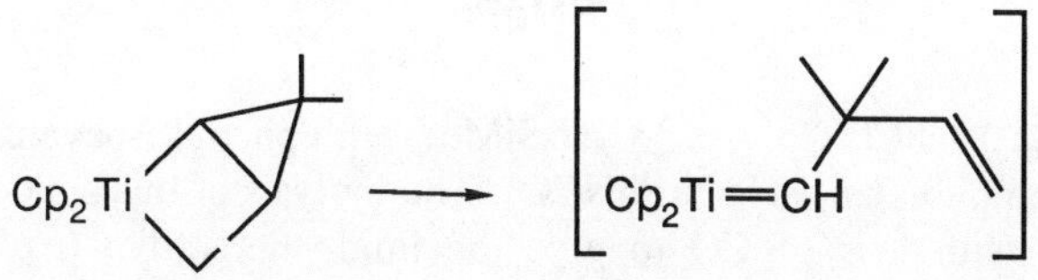

$$Cp_2Ti \longrightarrow [Cp_2Ti=CH] \qquad (41)$$

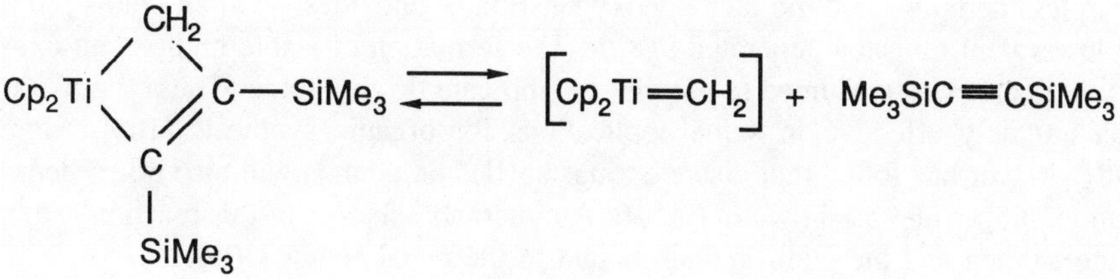

$$Cp_2Ti\diagup\overset{CH_2}{\underset{C}{\diagdown}}C-SiMe_3 \rightleftarrows [Cp_2Ti=CH_2] + Me_3SiC\equiv CSiMe_3$$

$$|\ SiMe_3 \qquad (42)$$

A striking example of an *oxidative* carbon–oxygen bond cleavage has been reported by Wolczanski [102]. It involves the cleavage of carbon monoxide by a tantalum(III) siloxide as shown in Eq. 43. The cleavage of CO resulting in an oxotungsten complex has also been reported as has cleavage of isonitriles to give the tungsten organoimido analogs [103]. A somewhat related example is the cleavage of nitriles by ditungsten hexa-*t*-butoxide according to Eq. 44 [104]. The resultant tungsten(VI)–nitrido complex precipitates as a nitrido-bridged polymer. (The cleavage of acetylenes by the same reagent is discussed in Section 3.2.5.) A particularly clean reaction of this type is the cleavage of carbon dioxide by tungsten(II) to afford a structurally characterized oxotungsten(IV) carbonyl complex (Eq. 45) [105]. The same tungsten complex will also cleave isocyanates RNCO and carbodiimides RN=C=NR to afford organoimido–

tungsten derivatives W(NR)Cl$_2$(CO)L$_2$ and W(NR)Cl$_2$(CNR)L$_2$, respectively. In very recent work, Mayer has demonstrated the cleavage of cyclopentanone to afford an oxotungsten alkylidene complex [360].

$$2\,(RO)_3Ta + CO \longrightarrow (RO)_3Ta=O + 1/2\,(RO)_3Ta=C=C=Ta(OR)_3$$

$$R = {}^tBu_3Si \tag{43}$$

$$({}^tBuO)_3W\equiv W(O{}^tBu)_3 + RC\equiv N \longrightarrow [({}^tBuO)_3W(N)]_n + ({}^tBuO)_3W\equiv CR \tag{44}$$

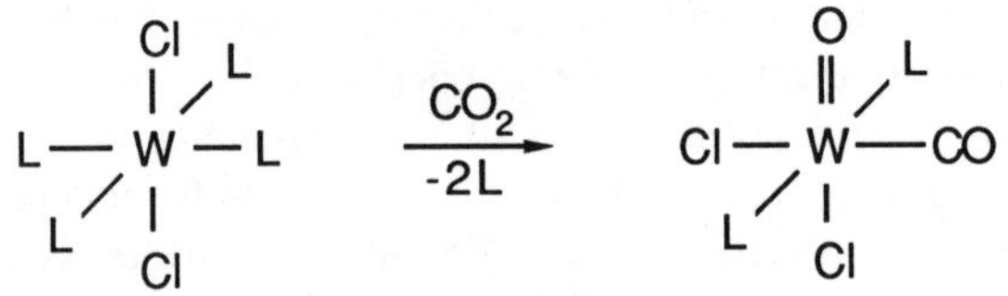

L = PMePh$_2$

Reaction of (Cp')$_2$NbCl (Cp' = η^5-C$_5$H$_4$SiMe$_3$) with phenyl isocyanate is reported to give a stable complex (Cp')$_2$NbCl(PhNCO). Thermolysis of this adduct in refluxing toluene proceeds with loss of CO to give the imido derivative (Cp')$_2$NbCl(NPh) [105a].

Sharpless demonstrated the facile deoxygenation of epoxides [106] and diols [107] using low-valent tungsten generated *in situ*. The formation of stable high-valent oxo-tungsten species was presumed to provide a significant driving force. The reaction has proven uniquely effective in some applications for organic synthesis [108]. More recently, Mayer has found that discrete tungsten(II) compounds will also effect deoxygenation of epoxides as shown in Eq. 46. A remarkable feature of this reaction is that both the oxygen and the olefin remain bound to the metal center [105].

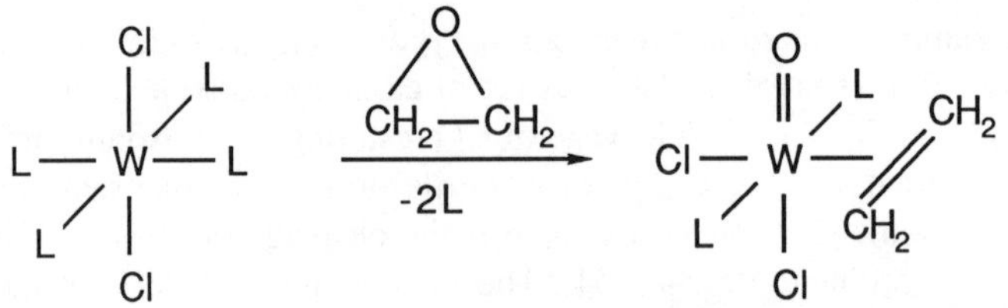

L = PMePh$_2$

A clever modification of the tungsten-mediated de-epoxidation involves the use of tungstenocene as the oxo acceptor, as reported by Green [109]. The oxo ligand in

$Cp_2W=O$ is apparently more labile than in other monooxotungsten species and can be reduced back to tungstenocene with sodium amalgam. (Note that the usual triple-bonded oxo would result in a 20-electron complex.) Thus, the combination of tungstenocene and excess amalgam allows the catalytic deoxygenation of epoxides. No rate data are provided for the catalytic reaction, which apparently is quite slow.

3.2.4 Routes Involving Cleavage of an Electronegative Leaving Group from the α Atom

This class of reaction usually results in a net oxidation of the metal, since the electronegative leaving group tends to depart with the electron pair that was initially involved in bonding to the α atom. In known examples the atom initially bonded to the α atom may be a halogen, oxygen, sulfur, nitrogen, phosphorus, or other pnictogen. These will be discussed in Sections 3.2.4.1–3.2.4.5. In Section 3.2.4.6 we mention some formally related examples in which the departing atom is a more electronegative transition metal. A special case of these reactions is cleavage of a symmetrical precursor such as dioxygen. Because of their importance these reactions are treated separately in Section 3.2.5.

3.2.4.1 Cleavage of α Halogen.

A variety of catalytic systems for olefin oxidation utilize the hypochlorite ion as the stoichiometric oxidant to generate reactive oxo species *in situ*. The oxo species involved include RuO_4 [110], ruthenium(IV) derivatives [111], and porphyrin-containing complexes of iron [112] or manganese [113]. In general it is not known whether these reactions proceed by cleavage of a coordinated hypochlorite moiety or, for example, by oxidation of the metal via outer-sphere electron transfer followed by deprotonation of coordinated water. However, in the case of the manganese porphyrin systems, labeling studies and kinetics both point to an O—Cl bond cleavage is indicated in Eq. 47 [113,114]. In fact, a high-valent "oxo-like" manganese porphyrin complex prepared via NaOCl oxidation has been isolated and shown to epoxidize styrene both stoichiometrically and catalytically [113].

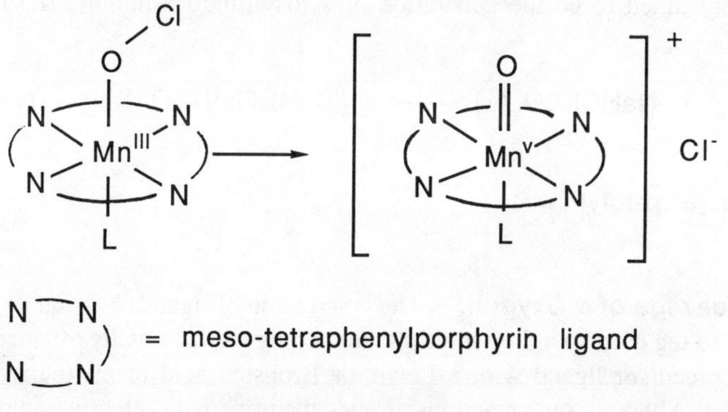

(47)

Periodate ion, IO_4^-, and especially iodosobenzene, "PhIO," have proven to be useful stoichiometric oxidants in biomimetic oxidations utilizing porphyrin complexes of iron [115,116], chromium [117], and manganese [118,119]. In each case the reaction is presumed to proceed by transfer of the oxo moiety to the porphyrin-bound metal; in the case of chromium, the oxo-containing product has been structurally characterized. Iodosobenzene has also been utilized by Kochi to prepare structurally characterized oxochromium(V) complexes with salicylidene ligands [120,121]. The use of the more soluble perfluorinated analog C_6F_5IO has been suggested to be advantageous in some applications [116]. Periodate is frequently used as a secondary oxidant in catalytic RuO_4 and OsO_4 oxidations in organic chemistry [122,123].

Similarly, N–Cl bond cleavage can provide a convenient route to metal–nitrogen multiple bonds. For example, treatment of tungsten hexacarbonyl with N,N-dichlorophenylsulfonamide leads in good yield to the yellow polymeric compound $[(PhSO_2N)_2WCl_2]_x$, which may be converted to the octahedral complex $[(PhSO_2N)_2WCl_2(MeCN)_2]$ by recrystallization from acetonitrile [124]. The reaction of nitrogen trichloride with rhenium(V) chloride affords nitridorhenium(VII) tetrachloride, as shown in Eq. 48 [125]. Moreover, in certain cases NX-type ligands will undergo substitution reactions: The complex $Cl_3V{=}NI$, upon treatment with chlorine or bromine, is converted to $Cl_3V{=}NCl$ or $Cl_3V{=}NBr$, respectively [126]. (However, early claims that the N—Cl bond in $Cl_3V{=}NCl$ is readily cleaved by treatment with either Lewis acids or Lewis bases [127] subsequently have been revised [128].)

$$ReCl_5 + NCl_3 \longrightarrow \left[\begin{array}{c} Cl \\ | \hspace{-0.2em}\diagup Cl \\ -Re{\equiv}N- \\ \diagup | \\ Cl \hspace{0.5em} Cl \end{array} \right]_n + 2\,Cl_2 \qquad (48)$$

Sharpless and coworkers utilized an N-chlorinated reagent, "chloramine-T", in the synthesis of the only known homoleptic imido complex as shown in Eq. 49 [129]. This same reagent was utilized as stoichiometric oxidant in the catalytic version of the osmium-mediated "oxyamination" reaction [130,131]. Here again the role of chloramine-T is presumed to be the generation of a tosylimido osmium(VIII) intermediate.

$$[Os(N^tBu)_3]_n + NaNCl(SO_2Ar) \longrightarrow (^tBuN)_3Os(NSO_2Ar) + NaCl$$

$$Ar = p\text{-tolyl} \qquad (49)$$

3.2.4.2 Cleavage of α Oxygen.

The conversion of ligands bearing an α oxygen functionality to the corresponding multiply bonded ligands is usually effected by treatment of the precursor ligand with a Lewis or Bronsted acid or alternatively with a reducing agent. Although such reactions provide the principal synthetic route to Fischer carbyne complexes, they are far less common as routes to the high-valent systems that constitute the main topic of this volume. Nevertheless, peracids have been found to be effective reagents for *in situ* oxidation of iron(III) porphyrin species to formal oxoiron(V) intermediates [132]. Kinetic studies indicate that this process requires prior formation of an iron(III) perbenzoate intermediate and is first order in [H$^+$], suggesting a need for protonation of the benzoate leaving group. A related reaction involves the cleavage of a manganese(III) perbenzoate, generated by acylation of a manganese peroxo complex, to give a formal manganese(V) oxo [133]. (See Section 7.1.3.)

Solutions containing the ruthenium(III) nitrosyl species K$_2$[Ru(NO)Cl$_5$] upon treatment with formaldehyde or tin(II) chloride afford the bridging nitrido complex K$_3$[Ru$_2$Cl$_8$(H$_2$O)$_2$N] [134–136]. Treatment of certain molybdenum nitrosyl complexes with alkylaluminum produces an active methathesis catalyst [137]. An N—O bond cleavage resulting in a terminal nitrido ligand apparently occurs under these conditions. The resultant "intramolecular redox isomerization" has been represented schematically as Eq. 50.

$$[\text{Mo}^{\circ}\text{Cl(NO)} \cdot x\text{AlCl}_3 \cdot y\text{AlEtCl}_2] \longrightarrow [\text{NMo}^{\text{VI}}(\text{O})\text{Cl} \cdot \text{AlCl}_3 \cdot y\text{AlEtCl}_2] \quad (50)$$

Cotton [138] has demonstrated the formation of a product with a terminal imido group, [W(OCMe$_3$)$_2$(NPh)]$_2$(μ-O)(μ-OCMe$_3$)$_2$, upon treatment of triply bonded ditungsten hexa-*tert*-butoxide with nitrosobenzene. (Here presumably the incipient bonding of the oxygen to tungsten is promoting the N—O cleavage.) An interesting related reaction is the deoxygenation of a formal rhenium nitrosobenzene complex upon treatment with excess isonitrile or trialkylphosphine. In the case of triphenylphosphine the reaction follows Eq. 51 [139].

$$\text{Re(ONAr)Cl}_3(\text{OPPh}_3) \xrightarrow[\text{MeCN}]{\text{PPh}_3} \text{Re(NAr)Cl}_3(\text{PPh}_3)_2 + 2\,\text{Ph}_3\text{PO} \quad (51)$$

Ar = p-tolyl, p-methoxyphenyl

Until recently α oxygen cleavage had not been employed for the direct synthesis of d^0 alkylidene or alkylidyne derivatives. However, as noted above, the classic route to Fischer carbyne complexes [140] has been the Lewis-acid-promoted elimination of an alkoxide from the corresponding carbene (Eq.52). Mayr and coworkers have discovered that tungsten carbynes can be oxidized with elemental bromine to afford the tungsten(VI) alkylidyne [141]. Thus the combination of Eqs. 52 and 53 opens an important new route to the high-valent analogs. Very recently Mayer has demonstrated the oxidative cleavage of cyclopentanone by WCl$_2$L$_4$ which affords an oxotungsten cyclopentylidene complex [360].

$$(\text{CO})_5\text{M}=\text{C}\begin{smallmatrix}\text{OR'}\\\text{R}\end{smallmatrix} + \text{BX}_3 \xrightarrow{-(\text{BX}_2\text{OR'})} (\text{CO})_4\text{M}(\equiv\text{CR})(\text{X}) + \text{CO}$$

M = Cr, Mo, W ; X = Cl, Br, I $\quad (52)$

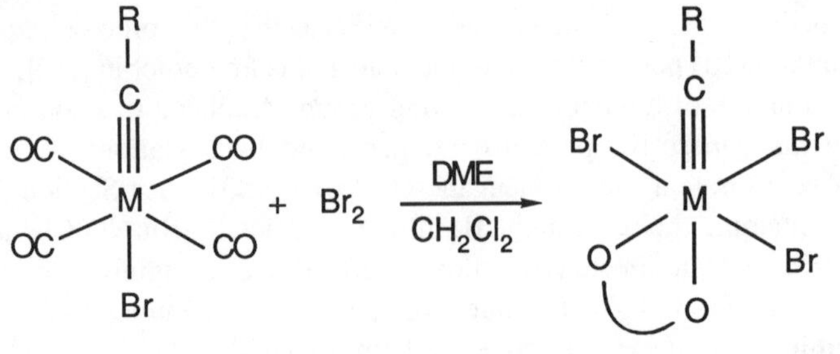

$$\text{OC}\underset{\underset{\text{Br}}{|}}{\overset{\overset{\text{R}}{\underset{|}{\text{C}}}}{\underset{|||}{\text{M}}}}\text{CO} + \text{Br}_2 \xrightarrow[\text{CH}_2\text{Cl}_2]{\text{DME}} \text{products} \quad (53)$$

M = Mo, W ; O⌒O = dimethoxyethane

3.2.4.3 Cleavage of α Sulfur.

By far the most common reactions in this category involve the cleavage of a sulfoxide to afford a metal–oxo complex and the dialkyl sulfide. It has long been known [142,143] that molybdenum(V) chloride would abstract oxygen from either dimethyl or diphenyl sulfoxide, as in Eq. 54. Similarly, it has been reported that Cp_2NbCl_2 and its analogs will react with dimethyl sulfoxide (DMSO) to afford Cp_2NbOCl containing a terminal oxo group [144]. Niobium(V) bromide reacts with DMSO to afford $NbOBr_3(DMSO)_2$ [145].

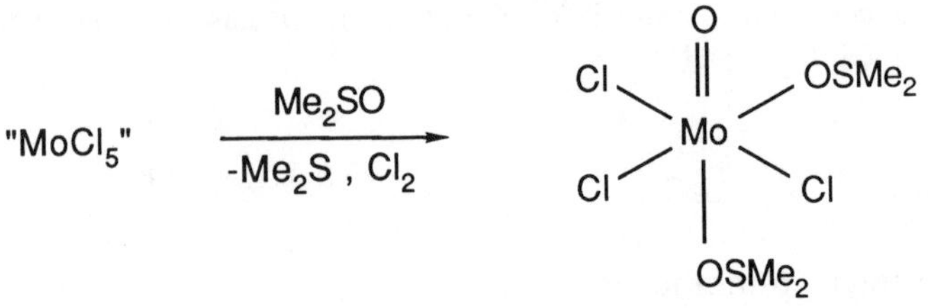

$$\text{"MoCl}_5\text{"} \xrightarrow[-\text{Me}_2\text{S, Cl}_2]{\text{Me}_2\text{SO}} \text{MoOCl}_3(\text{OSMe}_2)_2 \quad (54)$$

Interest in such reactions was greatly enhanced by the 1975 report of Mitchell and Scarle that sulfoxide deoxygenations could be carried out under mild conditions using sulfur-ligated molybdenum(IV) complexes (Eq.55) [146]. Such reactions are suggested to be relevant to the mode of action of the molybdenum-containing "oxo-transferase" enzymes and in particular of sulfite reductase [147]. The kinetics of deoxygenation of DMSO by $(dtc)_2MoO$ have been determined [148]. Noteworthy is the fact that the product molybdenum(VI) dioxo complexes can be reduced with trialkylphosphine; thus the reaction can be run with a catalytic amount of molybdenum complex in the presence of excess phosphine. Recently Holm has introduced the use of thiols as more biologically relevant reducing agents for this reaction [149].

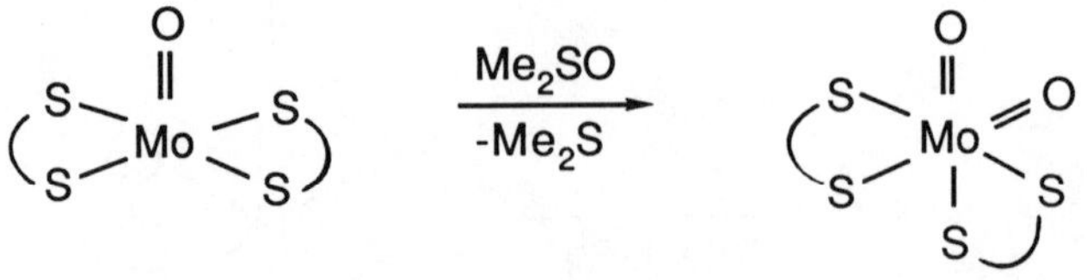

$$(dtc)_2\text{MoO} \xrightarrow[-\text{Me}_2\text{S}]{\text{Me}_2\text{SO}} (dtc)_2\text{MoO}_2 \quad (55)$$

S⌒S = diethyldithiocarbamate

Thionitrosyl complexes of molybenum, rhenium, and osmium upon treatment with tributylphosphine all afford the corresponding nitrido complex as represented schematically in Eq. 56 [150]. However, this interesting reaction has not proven particularly useful as a synthetic route to nitrides since the only known syntheses of thionitrosyls utilize nitrido complexes as starting materials. Of greater utility are reactions involving the scission of the M=NSCl moiety that have been developed by Dehnicke and coworkers as exemplified by Eq. 57 [151]. The (NSCl) ligand is extremely labile; in the case of $MoCl_4(NSCl)$ and $(Cl_3PO)Cl_4Re(NSCl)$, simply treating with Ph_4PCl is sufficient to form the $PPh_4[MNCl_4]$ derivative cleanly [152,153]. Another reaction involving an N—S cleavage is that between cyclopentadienylmolybdenum tricarbonyl dimer and *t*-butyl sulfur diimine, *t*-BuN=S=NBu-*t*. The product is a sulfur-bridged imidomolybdenum(V) dimer, $[CpMo(N^tBu)(\mu-S)]_2$ [154].

$$M-NS + PR_3 \longrightarrow M\equiv N + R_3PS \qquad (56)$$

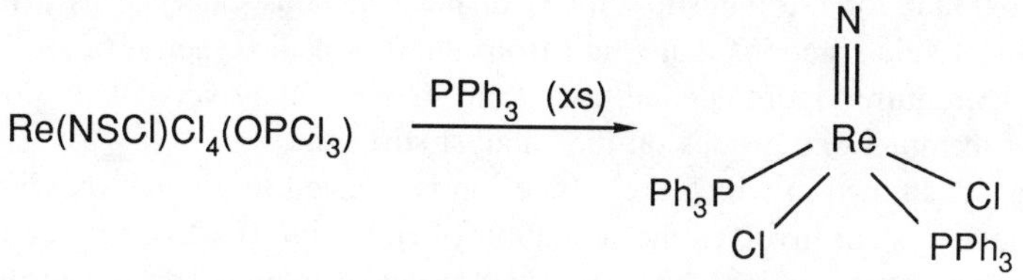

$$Re(NSCl)Cl_4(OPCl_3) \xrightarrow{PPh_3 \text{ (xs)}} \underset{\underset{Cl}{Ph_3P}}{\overset{\overset{N}{\overset{|||}{}}}{\underset{PPh_3}{Re}}}Cl \qquad (57)$$

3.2.4.4 Cleavage of α Nitrogen.

A variety of oxo nitrogen species have proven to be effective reagents for the preparation of oxo transition metal compounds. For example, the oxidation of either $ReMe_6$ or $Me_4Re=O$ with nitrogen monoxide (NO) affords *cis*-trimethyldioxorhenium(VII), ReO_2Me_3 [155]. In this and related reactions, the mechanistic details are typically obscure because of the radical nature of NO and because the nitrogen-containing products have not been identified. Treatment of $Cp_2^*VCl_2$ with NO gives monomeric $Cp^*VCl_2(O)$ as the major product [156]. Thermal decomposition of a discrete tungsten nitrosyl derivative has been shown to follow Eq. 58 [157], while oxotungsten acetylene complexes are formed in Eq. 59 [158].

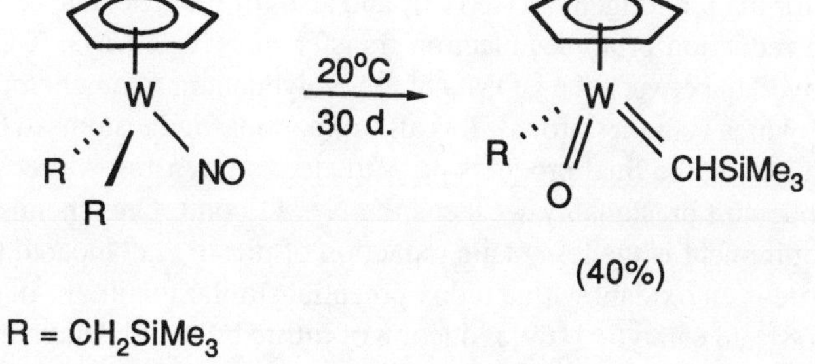

$$\text{[Cp-W(R)(R)(NO)]} \xrightarrow[30 \text{ d.}]{20°C} \text{[Cp-W(R)(O)(CHSiMe}_3\text{)]} \qquad (58)$$

(40%)

$R = CH_2SiMe_3$

Dinitrogen oxide (N_2O) is a more robust species than NO, with the result that no example of oxo transfer to form a discrete terminal oxo complex has been reported. Instead, a range of low-valent, early transition metal species were shown to react with N_2O to give oxo-bridged dimers or clusters [159,160]. A claim [146] that N_2O will transfer oxygen to a biologically relevant molybdenum(IV) system appears to be in error [161]. However, it should be noted that N_2O is increasingly utilized as a stoichiometric oxidant for mechanistic studies in heterogenous catalysis as a means to suppress radical side-reactions that result from the use of dioxygen [162,163]. While such reactions require vigorous conditions (300–600°C), they very likely do involve formation of terminal oxo groups on the catalyst surface.

Reactions are known in which the nitrate ion is cleaved to give an oxo ligand and NO_2. One such system involves the oxidation of ruthenium(II) to oxoruthenium(IV) [164]. Related reactions of molybdenum are of especial interest because of their relevance to the mechanism of the enzymes nitrite reductase and nitrate reductase. It has been suggested [165] that the reported [146] reduction of nitrate ion by a molybdenum(IV) complex, $MoO(S_2CNEt_2)_2$, is in error. In contrast, the reduction of nitrate by molydenum(V) has been known for some time [165,166]. Recent isotopic labeling studies demonstrate unambiguously that this reaction can proceed via an oxo transfer process to afford a terminal oxo ligand [167]. Stopped-flow kinetic studies [168] on the reduction of $[NEt_4][NO_3]$ by $MoOCl_3(OPPh_3)_2$ appear broadly relevant to other oxo transfer reactions as well: The mechanism is suggested to involve (1) dissociation of the phosphine oxide ligand trans to the oxo group, (2) addition of nitrate ion as a unidentate ligand at the vacant site, (3) intramolecular rearrangement moving the nitrato group cis to the oxo ligand, and (4) rapid electron transfer and concommitant oxygen atom transfer. This last step results in the expulsion of NO_2 and the introduction of the second oxo group cis to the original oxo ligand.

The necessity for the rearrangement step (3), and also for monodentate coordination of nitrate prior to reduction promotes electron transfer in several ways. The electron transfer requires overlap between the HOMO of the molybdenum fragment (d_{xy}) and the LUMO of nitrate, which is the π^* orbital. It is also important for the atoms to be located close to their positions in the final products prior to electron transfer. Moreover, coordination to molybdenum presumably weakens the N—O bond. One chemical consequence of this requirement is the less facile reduction of nitrite as compared to nitrate. Interestingly, nitrite is an oxidant with a redox potential similar to nitrate. In fact, both stoichiometric [168] and catalytic [169] reduction of nitrite by molybdenum complexes have been demonstrated. However, for MoOCl$_3$(OPPh$_3$)$_2$ the internal rearrangement step (3) is approximately 38 times slower than for nitrate. It has been suggested that this difference reflects the normal preference of nitrite to coordinate as an N-donor ligand.

Oxo transfer from organic nitrogen compounds is also known. Of practical importance is the discovery by Upjohn researchers that N-methylmorpholine-N-oxide is a convenient stoichiometric oxidant for use in catalytic osmium tetroxide oxidations [170]. This reagent was subsequently adopted by Holm and coworkers for studies on oxo transfer to biomimetic molybdenum(IV) compounds (Eq. 60) [148]. Similar oxidations of molybdenum have been achieved using pyridine N-oxide, azoxybenzene, and t-butyl nitrate [171,146]. N-Methylmorpholine-N-oxide will also convert iron(II) porphyrins to their oxoiron(IV) analogs [172].

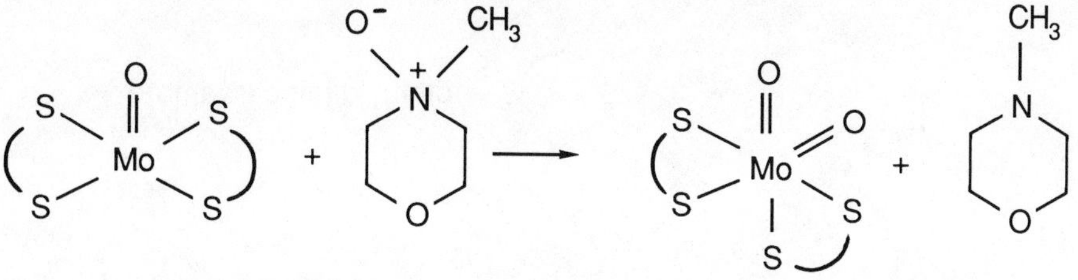

S⌢S = diethyldithiocarbamate   (60)

Another amine N-oxide, p-cyano-N,N-dimethylaniline-N-oxide, has been used as an oxo transfer agent to synthesize oxo–porphyrin complexes of (formally) iron(V), manganese(V), and (with photochemical activation) chromium(V) [173,174]. A feature of these reactions is that the dimethyl aniline that is produced is subsequently oxidized to a host of organic products and formaldehyde. A reaction in which both terminal and bridging oxo groups are formed is represented by Eq. 61 involving nitrobenzene and low-valent tungsten or molybdenum [175].

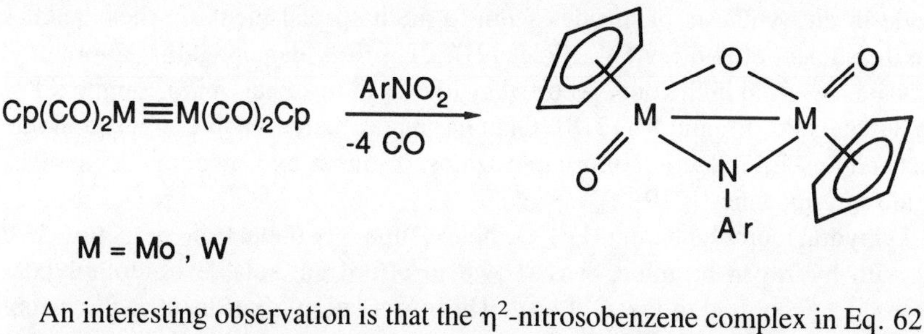

Cp(CO)$_2$M≡M(CO)$_2$Cp $\xrightarrow[-4\text{ CO}]{\text{ArNO}_2}$ (61)

M = Mo, W

An interesting observation is that the η^2-nitrosobenzene complex in Eq. 62 does not cleave to give the molybdenum(VI) valence isomer, despite the fact that the latter is a known stable complex [176]. In contrast, d^0 molybdenum metallooxaziridines are reported to thermolyze with extrusion of the elements of phenyl nitrene to afford oxo-molybdenum(VI) products [177,178]. In the presence of olefins the "phenyl nitrene" fragment is recovered as the allylic amine, in the presence of cylcohexanone as 2-

N-phenyliminocyclohexanone, and in the absence of additive as azobenzene. A similar pathway has been invoked by Wilkinson for the decomposition of an η^2-nitrosomethane intermediate in Eq. 63. When added to the reaction mixture of Eq. 63, styrene is converted to its *N*-methyl aziridine [179].

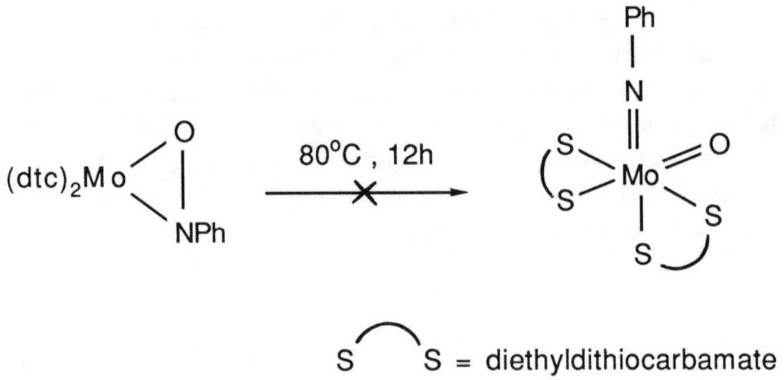

$$(dtc)_2Mo\overset{O}{\underset{NPh}{\diagup\!\!\!\diagdown}} \xrightarrow[\times]{80°C,\ 12h} (\underset{S}{\overset{S}{\diagup}})Mo(\overset{NPh}{\underset{S}{\diagdown}})=O \quad (62)$$

S⌒S = diethyldithiocarbamate

$$Cp_2Nb\overset{N(Me)}{\underset{CH_3}{\diagup\!\!\!\diagdown}}O \xrightarrow{25°C} Cp_2Nb\overset{O}{\underset{CH_3}{\diagdown}} + 1/2\ MeN=NMe \quad (63)$$

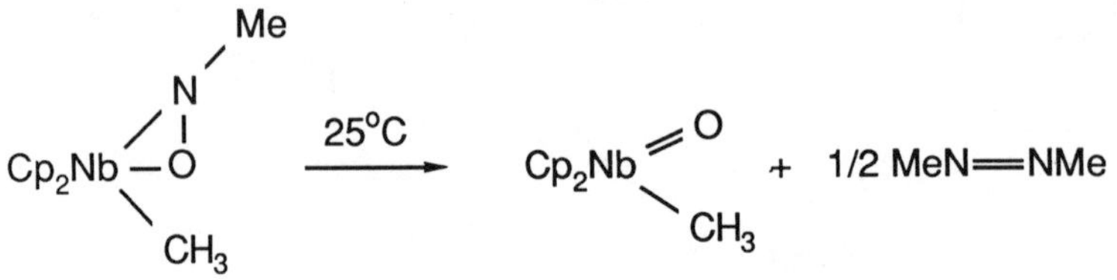

The decomposition of azides with formation of dinitrogen is a broadly useful synthetic route to both nitrido and imido complexes. A general review of the reactions of transition metal species with azides has appeared [139]. The extrusion of N_2 as a side-product results in an easy product isolation as well as providing a large driving force for such reactions. Applications of this approach for the synthesis of nitrido derivatives are summarized in Table 3.3 and for imido synthesis in Table 3.4. The extensive (and courageous!) studies of Dehnicke and coworkers on the use of explosive iodine azide in the synthesis of nitrides seem to merit special mention; these studies have been the subject of two review articles [216,217]. It is also noted that fluoroalkyl azide cleavage was used in the only reported syntheses of terminal imido complexes of the group 8 metals Pd, Pt, and Rh [218]. Olah has made the reasonable suggestion that the synthetically useful reduction of organic azides to amines by vanadium(II) involves organoimido intermediates [219].

A dialkylhydrazidomolybdenum(IV) complex, upon two-electron reduction and treatment with hydrogen bromide, was shown to afford an isolable imidomolybdenum(IV) product according to Eq. 64 [220]. This observation is pertinent to the mechanism of cleavage of dinitrogen by nitrogenase. (See Section 7.3; a variety of other reactions in which hydrazido complexes undergo N—N cleavage without formation of isolable imido complexes are surveyed in Section 6.4.) A series of organoimido complexes have been prepared by the cleavage of unsymmetrical acyl hydrazines according to Eq. 65. The other reaction products were not determined [221]. (For routes involving

TABLE 3.3 Syntheses of Nitrido Complexes Involving Azides

Starting Complex	Azide	Product	Reference[a]
VCl_3	NEt_4N_3	$NEt_4[VNCl_3]$	180
$Nb_2Cl_6(Me_2S)_3$	Me_3SiN_3	$[NbNCl_2(SMe_2)]_n$	181[b]
$Cr(salen)N_3(H_2O)$	—[c]	$CrN(salen)(H_2O)$	182
$MoCl_3(thf)_3$	Me_3SiN_3	$MoNCl_2(PPh_3)_2$	183
$MoCl_3(thf)_3$	Me_3SiN_3[d]	$[MoN(S_2P(OMe)_2)_2]_4$	184
$MoBr_4$	IN_3	$[MoNBr_3]_n$	185
$MoCl_4(NCMe)_2$	NaN_3[e]	$MoNCl_3(Ph_3PO)_2$	186 (183)
$MoCl_4(NCMe)_2$	Me_3SiN_3[f]	$MoN(N_3)_2Cl(terpy)$	187
$MoCl_4(bipy)$	Me_3SiN_3	$MoN(N_3)_3(bipy)$	188
$MoCl(dtc)_3$	NaN_3	$MoN(dtc)_3$	183
$MoCl_5$	ClN_3	$[MoNCl_3]_n$	189
$MoCl_5$	NEt_4N_3	$NEt_4[MoNCl_4]$	190 (183)
$M(N_2)_2(dppe)_2$ (M = Mo, W)	Me_3SiN_3	$MoN(N_3)(dppe)_2$	191, 192
WBr_5	PPh_4N_3	$PPh_4[WNBr_4]$	193
WCl_6	IN_3	$[WNCl_3]_n$	194 (189)
$M(N_3)(TPP)$ (M = Mn, Cr)	—[c]	$MN(TPP)$	195, 196
$[TcO_4]^-/HBr$	NaN_3	$[TcNBr_4]^-$	197
$ReCl_3(PMePh_2)_2$	NaN_3	$ReNCl_2(PMe_2Ph)_3$	198
$ReCl_5$	ClN_3	$[ReNCl_3]_n$	199
ReF_5	Me_3SiN_3	$ReNF_4$	200
$[MO_2Cl_4]^{2-}$ (M = Ru, Os)	NaN_3	$[MNCl_5]^{2-}$	201
$Os(\eta^4\text{–HBAB})(PPh_3)_2$[g]	Me_3SiN_3	$OsN(\eta^2\text{–HBAB})$	202

[a] References in parentheses use an alternative azide reagent.
[b] Product not characterized.
[c] Photochemical reaction.
[d] Subsequent addition of $[Et_2NH_2][S_2(P(OMe)_2]$.
[e] Subsequent addition of Ph_3PO.
[f] Subsequent addition of terpyridyl.
[g] HBAB = 1,2–bis(o–hydroxybenzamido)benzene ligand.

TABLE 3.4 Syntheses of Organoimido Complexes Using Organic Azides, RN_3

Starting Complex	R	Product	Reference
VCl_4	Me_3Si	$Cl_3V(NSiMe_3)$	203
Cp_2V	Me_3Si	$Cp_2V(NSiMe_3)$	204
Cp_2^*V	Ph	$Cp_2^*V(NPh)$	205, 206
$Nb_2Cl_6(SMe_2)_3$	Ph	$[Nb(NPh)Cl_3(SMe_2)]_2$	181
$Ta(CH^tBu)(PMe_3)_4Cl$	Me_3Si	$Ta(CH^tBu)(NSiMe_3)(PMe_3)_2Cl$	207
$Mo(CO)_4Cl_2$ [a]	Ph	$Mo(NPh)(S_2P(OEt)_2)_3$	208
$MoCl_4(thf)_2$	p-tolyl	$Mo(N-p-tolyl)Cl_4(thf)$	209
$Mo_2(O^tBu)_6$	Ph	$[Mo(NPh)(O^tBu)_2(\mu-NPh)]_2$	210
$Mo(CO)_3(S_2PPh_2)$	p-tolyl	$Mo(N-p-tolyl)_2(S_2PPh_2)_2$	211
$Mo(CO)_2(dtc)_2$	Ph	$Mo(NPh)_2(dtc)_2$	212
$[CpMo(CO)_2]_2$	Ph	$CpMo(NPh)(\mu-N_3RCO)Mo(CO)_2Cp$	213
$W(CO)_2(dtc)_2(PPh_3)$	p-tolyl	$W(N-p-tolyl)(CO)(dtc)_2$	214
$[Cp^*W(CO)_2]_2$	Et	$Cp^*W(NEt)(\mu-N_3RCO)W(CO)_2Cp^*$	215

[a] Reaction in presence of $NH_4[S_2P(OEt)_2]$.

cleavage of symmetrical hydrazines see Section 3.2.5.) Thermolysis of rhenium triazine complexes $ReCl_2(ArN_3Ar)(PPh_3)_2$ in carbon tetrachloride is also reported to yield an imido product $ReCl_3(NAr)(PPh_3)_2$ [222].

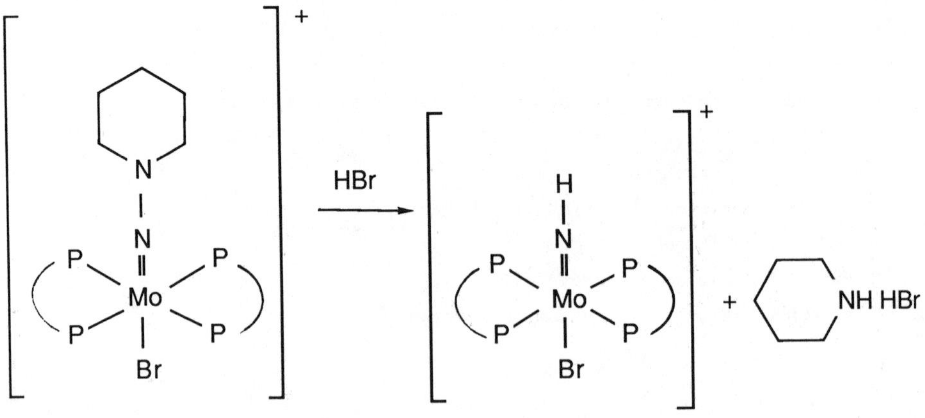

(64)

P⌢P = 1,2-bis(diphenylphosphino)ethane

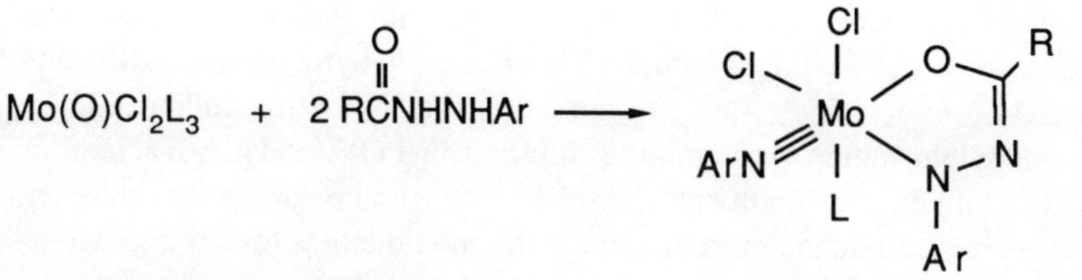

$$\text{Mo(O)Cl}_2\text{L}_3 \;+\; 2\,\text{RCNHNHAr} \longrightarrow \begin{array}{c}\text{complex shown}\end{array} \qquad (65)$$

L = PMePh$_2$ or PEt$_2$Ph

R = alkyl or aryl

Diazoalkanes have seen considerable use in the synthesis of Fischer-type carbenes. It can also be noted that catalytic cyclopropanation reactions utilizing diazo compounds are generally believed to involve transition metal carbene intermediates [223]. In contrast, this route has not been utilized for alkylidene synthesis. Despite earlier claims [214], Eq. 66 gives a diazoalkane complex, not the tungsten (IV) benzylidene [361]. Nevertheless, an important route to alkylidyne complexes via C—N bond cleavage is under development in the form of the nitrile cleavage reaction in Eq. 44 [224,104,225]. A useful feature of Eq. 44 as a route to alkylidynes is the precipitation of the side-product nitridotungsten complex as polymeric [(^{t}BuO)$_3$WN]$_x$.

$$\text{W(CO)}_2(\text{dtc})_2(\text{PPh}_3) \;+\; \text{PhCHN}_2 \;\xrightarrow[-\text{CO}]{37^\circ\text{C}}\; \text{W}(\equiv\text{NN}=\text{CHPh})(\text{dtc})_2(\text{CO}) \qquad (66)$$

3.2.4.5 Cleavage of α Phosphorus (and Other Pnictogens).

A report [146] that triphenylphosphine oxide can be deoxgenated by (dtc)$_2$MoO complexes subsequently has been shown to be in error [161]. It has, in fact, been pointed out that such a transformation is precluded on thermodynamic grounds; ΔH for Eq. 67 is -35 kcal/mole while ΔH for Eq. 68 is -67 kcal/mole [226].

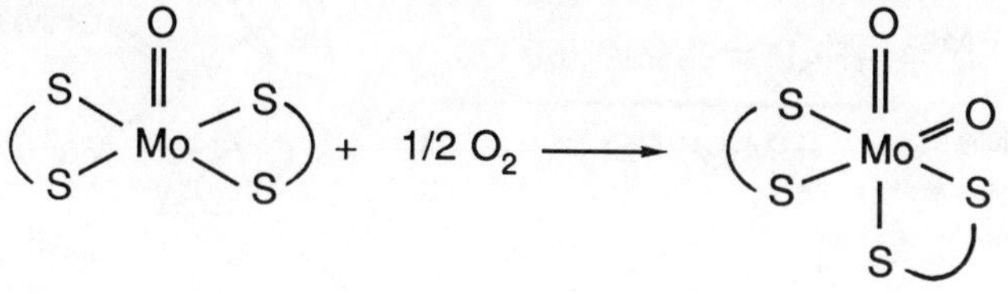

$$\left(\begin{array}{c}\text{S}\\\text{S}\end{array}\text{Mo}\begin{array}{c}\text{S}\\\text{S}\end{array}\right) \;+\; 1/2\,\text{O}_2 \;\longrightarrow\; \left(\begin{array}{c}\text{S}\\\text{S}\end{array}\text{Mo}\begin{array}{c}\text{O}\\\text{S}\\\text{S}\end{array}\right) \qquad (67)$$

S⌒S = diethyldithiocarbamate

$$\text{Ph}_3\text{P} \;+\; 1/2\,\text{O}_2 \;\longrightarrow\; \text{Ph}_3\text{PO} \qquad (68)$$

(Indeed, because of the great strength of the phosphorus–oxygen bond, phosphines are excellent reagents for deoxygenation of oxo complexes; see Section 6.3.2.) Nevertheless, there is some evidence that under appropriately forcing conditions the deoxygenation of phosphine oxides by transition metals can occur. The widely used syntheses of $(Ph_3PO)_2MoOCl_3$ and of $(Ph_3PO)_2MoO_2Cl_2$ from triphenylphosphine oxide and $MoCl_5$ are believed to involve such a transformation [142]. Niobium pentabromide will also cleave triphenylphosphine oxide to give $NbOBr_3(Ph_3PO)_2$ [145]. A fascinating feature of these reactions is that treatment of $MoCl_5$ with triphenylantimony oxide under the same conditions, despite the more favorable thermodynamics for deoxgenation of arsenic versus phosphorus, results in formation of $(Ph_3AsO)MoCl_5$ as the only isolated product [142].

In contrast, both triphenylarsenic oxide and triphenylantimony oxide will readily transfer oxygen to $(dtc)_2MoO$ to afford $(dtc)_2MoO_2$ [171]. This chemistry serves as the basis for a molybdenum-catalyzed deoxgenation system which employs triphenylphosphine as stoichiometric reductant [171]. This is shown in Eq. 69.

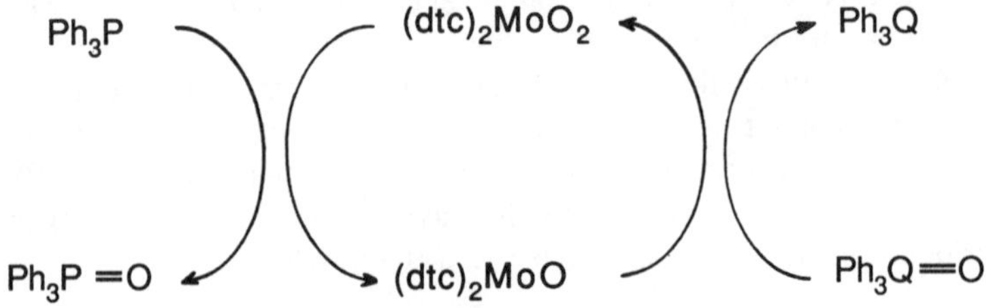

$$Ph_3P + (dtc)_2MoO_2 + Ph_3Q \longrightarrow Ph_3P=O + (dtc)_2MoO + Ph_3Q=O \qquad (69)$$

$$Q = As, Sb$$

Several examples have been reported in which a phosphorane ("Wittig reagent") undergoes transition metal-mediated P—C bond cleavage to afford an alkylidene complex [227,228,43]. Equations 70 and 71 are particularly noteworthy. The former provided the first example of an ethylidene complex, whereas the latter afforded the first group IV metal alkylidene compound.

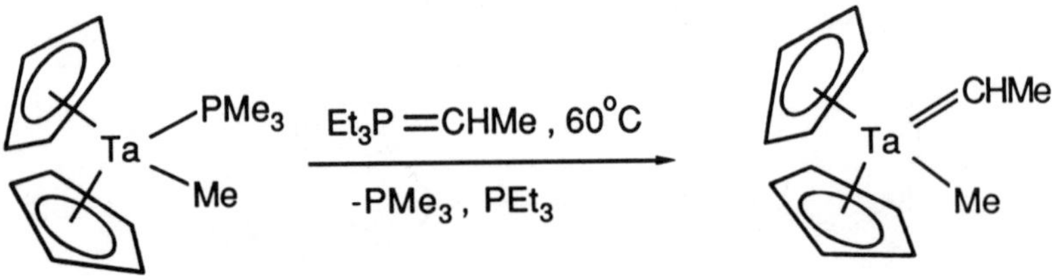

$$Cp_2Ta(PMe_3)(Me) \xrightarrow[-PMe_3, PEt_3]{Et_3P=CHMe, 60°C} Cp_2Ta(=CHMe)(Me) \qquad (70)$$

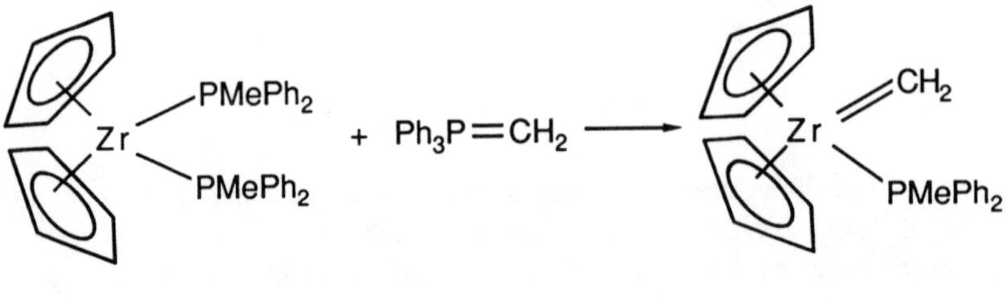

$$Cp_2Zr(PMePh_2)_2 + Ph_3P=CH_2 \longrightarrow Cp_2Zr(=CH_2)(PMePh_2) \qquad (71)$$

3.2.4.6 Transfer from More Electronegative Transition Metal.

A few reactions are known in which a multiply bonded ligand is transferred from a more electronegative transition metal to a more electropositive transition metal. Such transformations, at least formally, belong in the class of reactions currently under discussion, since they involve α cleavage and oxidation of the metal that accepts the multiply bonded ligand. For example, the oxomolybdenum(V) in Eq. 72 will cleanly transfer an oxo ligand to tungsten(IV) [229]. A somewhat related transformation is the reaction of $Cl_3V(NCl)$ with the pentahalides of molybdenum and rhenium which proceeds with the overall stoichiometry shown in Eq. 73 [230]. These reactions are said to involve initial formation of an bridged nitride species $Cl_4M—N\equiv VCl_3$; transfer of the nitride takes place in a subsequent redox step.

$$W(CO)(HC\equiv CH)(dtc)_2 \; + \; Mo_2O_3[S_2P(OEt)_2]_4 \; \xrightarrow{-CO}$$

$$W(O)(HC\equiv CH)(dtc)_2 \; + \; 2 \; Mo(O)[S_2P(OEt)_2]_2 \quad (72)$$

$$ReCl_5 \; + \; V(NCl)Cl_3 \; \xrightarrow[-Cl_2]{CCl_4,\;77°C} \; [Re(N)Cl_3]_n \; + \; VCl_4 \quad (73)$$

3.2.5 Cleavage of Symmetrical Ligand Precursor

In this section we discuss the formation of multiply bonded ligands by cleavage of symmetrical precursors: dioxygen, hydrogen peroxide, dinitrogen, azoalkanes (RN=NR), and symmetrical hydrazines and acetylenes. Although a case could be made for including these processes in previous sections, these reactions are collected here because of their special interest and because we wished to highlight, compare, and contrast some of this chemistry.

In reactions of dioxgen with low-valent metals in aqueous media, it is often unclear whether the oxo ligand is derived from dioxygen cleavage or outer-sphere oxidation followed by deprotonation of coordinated water. In one example where mechanistic studies have been conducted, the oxidation of aqueous vanadium(II), there is evidence for competing one-electron and two-electron pathways [231]. Within the last decade examples have begun to emerge in which reactions of dioxygen in nonaqueous solvents afford discrete oxo–metal complexes as products. In at least two systems isotopic labeling studies have confirmed that the terminal oxo ligand is derived from O_2 [232,233]. A number of these reactions are tabulated in Table 3.5 and others are discussed below. Inspection of Table 3.5 shows that the starting complexes in these transformations include not only low-valent metals but a variety of electron-rich species including σ-organometallics and even a catecholate complex.

Of prime interest because of their relevance as cytochrome P-450 model compounds

TABLE 3.5 Syntheses of Terminal Oxo Complexes by Cleavage of Dioxygen

Starting Complex	Solvent	Product	Reference
$Cp_2^*VI_2$	toluene	$[Cp^*V(O)I]_2(\mu\text{-}O)$	156
Cp_2NbCl	—[a]	$Cp_2Nb(O)Cl$	234
$(^tBu_3SiO)_3Ta$	—[a]	$(^tBu_3SiO)_3Ta(O)$	102
$Cp^*Cr(CO)_2(NO)$	toluene	$[Cp^*Cr(O)(\mu\text{-}O)]_2$	235
$Cp^*Mo(CO)_2(NO)$	toluene	$[Cp^*Mo(O)_2]_2(\mu\text{-}O)$	235
$Mo(DBCat)_3$[b]	toluene	$Mo_2O_2(DBCat)_4$	156
$(HB(pz)_3)MoCl_3$	CH_2Cl_2	$(HB(pz)_3)MoOCl_2$	236
$Mo(NAr)(dtc)_2$	toluene	$Mo(NAr)(O)(dtc)_2$	237
$CpW(NO)(CH_2SiMe_3)_2$	hexane	$CpW(O)_2(CH_2SiMe)_3$	157
$WCl_2(PMePh_2)_4$	benzene	$W(O)Cl_2(PMePh_2)_3$	238
$Mn(3\text{-}MeO\text{-}salen)$[c]	MeOH	$Mn(O)(3\text{-}MeO\text{-}salen)\cdot1.5MeOH$	239
$ReMe_6$	—[a]	$Re(O)Me_4$	155
$Re(O)Me_4$	none	$Re(O)_3Me$	240
$Cp^*Re(CO)_2(thf)$	thf	Cp^*ReO_3	241

[a] Not reported.
[b] DBCat = 3,5-di-*tert*-butylcatecholate.
[c] 3-MeO-salen = 3-methyoxysalicylideneimine.

has been the oxidation of iron(II) porphyrin complexes by molecular oxygen. It is believed that treatment of (TPP)Fe with O_2 at $-80°C$ results in formation of a peroxo-bridged iron(III) dimer, (TPP)Fe—O—O—Fe(TPP); on warming, this decomposes to the bridged oxo species (TPP)FeOFe(TPP) and dioxgen [242]. However, upon addition of nitrogen bases (L = *N*-methylimidazole or pyridine) to solutions of the peroxo dimer at $-80°C$, the terminal oxoiron(IV) is formed (Eq. 74) [233,243].

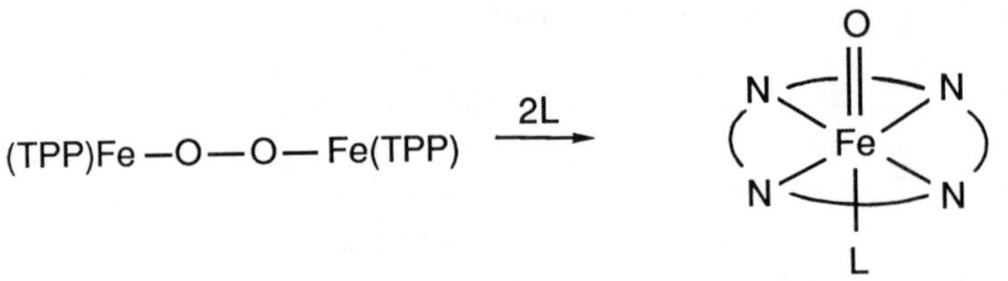

$$(TPP)Fe-O-O-Fe(TPP) \xrightarrow{2L} \begin{pmatrix} & O \\ N & \| & N \\ & Fe & \\ N & | & N \\ & L & \end{pmatrix} \quad (74)$$

Moreover, photolysis of the peroxo dimer at 15 K in the absence of added ligand using matrix isolation techniques also appears to produce a terminal oxo complex as product.

Evidence for this conclusion includes resonance Raman and ^{18}O labeling studies [233].

Oxygen activation in the P-450 enzyme necessarily differs from Eq. 74 in that it can involve only a single iron atom. The enzymatic process is believed to proceed by the formal heterolysis of the O—O bond in mononuclear peroxoiron species, which results in the formation of the reactive oxidant [FeO]$^{3+}$ and a molecule of water (see section 7.1.3).

Other metal–porphyrin systems undergo oxidation by O_2. Oxidation of a ruthenium(II)–porphyrin complex (TMP)Ru(CO) results in formation of a dioxoruthenium(VI) complex (TMP)RuO$_2$, a powerful oxidizing agent. The reaction is suggested to involve initial formation of a ruthenium(IV) intermediate, (TMP)RuO, which then disproportionates to Ru(II) and Ru(VI) [244]. There is also some circumstantial evidence for formation of oxomanganese products from O_2 and manganese porphyrin complexes: It has been shown that in protic media the combination of a manganese porphyrin complex, molecular oxygen, and a reducing agent (abscorbate or H$_2$/colloidal platinum) allows the catalytic epoxidation of olefins [245,246].

Oxomolybdenum(IV) complexes of the type (dtc)$_2$MoO are readily oxidized by molecular oxygen to the oxomolybenum(VI) analogs (dtc)$_2$MoO$_2$. (Eq. 67). (This observation is presumably relevant to biological oxidations mediated by the molybdenum-containing "oxo-transferase" family of enzymes.) Equation 67 was first noted by Barral who used it to effect the molybdenum-catalyzed oxidation of triphenylphosphine to triphenylphosphine oxide [247]. There is evidence that seemingly related phosphine oxidations involving S-deprotonated cysteine esters as ligands on molybdenum are, in fact, "fundamentally different" and require the presence of a small amount of water [248] (see Section 6.2.2).

Chisholm [232] has reported a series of oxidations of molybdenum alkoxides under aprotic conditions where oxo ligands were shown to be derived from molecular oxygen by ^{18}O labeling. Triply bonded Mo$_2$(OR)$_6$ compounds and O_2 react in hydrocarbon solvents to give MoO$_2$(OR)$_2$ compounds and alkoxy radicals. When R = t-butyl the reaction is rapid with no isolable intermediates; when R = i-propyl or neopentyl, oxomolybdenum clusters were formed as isolable intermediates. Also, several [Mo(OR)$_4$]$_x$ compounds react with O_2 to yield give (RO)$_4$Mo=O. A reaction between Mo(O^{t}Bu)$_4$ and O_2 yields MoO$_2$(O^{t}Bu)$_2$ and t-butoxy radicals in a 1:2 ratio. It is suggested that the initial reaction between Mo$_2$(OR)$_6$ and O_2 involves a facile cleavage of the Mo—Mo triple bond to produce MoO$_2$(OR)$_2$ and Mo(OR)$_4$. Of special interest is the proposal that subsequent reaction in the case of Mo(O^{t}Bu)$_4$ involves formation of a peroxy intermediate (Eq. 75), which may undergo either unimolecular homolytic decomposition via Eq. 76a or reaction with a second molybdenum(IV) with O—O cleavage according to Eq. 76b. (Such pathways involving bridged peroxo intermediates are being proposed with increasing frequency; see also Eq. 74 and ref. 235.)

$$\text{Mo(O}^t\text{Bu)}_4 + O_2 \longrightarrow [\text{Mo(O}^t\text{Bu)}_4(O_2)] \tag{75}$$

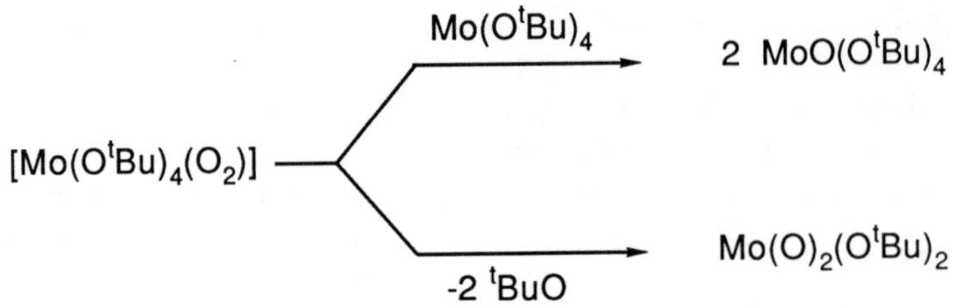

$$[\text{Mo(O}^t\text{Bu})_4(\text{O}_2)] \begin{cases} \xrightarrow{\text{Mo(O}^t\text{Bu})_4} 2\ \text{MoO(O}^t\text{Bu})_4 & (76a) \\ \xrightarrow{-2\ ^t\text{BuO}} \text{Mo(O)}_2(\text{O}^t\text{Bu})_2 & (76b) \end{cases}$$

Metal–oxo complexes can also be prepared from metal peroxo (η^2-O_2(2−)) complexes, which are in turn obtained by treating an appropriate starting complex with hydrogen peroxide. A remarkable case in point is the photolysis of bisperoxomolybdenum(VI) tetra-*p*-tolylporphyrin which is shown in Eq. 77. The product *cis*-dioxo–molybdenum(VI) complex is the only known example of a metalloporphyrin having *cis*-bis(monodentate) axial ligation. The highly strained product is reported to be unusually reactive toward triphenylphosphine oxidation [249].

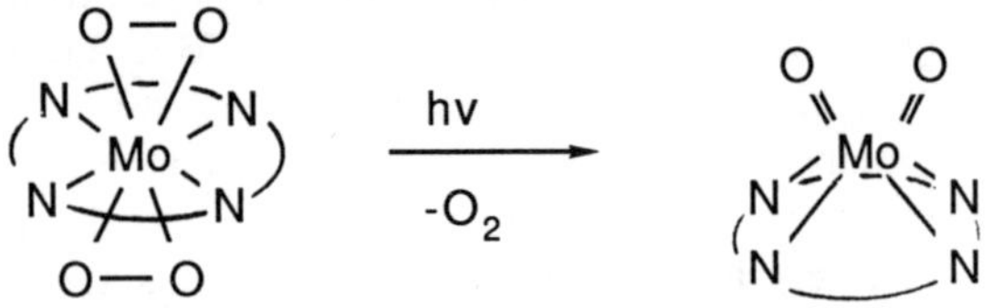

(77)

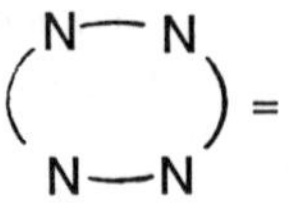

= tetra-p-tolylporphyrin

A more common pathway from metal peroxo compounds to their oxo analogs involves transfer of one oxygen atom to an external oxygen acceptor. A familiar example is Mimoun's diperoxomolybdenum complex, which reacts with olefins [250] according to Eq.78. A labeling study has confirmed that it is a peroxo oxygen rather than the oxo substituent which is transferred [251]. (In addition, this reagent will transfer oxygen to carbanions [252,253], also a useful reaction for organic synthesis.) Another nice example from the Mimoun group is Eq. 79 for which both the peroxo-vanadium starting complex and the dioxovanadium product have been characterized structurally [254].

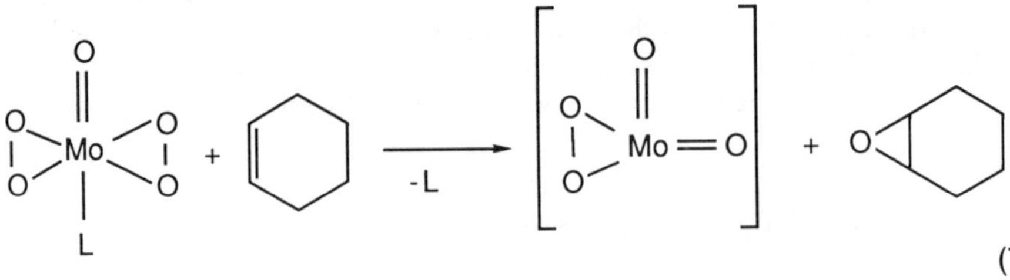

(78)

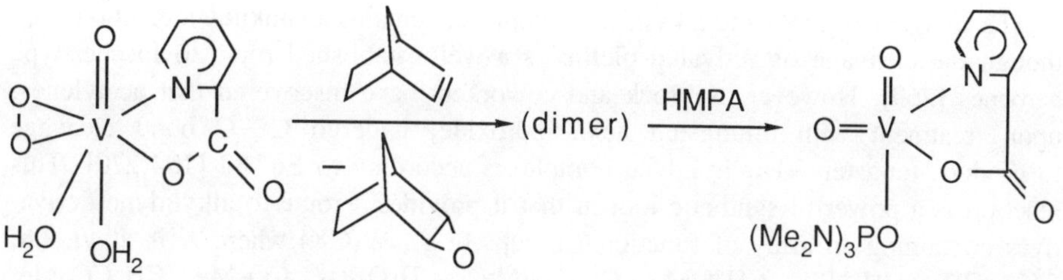

(79)

Azoalkanes (RN=NR) can be cleaved by low-valent metals to give organoimido complexes. For example, Eq. 80 proceeds for both M = niobium and tantalum to give chloride-bridged imido dimers as structurally characterized products [255,256]. Related azoalkane cleavages that afford terminal imido complexes are known for chromium and rhenium [257,258]. Moreover, Stone and coworkers have reported that treatment of an iridium(I) complex (PMePh$_2$)$_2$Ir(CO)Cl with hexafluoroazomethane yields the "mononuclear iridium nitrene complexes" cis- and trans-(PMePh$_2$)$_2$Ir(CO)Cl(NCF$_3$) [259,260].

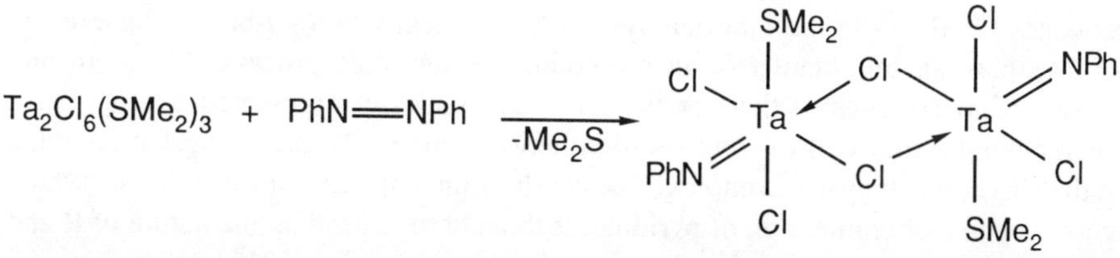

$$Ta_2Cl_6(SMe_2)_3 + PhN=NPh \xrightarrow{-Me_2S} \text{[dimer product]} \quad (80)$$

The cleavage of hydrazine provides an entry into nitrido complexes of rhenium and technetium. In the simplest case, a mixture of a rhenium(VII) compound (Re$_2$O$_7$ or KReO$_4$), a hydrazine salt, and a phosphine ligand is heated in ethanol. A wide range of rhenium(V) nitrido compounds ReNX$_2$(PR$_3$)$_n$, where n = 2 or 3 and X is halogen, have been prepared in this way [198,261,262]. Replacing perrhenate by pertechnetate in the preceding reaction gives the analogous nitridotechnetium(V) complexes [263]. Moreover, by substituting other ligands in place of the phosphine component, a range of other nitrides, including (dtc)$_2$TcN [264,265], K[ReN(H$_2$O)(CN)$_4$] [266], and [ReNCl(diphos)$_2$]Cl [267], can be obtained.

This synthetic approach was also extended to the synthesis of alkylimido rhenium complexes by the use of symmetrical 1,2-dialkylhydrazines [268]. Although the mechanism of the reaction is unknown, the reaction is suggested to follow Eq. 81.

$$\text{Re(O)Cl}_3(\text{PPh}_3)_2 + \text{MeNHNHMe·2HCl} \xrightarrow[-\text{Ph}_3\text{PO}]{\text{PPh}_3} \text{Re(NMe)Cl}_3(\text{PPh}_3)_2 + \text{MeNH}_3\text{Cl} \quad (81)$$

The cleavage of olefins to alkylidene complexes remains an unknown reaction even though the cleavage of activated olefins is a well-established route to Fischer-type carbenes [269]. However, Schrock and coworkers have discovered that acetylenes, upon treatment with ditungsten hexa-*t*-butoxide, undergo C—C bond cleavage to produce tungsten(VI) alkylidyne complexes according to Eq. 82 [104,270]. This reaction is a powerful synthetic tool in that it provides a route to alkylidyne derivatives containing an array of functional groups (e.g., W(CX) where X is alkyl, Ph, CH=CH$_2$, CH$_2$NR$_2$, CH$_2$OMe, CH$_2$OSiMe$_3$, CH(OEt)$_2$, CO$_2$Me, CH$_2$CO$_2$Me, C(O)Me, SCMe$_3$, or H). In some cases the compounds can be isolated only as adducts (^{t}BuO)$_3$W(CX)L, where L is pyridine or quinuclidine, and, in fact, it is suggested that added nitrogenous base can play a direct role in the scission reaction [270]. The reactions appear to proceed in high yields, although Cotton and coworkers have succeeded in isolating and characterizing several side-products [271,272].

$$(^tBuO)_3W \equiv W(O^tBu)_3 + RC \equiv CR \xrightarrow{25°C} 2\ (^tBuO)_3W \equiv CR \qquad (82)$$

Reactions analogous to Eq. 82 but using dimolybdenum hexa-*t*-butoxide afford molybdenum alkylidyne complexes; however, this reaction only proceeds for terminal acetylenes [273]. As in the tungsten system, the sterically bulky *t*-butoxy ligands are found to play an important role in promoting the cleavage process. Chisholm and coworkers have suggested that for the more general case represented by Eq. 83, a dynamic equilibrium will exist in solution between the μ-alkyne ditungsten complex and the alkylidyne tungsten complex. The equililbrium constant for this process, which involves uptake or elimination of pyridine, is thought to depend on the nature of R and R' with bulky combinations favoring the alkylidyne species. Labeling studies and experiments using various trapping reagents lend support to this hypothesis [274,275].

$$\text{[complex]} \underset{+2py}{\overset{-2py}{\rightleftharpoons}} 2\ (RO)_3W \equiv CR' \qquad (83)$$

3.2.6 Routes Involving "Wittig-Like" (2 + 2) Replacement of Existing Multiply Bonded Ligands

This class of reactions involves the direct replacement of an existing multiply bonded ligand by another upon treatment with an unsaturated reagent, as shown schematically in Eq. 84. Two reactions in this family that are of practical importance are the catalytic metathesis of olefins and of acetylenes; these we will discuss in greater detail in Sections 7.4.1 and 7.5. These two reactions have the added significance that they are the only two cases where serious mechanistic studies have been undertaken. Nevertheless, it is widely believed that all of these reactions involve a four-center mechanism as shown in Eq. 84.

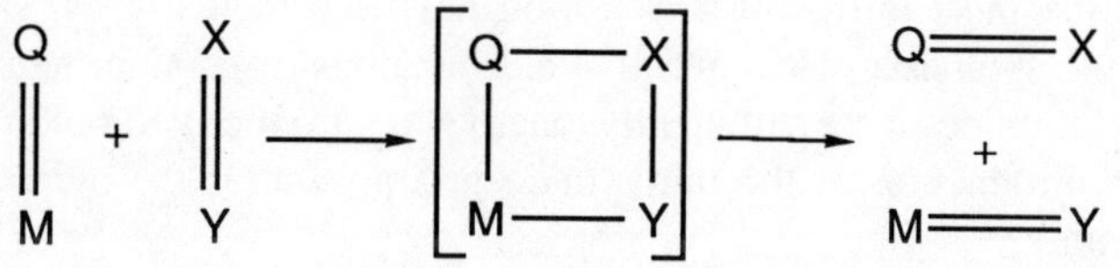

$$\begin{matrix} Q \\ \| \\ M \end{matrix} + \begin{matrix} X \\ \| \\ Y \end{matrix} \longrightarrow \begin{bmatrix} Q \text{---} X \\ | \quad | \\ M \text{---} Y \end{bmatrix} \longrightarrow \begin{matrix} Q \text{==} X \\ + \\ M \text{==} Y \end{matrix} \qquad (84)$$

Alkylidene complexes react with organic carbonyl compounds (aldehydes, ketones, esters) by such a process to afford an olefin and an oxo–metal complex. An early example was Eq. 85 [276]. (Extensions of this chemistry [277,278] having potential utility in organic synthesis are discussed in section 6.2.4.) Related reactions of alkylidyne complexes that afford oxo–vinyl complexes as products have been demonstrated (Eq. 86) [224]. Similarly, reactions of organoimido complexes with aldehydes and ketones afford oxo complexes as products [31,279,207]. However, none of these reactions has proven widely useful as a synthetic route to oxo complexes: Synthesis of the starting alkylidene and organoimido derivatives is generally more difficult than other more direct routes to the oxo compound. Nevertheless, in one case this approach provides a route to an oxo complex that is not available by other means [66]. Reaction of the diiimido chromium complex in Eq. 87 stops cleanly after introduction of a single oxo moiety; the Schiff's base side-product can be precipitated by treatment with methyl triflate.

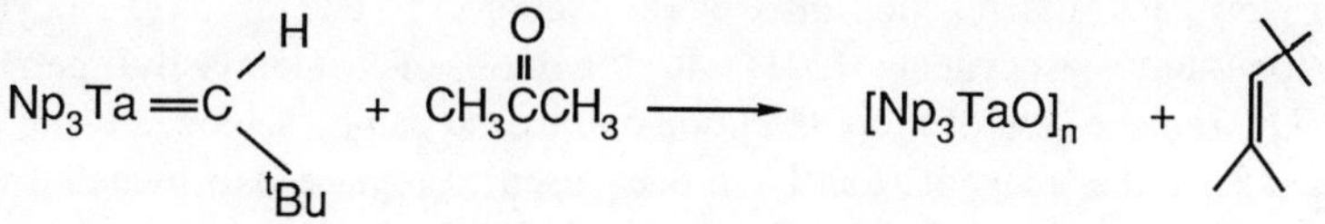

$$Np_3Ta=C\begin{matrix}H\\ \\^tBu\end{matrix} + CH_3\overset{O}{\overset{\|}{C}}CH_3 \longrightarrow [Np_3TaO]_n + \text{(isobutylene)} \qquad (85)$$

$$(ArO)_3W\equiv C^tBu + \text{PhCHO} \longrightarrow \begin{matrix}ArO\diagdown\overset{O}{\overset{\|}{W}}\diagup OAr \\ ArO\diagup\quad\diagdown{^tBu}\\ H\diagdown \\ Ph \end{matrix} \qquad (86)$$

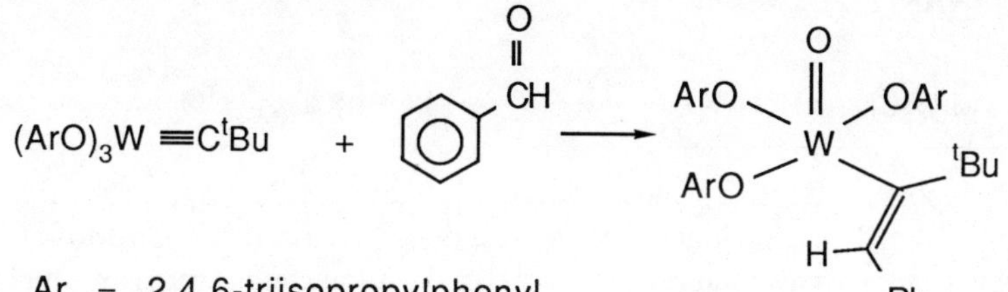

Ar = 2,4,6-triisopropylphenyl

$$\begin{matrix}Me_3SiO\diagdown\quad\diagup N^tBu\\ Cr\\ Me_3SiO\diagup\quad\diagdown N^tBu\end{matrix} \xrightarrow[-PhCH=N^tBu]{PhCH=O} (Me_3SiO)_2Cr(O)(N^tBu) \qquad (87)$$

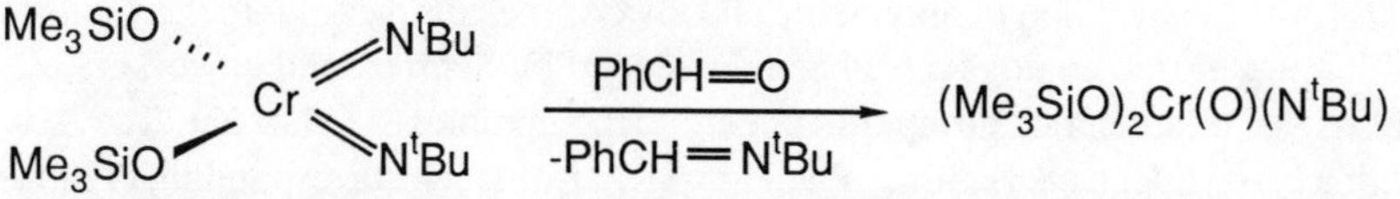

Related (2 + 2) exchange reactions have proven useful for the synthesis of organoimido complexes starting from either oxo or alkylidene complexes. Two types of reagent have proven particularly valuable for replacing an oxo by an organoimido ligand. These are the phosphinimines, $R_3P{=}NR'$, and isocyanates, $RN{=}C{=}O$. Phosphinimines were first applied to the synthesis of imidorhenium(V) species $L_2Cl_3Re{=}NPh$ [280] and were subsequently used for the synthesis of $(dtc)_2Mo{=}NPh$ [281]. Perhaps the most intriguing application has been their use by Sharpless and coworkers for the synthesis of di- and tri-imido analogs of osmium tetroxide [282]. Triphenylphosphinimines were sufficiently reactive to introduce two imido substituents in Eq. 88, but introduction of the third imido group requires the more reactive tri-butylphosphinimine.

$$OsO_4 \xrightarrow[-2Ph_3PO]{2Ph_3P=NR} OsO_2(NR)_2 \xrightarrow[Bu_3PO]{Bu_3P=NR} OsO(NR)_3 \quad (88)$$

R = 3° alkyl

(However, published reports [280,283] regarding the synthesis of osmium(VI)–phenylimido complexes using N-benzoyl phosphinimines are in error. The "Os-OCl$_3$(PPh$_3$)$_2$" starting material [198] is, in fact, a mixture of osmium(IV) and (VI) [284] and the products are actually benzonitrile complexes [285].)

The isocyanate route was first used by Soviet workers to convert $L_2Cl_3Re{=}O$ to $L_2Cl_3Re{=}NPh$ [286,287]. Subsequently this has developed into an important synthetic method, since the products of Eq. 89 are themselves versatile starting materials for a wide range of other imido species. In refluxing hydrocarbon solvents, Eq. 89 proceeds for M = W, n = 4 [12,288,289]; for M = Re, n = 4 [289]; and for M = V, n = 3 [290]. In the tungsten case, the reaction also works well for 1° and 2° alkyl isocyanates, but fails for t-butyl isocyanate [291]. In the case of trimethylsilyl perrhenate, Me$_3$SiOReO$_3$, replacement of more than one oxo ligand can be achieved using phenyl isocyanate [13]. Other reagents that have been used to replace oxo ligands by organoimido ligands include arylsulfynilamines ArNSO [292] and aryl formamidines ArNHCH=NAr [222].

$$M(O)Cl_n + PhNCO \xrightarrow{\Delta} M(NPh)Cl_n + CO_2 \quad (89)$$

A recent report by Geoffroy and coworkers [292a] provides support for the intermediacy of a [2 + 2] cycloadduct in these reactions. Treatment of Cp$_2$Mo=O with phenyl isocyanate gives a structurally characterized cyclometallacarbamate which does not lose CO$_2$ even at reflux in THF.

Methods for the conversion of alkylidene and alkylidyne complexes into their imido, μ-hydrazido, and nitrido analogs have been developed by Schrock and coworkers. As shown in Eq. 90, treatment of group 5 neopentylidene complexes (M = Nb, Ta; X = Cl, Br) with a benzylidene alkylamine, PhCH=NR, affords the corresponding organoimido complexes [207,293]. By replacing the Schiff's base component in Eq. 90 with PhCH=NN=CHPh, the same type of reaction can be used to prepare μ-N$_2$ derivatives which are best regarded as containing a bridging hydrazido(4−) ligand [294,293].

The somewhat related reaction of alkylidynes with nitriles to afford a nitridotungsten product (Eq. 91, Ar = 2,6-diisopropylphenyl) has also been reported [224].

$$\text{THF}\diagdown\underset{\underset{\text{Cl}}{|}}{\overset{\overset{\text{Cl}}{|}}{\text{Ta}}}\diagup\overset{\text{Cl}}{\underset{\text{CH}^t\text{Bu}}{\lessapprox}} \xrightarrow[-^t\text{BuCH}=\text{CHPh}]{\text{PhCH}=\text{NR}} \text{THF}\diagdown\underset{\underset{\text{Cl}}{|}}{\overset{\overset{\text{Cl}}{|}}{\text{Ta}}}\diagup\overset{\text{Cl}}{\underset{\text{NR}}{\lessapprox}} \qquad (90)$$

$$(\text{ArO})_3\text{W} \equiv \text{C}^t\text{Bu} + \text{MeCN} \longrightarrow {}^t\text{BuC} \equiv \text{CMe} + [(\text{ArO})_3\text{WN}]_n \qquad (91)$$

A potentially very useful reaction that has yet to be demonstrated, at least in solution phase, is the replacement of an oxo ligand by an alkylidene ligand. Attempts to achieve this transformation by treating oxo complexes with Wittig reagents have either afforded stable adducts [295] or alternatively have resulted in a variety of interesting rearrangements [296,297] rather than discrete alkylidene complexes. Another approach that has yet to be achieved in solution is the reaction between an oxo ligand and an olefin to afford an alkylidene and an organic carbonyl compound. (Such reactions might explain the initial generation of active alkylidene species upon treatment of metal oxides with olefins during catalytic olefin metathesis.) In fact, in the gas phase, such reactions involving naked oxometal cations and ethylene (Eq. 92) have been observed for both M = Mn [298] and M = Cr [299].

$$[\text{MO}]^+ + \text{CH}_2=\text{CH}_2 \longrightarrow [\text{MCH}_2]^+ + \text{CH}_2=\text{O} \qquad (92)$$

The aforementioned gas-phase reactions involve severely coordinatively unsaturated low-valent oxo species; one may question their relevance to the high-valent oxo complexes one encounters in solution chemistry. In fact, one example of a gas-phase reaction of a d^0 oxo species and an olefin has been reported [300]. In the event, the ClCrO_2^+ ion reacted with ethylene to afford both products from oxygen transfer ($\text{C}_2\text{H}_3\text{O}^+$ and ClCrO^+) and a product containing both carbon and oxygen, ClCrOCH_2^+. Although the connectivity of this product is unknown, it was suggested to be as shown in Eq. 93. (Another possibility would be an η^2-formaldehyde complex.) The interpretation of Eq. 93 is also in accord with *ab initio* calculations on the reaction between CrO_2Cl_2 and ethylene that show that such an alkylidene-forming pathway would be exothermic by about 18 kcal/mole [301,302]. (Some caution is required in reading ref. 301 and other papers in this series despite the fact that the calculations are of excellent quality and, importantly, include electron correlation. As one example, the initial formation of a π-complex between d^0 MO_2Cl_2 and ethylene in ref. 301 is said to be exothermic by 20 kcal/mole. However, this energy was not calculated and is based on "analogy to similar systems," which in this case refers to an olefin complex of palladium(II). See also Sections 2.2.4 and 7.4.1.)

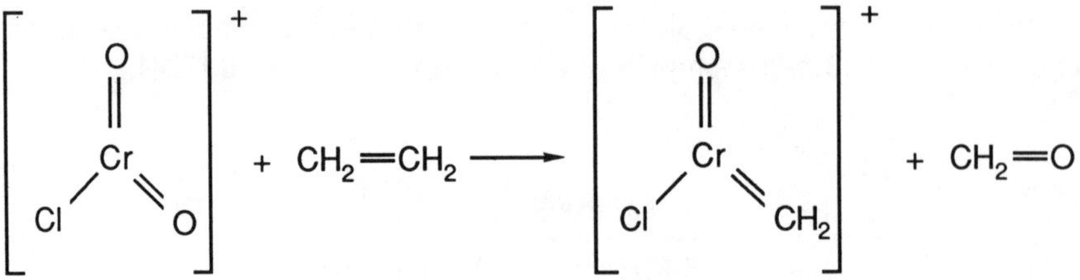

$$\left[\begin{array}{c}O\\||\\Cl\diagup Cr\diagdown O\end{array}\right]^{+} + CH_2=CH_2 \longrightarrow \left[\begin{array}{c}O\\||\\Cl\diagup Cr\diagdown CH_2\end{array}\right]^{+} + CH_2=O \quad (93)$$

We should not close this section without reiterating that the olefin metathesis reaction represents an example of Eq. 84, an example in which both the starting complex and the metal-containing product are alkylidene derivatives. While we defer most of our discussion of this reaction until later sections, it is noted that Osborn has used this transformation as a synthetic route to alkylidene species. Thus, readily available neopentylidene complexes react with terminal olefins with extrusion of neohexene according to Eq. 94. This approach has allowed the synthesis of the first non-18 electron metal alkylidene complexes containing β hydrogen [303].

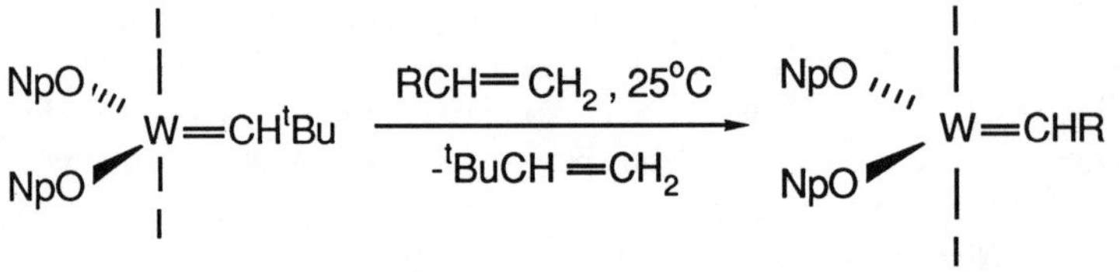

$$\begin{array}{c}NpO_{\prime\prime\prime}\\NpO\end{array}\!\!W\!=\!CH^tBu \xrightarrow[-^tBuCH=CH_2]{RCH=CH_2,\ 25°C} \begin{array}{c}NpO_{\prime\prime\prime}\\NpO\end{array}\!\!W\!=\!CHR \quad (94)$$

3.3 BY FORMATION OF A BOND TO THE α ATOM

Although, for example, the protonation and alkylation of nitrido complexes are known reactions, such processes are far from common. The reasons for the limited nucleophilicity of the nitrogen atom in tightly bound nitrido complexes were delineated in Chapter 2 and ultimately reflect the absence of a high-lying (easily oxidized), nonbonding electron pair on nitrogen. This is particularly true for the more covalent nitrido complexes of the later transition metals. Analogy might be made to the difficulty in N-alkylation of acetonitrile or of dinitrogen. As we shall see, similar considerations apply to ligands multiply bonded through carbon and through oxygen (Section 6.2) as well.

3.3.1 Imido and Alkylidene Ligands via Protonation

Group 6 nitrido complexes with chelating phosphine ligands can be protonated to afford cationic imido complexes. For example, Eq. 95 has been demonstrated for both M = molybenum and tungsten [192,304].

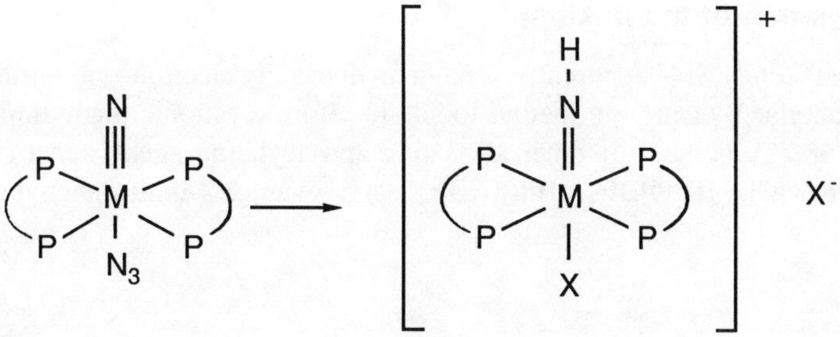

$$\begin{pmatrix} P & \underset{\underset{N_3}{|}}{\overset{\overset{N}{|||}}{M}} & P \\ P & & P \end{pmatrix} \longrightarrow \begin{bmatrix} P & \underset{\underset{X}{|}}{\overset{\overset{H-N}{||}}{M}} & P \\ P & & P \end{bmatrix}^+ X^- \quad (95)$$

P⌒P = 1,2-bis(diphenylphosphino)ethane

X = Cl , Br , I

Earlier we had discussed (Eq. 16 above) intramolecular hydrogen transfer to alkylidyne ligands as a route to oxo and organoimido ligands. Obviously this reaction also results in formation of an alkylidene ligand [46]. Indeed, this approach recently has been applied to the synthesis of a well-characterized, highly active, Lewis acid-free olefin metathesis catalyst as shown in Eq. 96 [305]. Intermolecular analogs are also known. Thus treatment of W(CR)(O-t-Bu)$_3$ with hydrogen halides, phenols, or carboxylic acids results in formation of an alkylidene complex as in Eq. 97 [47]. In some cases, even alkylidenes containing β hydrogen could be prepared by this route. An interesting related example is the protonation of W(CH)(PMe$_3$)$_4$Cl by triflic acid. Extensive ^{1}H and ^{13}C NMR studies suggest that the product methylidene complex [W(CH$_2$)(PMe$_3$)$_4$Cl]OTf is better described as a "face-protonated methylidyne" species [306].

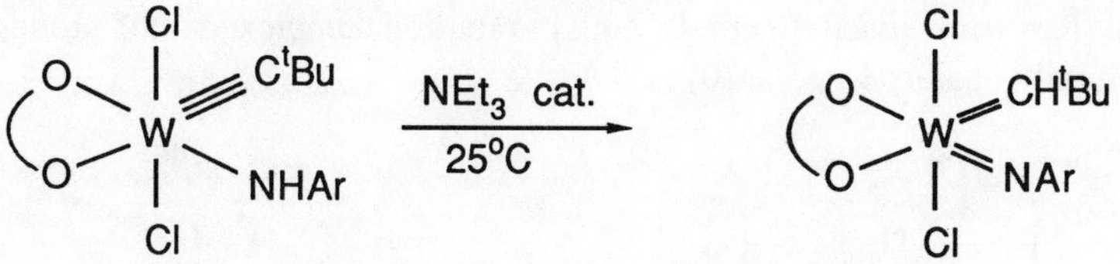

$$\begin{pmatrix} O & \underset{\underset{Cl}{|}}{\overset{\overset{Cl}{|}}{W}} & \equiv C^tBu \\ O & & NHAr \end{pmatrix} \xrightarrow[25°C]{NEt_3 \text{ cat.}} \begin{pmatrix} O & \underset{\underset{Cl}{|}}{\overset{\overset{Cl}{|}}{W}} & =CH^tBu \\ O & & =NAr \end{pmatrix} \quad (96)$$

O⌒O = dimethoxyethane

Ar = 2,6-diisopropylphenyl

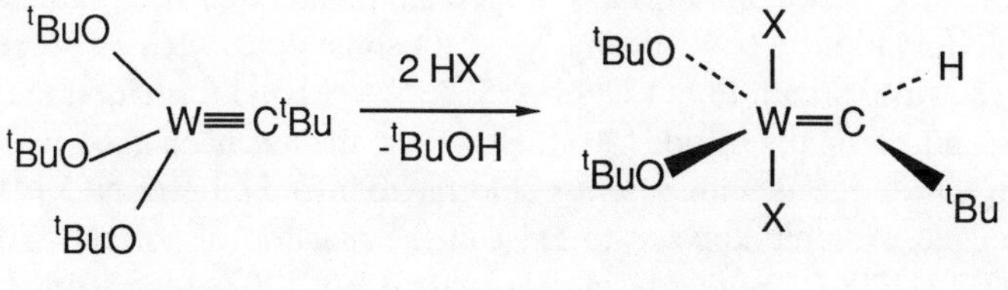

$$\begin{array}{c} ^tBuO \\ ^tBuO-W\equiv C^tBu \\ ^tBuO \end{array} \xrightarrow[-^tBuOH]{2 \text{ HX}} \begin{array}{c} ^tBuO \cdots \overset{X}{\underset{X}{|}} \cdots H \\ {}^tBuO-W=C \\ {}^tBu \end{array} \quad (97)$$

3.3.2 Alkylation of the α Atom

The complexes $(dtc)_3MoN$ apparently contain an unusually electron-rich nitrido ligand that can be alkylated even with methyl iodide to afford a cationic methylimido compound (Eq. 98). A number of other alkylating and acylating agents react similarly, including PhCOCl, $[R_3O]BF_4$, $[Ph_3C]BF_4$, and even 2,4-dinitrophenyl chloride [307].

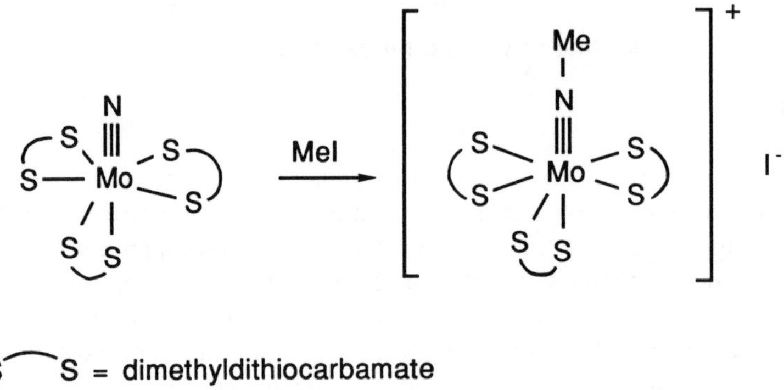

(98)

S⌒S = dimethyldithiocarbamate

Other examples are known. The trityl cation Ph_3C^+ has been found to react with nitrido complexes of rhenium and of tungsten with C—N bond formation [307,308]. The acylation of a nitridomanganese(V)–porphyrin complex with trifluoracetic anhydride has been used to promote its reaction with olefins (the aza analog of epoxidation, see Section 6.2.4.1) [309]. A particularly nice example is Eq. 99, reported by Shapley: In one case (R = trimethylsilylmethyl), both the starting complex and the product have been structurally characterized [310].

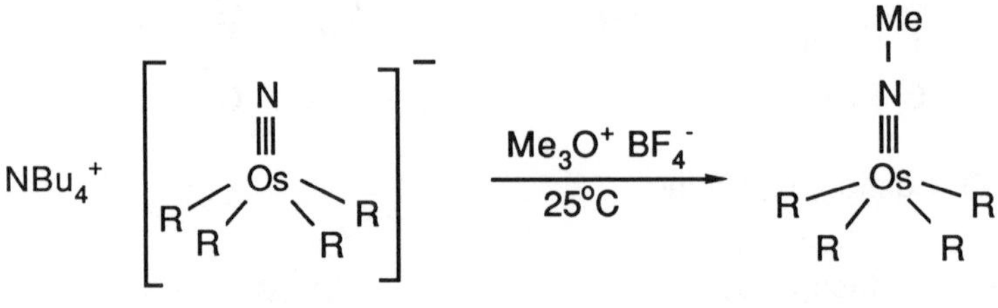

(99)

α-Bond formation between a nitrido nitrogen atom and other heteroatoms can also serve as a route to various "NX" ligand types. In some cases such as the reaction of $(dtc)_2ReN$ or $(dtc)_3MoN$ with $PhSO_2Cl$ or $ArSCl$, N—S bond formation may proceed by electrophilic attack on the nitride [307]. However, the mechanism by which refractory vanadium nitride reacts with gaseous chlorine to afford $Cl_3V\equiv NCl$ [216] is less obvious. A related example appears to exist in the reaction of F_4ReN with ClF_3 to produce $F_5Re(NCl)$ [200]. Another fascinating case is Eq. 100, which affords a product containing a metallaheteroaromatic ring (a "cyclothiazeno ligand"). In this example it is not even known whether the nitrido nitrogen atom is retained in the metal-containing product [311].

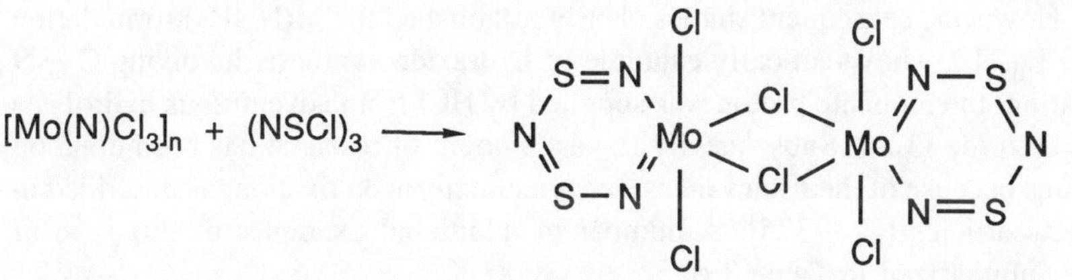

$$[Mo(N)Cl_3]_n + (NSCl)_3 \longrightarrow \text{[dimeric Mo complex]} \quad (100)$$

The reaction of an alkylidyne species with an electrophilic alkylating agent to afford an alkylidene has not yet been reported. A reaction that does result in carbon–carbon bond formation to an alkylidyne ligand is that which occurs upon treatment with acetylenes, one example being Eq. 101. The resultant "metallacyclobutadienes" contain the alkylidene M=C bond. Although such species tend to be unstable toward β-deprotonation processes when prepared from terminal acetylenes [18,312,313], several examples derived from internal acetylenes have now been structurally characterized [314–316]. The position of the equilibrium in these reactions depends on the steric bulk and electronic properties of the ancillary ligands. In the particular case of Eq. 101 (R = 2,6-diisopropylphenyl), the product can be isolated in the solid state but at room temperature in toluene-d_8 it is virtually totally dissociated to the propylidyne complex and 3-hexyne [18]. Catalytic acetylene metathesis is based on the existence of such equilibria; this reaction will be discussed in detail in Section 7.5.

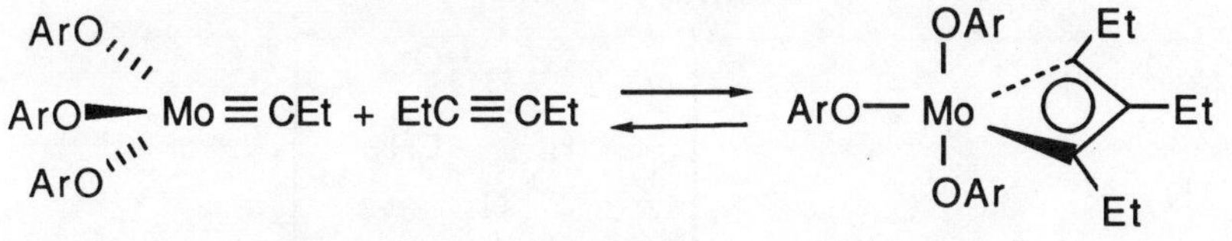

$$(ArO)_3Mo\equiv CEt + EtC\equiv CEt \rightleftharpoons (ArO)_3Mo[\text{metallacyclobutadiene with Et groups}] \quad (101)$$

Ar = 2,6-diisopropylphenyl

3.4 BY FORMATION OF A BOND TO THE β ATOM

This family of reactions involves formation of a new bond (typically to C or H) to the eventual β atom of a hydrazido, organoimido, or alkylidene ligand. The ligand precursor is invariably an unsaturated species and this provides a common denominator for this class of reactions. However, formation of the β bond may proceed via either formal nucleophilic or electrophilic attack on that unsaturated precursor, as illustrated below.

Hydrazido(2-) ligands are frequently prepared by the protonation or alkylation of a dinitrogen ligand. In some cases, the intermediate diazenido(1-) complex, M(N=NR), is isolated while in others the dinitrogen ligand is converted directly to the hydrazido species. The first examples of this type of reaction were reported by the Chatt group in 1972 and involved the use of chelating diphosphine ligands [317,318]. IR spectral data for the protonated nitrogen ligand from Eq. 102 initially suggested a M(NHNH) structure [318]. However, subsequent studies clearly established the M(NNH$_2$) formulation [319–321]. Eq. 103 shows an early example of hydrazido synthesis involving C—N bond formation; the requisite proton was supplied by HCl from adventitious hydrolysis of the acid chloride [322]. Subsequently, a vast amount of research has been done on such reactions because of their relevance to enzymatic nitrogen fixation, as described in recent review articles [323–325]. A number of additional examples of this type of process are summarized in Table 3.6.

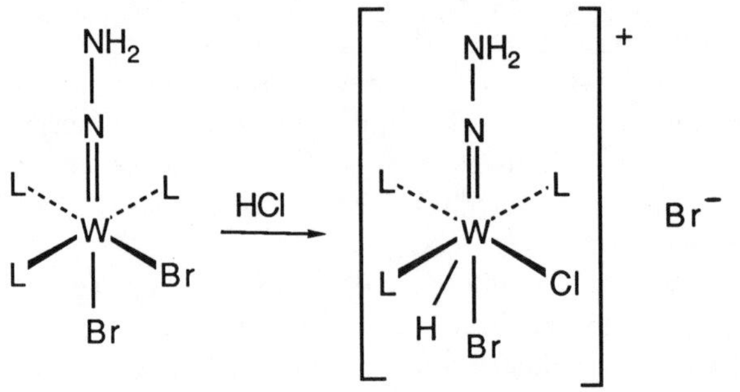

$$\text{(102)}$$

L = PMePh$_2$; position of hydride not determined

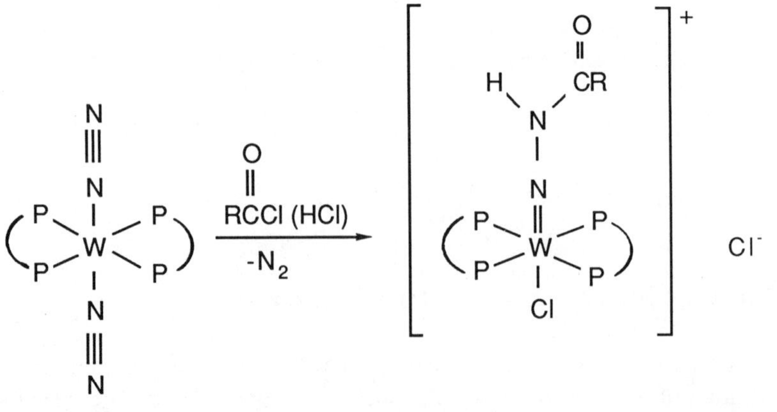

$$\text{(103)}$$

Henderson [340] has studied the mechanism of formation of [M(NNH$_2$)(OMe)$_2$L$_3$] from cis-[M(N$_2$)$_2$L$_4$] in acidic methanol where M = Mo or W and L = dimethylphenylphosphine. The reaction was found to be first order in metal complex and second order in acid concentration. It is suggested that diprotonation of one coordinated dinitrogen ligand labilizes the remaining N$_2$ ligand which dissociates to yield a five-coordinate intermediate [M(NNH$_2$)L$_4$]$^{2+}$. Rapid attack of methanol gives cis-

TABLE 3.6 Formation of Hydrazido(2−) Ligands by Alkylation or Protonation of a β-Nitrogen Atom[a]

Starting Complex	Electrophile	Product	Reference
$(dtc)_3Mo(NNPh)$	MeI	$[(dtc)_3Mo(NNMePh)]I$	326
$Mo(NNBu)Br(dppe)_2$	MeOTf	$[Mo(NNMeBu)Br(dppe)_2]OTf$	327
$Mo(NNMe)Br(dppe)_2$	HBF_4	$[Mo(NNHMe)Br(dppe)_2]BF_4$	328
$Mo(N_2)_2(triphos)(PPh_3)$	HCl	$[Mo(NNH_2)Cl(triphos)(PPh_3)]Cl$	329
$Mo(N_2)_2(dppe)_2$	HBF_4	$[Mo(NNH_2)F(dppe)_2]BF_4$	330
$Mo[NNC(O)Ph]Cl(dppe)_2$	HCl	$Mo[NNH(COPh)]Cl_2(dppe)_2$	331
$Mo(NNCH_2CO_2Et)Cl(dppe)_2$	HBF_4	$[Mo\{NNH(CH_2CO_2Et)\}Cl(dppe)_2]^+$	332
$M(N_2)_2(dppe)_2$ (M = Mo, W)	$Br(CH_2)_4Br$	$[M\{NN(CH_2)_4\}Br(dppe)_2]Br$	333
$M(N_2)_2(PMe_2Ph)_4$ (M = Mo, W)	HCl	$M(NNH_2)Cl_2(PMe_2Ph)_3$	334
$W(NNH)Br(dppe)_2$	MeBr	$[W(NNHMe)Br(dppe)_2]Br$	335
$W(NN^iPr)Br(dppe)_2$	HBr	$[W(NNH^iPr)Br(dppe)_2]Br$	322
$W(N_2)_2(PMe_2Ph)_4$	Me_3SiI	$W[NNH(SiMe_3)]I_2(PMe_2Ph)_3$	336
$W(N_2)_2(dppe)_2$	TsOH	$[W(NNH_2)(OTs)(dppe)_2]OTs$	337
$Re(NNPh)_2Br(PPh_3)_2$	HBr	$Re(NNHPh)(NNPh)Br_2(PPh_3)_2$	338
$Re(NNPh)Cl_2(NH_3)(PMe_2Ph)_2$	HCl	$[Re(NNHPh)Cl_2(NH_3)(PMe_2Ph)_2]Cl$	339

[a] Abbreviations: OTf = triflate; OTs = tosylate; triphos = $(Ph_2PCH_2CH_2)_2PPh$.

$[M(NNH_2)(OMe)L_4]^+$ and a mole-equivalent of protons. Subsequent dissociation of phosphine and rapid addition of methanol then accounts for the observed product.

The examples in Table 3.6 involve electrophilic attack on the β-nitrogen atom. Sutton and coworkers have demonstrated that the opposite tack can also be used to prepare hydrazido(2-) complexes. For example, a hydride ligand of Cp_2WH_2 was found to add to arenediazonium salts to afford a tungsten(VI) product as shown in Eq. 104 [341,342]. Moreover, treatment of $CpRe(CO)_2(THF)$ with an arenediazonium tetrafluoroborate resulted in an aryldiazenido complex, $[CpRe(CO)_2(NNAr)]BF_4$. Subsequent addition of methyllithium occurred at the β-nitrogen atom to yield a rhenium hydrazido derivative, $CpRe(CO)_2(NNMeR)$ [342].

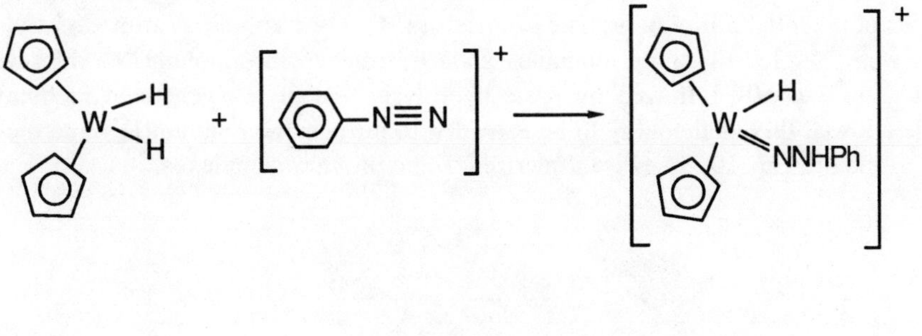

$$Cp_2WH_2 + [Ph\text{-}N{\equiv}N]^+ \longrightarrow [Cp_2W(H)(NNHPh)]^+ \quad (104)$$

Tungsten(VI) chloride reacts with trichloroacetonitrile as shown in Eq. 105 to form a product containing the pentachloroethylimido ligand [343,344]. The product has also been obtained free of nitrile ligand as a chloride-bridged dimer, [Cl$_4$W(NR)]$_2$ [345]. This chemistry can be extended to some nonchlorinated (aliphatic and aromatic) nitriles. However, nitriles containing hydrogen on the cyanide-bearing carbon result in imido complexes that can decompose with evolution of hydrogen halide [346]. Such a pathway apparently accounts for the well-known reduction of WCl$_6$ to WCl$_4$(NCMe)$_2$ in acetonitrile [346]. Dehnicke and coworkers have extended the scope of this chemistry by the introduction of Cl$_3$PO as a stabilizing ligand and in particular by the use of cyanogen chloride, ClCN, as the nitrile component. In this way trichloromethylimido complexes (Cl$_3$PO)M(NCCl$_3$)Cl$_4$ have been isolated and characterized for M = Mo, W, and Re [347–349].

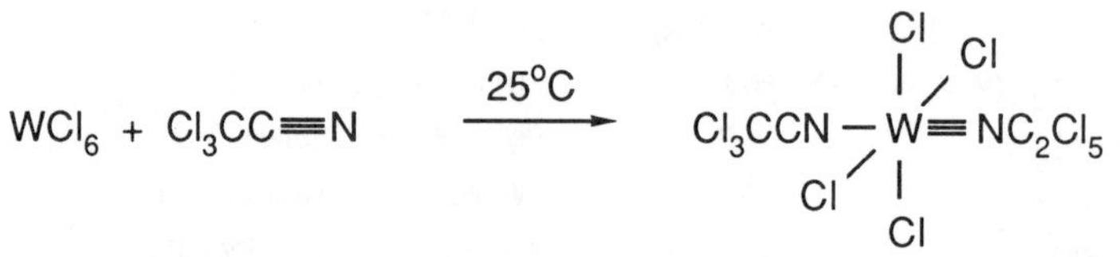

$$WCl_6 + Cl_3CC\equiv N \xrightarrow{25°C} Cl_3CCN-\underset{\underset{Cl}{|}}{\overset{\overset{Cl}{|}}{W}}\!\!\diagup^{Cl}_{\diagdown Cl}\!\!\equiv NC_2Cl_5 \quad (105)$$

Tantalum(III) and niobium(III) chloride species, generated by reduction *in situ*, react with acetonitrile to afford interesting imido-bridged dimers as indicated in Eq. 106. (Earlier reports of simple nitrile adduct formation [350,351] appear to be in error.) Derivatives containing this bridging "1,2-dimethyl-1,2-diimidoethene" ligand have been characterized structurally for both niobium [352] and for tantalum [353].

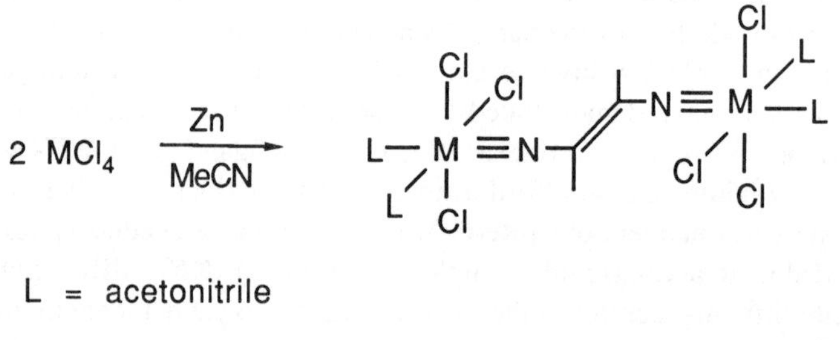

$$2\ MCl_4 \xrightarrow[\text{MeCN}]{\text{Zn}} L-M\equiv N{-}\!\!\diagup\!\!{-}N\equiv M-L \quad (106)$$

L = acetonitrile

Pedersen has harnessed the driving force for imide formation to promote coupling reactions of potential utility in organic synthesis [354]. For example, *N*-trimethylsilylimines were coupled to vicinal diamines upon treatment with niobium(IV) chloride according to Eq. 107 followed by basic hydrolysis. An *in situ*-generated niobium hydride allowed this methodolgy to be extended to nitriles; the niobium(IV) intermediates formed in Eq. 108 likewise dimerized to the diimido complexes.

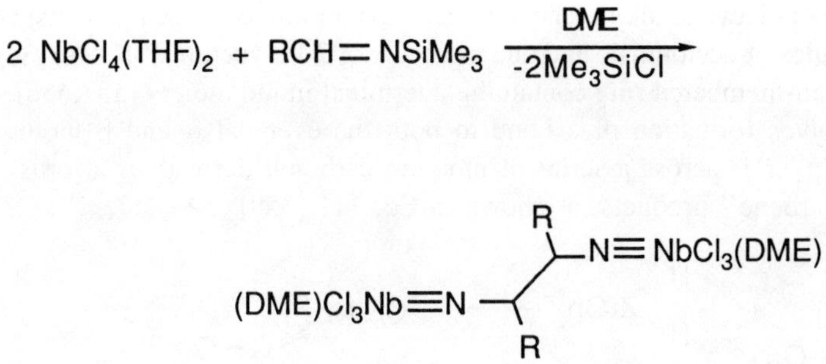

$$2\ NbCl_4(THF)_2 + RCH=NSiMe_3 \xrightarrow[-2Me_3SiCl]{DME}$$

(107)

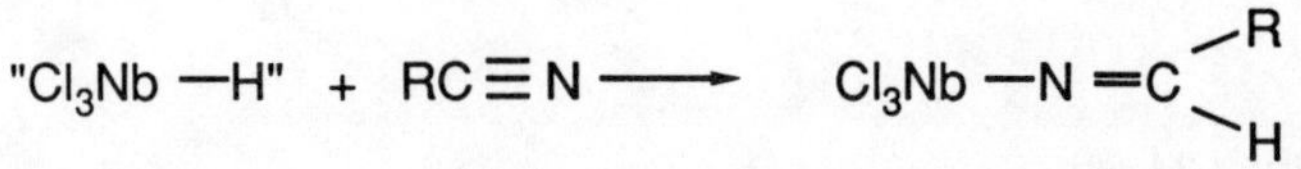

$$(DME)Cl_3Nb\equiv N-\underset{R}{\overset{R}{C}}-N\equiv NbCl_3(DME)$$

$$\text{"}Cl_3Nb-H\text{"} + RC\equiv N \longrightarrow Cl_3Nb-N=C\underset{H}{\overset{R}{\diagup}} \qquad (108)$$

Tantalum alkylidene complexes react with nitriles with formation of an imido ligand. This reaction is exemplified by Eq. 109 [355]. CpTaCl$_2$(CHCMe$_3$) reacts similarly with acetonitrile or benzonitrile [38]. The latter complex was also shown to react with diphenylacetylene to afford a new alkylidene species according to Eq. 110. No mechanistic studies have been reported for either Eqs. 109 or 110; however the proposal [38] that they involve an initial (2 + 2) cycloaddition process analogous to Eq. 84 seems quite reasonable.

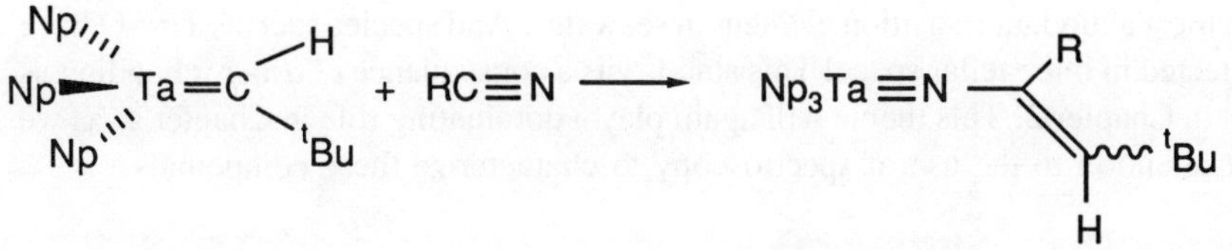

$$Np_3Ta=C\underset{^tBu}{\overset{H}{\diagup}} + RC\equiv N \longrightarrow Np_3Ta\equiv N-\overset{R}{\underset{H}{C}}=\underset{^tBu}{\overset{}{C}}\text{H} \qquad (109)$$

$$Cp(Cl)_2Ta=CH^tBu + PhC\equiv CPh \xrightarrow{25°C} \text{product} \qquad (110)$$

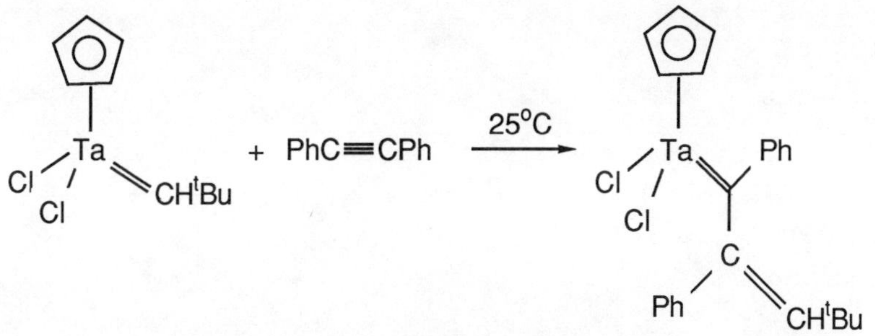

Other reactions are known in which unsaturated molecules are ultimately converted to multiply bonded ligands. Curtis [356] has reported an example in which coupling of a coordinated alkyne and an η^2-iminoacyl on a tantalum center results in a nonplanar,

6π-electron metallacyclic alkylidene complex. Even more complicated is the coupling of two molecules of acetonitrile and one molecule of 2-butyne on $W_2(OR)_6(py)_2$ which leads to a seven-membered ring containing a terminal imido moiety [357,358]. A final example involves formation of a bond to both the eventual α and β atoms. Thus, addition of $Cp_2^*ZrH_2$ across a series of niobium carbonyl derivatives affords unusual "zirconoxy carbene" products, as shown in Eq. 111 [359].

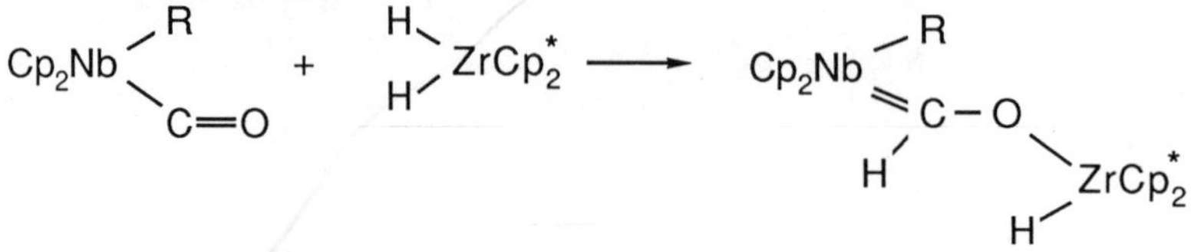

$$Cp_2Nb\begin{matrix}R\\C=O\end{matrix} + \begin{matrix}H\\H\end{matrix}ZrCp_2^* \longrightarrow Cp_2Nb\begin{matrix}R\\=C-O\\H\end{matrix}\begin{matrix}\\\\ZrCp_2^*\\H\end{matrix} \quad (111)$$

R = H , Me , Ph , CH_2Ph

3.5 CONCLUDING REMARKS

The wide range of synthetic procedures that can be used to introduce multiply bonded ligands underscores the fact that, as a class, these compounds are by no means fragile laboratory curiosities. Indeed it is the tendency of molybdenum to form terminal oxo ligands spontaneously—as molybdate and isopolymolybdates—that makes molybdenum the most abundant transition element in seawater. And species such as Ti=O have been detected in interstellar space! This stability is a consequence of $d\pi$–$p\pi$ bonding as set forth in Chapter 2. This theme will again play a dominating role in Chapter 4, as we turn our attention to the use of spectroscopy to characterize these compounds.

REFERENCES

1. Gamsjaeger, H.; Murmann, R.K. *Adv. Inorg. Bioinorg. Mech.* **1983**, *2*, 317-380.
2. Schrock, R.R.; Messerle, L.W.; Clayton, C.D.; Guggenberger, L.J. *J. Am. Chem. Soc.* **1978**, *100*, 3793-3800.
3. Turner, H.W.; Schrock, R.R.; Fellman, J.D.; Holmes, S.J. *J. Am. Chem. Soc.* **1983**, *105*, 4942-4950.
4. Messerle, L.W.; Jennische, P.; Schrock, R.R.; Stucky, G. *J. Am. Chem. Soc.* **1980**, *102*, 6744-6752.
5. Andersen, R.A. *Inorg. Chem.* **1979**, *18*, 3622-3623.
6. Chamberlain, L.R.; Rothwell, I.P.; Huffman, J.C. *J. Am. Chem. Soc.* **1986**, *108*, 1502-1509.
7. Fellmann, J.D.; Schrock, R.R.; Traficante, D.D. *Organometallics* **1982**, *1*, 481-484.
8. Fellmann, J.D.; Schrock, R.R.; Rupprecht, G.A. *J. Am. Chem. Soc.* **1981**, *103*, 5752-5758.
9. Ehrenfeld, D.; Kress, J.; Moore, B.D.; Osborn, J.A.; Schoettel, G. *J. Chem. Soc., Chem. Comm.* **1987**, 129-131.
10. Andersen, R.A.; Chisholm, M.H.; Gibson, J.F.; Reichert, W.W.; Rothwell, I.P.; Wilkinson, G. *Inorg. Chem.* **1981**, *20*, 3934-3936.
11. Quignard, F.; Leconte, M.; Basset, J.-M. *J. Chem. Soc., Chem. Comm.* **1985**, 1816-1817.
12. Pedersen, S.F.; Schrock, R.R. *J. Am. Chem. Soc.* **1982**, *104*, 7483-7491.
13. Horton, A.D.; Schrock, R.R.; Freudenberger, J.H. *Organometallics* **1987**, *6*, 893-894.
14. Edwards, D.S.; Biondi, L.V.; Ziller, J.W.; Churchill, M.R.; Schrock, R.R. *Organometallics* **1983**, *2*, 1505-1513.
15. Mowat, W.; Wilkinson, G. *J. Chem. Soc., Dalton Trans.* **1973**, 1120-1124.
16. McLain, S.J.; Wood, C.D.; Messerle, L.W.; Schrock, R.R.; Hollander, F.J.; Youngs, W.J.; Churchill, M.R. *J. Am. Chem. Soc.* **1978**, *100*, 5962-5964.
17. Fellmann, J.D.; Turner, H.W.; Schrock, R.R. *J. Am. Chem. Soc.* **1980**, *102*, 6608-6609.
18. McCullough, L.G.; Schrock, R.R.; Dewan, J.C.; Murdzek, J.C. *J. Am. Chem. Soc.* **1985**, *107*, 5987-5998.
19. Ahmed, K.J.; Chisholm, M.H.; Huffman, J.C. *Organometallics* **1985**, *4*, 1168-1174.
20. Chisholm, M.H.; Cotton, F.A.; Extine, M.W.; Murillo, C.A. *Inorg. Chem.* **1978**, *17*, 696-698.
21. Schrock, R.R.; Clark, D.N.; Sancho, J.; Wengrovius, J.H.; Rocklage, S.M.; Pedersen, S.F. *Organometallics* **1982**, *1*, 1645-1651.
22. Sharp, P.R.; Holmes, S.J.; Schrock, R.R.; Churchill, M.R.; Wasserman, H.J. *J. Am. Chem. Soc.* **1981**, *103*, 965-966.
23. Mayr, A.; Asaro, M.F.; Kjelsberg, M.A.; Lee, S.L.; Van Engon, D. *Organometallics* **1987**, *6*, 432-434.
24. Henderson, R.A.; Davies, G.; Dilworth, J.R.; Thorneley, R.N.F. *J. Chem. Soc., Dalton Trans.* **1981**, 40-50.

25. Buchler, J.W.; Dreher, C.; Lay, K.L.; Raap, A.; Gersonde, K. *Inorg. Chem.* **1983**, *22*, 879–884.
26. Buhr, J.D.; Winkler, J.R.; Taube, H. *Inorg. Chem.* **1980**, *19*, 2416–2425.
27. Jernakoff, P.; Cooper, N.J. *Organometallics* **1986**, *5*, 747–751.
28. Moyer, B.A.; Meyer, T.J. *Inorg. Chem.* **1981**, *20*, 436–444.
29. Casey, A.T.; Vecchio, A.M. *Trans. Met. Chem.* **1986**, *11*, 366–368.
30. Mayer, J.M.; Curtis, C.J.; Bercaw, J.E. *J. Am. Chem. Soc.* **1983**, *105*, 2651–2660.
31. Nugent, W.A.; Harlow, R.L. *J. Chem. Soc., Chem. Comm.* **1978**, 579–580.
32. Murray, R.C.; Schrock, R.R. *J. Am. Chem. Soc.* **1985**, *107*, 4557–4558.
33. Nugent, W.A.; Harlow, R.L. *Inorg. Chem.* **1980**, *19*, 777–779.
34. Clifford, A.F.; Kobayashi, C.S. *Inorg. Synth.* **1960**, *6*, 204–208.
35. Schrock, R.R.; Sharp, P.R. *J. Am. Chem. Soc.* **1978**, *100*, 2389–2399.
36. Liu, A.H.; Murray, J.C.; Dewan, J.C.; Santarsiero, B.D.; Schrock, R.R. *J. Am. Chem. Soc.* **1987**, *109*, 4282–4291.
37. Schrock, R.R. *J. Am. Chem. Soc.* **1974**, *96*, 6796–6797.
38. Wood, C.D.; McLain, S.J.; Schrock, R.R. *J. Am. Chem. Soc.* **1979**, *101*, 3210–3222.
39. Rupprecht, G.A.; Messerle, L.W.; Fellmann, J.D.; Schrock, R.R. *J. Am. Chem. Soc.* **1980**, *102*, 6236–6244.
40. Clark, D.N.; Schrock, R.R. *J. Am. Chem. Soc.* **1978**, *100*, 6774–6776.
41. Fellmann, J.D.; Rupprecht, G.A.; Wood, C.D.; Schrock, R.R. *J. Am. Chem. Soc.* **1978**, *100*, 5964–5966.
42. Churchill, M.R.; Wasserman, H.J.; Turner, H.W.; Schrock, R.R. *J. Am. Chem. Soc.* **1982**, *104*, 1710–1716.
43. van Asselt, A.; Burger, B.J.; Gibson, V.C.; Bercaw, J.E. *J. Am. Chem. Soc.* **1986**, *108*, 5347–5349.
44. Cooper, N.J.; Green, M.L.H. *J. Chem. Soc., Chem. Comm.* **1974**, 761–762.
45. Erickson, T.K.G.; Mayer, J.M. submitted to *Angew. Chem.*.
46. Rocklage, S.M.; Schrock, R.R.; Churchill, M.R.; Wasserman, H.J. *Organometallics* **1982**, *1*, 1332–1338.
47. Freudenberger, J.H.; Schrock, R.R. *Organometallics* **1985**, *4*, 1937–1944.
48. Chatt, J.; Rowe, G.A. *J. Chem. Soc.* **1962**, 4019–4033.
49. Lappert, M.F.; Patil, D.S.; Pedley, J.B. *J. Chem. Soc., Chem. Comm.* **1975**, 830–831.
50. Bishop, M.W.; Chatt, J.; Dilworth, J.R.; Hursthouse, M.B.; Motevalli, M. *J. Chem. Soc., Dalton Trans.* **1979**, 1600–1602.
51. Hsieh, T.-C.; Gebreyes, K.; Zubieta, J. *J. Chem. Soc., Chem. Comm.* **1984**, 1172–1174.
52. Sakharov, S.G.; Kokunov, Yu.V.; Gustyakova, M.P.; Buslaev, Yu.A. *Koord. Khim.* **1982**, *8*, 1669–1672.
53. Chan, D.M.-T.; Fultz, W.C.; Nugent, W.A.; Roe, D.C.; Tulip, T.H. *J. Am. Chem. Soc.* **1985**, *107*, 251–253.
54. Schrock, R.R.; Clark, D.N.; Sancho, J.; Wengrovius, J.H.; Rocklage, S.M.; Pedersen, S.F. *Organometallics* **1982**, *1*, 1645–1651.

55. Chambers, O.R.; Harman, M.E.; Rycroft, D.S.; Sharp, D.W.A.; Winfield, J.M. *J. Chem. Res. (M)* **1977**, 1849-1876.
56. Viswanathan, N.; VanDyke, C.H. *J. Organomet. Chem.* **1968**, *11*, 181-184.
57. Godemeyer, T.; Berg, A.; Gross, H.D.; Müller, U.; Dehnicke, K. *Z. Naturforsch.* **1985**, *40B*, 999-1004.
58. Chatt, J.; Crichton, B.A.L.; Dilworth, J.R.; Dahlstrom, P.; Gutkowska, R.; Zubieta, J. *Inorg. Chem.* **1982**, *21*, 2383-2391.
59. Becker, F. *J. Organomet. Chem.* **1973**, *51*, C9-C10.
60. Wiberg, N.; Häring, H.W.; Schieda, O. *Angew. Chem., Int. Ed. Engl.* **1976**, *15*, 386-387.
61. Harmon, M.; Sharp, D.W.A.; Winfield, J.M. *Inorg. Nucl. Chem. Lett.* **1974**, *10*, 183-185.
62. Preuss, F.; Towae, W. *Z. Naturforsch.* **1981**, *36B*, 1130-1135.
63. Bradley, D.C.; Hursthouse, M.B.; Jelfs, A.N. de M.; Short, R.L. *Polyhedron* **1983**, *2*, 849-852.
64. Bates, P.A.; Nielson, A.J.; Waters, J.M. *Polyhedron* **1985**, *4*, 1391-1401.
65. Jones, T.C.; Nielson, A.J.; Ricard, C.E.F. *J. Chem. Soc., Chem. Comm.* **1984**, 205-206.
66. Nugent, W.A. *Inorg. Chem.* **1983**, *22*, 965-969.
67. Ashcroft, B.R.; Bradley, D.C.; Clark, G.R.; Errington, R.J.; Nielson, A.J.; Rickard, C.E.F. *J. Chem. Soc., Chem. Comm.* **1987**, 170-171.
68. Bradley, D.C.; Errington, R.J.; Hursthouse, M.B.; Nielson, A.J.; Short, R.L. *Polyhedron* **1983**, *2*, 843-847.
69. Nugent, W.A.; Harlow, R.L.; McKinney, R.J. *J. Am. Chem. Soc.* **1979**, *101*, 7265-7268.
70. Mowat, W.; Shortland, A.; Yagupsky, G.; Hill, N.J.; Yagupsky, M.; Wilkinson, G. *J. Chem. Soc., Dalton Trans.* **1972**, 533-542.
71. Hug, F.; Mowat, W.; Skapski, A.C.; Wilkinson, G. *J. Chem. Soc., Chem. Comm.* **1971**, 1477-1478.
72. Dolgoplosk, B.A.; Oreshkin, I.A.; Makovetsky, K.L.; Tinyakova, E.I.; Ostrovskaya, I.Ya.; Kershenbaum, I.L.; Chernenko, G.M. *J. Organomet. Chem.* **1977**, *128*, 339-344.
73. Vedejs, E.; Martinez, G.R. *J. Am. Chem. Soc.* **1979**, *101*, 6452-6454.
74. Tebbe, F.N.; Parshall, G.W.; Reddy, G.S. *J. Am. Chem. Soc.* **1978**, *100*, 3611-3613.
75. McDermott, G.A.; Mayr, A. *J. Am. Chem. Soc.* **1987**, *109*, 580-582.
76. Hartner, F.W., Jr.; Schwartz, J. *J. Am. Chem. Soc.* **1981**, *103*, 4979-4981.
77. Wengrovius, J.H.; Schrock, R.R. *Organometallics* **1982**, *1*, 148-155.
78. Noble, A.M.; Winfield, J.M. *Inorg. Nucl. Chem. Lett.* **1968**, *4*, 339-342.
79. Kokunov, Yu.V.; Chubar, Yu.D.; Bochkareva, V.A.; Buslaev, Yu.A. *Soviet J. Coord. Chem.* **1978**, *4*, 162-167.
80. Kokunov, Yu.V.; Bochkareva, V.A.; Chubar, Yu.D.; Buslaev, Yu.A. *Koord. Khim.* **1980**, *6*, 1213-1216.
81. Bunker, M.J.; DeCian, A.; Green, M.L.H. *J. Chem. Soc., Chem. Comm.* **1977**, 59.

82. Drew, M.G.B.; Fowles, G.W.A.; Rice, D.A.; Shanton, K.J. *J. Chem. Soc., Chem. Comm.* **1974**, 614–615.
83. Bradley, D.C.; Chisholm, M.H.; Extine, M.W.; Stager, M.E. *Inorg. Chem.* **1977**, *16*, 1794–1801.
83a Barner, C.J.; Collins, T.J.; Mapes, B.E.; Santarsiero, B.D. *Inorg. Chem.* **1986**, *25*, 4322–4323.
84. Deutscher, R.L.; Kepert, D.L. *Inorg. Chim. Acta* **1970**, *4*, 645–650.
85. Noble, A.M.; Winfield, J.M. *J. Chem. Soc. A.* **1970**, 501–506.
86. Bradley, D.C.; Thomas, I.M. *Proc. Chem. Soc., London* **1959**, 225–226.
87. Bradley, D.C.; Thomas, I.M. *Can. J. Chem.* **1962**, *40*, 1355–1360.
88. Airoldi, C.; Bradley, D.C.; Vuru, G. *Trans. Met. Chem.* **1979**, *4*, 64.
89. Bradley, D.C.; Chisholm, M.H.; Extine, M.W. *Inorg. Chem.* **1977**, *16*, 1791–1794.
90. Takahashi, Y.; Onoyama, N.; Ishikawa, Y.; Motojima, S.; Sugiyama, K. *Chem. Lett.* **1978**, 525–528.
91. Chamberlain, L.R.; Rothwell, I.P.; Huffman, J.C. *J. Chem. Soc., Chem. Comm.* **1986**, 1203–1205.
92. Chiu, K.W.; Jones, R.A.; Wilkinson, G.; Galas, A.M.R.; Hursthouse, M.B. *J. Chem. Soc., Dalton Trans.* **1981**, 2088–2097.
93. Straus, D.A.; Grubbs, R.H. *Organometallics* **1982**, *1*, 1658–1661.
94. Clawson, L.; Buchwald, S.L.; Grubbs, R.H. *Tetrahedron Lett.* **1984**, 5733–5736.
95. Wallace, K.C.; Dewan, J.C.; Schrock, R.R. *Organometallics* **1986**, *5*, 2162–2164.
96. Brown-Wensley, K.A.; Buchwald, S.L.; Cannizzo, L.; Clawson, L.; Ho, S.; Meinhardt, D.; Stille, J.R.; Straus, D.; Grubbs, R.H. *Pure Appl. Chem.* **1983**, *55*, 1733–1744.
97. Ikariya, T.; Ho, S.C.H.; Grubbs, R.H. *Organometallics* **1985**, *4*, 199–200.
98. Gilliom, L.R.; Grubbs, R.H. *Organometallics* **1986**, *5*, 721–724.
99. Gilliom, L.R.; Grubbs, R.H. *J. Am. Chem. Soc.* **1986**, *108*, 733–742.
100. Tebbe, F.N.; Harlow, R.L. *J. Am. Chem. Soc.* **1980**, *102*, 6151–6153.
101. McKinney, R.J.; Tulip, T.H.; Thorn, D.L.; Coolbaugh, T.S.; Tebbe, F.N. *J. Am. Chem. Soc.* **1981**, *103*, 5584–5586.
102. LaPointe, R.E.; Wolczanski, P.T.; Mitchell, J.F. *J. Am. Chem. Soc.* **1986**, *108*, 6382–6384.
103. Chisholm, M.H.; Heppert, J.A.; Huffman, J.C.; Streib, W.E. *J. Chem. Soc., Chem. Comm.* **1985**, 1771–1773.
104. Schrock, R.R.; Listemann, M.L.; Sturgeoff, L.G. *J. Am. Chem. Soc.* **1982**, *104*, 4291–4293.
105. Bryan, J.C.; Geib, S.J.; Rheingold, A.L.; Mayer, J.M. *J. Am. Chem. Soc.* **1987**, *109*, 2826–2828.
105a Antiñolo, A.; Garcia-Lledó, S.; Martinez de Ilarduya, J.; Otero, A. *J. Organomet. Chem.* **1987**, *335*, 85–90.
106. Sharpless, K.B.; Umbreit, M.A.; Nieh, M.T.; Flood, T.C. *J. Am. Chem. Soc.* **1972**, *94*, 6538–6540.
107. Sharpless, K.B.; Flood, T.C. *J. Chem. Soc., Chem. Comm.* **1972**, 370–371.

108. Sattar, A.; Forrester, J.; Moir, M. Roberts, J.S.; Parker, W. *Tetrahedron Lett.* **1976**, 1405-1406.
109. Berry, M.; Davies, S.G.; Green, M.L.H. *J. Chem. Soc., Chem. Comm.* **1978**, 99-100.
110. Courtney, J.L.; Swansborough, K.F. *Rev. Pure Appl. Chem.* **1972**, *22*, 47-54.
111. Dobson, J.C.; Seok, W.K.; Meyer, T.J. *Inorg. Chem.* **1986**, *25*, 1513-1514.
112. Takagi, S.; Takahashi, E.; Miyamoto, T.K.; Sasaki, Y. *Chem. Lett.* **1986**, 1275-1278.
113. Bortolini, O.; Meunier, B. *J. Chem. Soc., Chem. Comm.* **1983**, 1364-1366.
114. Collman, J.P.; Brauman, J.I.; Meunier, B.; Raybuck, S.A.; Kodadek, T. *Proc. Natl. Acad. Sci.* **1984**, *81*, 3245-3248.
115. Groves, J.T.; Nemo, T.E.; Myers, R.S. *J. Am. Chem. Soc.* **1979**, *101*, 1032-1033.
116. Traylor, P.S.; Dolphin, D.; Traylor, T.G. *J. Chem. Soc., Chem. Comm.* **1984**, 279-280.
117. Groves, J.T.; Kruper, W.J. *J. Am. Chem. Soc.* **1979**, *101*, 7613-7615.
118. Groves, J.T.; Kruper, W.J.; Haushalter, R.C. *J. Am. Chem. Soc.* **1980**, *102*, 6375-6377.
119. Hill, C.L.; Schardt, B.C. *J. Am. Chem. Soc.* **1980**, *102*, 6375-6377.
120. Srinivasan, K.; Kochi, J.K. *Inorg. Chem.* **1985**, *24*, 4671-4679.
121. Samsel, E.G.; Srinivasan, K.; Kochi, J.K. *J. Am. Chem. Soc.* **1985**, *107*, 7606-7617.
122. Carlsen, P.H.J.; Katsuki, T.; Martin, V.S.; Sharpless, K.B. *J. Org. Chem.* **1981**, *46*, 3936-3938.
123. Schroeder, M. *Chem. Rev.* **1980**, *80*, 187-213.
124. Roesky, H.W.; Sundermeyer, J., Schimkowiak, J.; Jones, P.G.; Noltemeyer, M.; Schroeder, T.; Sheldrick, G.M. *Z. Naturforsch.* **1985**, *40B*, 736-739.
125. Liese, W.; Dehnicke, K.; Walker, I.; Strähle, J. *Z. Naturforsch.* **1979**, *34B*, 693-696.
126. Dehnicke, K.; Liebelt, W. *Z. Anorg. Allg. Chem.* **1979**, *453*, 9-13.
127. Strähle, J.; Dehnicke, K. *Z. Anorg. Chem.* **1965**, *338*, 287-298.
128. Lorcher, K.-P.; Strähle, J.; Walker, I. *Z. Anorg. Allg. Chem.* **1979**, *452*, 123-140.
129. Hentges, S.G.; Sharpless, K.B.; Tulip, T.H. unpublished results.
130. Sharpless, K.B.; Chong, P.O.; Oshima, K. *J. Org. Chem.* **1976**, *41*, 177-179.
131. Herranz, E.; Sharpless, K.B. *J. Org. Chem.* **1978**, *43*, 2544-2548.
132. Groves, J.T.; Watanabe, Y. *J. Am. Chem. Soc.* **1986**, *108*, 7834-7836.
133. Groves, J.T.; Watanabe, Y.; McMurry, T.J. *J. Am. Chem. Soc.* **1983**, *105*, 4489-4490.
134. Cleare, M.J.; Griffith, W.P. *J. Chem. Soc., Chem. Comm.* **1968**, 1302.
135. Mukaida, M. *Bull. Chem. Soc. Jpn.* **1970**, *43*, 3805-3813.
136. Ciechanowicz, M.; Skapski, A.C. *J. Chem. Soc. A* **1971**, 1792-1794.
137. Seyferth, K.; Taube, R. *J. Mol. Catal.* **1985**, *28*, 53-69. See also Sellmann, D.; Binker, G. *Z. Naturforsh* **1987**, *42B*, 341-347.

138. Cotton, F.A.; Shamshoum, E.S. *J. Am. Chem. Soc.* **1984**, *106*, 3222-3225.
139. LaMonica, G.; Cenini, S. *J. Chem. Soc., Dalton Trans.* **1980**, 1145-1149.
140. Fischer, E.O.; Schubert, U. *J. Organomet. Chem.* **1975**, *100*, 59-81.
141. Mayr, A.; McDermott, G.A. *J. Am. Chem. Soc.* **1986**, *108*, 548-549.
142. Horner, S.M.; Tyree, Jr., S.Y. *Inorg. Chem.* **1962**, *1*, 122-127.
143. Behzadi, K.; Baghlaf, A.O.; Thompson, A. *J. Less-Common Met.* **1978**, *57*, 103-110.
144. Broussier, R.; Normand, H.; Gautheron, B. *J. Organomet. Chem.* **1976**, *120*, C28-C30.
145. Behzadi, K.; Ahwaz Iran, A.I.T.; Thompson, A. *J. Less Common Met.* **1986**, *124*, 135-139.
146. Mitchell, P.C.H.; Scarle, R.D. *J. Chem. Soc., Dalton Trans.* **1975**, 2552-2555.
147. Holm, R.H.; Berg, J.M. *Acc. Chem. Res.* **1986**, *19*, 363-370.
148. Reynolds, M.S.; Berg, J.M.; Holm, R.H. *Inorg. Chem.* **1984**, *23*, 3057-3062.
149. Caradonna, J.P.; Harlan, E.W.; Holm, R.H. *J. Am. Chem. Soc.* **1986**, *108*, 7856-7858.
150. Bishop, M.W.; Chatt, J.; Dilworth, J.R. *J. Chem. Soc., Dalton Trans.* **1979**, 1-5.
151. Hauck, H.G.; Klingelhöfer, P.; Müller, U.; Dehnicke, K. *Z. Anorg. Allg. Chem.* **1984**, *510*, 180-188.
152. Müller, U.; Schweda, E.; Strähle, J. *Z. Naturforsch.* **1983**, *38B*, 1299-1300.
153. Müller, U.; Kafitz, W.; Dehnicke, K. *Z. Anorg. Allg. Chem.* **1983**, *501*, 69-78.
154. Dahl, L.F.; Frisch, P.D.; Gust, G.R. *J. Less-Common Met.* **1974**, *36*, 255-264.
155. Mertis, K.; Wilkinson, G. *J. Chem. Soc., Dalton Trans.* **1976**, 1488-1492.
156. Bottomley, F.; Darkwa, J.; Sutin, L.; White, P.S. *Organometallics* **1986**, *5*, 2165-2171.
157. Legzdins, P.; Rettig, S.J.; Sánchez, L. *Organometallics* **1985**, *4*, 1470-1471.
158. Alt, H.G.; Hayen, H.I. *Angew. Chem., Int. Ed. Engl.* **1985**, *24*, 497-498.
159. Bottomley, F.; Lin, I.J.B.; Mukaida, M. *J. Am. Chem. Soc.* **1980**, *102*, 5238-5242.
160. Bottomley, F.; Paez, D.E.; White, P.S. *J. Am. Chem. Soc.* **1982**, *104*, 5651-5657.
161. Chen, G.J.-J.; McDonald, J.W.; Newton, W.E. *Inorg. Chem.* **1976**, *15*, 2612-2615.
162. Iwasawa, Y.; Nakamura, T.; Takamatsu, K.; Ogasawara, S. *J. Chem. Soc., Faraday Trans. 1* **1980**, *76*, 939-951.
163. Liu, H.-F.; Liu, R.-S.; Liew, K.Y.; Johnson, R.E.; Lunsford, J.H. *J. Am. Chem. Soc.* **1984**, *106*, 4117-4121.
164. Moyer, B.A.; Meyer, T.J. *J. Am. Chem. Soc.* **1979**, *101*, 1326-1328.
165. Durant, R.; Garner, C.D.; Hyde, M.R.; Mabbs, F.E.; Parsons, J.R.; Richens, D. *J. Less-Common Met.* **1977**, *54*, 459-464.
166. Garner, C.D.; Hyde, M.R.; Mabbs, F.E.; Routledge, V.I. *J. Chem. Soc., Dalton Trans.* **1975**, 1180-1186.

167. Wieghardt, K.; Woeste, M.; Roy, P.S.; Chaudhuri, P. *J. Am. Chem. Soc.* **1985**, *107*, 8276-8277.
168. Hyde, M.R.; Garner, C.D. *J. Chem. Soc., Dalton Trans.* **1975**, 1186-1191.
169. Tanaka, K.; Honjo, M.; Tanaka, T. *Inorg. Chem.* **1985**, *24*, 2662-2665.
170. VanRheenan, V.; Kelly, R.C.; Cha, D.Y. *Tetrahedron Lett.* **1976**, 1973-1976.
171. Lu, X.; Sun, J.; Tao, X. *Synthesis* **1982**, 185-187.
172. Shin, K.; Goff, H.M. *J. Am. Chem. Soc.* **1987**, *109*, 3140-3142.
173. Woon, T.C.; Dicken, C.M.; Bruice, T.C. *J. Am. Chem. Soc.* **1986**, *108*, 7990-7995.
174. Yuan, L.-C.; Calderwood, T.S.; Bruice, T.C. *J. Am. Chem. Soc.* **1985**, *107*, 8273-8274.
175. Alper, H.; Petrignani, J.-F.; Einstein, F.W.B.; Willis, A.C. *J. Am. Chem. Soc.* **1983**, *105*, 1701-1702.
176. Maatta, E.A.; Wentworth, R.A.D. *Inorg. Chem.* **1980**, *19*, 2597-2599.
177. Liebeskind, L.S.; Sharpless, K.B.; Wilson, R.D.; Ibers, J.A. *J. Am. Chem. Soc.* **1978**, *100*, 7061-7063.
178. Muccigrosso, D.A.; Jacobson, S.E.; Apgar, P.A.; Mares, F. *J. Am. Chem. Soc.* **1978**, *100*, 7863-7865.
179. Middleton, A.R.; Wilkinson, G. *J. Chem. Soc., Dalton Trans.* **1980**, 1888-1892.
180. Kasper, M.; Bereman, R.D. *Inorg. Nucl. Chem. Lett.* **1974**, *10*, 443-447.
181. Hubert-Pfalzgraf, L.G.; Aharonian, G. *Inorg. Chim. Acta* **1985**, *100*, L21-L22.
182. Arshankow, V.S.I.; Poznjak, A.L. *Z. Anorg. Allg. Chem.* **1981**, *481*, 201-206.
183. Chatt, J.; Dilworth, J.R. *J. Indian Chem. Soc.* **1977**, *54*, 13-18.
184. Noble, M.E.; Folting, K.; Huffman, J.C.; Wentworth, R.A.D. *Inorg. Chem.* **1982**, *21*, 3772-3776.
185. Dehnicke, K.; Krüger, N. *Z. Naturforsch.* **1978**, *33B*, 1242-1244.
186. Seyferth, K.; Taube, R. *J. Organomet. Chem.* **1982**, *229*, C19-C23.
187. Beck, J.; Schweda, E.; Strähle, J. *Z. Naturforsch.* **1985**, *40B*, 1073-1076.
188. Schweda, E.; Strähle, J. *Z. Naturforsch.* **1980**, *35B*, 1146-1149.
189. Dehnicke, K.; Strähle, J. *Z. Anorg. Allg. Chem.* **1965**, *339*, 171-181.
190. Bereman, R.D. *Inorg. Chem.* **1972**, *11*, 1149-1150.
191. Chatt, J.; Dilworth, J.R. *J. Chem. Soc., Chem. Comm.* **1975**, 983-984.
192. Bevan, P.C.; Chatt, J.; Dilworth, J.R.; Henderson, R.A.; Leigh, G.J. *J. Chem. Soc., Dalton Trans.* **1982**, 821-824.
193. Rushke, P.; Dehnicke, K. *Z. Naturforsch.* **1980**, *35B*, 1589-1591.
194. Musterle, W.; Strähle, J.; Liebolt, W.; Dehnicke, K. *Z. Naturforsch.* **1979**, *34B*, 942-948.
195. Buchler, J.W.; Dreher, C. *Z. Naturforsch.* **1984**, *39B*, 222-230.
196. Groves, J.T.; Takahashi, T.; Butler, W.M. *Inorg. Chem.* **1983**, *22*, 884-887.
197. Baldas, J.; Boas, J.F.; Bonnyman, J.; Williams, G.A. *J. Chem. Soc., Dalton Trans.* **1984**, 2395-2400.

198. Chatt, J.; Falk, C.D.; Leigh, G.J.; Paske, R.J. *J. Chem. Soc. A* **1969**, 2288–2293.
199. Dehnicke, K.; Liese, W.; Köhler, P. *Z. Naturforsch.* **1977**, *32B*, 1487.
200. Fawcett, J.; Peacock, R.D.; Russell, D.R. *J. Chem. Soc., Chem. Comm.* **1982**, 958–959; *J. Chem. Soc., Dalton Trans.* **1987**, 567.
201. Griffith, W.P.; Pawson, D. *J. Chem. Soc., Dalton Trans.* **1973**, 1315–1320.
202. Barner, C.J.; Collins, T.J.; Mapes, B.E.; Santarsiero, B.D. *Inorg. Chem.* **1986**, *25*, 4322–4323.
203. Schweda, E.; Scherfise, K.D.; Dehnicke, K. *Z. Anorg. Allg. Chem.* **1985**, *528*, 117–124.
204. Wiberg, N.; Häring, H.W.; Schubert, U. *Z. Naturforsch.* **1980**, *35B*, 599–603.
205. Gambarotta, S.; Chiesi-Villa, A.; Guastini, C. *J. Organomet. Chem.* **1984**, *270*, C49–C52.
206. Osborne, J.H.; Rheingold, A.L.; Trogler, W.C. *J. Am. Chem. Soc.* **1985**, *107*, 7945–7952.
207. Rocklage, S.M.; Schrock, R.R. *J. Am. Chem. Soc.* **1980**, *102*, 7808–7809.
208. Edelblut, A.W.; Wentworth, R.A.D. *Inorg. Chem.* **1980**, *19*, 1110–1117.
209. Chou, C.Y.; Huffman, J.C.; Maatta, E.A. *J. Chem. Soc., Chem. Comm.* **1984**, 1184–1185.
210. Chisholm, M.H.; Folting, K.; Huffman, J.C.; Ratermann, A.L. *Inorg. Chem.* **1982**, *21*, 978–982.
211. Chou, C.Y.; Maatta, E.A. *Inorg. Chem.* **1984**, *23*, 2912–2914.
212. Haymore, B.L.; Maatta, E.A.; Wentworth, R.A.D. *J. Am. Chem. Soc.* **1979**, *101*, 2063–2068.
213. D'Erries, J.J.; Messerle, L.; Curtis, M.D. *Inorg. Chem.* **1983**, *22*, 849–851.
214. Hillhouse, G.L.; Haymore, B.L. *J. Organomet. Chem.* **1978**, *162*, C23–C26.
215. Herrmann, W.A.; Kriechbaum, G.W.; Dammel, R.; Bock, H.; Ziegler, M.L.; Pfisterer, H. *J. Organomet. Chem.* **1983**, *254*, 219–241.
216. Dehnicke, K.; Strähle, J. *Angew. Chem., Int. Ed. Engl.* **1981**, *20*, 413–426.
217. Dehnicke, K. *Angew. Chem., Int. Ed. Engl.* **1979**, *18*, 507–514.
218. McGlinchey, M.J.; Stone, F.G.A. *J. Chem. Soc., Chem. Comm.* **1970**, 1265.
219. Ho, T.-L.; Henninger, M.; Olah, G.A. *Synthesis* **1976**, 815–816.
220. Hussain, W.; Leigh, G.J.; Pickett, C.J. *J. Chem. Soc., Chem. Comm.* **1982**, 747–748.
221. Bishop, M.W.; Chatt, J.; Dilworth, J.R.; Hursthouse, M.B.; Jayaweera, S.A.A.; Quick, A. *J. Chem. Soc., Dalton Trans.* **1979**, 914–920.
222. Rossi, R.; Marchi, A.; Duatti, A.; Magon, L.; DiBernardo, P. *Trans. Met. Chem.* (Weinheim, Ger.) **1985**, *10*, 151–153.
223. Doyle, M.P. *Chem. Rev.* **1986**, *86*, 919–939.
224. Freudenberger, J.H.; Schrock, R.R. *Organometallics* **1986**, *5*, 398–400.
225. Chisholm, M.H.; Huffman, J.C.; Marchant, N.S. *J. Am. Chem. Soc.* **1983**, *105*, 6162–6163.

226. Harlan, E.W.; Berg, J.M.; Holm, R.H. *J. Am. Chem. Soc.* **1986**, *108*, 6992–7000.
227. Schwartz, J.; Gell, K.I. *J. Organomet. Chem.* **1980**, *184*, C1–C2.
228. Sharp, P.R.; Schrock, R.R. *J. Organomet. Chem.* **1979**, *171*, 43–51.
229. Templeton, J.L.; Ward, B.C.; Chen, G.J.-J.; McDonald, J.W.; Newton, W.E. *Inorg. Chem.* **1981**, *20*, 1248–1253.
230. Liese, W.; Dehnicke, K. *Z. Naturforsch.* **1978**, *33B*, 1061–1062.
231. Swinehart, J.H. *Inorg. Chem.* **1965**, *4*, 1069–1070.
232. Chisholm, M.H.; Folting, K.; Huffman, J.C.; Kirkpatrick, C.C. *Inorg. Chem.* **1984**, *23*, 1021–1037.
233. Bajdor, K.; Nakamoto, K. *J. Am. Chem. Soc.* **1984**, *106*, 3045–3046.
234. Lemenovskii, D.A.; Baukova, T.V.; Fedin, V.P. *J. Organomet. Chem.* **1977**, *132*, C14–C16.
235. Herberhold, M.; Kremnitz, W.; Razavi, A.; Schöllhorn, H.; Thewalt, U. *Angew. Chem., Int. Ed. Engl.* **1985**, *24*, 601.
236. Lincoln, S.; Koch, S.A. *Inorg. Chem.* **1986**, *25*, 1594–1602.
237. Devore, D.D.; Maatta, E.A. *Inorg. Chem.* **1985**, *24*, 2846–2849.
238. Bryan, J.C.; Mayer, J.M. unpublished results.
239. Yarino, T.; Matsushita, T.; Masuda, I.; Shinra, K. *J. Chem. Soc., Chem. Comm.* **1970**, 1317–1318.
240. Beattie, I.R.; Jones, P.J. *Inorg. Chem.* **1979**, *18*, 2318–2319.
241. Herrmann, W.A.; Serrano, R.; Bock, H. *Angew. Chem., Int. Ed. Engl.* **1984**, *23*, 383–385.
242. Chin, D.-H.; LaMar, G.N.; Balch, A.L. *J. Am. Chem. Soc.* **1980**, *102*, 4344–4350.
243. Simonneaux, G.; Scholz, W.F.; Reed, C.A.; Lang, G. *Biochim. Biophys. Acta* **1982**, *716*, 1–7.
244. Groves, J.T.; Quinn, R. *J. Am. Chem. Soc.* **1985**, *107*, 5790–5792.
245. Mansuy, D.; Fontecave, M.; Bartoli, J.-F. *J. Chem. Soc., Chem. Comm.* **1983**, 253–254.
246. Tabushi, I.; Yazaki, A. *J. Am. Chem. Soc.* **1981**, *103*, 7371–7373.
247. Barral, R.; Bocard, C.; Sérée de Roch, I.; Sajus, L. *Tetrahedron Lett.* **1972**, 1693–1696.
248. Ueyama, N.; Yano, M.; Miyashita, H.; Nakamura, A.; Kamachi, M.; Nozakura, S.-I. *J. Chem. Soc., Dalton Trans.* **1984**, 1447–1451.
249. Ledon, H.; Bonnet, M.; Lallemand, J.-Y. *J. Chem. Soc., Chem. Comm.* **1979**, 702–704.
250. Mimoun, H.; Seree de Roch, I.; Sajus, L. *Tetrahedron* **1970**, 37–50.
251. Sharpless, K.B.; Townsend, J.M.; Williams, D.R. *J. Am. Chem. Soc.* **1972**, *94*, 295–296.
252. Vedejs, E.; Engler, D.A.; Telschow, J.E. *J. Org. Chem.* **1978**, *43*, 188–196.
253. Vedejs, E.; Telschow, J.E. *J. Org. Chem.* **1976**, *41*, 740–741.
254. Mimoun, H.; Saussine, L.; Daire, E.; Postel, M.; Fischer, J.; Weiss, R. *J. Am. Chem. Soc.* **1983**, *105*, 3101–3110.
255. Cotton, F.A.; Duraj, S.A.; Roth, W.J. *J. Am. Chem. Soc.* **1984**, *106*, 4749–4751.
256. Canich, J.A.M.; Cotton, F.A.; Duraj, S.A.; Roth, W.J. *Polyhedron* **1986**, *5*, 895–898.
257. Wiberg, N.; Häring, H.-W.; Schubert, U. *Z. Naturforsch.* **1978**, *33B*, 1365–1369.

258. Lahiri, G.K.; Goswami, S.; Falvello, L.R.; Chakravorty, A. *Inorg. Chem.* **1987**, *26*, 3365–3370.
259. Ashley-Smith, J.; Green, M.; Mayne, N.; Stone, F.G.A. *J. Chem. Soc., Chem. Comm.* **1969**, 409.
260. Ashley-Smith, J.; Green, M.; Stone, F.G.A. *J. Chem. Soc., Dalton Trans.* **1972**, 1805–1809.
261. Chatt, J.; Garforth, J.D.; Johnson, N.P.; Rowe, G.A. *J. Chem. Soc.* **1964**, 1012–1020.
262. Chatt, J.; Garforth, J.D.; Johnson, N.P.; Rowe, G.A. *J. Chem. Soc.* **1964**, 1012–1020.
263. Kaden, L.; Lorenz, B; Schmidt, K.; Sprinz, H.; Wahren, M. *Isotopenpraxis* **1981**, *17*, 174–175.
264. Abram, U.; Spies, H. *Inorg. Chim. Acta* **1984**, *94*, L3–L6.
265. Baldas, J.; Bonnyman, J.; Pojer, P.M.; Williams, G.A.; Mackay, M.F. *J. Chem. Soc., Dalton Trans.* **1981**, 1798–1801.
266. Lock, C.J.L.; Wilkinson, G. *J. Chem. Soc.* **1964**, 2281–2285.
267. Jabs, W.; Herzog, S. *Z. Chem.* **1972**, *12*, 268–269.
268. Chatt, J.; Dilworth, J.R.; Leigh, G.J. *J. Chem. Soc. A* **1970**, 2239–2243.
269. Cardin, D.J.; Cetinkaya, B.; Lappert, M.F.; Manojlovic-Muir, L.; Muir, K.W. *J. Chem. Soc., Chem. Comm.* **1971**, 400–401.
270. Listemann, M.L.; Schrock, R.R. *Organometallics* **1985**, *4*, 74–83.
271. Cotton, F.A.; Schwotzer, W.; Shamshoum, E.S. *Organometallics* **1983**, *2*, 1167–1171.
272. Cotton, F.A.; Schwotzer, W.; Shamshoum, E.S. *J. Organomet. Chem.* **1985**, *296*, 55–68.
273. Strutz, H.; Schrock, R.R. *Organometallics* **1984**, *3*, 1600–1601.
274. Chisholm, M.H.; Conroy, B.K.; Huffman, J.C.; Marchant, N.S. *Angew. Chem., Int. Ed. Engl.* **1986**, *25*, 446–447.
275. Chisholm, M.H.; Folting, K.; Hoffman, D.M.; Huffman, J.C. *J. Am. Chem. Soc.* **1984**, *106*, 6794–6805.
276. Schrock, R.R. *J. Am. Chem. Soc.* **1976**, *98*, 5399–5400.
277. Aguero, A.; Kress, J.; Osborn, J.A. *J. Chem. Soc., Chem. Comm.* **1986**, 531–533.
278. Kauffmann, T.; Ennen, B.; Sander, J.; Wieschollek, R. *Angew. Chem., Int. Ed. Engl.* **1983**, *22*, 244–245.
279. Cotton, F.A.; Hall, W.T. *J. Am. Chem. Soc.* **1979**, *101*, 5094–5095.
280. Chatt, J.; Dilworth, J.R. *J. Chem. Soc., Chem. Comm.* **1972**, 549.
281. Maatta, E.A.; Haymore, B.L.; Wentworth, R.A.D. *Inorg. Chem.* **1980**, *19*, 1055–1059.
282. Chong, A.O.; Oshima, K.; Sharpless, K.B. *J. Am. Chem. Soc.* **1977**, *99*, 3420–3426.
283. Bell, B.; Chatt, J.; Dilworth, J.R.; Leigh, G.J. *Inorg. Chim. Acta* **1972**, *6*, 635–638.
284. Salmon, D.J.; Walton, R.A. *Inorg. Chem.* **1978**, *17*, 2379–2382.
285. Goeden, G.V.; Haymore, B.L. *Inorg. Chim. Acta* **1983**, *71*, 239–249.
286. Kolomnikov, I.S.; Koreshkov, Yu.D.; Lobeeva, T.S.; Vol'pin, M.E. *Akad. Nauk SSSR, Ser. Khim.* **1971**, 2065–2066.
287. Kolomnikov, I.S.; Koreshkov, Yu.D.; Lobeeva, T.S.; Volpin, M.E. *J. Chem. Soc., Chem. Comm.* **1970**, 1432.
288. Bradley, D.C.; Hursthouse, M.B.; Malik, K.M.A.; Nielson, A.J.; Short, R.L. *J. Chem. Soc., Dalton Trans.* **1983**, 2651–2666.

289. Nielson, A.J. *Inorg. Synth.* **1986**, *24*, 194-200.
290. Maatta, E.A. *Inorg. Chem.* **1984**, *23*, 2560-2561.
291. Ashcroft, B.R.; Clark, G.R.; Nielson, A.J.; Rickard, C.E.F. *Polyhedron* **1986**, *5*, 2081-2091.
292. LaMonica, G.; Cenini, S. *Inorg. Chim. Acta* **1973**, *29*, 183-187.
292a Jernakoff, P.; Geoffroy, G.L.; Rheingold, A.L.; Geib, S.J. *J. Chem. Soc., Chem. Comm.* **1987**, 1610-1611.
293. Rocklage, S.M.; Schrock, R.R. *J. Am. Chem. Soc.* **1982**, *104*, 3077-3081.
294. Turner, H.W.; Fellmann, J.D.; Rocklage, S.M.; Schrock, R.R.; Churchill, M.R.; Wasserman, H.J. *J. Am. Chem. Soc.* **1980**, *102*, 7809-7811.
295. Lai, R.; LeBot, S.; Baldy, A.; Pierrot, M.; Arzoumanian, H. *J. Chem. Soc., Chem. Comm.* **1986**, 1208-1209.
296. Arzoumanian, H.; Baldy, A.; Lai, R.; Odreman, A.; Metzger, J.; Pierrot, M. *J. Organomet. Chem.* **1985**, *295*, 343-352.
297. Arzoumanian, H.; Baldy, A.; Lai, R.; Metzger, J.; Peh, M.-L.N.; Pierrot, M. *J. Chem. Soc., Chem. Comm.* **1985**, 1151-1152.
298. Stevens, A.E.; Beauchamp, J.L. *J. Am. Chem. Soc.* **1979**, *101*, 6449-6450.
299. Kang, H.; Beauchamp, J.L. *J. Am. Chem. Soc.* **1986**, *108*, 5663-5668.
300. Walba, D.M.; DePuy, C.H.; Grabowski, J.J.; Bierbaum, V.M. *Organometallics* **1984**, *3*, 498-499.
301. Rappé, A.K.; Goddard, III, W.A. *J. Am. Chem. Soc.* **1980**, *102*, 5114-5115.
302. Rappé, A.K.; Goddard, III, W.A. *J. Am. Chem. Soc.* **1982**, *104*, 448-456.
303. Aguero, A.; Kress, J.; Osborn, J.A. *J. Chem. Soc., Chem. Comm.* **1985**, 793-794.
304. Henderson, R.A.; Davies, G.; Dilworth, J.R.; Thorneley, R.N.F. *J. Chem. Soc., Dalton Trans.* **1981**, 40-50.
305. Schaverien, C.J.; Dewan, J.C.; Schrock, R.R. *J. Am. Chem. Soc.* **1986**, *108*, 2771-2773.
306. Holmes, S.J.; Clark, D.N.; Turner, H.W.; Schrock, R.R. *J. Am. Chem. Soc.* **1982**, *104*, 6322-6329.
307. Bishop, M.W.; Chatt, J.; Dilworth, J.R.; Neaves, B.D.; Dahlstrom, P.; Hyde, J.; Zubieta, J. *J. Organomet. Chem.* **1981**, *213*, 109-124.
308. Fenske, D.; Kujanek, R.; Dehnicke, K. *Z. Anorg. Allg. Chem.* **1983**, *507*, 51-58.
309. Groves, J.T.; Takahashi, T. *J. Am. Chem. Soc.* **1983**, *105*, 2073-2074.
310. Shapley, P.A.B.; Own, Z.-Y.; Huffman, J.C. *Organometallics* **1986**, *5*, 1269-1271. Shapley, P.A.B.; Own, Z.-Y. *J. Organomet. Chem.* **1987**, *335*, 269-276.
311. Kynast, U.; Conradi, E.; Müller, U.; Dehnicke, K. *Z. Naturforsch.* **1984**, *39B*, 1680-1685.
312. Strutz, H.; Dewan, J.C.; Schrock, R.R. *J. Am. Chem. Soc.* **1985**, *107*, 5999-6005.
313. Freudenberger, J.H.; Schrock, R.R. *Organometallics* **1986**, *5*, 1411-1417.

314. Churchill, M.R.; Ziller, J.W.; Freudenberger, J.H.; Schrock, R.R. *Organometallics* **1984**, *3*, 1554-1562.
315. Churchill, M.R.; Ziller, J.W.; McCullough, L.; Pedersen, S.F.; Schrock, R.R. *Organometallics* **1983**, *2*, 1046-1048.
316. Freudenberger, J.H.; Schrock, R.R.; Churchill, M.R.; Rheingold, A.L.; Ziller, J.W. *Organometallics* **1984**, *3*, 1563-1573.
317. Chatt, J.; Heath, G.A.; Leigh, G.J. *J. Chem. Soc., Chem. Comm.* **1972**, 444-445.
318. Chatt, J.; Heath, G.A.; Richards, R.L. *J. Chem. Soc., Chem. Comm.* **1972**, 1010-1011.
319. Chisholm, M.G.; Folting, K.; Huffman, J.C.; Ratermann, A.L. *J. Chem. Soc., Chem. Comm.* **1981**, 1229-1231.
320. Heath, G.A.; Mason, R.; Thomas, K.M. *J. Am. Chem. Soc.* **1974**, *96*, 259-260.
321. Henderson, R.A. *J. Chem. Soc., Dalton Trans.* **1982**, 917-925.
322. Chatt, J.; Diamantis, A.A.; Heath, G.A.; Hooper, N.E.; Leigh, G.J. *J. Chem. Soc., Dalton Trans.* **1977**, 688-697.
323. Chatt, J.; Dilworth, J.R.; Richards, R.L. *Chem. Rev.* **1978**, *78*, 589-625.
324. Chatt, J.; da Camera Pina, L.M.; Richards, R.L. eds. "New Trends in the Chemistry of Nitrogen Fixation" Academic Press, London, 1980.
325. Hidai, M. in "Molybdenum Enzymes"; Spiro, T.G., ed. Wiley, New York, 1985.
326. Bishop, M.W.; Butler, G.; Chatt, J.; Dilworth, J.R.; Leigh, G.J. *J. Chem. Soc., Dalton Trans.* **1979**, 1843-1850.
327. Bossard, G.E.; George, T.A.; Lester, R.K.; Tisdale, R.C.; Turcotte, R.L. *Inorg. Chem.* **1985**, *24*, 1129-1132.
328. Bossard, G.E.; Busby, D.C.; Chang, M.; George, T.A.; Iske, Jr., S.D.A. *J. Am. Chem. Soc.* **1980**, *102*, 1001-1008.
329. Baumann, J.A.; Bossard, G.E.; George, T.A.; Howell, D.B.; Koczon, L.M.; Lester, R.K.; Noddings, C.M. *Inorg. Chem.* **1985**, *24*, 3568-3578.
330. Hidai, M.; Kodama, T.; Sato, M.; Harakawa, M.; Uchida, Y. *Inorg. Chem.* **1976**, *15*, 2694-2697.
331. Tatsumi, T.; Hidai, M.; Uchida, Y. *Inorg. Chem.* **1975**, *14*, 2530-2534.
332. Busby, D.C.; George, T.A. *Inorg. Chem.* **1979**, *18*, 3164-3167.
333. Chatt, J.; Hussian, W.; Leigh, G.J.; Terreros, F.P. *J. Chem. Soc., Dalton Trans.* **1980**, 1408-1415.
334. Chatt, J.; Pearman, A.J.; Richards, R.L. *J. Chem. Soc., Dalton Trans.* **1978**, 1766-1776.
335. Chatt, J.; Pearman, A.J.; Richards, R.L. *J. Chem. Soc., Dalton Trans.* **1976**, 1520-1524.
336. Hidai, M.; Komori, K.; Kodama, T.; Jin, D.-M.; Takahashi, T.; Sugiura, S.; Uchida, Y.; Mizobe, Y. *J. Organomet. Chem.* **1984**, *272*, 155-167.
337. Pickett, C.J.; Ryder, K.S.; Talarmin, J. *J. Chem. Soc., Dalton Trans.* **1986**, 1453-1457.
338. Dilworth, J.R.; Harrison, S.A.; Walton, D.R.M.; Schweda, E. *Inorg. Chem.* **1985**, *24*, 2594-2595.
339. Douglas, P.G.; Galbraith, A.R.; Shaw, B.L. *Trans. Met. Chem.* **1975**, *1*, 17-20.

340. Henderson, R.A. *J. Organomet. Chem.* **1981**, *208*, C51–C54.
341. Carroll, J.A.; Sutton, D. *Inorg. Chem.* **1980**, *19*, 3137–3142.
342. Barrientos-Penna, C.F.; Einstein, F.W.B.; Jones, T.; Sutton, D. *Inorg. Chem.* **1982**, *21*, 2578–2585.
343. Drew, M.G.B.; Moss, K.C.; Rolfe, N. *Inorg. Nucl. Chem. Lett.* **1971**, *7*, 1219–1222.
344. Fowles, G.W.A.; Moss, K.C.; Rice, D.A.; Rolfe, N. *J. Chem. Soc., Dalton Trans.* **1973**, 1871–1873.
345. Drew, M.G.B.; Fowles, G.W.A.; Rice, D.A.; Rolfe, N. *J. Chem. Soc., Chem. Comm.* **1971**, 231–232.
346. Fowles, G.W.A.; Rice, D.A.; Shanton, K.J. *J. Chem. Soc., Dalton Trans.* **1977**, 1212–1214.
347. Dehnicke, K.; Weiher, U.; Fenske, D. *Z. Anorg. Allg. Chem.* **1979**, *456*, 71–80.
348. Weiher, U.; Dehnicke, K.; Fenske, D. *Z. Anorg. Allg. Chem.* **1979**, *457*, 105–114.
349. Weiher, U.; Dehnicke, K.; Fenske, D. *Z. Anorg. Allg. Chem.* **1979**, *457*, 115–122.
350. Blight, D.G.; Deutscher, R.L.; Kepert, D.L. *J. Chem. Soc., Dalton Trans.* **1972**, 87–89.
351. Gert, R.; Perron, W. *J. Less-Common Met.* **1972**, *26*, 369–379.
352. Finn, P.A.; King, M.S.; Kilty, P.A.; McCarley, R.E. *J. Am. Chem. Soc.* **1975**, *97*, 220–221.
353. Cotton, F.A.; Hall, W.T. *Inorg. Chem.* **1978**, *17*, 3525–3528.
354. Roskamp, E.J.; Pedersen, S.F. *J. Am. Chem. Soc.* **1987**, *109*, 3152–3154.
355. Schrock, R.R.; Fellmann, J.D. *J. Am. Chem. Soc.* **1978**, *100*, 3359–3370.
356. Curtis, M.D.; Real, J. *J. Am. Chem. Soc.* **1986**, *108*, 4668–4669.
357. Chisholm, M.H.; Hoffman, D.M.; Huffman, J.C. *J. Chem. Soc., Chem. Comm.* **1983**, 967–968.
358. Chisholm, M.H.; Hoffman, D.M.; Huffman, J.C. *J. Am. Chem. Soc.* **1984**, *106*, 6815–6826.
359. Threlkel, R.S.; Bercaw, J.E. *J. Am. Chem. Soc.* **1981**, *103*, 2650–2659.
360. Bryan, J.C.; Mayer, J.M. *J. Am. Chem. Soc.* **1987**, *109*, 7213–7214.
361. Hillhouse, G.L.; Haymore, B.L. *J. Am. Chem. Soc.* **1982**, *104*, 1537–1548.

CHAPTER 4

VIBRATIONAL AND NMR SPECTROSCOPY OF MULTIPLY BONDED LIGANDS

Spectroscopic techniques are, of course, indispensable to the inorganic chemist. Space will not allow us to discuss all of the ways in which spectroscopy has been used to elucidate the overall structure of complexes containing multiply bonded ligands. Instead, our discussion must focus on the spectroscopic properties of the ligands themselves. Two techniques that provide particular insight into the nature of bonding in such ligands, ESR and UV-visible spectroscopy, were covered in Chapter 2. The current chapter is primarily concerned with vibrational (IR and Raman) and NMR spectroscopy.

The vibrational spectra of complexes containing the various ligands (with a heavy emphasis on metal–ligand stretching vibrations) are considered in Sections 4.1–4.5. Relevant aspects of oxygen, nitrogen, proton, and carbon NMR are covered in Sections 4.6–4.9. Section 4.10 contains some comments on the particular insights available from the NMR of the metal nucleus.

4.1 IR SPECTRA OF TERMINAL OXO AND NITRIDO COMPLEXES

4.1.1 General Considerations

The oxo and nitrido groups are ideal chromophores for IR spectroscopy. Because of the large change in bond dipole for these ligands, absorbance bands due to M=O and M≡N stretching modes are generally intense. Moreover, since the stretching vibrations in these terminal ligands are not strongly coupled to other ligand oscillations (cf. organoimido ligands, Section 4.3), the bands are also often sharp. Observation of such a band at an appropriate frequency is commonly considered diagnostic for the presence of a terminal ligand. In addition, force constants derived from such stretching frequencies provide a measure of the metal–ligand bond strength. The corresponding deformation modes occur at a much lower frequency than the stretching bands and tend to be broader and weaker. As a diagnostic tool the bending bands have proven much less useful than the stretching modes and therefore we will focus our attention on the latter.

The techniques of IR and Raman spectroscopy tend to be complementary in some respects. Used in combination, the different selection rules for these two types of

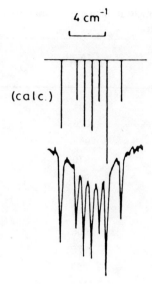

Figure 4.1 Observed and calculated isotope pattern in the Mo≡O stretching region of MoOF$_4$ in a dinitrogen matrix. (Adapted from ref. 39 with permission of the Royal Society of Chemistry.)

spectroscopy often allow the ready assignment of symmetric versus asymmetric stretching modes in polyoxo species. Raman spectroscopy is particularly useful for studying oxo species in an aqueous solution since water has only a very weak Raman spectrum. On the other hand, Raman spectroscopy has seen limited use for study of nitrido complexes, which are typically intensely colored.

In recent years, the combination of low temperatures, matrix isolation techniques, and improved instrumentation have allowed remarkable improvements in high-resolution spectra, especially of metal–oxo complexes. The spectrum of the oxo-stretching region of MoOF$_4$ in a dinitrogen matrix shown in Figure 4.1 was reported by J.S. Ogden and coworkers [1]. Note the complete resolution of the bands due to the seven isotopes in natural abundance molybdenum. The use of matrix isolation has also allowed the observation of a number of otherwise unaccessible oxo species such as CrOCl$_3$[2], OsO$_3$F$_2$[3], and ReO$_2$F$_3$[4].

4.1.2 Assignment of M—N and M—O Stretching Frequencies

In 1959 Barraclough, Lewis, and Nyholm published a classic study in which they suggested that bands in the region of 900–1100 cm^{-1} of oxometal species could be assigned as metal–oxygen stretching modes of terminal oxo ligands [5]. Bands due to bridging M—O—M systems were said to occur at lower frequencies (800–900 cm^{-1}) In his 1972 review, Griffith [6] proposed the range 1020–1150 cm^{-1} for the M—N stretching mode of terminal nitrides. Although the wealth of spectral data that have

TABLE 4.1 Effect of ^{18}O Substitution[a] on Terminal Metal–Oxo Stretching Frequencies (Selected Examples)

Complex	$\nu(^{16}O)$[b]	$\nu(^{18}O)$[b]	Reference
[Cr(O)(salen)]OTf	1004	965	7
Cr(O)Cl(TPP)	1026	982	8
Mo(O)(dtc)$_2$	962	914	9
Mo(O)(O$_2$)$_2$(HMPA)	970	920	10
Mo(O)Cl$_2$(dtc)$_2$	946	898	11
[MoO(μ–O)(cyclen)]$_2$	918	868	12
Mn(O)(TPP)	1060	1000	13
[Mo(O)Cl$_4$]Ph$_4$As	1008	962	14
Re(O)I(MeC≡CMe)$_2$	975	926	15
Fe(O)(TPP)	852	818	16
[Ru(O)(bipy)$_2$(py)](ClO$_4$)$_2$	792	752	17
Ru(O)$_2$(TMP)	821	785	18
Mo(O)$_2$(O^{t}Bu)$_2$	968, 930	920, 887	19
Mo(O)$_2$(O^{t}Bu)$_2$(bipy)	912, 888	863, 843	19
Mo(O)$_2$(dtc)$_2$	910, 877	865, 836	9
Np$_3$(O)WOW(O)Np$_3$	962, 942[c]	905, 894[c]	20
[MnO$_4$]K[d]	910, 844	873, 806	20
ReO$_2$F$_3$	1024, 988	970, 939	4
[Os(N)O$_3$]K	892, 872	844, 826	21

[a] Data are for complexes with complete ^{18}O substitution of all terminal oxo groups, except as indicated.
[b] Frequencies in wavenumbers (cm^{-1}).
[c] Raman band.
[d] Isotopic enrichment 86%.

appeared in intervening years allows us to identify a few exceptions and refinements, these basic proposals have stood up remarkably well.

The task of assigning stretching frequencies in particular oxo and nitrido complexes is abetted by the availability of data on simple systems where assignments are unambiguous. Oxo–halide and nitrido–halide complexes have been especially valuable in this regard. The metal–halogen modes occur at low frequencies (typically <450 cm^{-1} for Cl and Br), providing an unobstructed view of bands due to the multiply bonded ligand.

Isotopic labeling is an invaluable tool for identifying M—N and M—O stretching

TABLE 4.2 Effect of Isotopic Substitution on the M–N Stretching Frequency of Terminal Nitride Complexes

Complex	$\nu(^{14}N)$[a]	$\nu(^{15}N)$[a]	Reference
Cr(N)(TPP)	1017	991	22
Mn(N)(TPP)	1036	1008	23
Tc(N)(phthalocyanine)	1078	1049	24
[Os(N)O$_3$]K	1026	995	21
Os(N)Cl$_3$(py)$_2$	1060	1030	25
Os(N)Cl$_3$(PPh$_3$)$_2$	1058	1027	26
Os(N)Br$_3$(PPh$_3$)$_2$	1068	1035	26
Os(N)Cl$_3$(SbPh$_3$)$_2$	1065	1032	26
[Os(N)Cl$_4$]Ph$_4$As	1123	1087	14
[Os(N)Br$_4$]Ph$_4$As	1119	1082	14
[Os(N)Cl$_5$]K$_2$	1073	1041	27
[Os(N)Cl$_5$]Cs$_2$	1073	1040	28
[Os(N)Br$_5$]Cs$_2$	1073	1040	28

[a] Frequencies in cm^{-1}.

modes in oxides and nitrides in molecules that additionally contain complex organic ligands. Replacement of naturally occurring ^{16}O by ^{17}O or ^{18}O shifts the M=O stretch to a lower frequency; a similar downward shift is observed on replacing ^{14}N in nitrido complexes with ^{15}N. In both cases the magnitude of the change can be calculated to a good approximation from the reduced mass of the diatomic oscillator. In Table 4.1 we have collected the reported labeling studies on metal–oxo species. Table 4.2 provides a similar tabulation for nitrides.

4.1.3 Stretching Frequencies for Monooxo and Nitrido Complexes

In Table 4.3 we show the range of reported values for the IR stretching frequencies of mononuclear monooxo and nitrido complexes. The indicated references are those containing the limiting (highest and lowest frequency) examples. Stretching frequencies tend to be significantly higher for square pyramidal complexes than in other coordination geometries.

If we focus our attention on a single metal, several trends within the indicated ranges can be identified. In general, adding a trans ligand to a given square pyramidal complex decreases the stretching frequency of an oxo or nitrido ligand by 25–75 cm^{-1}. For example, coordination of pyridine N-oxide to the chromium(V) complex [(salen)CrO] decreases ν(Cr=O) from 1004 to 939 cm^{-1} [7]. Similarly, upon addition of chloride

TABLE 4.3   Range of Reported Stretching Frequencies for Monooxo and Nitrido Complexes

Metal	$\nu(M-O)^a$	$\nu(M-N)^a$	References
V	875 – 1035	970 – 1033	29,30,31
Nb	835 – 1020	—	32,33
Ta	905 – 935	—	34,35
Cr	930 – 1028	1012 – 1017	1,22,36,37
Mo	922^b – 1050	948 – 1109	38,39,40,41
W	922^b – 1058	980 – 1135	39,42,43,44
Mn	950 – 1060	1036 – 1052	15,23,45,46
Tc	882 – 1020	1027 – 1089	47,48,49,50
Re	945 – 1067	974 – 1099	47,51,52,53
Ru	—^{c}	1023 – 1092	14,54
Os	960 – 1040	1008 – 1125	55,56,57,58

^{a}Frequencies in wavenumbers (cm^{-1}).
^{b}M–O stretch for $Cp_2M=O$ comes at lower frequency (see text).
^{c}Oxoruthenium(IV) complexes show unusually low stretching frequencies (see text).

ion to [OsNCl$_4$]$^-$ to afford octahedral [OsNCl$_5$]$^{2-}$ the Os—N stretching frequency drops from 1123 to 1073 cm^{-1} [28]. In this regard, good π-donor ligands have the largest effect. For example, a trans-chloro ligand in *trans*-mer-ReOCl$_3$L$_2$ complexes (L = phosphine ligand) can be replaced by methoxide to give the *trans*-ReO(OMe)Cl$_2$L$_2$ analogs. In a series of such substitutions, the Re=O consistently dropped 20–30 cm^{-1} [59]. Extreme examples of this effect are the *trans*-dioxo complexes discussed below (Section 4.1.4).

It has been suggested [60] that strongly π-donating substituents located cis to the multiply bonded ligand can also sharply decrease the stretching frequency. This argument has been applied to the series of complexes AsPh$_4$[MoNX$_4$]. When X is Br or Cl, ν(Mo—N) is 1060 and 1054 cm^{-1}, respectively, but when X is the strongly π-donating F, the frequency drops to 969 cm^{-1} [60]. However, some caution is indicated as regards the generality of this proposal. Note that in the related series of oxo compounds WOX$_4$, the exact opposite pattern is observed: ν(W=O) actually increases 1025 < 1032 < 1058 cm^{-1} as X is varied from Br to Cl to F [39].

It can be seen in Table 4.3 that some of the data lie somewhat below the 900–1100 range originally suggested [5] for M=O stretching. Although all of these complexes have been proposed to be mononuclear, it is conceivable that some unrecognized bridging oxos are included. However, a clear-cut example appears to exist in the case of bis(N-phenylsalicylideneiminato)oxovanadium(IV) complexes containing electron-withdrawing para substituents on the N-phenyl group. In such complexes V=O is observed in the range 875–885 cm^{-1}. Moreover, an X-ray crystal structure of the p-chloro derivative unambiguously shows it to contain a terminal oxo moiety [29]. It is fascinating that analogous complexes containing an electron-releasing para substituent, or even an electron-withdrawing substituent in the meta position, exhibit a "normal" V=O stretch in the 940–980 cm^{-1} range [29]. These results have caused a reevaluation the earlier proposal [61] that stretching frequencies in the 800 cm^{-1} range are diagnostic for solid-state polymerization to V—O—V bridged structures.

We have excluded from Table 4.3 the M=O stretching frequencies for three series of compounds where the data fall significantly outside the range of expected values. The first series includes Meyer's ruthenium oxo derivative [(bipy)$_2$(py)RuO]$^{2+}$ and several other Ru(IV) oxo species, all of which fall in the range 790–805 cm^{-1} [62–64]. The second example is the oxoiron(IV) porphyrin complex (TPP)FeO where, despite a square pyramidal structure, ν(M=O) is 852 cm^{-1} [16]. In both cases the assignment has been confirmed by an isotopic labeling experiment (Table 4.1). The low frequencies may be attributed in part to the low oxidation state of these d^4 complexes. However, it must be noted that in a series of d^4 oxorhenium compounds, Re(O)X(RC≡CR)$_2$, ν(Re=O) is observed at 971–980 cm^{-1}, well within the "normal" range and comparable to that (992 cm^{-1}) for the d^0 complex ReOF$_5$ [55]. It is interesting to speculate that the unusually low stretching frequencies of the Fe and Ru complexes may be related to the presence of electrons in metal–oxygen antibonding orbitals in these two species. (The complexes contain formal metal oxygen double rather than triple bonds.) Noteworthy in this regard are the frequencies of the first excited states of oxo and nitrido species which can occasionally be discerned from the vibronic structure of UV-visible spectra. For example, in both AsPh$_4$[OsNBr$_4$] and trans-[Os(TMC)O$_2$]$^{2+}$ a decrease in stretching frequency of 100–150 cm^{-1} is observed upon comparing the ground-state molecule with the excited state [65,66]. The first excited state in both cases has one electron in a metal–oxygen π^* level.

The third series of "deviant" complexes are the diamagnetic complexes Cp$_2$M=O, where M=Mo and W. Green has reported ν(M=O) for the molybdenum complex as 793–868 cm^{-1} and for the tungsten analog as 789–879 cm^{-1}. The unusual properties of these complexes has provoked some question as to whether they might in fact contain bridging oxo ligands. However, an X-ray crystal structure has confirmed the presence of a terminal oxo for the methylcyclopentadienyl analog (MeCp)$_2$Mo=O [67]. (The Mo—O bond length of 1.721(2) Å clearly lies at the end of the range for monooxo molybdenum complexes in Fig. 5.2.) A tentative explanation for the low M=O frequency in this case is that the complexes are again restricted to a double-bonded oxo ligand, in this case by electron counting considerations. (A M—O triple bond would result in a 20-electron complex.)

At first glance, the frequency ranges in Table 4.3 seem to be remarkably independent of the nature of the metal that is present. (This is partly coincidence, the result of compensating trends in stretching force constants and the weight of the metal atoms.) Focusing our attention on a structurally similar series of compounds, the nitrido complexes AsPh$_4$[MNBr$_4$] [49,53], allows us to discern some systematic changes. When arranged in periodic fashion the M—N stretching data appear as follows:

MoBr$_4$N$^-$	TcBr$_4$N$^-$	RuBr$_4$N$^-$
1060	1074	1088
WBr$_4$N$^-$	ReBr$_4$N$^-$	OsBr$_4$N$^-$
1045	1099	1119

The wavenumbers are seen to increase from left to right along a period. (In making comparisons among metals in different triads, one must necessarily choose between holding the charge of the species constant or holding the d electron count constant. In this case the d electron count increases from left to right. However, the same trend can be seen in Table 4.6, where the comparison involves only d^0 complexes.) In general, it is expected that the frequency will also increase on proceeding down a given triad. However, as exemplified by Mo and W in the current example, there are exceptions. In the oxo complexes MOF$_4$ the more usual sequence is observed; ν(M=O) increases 1028<1050<1058 cm^{-1} along the series M = Cr, Mo, W [1,39].

4.1.4 Stretching Modes of Polyoxo Complexes

The presence of additional multiply bonded ligands in an oxo complex invariably shifts ν(M—O) to lower frequency when compared with monooxo species. The reasons for this weakening of the metal oxygen bond—competition for a few π symmetry orbitals and consequent reduction in bond order—were delineated in Chapter 2. The number of stretching bands in the IR and Raman for di-, tri-, and tetraoxo species in various coordination geometries can be predicted by the straightforward application of group theory [68]. This exercise will not be reproduced here.

With the exception of trans dioxo derivatives, 4-, 5-, and 6-coordinate dioxo complexes are expected to exhibit two M—O stretching bands in both the IR and Raman. The two bands correspond to the symmetric (A$_1$) and antisymmetric (B$_2$) modes of the MO$_2$ unit. In contrast, *trans*-dioxo complexes exhibit an A$_{2u}$ infrared active antisymmetric stretching mode and an A$_{1g}$ Raman active symmetric stretching mode. The stretching frequencies for a number of dioxo complexes are collected in Table 4.4. It can be seen that these fall in the range of 768–895 cm^{-1} for the *trans*-dioxo complexes and 862–1025 cm^{-1} for the other dioxo species.

The pseudo-C$_{3v}$ symmetry of most trioxo complexes dictates that the ν(M—O) stretching vibrations will group as a symmetric (A$_1$) and antisymmetric (E) mode and each mode will possess infrared and Raman activity. Stretching frequency data for a number of four- and six-coordinate trioxo complexes is shown in Table 4.5. A different situation pertains to five-coordinate OsO$_3$F$_2$. The three oxo ligands in this trigonal bipyramidal d^0 complex all lie in the equatorial plane of the molecule. Consistent with this D_{3h} symmetry the E' band at 929 cm^{-1} is observed in both the IR and Raman, while the A$_1$' Raman band at 946 cm^{-1} has no counterpart in the IR [3].

The vibrational spectra of the [MO$_4$]$^{n-}$ species have been studied extensively. The T_d symmetry of these species leads to the prediction of two stretching modes ν_1(A$_1$) and ν_3(T$_2$) and two deformation modes ν_2(E) and ν_4(T$_2$). In practice, the two bending

TABLE 4.4 Oxo Stretching Modes for Dioxo Complexes (Selected Examples)

Complex	ν(symmetric)[a]	ν(asymmetric)[a]	Reference
Trans Dioxo Complexes[b]			
$K_4[Mo(O)_2(CN)_4]$	779	826	69
$[Tc(O)_2(en)_2]I$	n.d.[c]	833	70
$K_3[Re(O)_2(CN)_4]$	871	768	69
$Ru(O)_2(OH)_2(py)_2$	850	790	71
$Ru(O)_2(TMP)$[d]	n.d.[c]	821	18
$Os(O)_2(OCH_2CH_2O)(py)_2$	870	824	71
$Os(O)_2(OH)_2(phen)$	895[e]	850	72
$[Os(O)_2(bipy)_2]^{2+}$ (trans)	n.d.[c]	872	73
Other Dioxo Complexes			
$K_2[V(O)_2F_3]$	925	887	74
$Cs[V(O)_2(dipic)]$[f]	956	947	75
$Cr(O)_2Cl_2$[g]	984	994	76
$Mo(O)_2F_2$[g]	1009	987	76
$Mo(O)_2Cl_2$[g]	994	972	76
$Mo(O)_2Br_2$[g]	991	969	76
$Mo(O)_2Br_2(Ph_3PO)_2$	944	903	77
$Mo(O)_2Me_2(bipy)$	(934)[h]	(905)[h]	78
$W(O)_2Cl_2$[g]	992	985	76
$Na_2[W(O)_2F_4]$	958	904	69
$Re(O)_2F_3$	1025	989	4
$Re(O)_2Me_3$	(992)[h]	(951)[h]	79
$[Os(O)_2(bipy)_2]^{2+}$ (cis)	883	862	73

[a] Frequencies in wavenumbers (cm^{-1}).
[b] Data for ν(symmetrical) from Raman spectrum and ν(asymmetric) from IR spectrum.
[c] Raman spectrum not reported.
[d] TMP = 5,10,15,20–tetramesitylporphyrin.
[e] Shoulder at 878 cm^{-1}.
[f] dipic = pyridine–2,6–dicarboxylate.
[g] Gas phase IR data.
[h] Asymmetric and symmetric modes not assigned.

TABLE 4.5 Oxo Stretching Modes for 4- and 6-coordinate Trioxo Complexes[a]

Complex	Conditions[b]	ν(symmetric)[c]	ν(asymmetric)[c]	Ref.
K[Cr(O)$_3$Cl]	R, H$_2$O	907	954	80
[Mo(O)$_3$Cl]$^-$	ir, solid	934	894	81
[Mo(O)$_3$(OSi^{t}Bu$_3$)]$^-$	ir, nujol	901	878	82
K$_3$[Mo(O)$_3$F$_3$]	ir, mull	892	839	83
(NH$_4$)$_3$[Mo(O)$_3$F$_3$]	ir, mull	900	824	83
Mo(O)$_3$(dien)	ir, mull	911	872	83
Mn(O)$_3$F	ir, gas	905.2	952.5	84
Tc(O)$_3$F	R, liquid	965	951	85
Tc(O)$_3$Cl(bipy)	R, solid	(902)[d]	(882)[d]	47
Re(O)$_3$F	ir, Ar matrix	1013.2	978.3	4
Re(O)$_3$Cl	ir, liquid	1001	960	86
Re(O)$_3$Br	ir, liquid	997	963	86
Re(O)$_3$Me	ir, CS$_2$	999	960	87
Cp*Re(O)$_3$[e]	ir, KBr	910	881	88
[(CTAN)Re(O)$_3$]Cl[f]	ir, KBr	945	920	89
K[Os(N)O$_3$]	R, H$_2$O	897	871	90

[a]For additional examples see refs. 47 and 91.
[b]R = Raman; ir = infrared.
[c]Frequencies in wavenumbers (cm^{-1}).
[d]Asymmetric and symmetric modes not assigned.
[e]However ref. 92 reports three bands [Raman, solid], at 907, 889, 874 cm^{-1}, said to reflect C_s symmetry.
[f]CTAN = 1,4,7-triazacyclononane.

modes are sometimes difficult to resolve. The assignments tabulated in Table 4.6 are those suggested in the review of Diemann and Mueller [93]. Also included are M—O stretching force constants. (The values cited are those for aqueous solution where crystal splitting effects are avoided.)

In 1965 Cotton and Wing [94] predicted that the force constants for [MO$_4$]$^{n-}$ species should increase monotonically both on proceeding down a given triad or on proceeding left to right across a given period of the periodic table. Although there were some inconsistencies in the available data in 1966, it is interesting to note that the data of Table 4.6 now support their original contention. (Note that the increasing force constants on proceeding left to right must be attributed partly to the decrease in ionic charge, for example in [CrO$_4$]$^{3-}$ versus [MnO$_4$]$^-$; compare section 4.1.3 above.)

TABLE 4.6 Vibrational Frequencies and Stretching Force Constants for d^0 Tetraoxo Species[a]

Complex	$\nu_1(A_1)$	$\nu_2(F_2)$	$\nu_3(E)$	$\nu_4(F_2)$	K_{FC}
$[VO_4]^{3-}$	826	336[b]	804	336[b]	4.80
$[CrO_4]^{2-}$	846	349	890	378	5.65
$[MoO_4]^{2-}$	897	317[b]	837	317[b]	5.93
$[WO_4]^{2-}$	931	325[b]	838	325[b]	6.48
$[MnO_4]^-$	839	360	914	430	5.92
$[TcO_4]^-$	912	325	912	336	6.78
$[ReO_4]^-$	971	332[b]	920	332[b]	7.56
RuO_4	882	323	914	334	6.96
OsO_4	965	333	960	323	8.29

[a]Raman frequencies (cm^{-1}) and stretching force constants (mdyn/Å) from ref. 93.
[b]The ν_2 and ν_4 bands cannot be resolved.

4.2 VIBRATIONAL SPECTRA OF BRIDGING NITRIDES

Some years ago Chatt and Heaton examined the effect of coordinating Lewis acids, such as boron trihalides, to the nitrido ligand in $(PPhEt_2)_3Re(N)X_2$ complexes (X = halogen) [95]. In all cases the M—N stretching frequency surprisingly increased upon complexation; in the case of boron trihalides, the increase was in the range of 80–120 cm^{-1}. Similarly, Dehnicke et al. have noted that $\nu(Re—N)$ in the $[ReBr_4N]^-$ anion shifts from 1099 to 1170 cm^{-1} upon complexation to BBr_3 [53].

In unsymmetrically bridging transition metal nitrides of the type M≡N—M, the absorption due to the short MN bond is found in roughly the same range as in terminal nitrides. Several examples are shown in Table 4.7. Occasionally this band is split. Data for the vibrational frequency of the long MN bond have not been reported; it is expected to occur at very low (<200 cm^{-1}) wavenumbers.

Symmetrically bridging nitrido derivatives of the type M=N=M must be treated somewhat differently. In this case, the two stretching vibrations should be regarded as $\nu_{sym}(M=N=M)$ and $\nu_{asy}(M=N=M)$. The symmetric oscillation is expected to fall at very low frequency, since it leaves the nitrogen atom unmoved. Pioneering studies on such systems were carried out by Cleare and Griffith [104], who examined dozens of ruthenium and osmium complexes of this type. In all cases, the asymmetric mode was observed in the IR in the range of 1013–1137 cm^{-1} while the symmetric mode appeared in the Raman at 263–360 cm^{-1}. Subsequently, vibrational spectra have been reported for symmetrically bridging nitrides of a variety of other transition metals as shown in Table 4.7.

TABLE 4.7 Vibrational Modes for Symmetrically and Unsymmetrically Bridging Nitrides

Complex	ν(symmetric)	ν(asymmetric)	Reference
Symmetrically Bridging[a]			
(DTTAA)VNV(DTTAA)[b]	890	—	96
$(NH_4)_3[Br_5TaNTaBr_5]$[c]	985	228	97
$PPh_4[Cl_5WNWCl_5]$[d]	945	—	98
(TPP)FeNFe(TPP)	910	424[e]	99
[(TPP)FeNFe(TPP)]ClO_4	~1000	465[e]	100
(OEP)FeNFe(OEP)	940	439[e]	101
(Pc)FeNFe(Pc)	915	—	102
$K_3[(H_2O)Br_4RuNRuBr_4(H_2O)]$	1050	392	103
$K_3[(H_2O)Cl_4RuNRuCl_4(H_2O)]$	1080[f]	402	103
$[Br(NH_3)_4RuNRu(NH_3)_4Br]Br_3$	1039	348	104
$K_3[(H_2O)Cl_4OsNOsCl_4(H_2O)]$	1137	267	104
$[Br(NH_3)_4OsNOs(NH_3)_4]Br_3$	1095[g]	—[h]	104
Unsymmetrically Bridging[i]			
$[Mo(N)Cl_3]_4$		1045	105
$[Mo(N)Cl_3(OPCl_3)]_4$		1042, 1049	106
$[Mo(N)Br_2]_n$		951	107
$[W(N)Cl_3]_n$		1068, 1084	105
$[Re(N)Cl_3]_n$		1080	108
$[Re(N)Cl_4]_n$		944, 995, 1011	109

[a] Data (cm^{-1}) for ν(asymmetric) from ir and for ν(symmetric) from Raman spectra. Additional examples in ref. 104.
[b] DTTAA = dibenzotetramethyltetraaza[14]annulene ligand.
[c] A band at 375 cm^{-1} was assigned as δ M–N–M.
[d] A band at 840 cm^{-1} was assigned as δ M–N–M.
[e] Resonance Raman measurement; shifted to higher frequency by ^{54}Fe substitution.
[f] Shifted to 1047 cm^{-1} by ^{15}N substitution.
[g] Shifted to 1063 cm^{-1} by ^{15}N substitution.
[h] Reported as 229 cm^{-1} in ref. 110.
[i] Indicated frequencies (cm^{-1}) are M≡N stretching modes.

In a single case the IR spectrum of a bridging hydrazido(4−) complex has been reported. The IR of [Ta(CH^{t}Bu)Cl(PMe$_3$)$_2$]$_2$(μ-N$_2$) exhibits a medium strength absorbance at 847 cm^{-1}, which moves to 820 cm^{-1} in the ^{15}N$_2$ substituted analog [111]. It is not known if this is an N—N stretching mode, a Ta—N stretching mode, or a mode characteristic of the entire Ta$_2$N$_2$ linkage.

4.3 IR SPECTRA OF ALKYLIDYNE, ORGANOIMIDO, AND RELATED NX COMPOUNDS

The assignment of the metal–ligand stretching vibrations in both alkylidyne and organoimido complexes is problematic. In such M(NR) or M(CR) systems, coupling of the M—C or M—N vibrational mode with other metal ligand modes, the C—R or N—R stretching modes, and/or with vibrations of the R group appear to be significant complications. As one consequence, no assignments of ν(MC) in high-valent alkylidynes have been reported to date. [In contrast ν(MC) for several carbynes are reported to lie in the range 1270–1420 cm^{-1} [112,113].] The assignment of ν(MN) in organoimido compounds has become somewhat controversial. Therefore, we provide some historical background relating to this problem.

In their original studies on the IR spectra of phenylimido complexes, Chatt and coworkers suggested that the M—N stretching band should be found at a lower frequency than that in oxo or nitrido complexes [114]. For example, they assigned ν(Re—N) in ReCl$_3$(NPh)(PEt$_3$)$_2$ to a very strong band at 780 cm^{-1}. However, the Chatt group subsequently revised this position and suggested that ν(M—N) in organoimido derivatives should in general lie at higher frequency than their oxo or nitrido analogs [115]. This was exemplified by the assignment of the Re—N stretching mode in Re(NMe)Cl$_3$(PPh$_2$Et)$_2$ as a band at 1096 cm^{-1}.

Three factors were cited in support of this revision.

1. The M—N stretching frequencies of nitrides were known to increase upon complexation of Lewis acids (see Section 4.2).
2. In the complex OsO$_3$(N^{t}Bu) the ν(Os—N) had been assigned as 1184 cm^{-1}, some 160 cm^{-1} higher than in the nitride [OsO$_3$N]$^-$ [116].
3. The Re—N distance of 1.788 Å in the nitrido complex ReNCl$_2$(PEt$_2$Ph)$_3$ is in fact longer than Re—N in several rhenium imido analogs that fall in the range of 1.68–1.70 Å. (In retrospect, we now recognize that the structure of this particular complex is anomalous and best regarded as a "distortional isomer" of the more usual nitride structures in which Re—N bond lengths of 1.59–1.66 Å are typical (see Section 5.3.1).

^{15}N isotopic labeling studies have been reported for a dozen phenylimido complexes, and the results are shown in Table 4.8. With the exception of the last two compounds, all of the complexes show a band in the 1310–1360 cm^{-1} region, which is shifted to lower frequency by 20–30 cm^{-1} upon ^{15}N substitution. The authors of these studies have, for the most part, taken the position that this band is associated with the

TABLE 4.8   Effect of ^{15}N Substitution on Characteristic IR Bands of Phenylimido Complexes

Complex	$\nu(^{14}N)^a$	$\nu(^{15}N)^a$	Reference
$Cp_2^*V(NPh)$	1330^b	1307^b	117
$Ta(NPh)(THF)_2Cl_3$	1360	1335	118
$Ta(NPh)(PMe_3)_2Cl_3$	1345	1325	118
$Ta(NPh)(PMe_3)_4Cl$	1340	1318	118
$Ta(NPh)(dmpe)_2Cl$	1350	1328	118
$Ta(NPh)(C_2H_4)(PMe_3)_3Cl$	1355	1332	118
$Mo(NPh)_2(O^tBu)_2$	1310	1280	119
$W(NPh)(S_2CNMe_2)_2(CO)$	1344	$\sim1316^c$	120
$Re(NPh)Cl_3(PPh_3)_2$	1347	1317	121
$Re(NPh)Cl(S_2CNMe_2)_2$	1353^d	1328^d	121
$Mo(NPh)[S_2P(OEt)_2]_3$	539^e	534^e	122
$Mo(NPh)Cl[S_2P(OEt)_2]_2$	$542^{e,f}$	$533^{e,f}$	122

[a] Frequencies in wavenumbers (cm^{-1}).
[b] In addition a band at 934 shifts to 923 cm^{-1}.
[c] The 1344 cm^{-1} band is "seen to shift 25–30 cm^{-1} upon ^{15}N substitution."
[d] In addition, bands at 1025, 1009, and 991 shifted to 1019, 1004, and 986 cm^{-1}.
[e] The higher frequency bands do not shift on ^{15}N substitution.
[f] In addition a weak band at 385 shifted to 376 cm^{-1}.

phenylimido ligand, but that it is not clear whether it represents a M—N or C—N stretching mode or some combination of the two. (It can be noted that aryl amines exhibit a C—N stretching mode in the range of 1250–1360 cm^{-1}. However, the high frequency of this band is said to reflect conjugation of the nitrogen lone-pair with the ring; it is unclear how this relates to the phenylimido derivatives where this electron pair is presumably involved in π-bonding with the metal. One might, for example, compare with the C—N stretching mode in nitromethane which occurs at 918 cm^{-1} [123].)

The view that the M—N stretch in organoimido complexes will occur at a considerably higher frequency than the C—N stretch and that this is due in part to vibrational coupling has been espoused by Dehnicke [60]. It was suggested that the decreasing influence of coupling can be seen in the series of compounds $Cl_3V(NX)$, all of which have a linear VNX arrangement.

	$Cl_3V(N-Cl)$	$Cl_3V(N-Br)$	$Cl_3V(N-I)$
$\nu(VN)$	1107	1032	963
$\nu(NX)$	510	435	390

It was further argued that when X is an organic group ν(MN) appears at even higher wavenumbers, generally between 1200 and 1300 cm^{-1}. On the other hand, ν(NC) in such compounds is said to be shifted to lower wavenumbers by more than 100 cm^{-1} relative to uncoupled ν(NC) stretching vibrations [60]. As an example, the M—N stretch in AsPh$_4$[Cl$_5$ReNCCl$_3$] is said to fall at 1240 cm^{-1}, while the C—N stretch is observed at 851 cm^{-1}. Similarly, ν(Mo—N) for AsPh$_4$[Cl$_5$MoNCCl$_3$] is assigned as 1200 cm^{-1} and ν(C—N) as 928 cm^{-1}. (Similar arguments have been applied to carbyne complexes [113].)

However, an essentially opposite viewpoint has been given by Osborn and Trogler [117]. They note that the IR spectrum of Cp*$_2$V(NPh) exhibits two bands that are sensitive to ^{15}N substitution (see Table 4.8). It was suggested that the band at 1330 cm^{-1} corresponds to ν(C—N) and that at 934 cm^{-1} is due to V—N stretching. The assignment of ν(M—N) below the usual range for nitrido complexes and at the low end of the range for oxo derivatives was said to be consistent with the relative bond strengths M(N) > M(O) > M(NR) suggested by bond length data. Moreover, when the phenyl group in Cp*$_2$V(NPh) was replaced by 2,6-dimethylphenyl, causing an increase in the C—N bond length and a decrease in V—N bond length, the 1330-cm^{-1} band dropped in frequency to 1293 cm^{-1}. This is said to be consistent with the proposed assignment as ν(C—N). On the basis of their studies, these authors proposed a range of 850–1150 cm^{-1} for the stretching frequencies in organoimido compounds.

It seems evident that the resolution of these conflicting proposals must await labeling studies utilizing both ^{13}C and ^{15}N substitution. Furthermore, the current authors again emphasize that the observed stretches need not be due to diatomic oscillators at all. At any rate, a cautious approach seems in order when making assignments of the IR spectra of organoimido complexes.

4.4 IR DATA AS A MEASURE OF THE RELATIVE π-DONATING ABILITY OF VARIOUS MULTIPLY BONDED LIGANDS

The widely cited assertion of Griffith [6] that nitride ligands are more effective π-donors than oxo ligands is based on the force constants for [ReO$_3$N]$^{2-}$ and [OsO$_3$N]$^-$ ions. In particular, he noted that the metal oxo force constant for these ions is significantly smaller than those in [ReO$_4$]$^-$ and OsO$_4$, respectively. For example, in the rhenium compounds K(ReO) drops from 7.50 to 6.24 mdyne/Å upon replacement of oxo by nitride [124]. In the years since Griffith's original proposal, a variety of additional data on the osmium systems have become available as summarized in Table 4.9. Despite the ambiguities surrounding the assignment of ν(MN) in organoimido complexes, these data do appear to have some interesting implications.

Replacement of an oxo substituent of osmium tetroxide with either a nitrido or an imido ligand leads to a decrease in ν(OsO) and an increase in the Os—O bond length. As additional oxo groups are replaced by imido ligands, the stretching frequencies of the remaining oxo groups continue to drop by an average of 40 cm^{-1} per imido ligand [127]. In the four cases where structural data are available, the Os—O bond length increases monotonically with decreasing stretching frequency.

Table 4.9 shows that the effect of an imido ligand on ν(OsO) is less than that of a

TABLE 4.9 Intramolecular Comparison of Oxo, Nitrido, and Organoimido Ligands

Complex	Y	Os–Y Bond Length[a]	ν(Os–Y)[b]	K[c]	References
OsO_4	O	1.697(7)[d]	963[e], 956	8.29	3,125
$OsO_3(NR)$[f]	O	1.715(4)	925, 912		126,127
	N	1.697(4)	1184		
$Os(O)_2(N^tBu)_2$	O	1.744(6)	888, 878		126,127
	N	1.715(8)	1200		
$Os(O)(N^tBu)_3$	O		838		127
	N		1190, (1100)[g]		
$M[OsO_3N]$[h]	O	1.740(10)	897, 871	6.76	90,128
	N	1.676(15)	1073	7.95	

[a] Average value, in Å.
[b] Frequencies in wavenumbers (cm^{-1}).
[c] Stretching force constant, in units of mdyne/Å.
[d] X-ray structure. Electron diffraction gives 1.712 Å [129].
[e] From Raman spectrum.
[f] Structural data for R = 1-adamantyl; IR data for R = t-butyl.
[g] Additional strong sharp absorption.
[h] X-ray structual data for $Cs[OsO_3N]$, Raman spectrum of aqueous $K[OsO_3N]$.

nitrido ligand. The data suggest that for such osmium(VIII) complexes π-donation decreases along the series nitrido > imido > oxo. Generalization of this sequence for systems other than osmium(VIII) is dangerous. There are indications that osmium differs from early transition metals in its relative affinity for oxygen and nitrogen. Thus OsO_4, unlike other metal oxides, reacts spontaneously with amines to afford the imido compound, which is then not hydrolytically sensitive. (It should also be noted that comparison of ν(Os—O) in OsO_4 and $[OsO_3N]^-$ is complicated by the different ionic charges involved.)

4.5 HYDROGEN STRETCHING MODES

For a number of cationic terminal imido complexes of the type $[(dppe)_2W(NH)X]^+$, ν(NH) is reported to lie in the range of 3220–3420 cm^{-1} [43]. In hydrazido(2−) complexes, ν(NH) for the β hydrogen usually lies in the range of 2800–3300 cm^{-1} [130]. Some caution is required in interpreting such bands; their frequencies are subject to large shifts due to hydrogen bonding, especially in the solid state. For example, the complexes [(dppe)$_2$M(N$_2$H$_2$)X]$^+$ were originally thought to contain the M—NH=NH moiety on the basis of the large splitting of their N—H stretching IR band [131]. It was subsequently shown that the splitting results from hydrogen bonding with the halide ion and that the complexes in fact contain the hydrazido(2−) ligand [130].

No data for ν(CH) have been reported for any of the known methylidyne species. However, ν(CH) has been reported for a number of alkylidene derivatives and provides both a useful diagnostic as well as structural insight. It has been noted that for electron-deficient 14-electron alkylidene complexes ν(CH) lies at a surprisingly low frequency. For instance, in a series of tantalum complexes Ta(CH^{t}Bu)X$_3$L$_2$ (X = halogen, L = neutral donor ligand), ν(CHα) usually occurs at 2400–2600 cm^{-1} [132]. Stretching frequencies in this range are taken as an indication of a distorted "T-shaped" alkylidene with a large M—C—C angle. Examples include CpTa(CH^{t}Bu)Cl$_2$ (2510 cm^{-1}, 165.0°) [133], Cp*Ta(CH^{t}Bu)(C$_2$H$_4$)(PMe$_3$) (2520 cm^{-1}, 170.0°) [134], and [Ta(CH^{t}Bu)Cl$_3$(PMe$_3$)]$_2$ (2605 cm^{-1}, 161.2°) [134]. For the d^2 alkylidenes Ta(CH^{t}Bu)ClL$_4$ (L = PMe$_3$ or dmpe) ν(CHα) drops to 2200 cm^{-1} [135, 136].

The presence of competitive π donors causes the C—H stretching frequency to rise. The π-donor alkoxide ligands in Ta(CH^{t}Bu)(O^{t}Bu)$_2$(PMe$_3$)Cl result in a ν(CHα) of 2720 cm^{-1} [137]. Higher ν(CHα) frequencies are an indication of a bent alkylidene structure. For example, in the 18-electron complex Cp$_2$Ta(CH^{t}Bu)Cl the Ta—C—C angle is 150.4°. Although ν(CHα) cannot be observed directly (due to overlap with other C—H stretching modes) it is inferred from ^{2}H labeling studies to occur at 2900 cm^{-1} [133]. [Because of such overlap problems, the NMR coupling constant J(CHα) provides a more broadly applicable diagnostic for alkylidene structure; see Section 4.9.]

4.6 ^{17}O NMR OF OXO COMPLEXES

The ^{17}O NMR of transition metal–oxo compounds was first studied by Figgis, Kidd, and Nyholm [138]. It was discovered that for the d^0 tetraoxometallates of V, Cr, Mo, W, Tc, Re, Ru, and Os there is a linear relationship between the lowest energy ($t_1 \rightarrow e$) electronic transition and the ^{17}O chemical shift. This striking observation was explained in terms of Ramsay's general equation for nuclear shielding, which contains both a diamagnetic and a paramagnetic term. The magnitude of the chemical shifts observed in ^{17}O spectra requires that they be dominated by the paramagnetic term. This term results from the nonspherical distribution of electronic charge surrounding the oxygen nucleus and therefore is very sensitive to orbital mixing of the excited states in metal–oxo species. As compared with the ground state, such excited states involve placing an electron in an oxygen $p_x(\pi)$ or $p_y(\pi)$ orbital which introduces orbital angular momentum. The contribution from such excited states will be inversely proportional to the energy of the electronic transition. In fact, using this model and approximate energies for the relevant electronic transitions, it was possible to calculate an ^{17}O chemical shift for [MnO$_4$]$^-$ having the correct order of magnitude [138].

Large linewidths resulting from the oxygen quadrupole are a significant complication in ^{17}O NMR. Therefore, it is desirable to run ^{17}O spectra at elevated temperatures and at low concentration in a solvent of low viscosity to increase quadrupole relaxation rates. Fortunately, the broad resonances are compensated by the wide range of chemical shifts [139]. Moreover, improvements in instrumentation make it possible in certain cases to obtain useful spectra on unenriched samples utilizing natural abundance (0.037%) ^{17}O [140]. In general, the broad linewidths preclude the observation of coupling to the metal nucleus, but occasionally (e.g., MnO$_4^-$) such coupling can be observed [141].

^{17}O NMR has proven particularly useful for distinguishing between terminal and bridging oxygen atoms. This was first demonstrated on a transition metal system for [Cr$_2$O$_7$]$^{2-}$. As is generally the case, the chemical shift (relative to H$_2$^{17}O) is greater for terminal (1090 ppm) than for the bridging oxo groups (338 ppm) [142]. In molybdenum complexes, for which considerable data are available, μ_2 oxygen generally resonates in the 300- to 700-ppm region while terminal oxygen is found in the 700- to 1100-ppm region. This approach has been used to demonstrate the occurrence of a bridging oxo group in [Mo(NAr)(dtc)$_2$]$_2$O, δ = 643 ppm [143] and to show the presence of terminal oxo groups in (^{t}BuO)$_2$MoO$_2$, δ = 862 ppm [140]. The extension of this approach has proven extremely powerful as a structural probe of polyoxoanions [144,145]. The interested reader should consult the review of Klemperer in which characteristic ranges are identified for each type of bridging oxygen from μ_2 through μ_6 in molybdates [139].

Because ^{17}O chemical shifts derive from the paramagnetic term of the Ramsay equation, they are extremely sensitive to the π-bonding environment of the oxometal complexes. (σ Bonds, by definition, have zero angular momentum about the direction of the bond axis and therefore are unimportant in this respect.) Attempts have been made to correlate the ^{17}O chemical shifts with π-bond orders in complexes of chromi-

TABLE 4.10 ^{17}O Chemical Shifts for Some Oxochromium and Oxomolybdenum Species[a]

Complex	π-Bond Order	δ (M = Cr)	δ (M = Mo)
[O$_3$MOMO$_3$]$^{2-}$ (bridging)	0	345	248
[MO$_4$]$^{2-}$	0.75	835	532
[O$_3$MOMO$_3$]$^{2-}$ (terminal)	1.0	1129	715
MO$_2$X$_2$[b]	1.5	1460	921

[a] ^{17}O chemical shift (versus H$_2$^{17}O) from refs. 11, 146, and 147.
[b] For Cr, X = Cl; for Mo, X = ethylcysteinyl.

um [146] and of molybdenum [11]. As seen in Table 4.10, for a closely related series of compounds, the chemical shift does fall off monotonically with increasing π-bond order. (In assigning the bond orders in Table 4.10, it is assumed that the total number of π bonds formed by a metal in either octahedral [94] or tetrahedral [11] geometry is three. Wentworth has defended such a model using perturbation molecular orbital arguments. We further assume with Kidd [146] that the bridging oxo in M_2O_7 species has a π-bond order of 0. This is undoubtedly an oversimplification.)

For each series of closely related compounds, a plot of the chemical shift versus π-bond order is fairly linear with a correlation coefficient $r^2 = 0.99$. Obviously, one would have to be cautious in using this approach to predict chemical shifts in more complex systems containing competitive π-donor ligands. (Interestingly, the Cr and Mo data in Table 4.10 correlate linearly with one another with a correlation coefficient $r^2 = 0.9996$. A similar correlation has been noted in the ^{15}N chemical shifts of homologous tungsten and molybdenum compounds [148].)

Miller and Wentworth have also investigated the relationship between the ^{17}O chemical shift in oxomolybdenum complexes and the Mo—O bond length [11]. In fact, based on a limited series of compounds, a plot comparing these two parameters was strikingly linear. A subsequent study using a larger set of compounds revealed a considerable amount of scatter in this type of plot [149]. It appears that this interesting relationship is best viewed as having qualitative rather than quantitative significance.

Dynamic ^{17}O NMR is also a useful technique. Early examples included two independent studies of oxygen exchange in the $[CrO_4]^{2-}/[Cr_2O_7]^{2-}$ system. In the first study [150], evidence was presented for nucleophilic attack of $[CrO_4]^{2-}$ on $[Cr_2O_7]^{2-}$ leading to substitution at Cr(VI):

$$[Cr^*O_4]^{2-} + [O_3CrOCrO_3]^{2-} \rightarrow [O_3Cr^*OCrO_3]^{2-} + [CrO_4]^{2-}$$

However, as noted by Klemperer [139], this interpretation is dubious since the investigators incorrectly assumed equal linewidths for the terminal and bridging oxygen resonances in dichromate in the absence of oxygen exchange. In contrast, a study [142] of the effect of acidification on the observed linewidths for dichromate and solvent resonances in aqueous $[Cr_2O_7]^{2-}$ provided solid evidence for the acid catalyzed equilibrium

$$[Cr_2O_7]^{2-} + H_2O \rightleftarrows 2[HCrO_4]^-$$

4.7 NITROGEN NMR OF MULTIPLY BONDED LIGANDS

Both ^{14}N and ^{15}N have magnetic moments and can give NMR spectra [151]. The more abundant ^{14}N has a spin of 1 and, therefore, a quadrupole moment. Nevertheless, useful ^{14}N spectra can be obtained for highly symmetrical molecules where quadrupole broadening is minimized. The first demonstration of this approach for a transition metal system involved the tungsten(VI) imido complexes trans-$[WF_4(NMe)L]^{n-}$ [152]. Shielding of the nitrogen nucleus was shown to decrease by 26 ppm as the trans ligand

TABLE 4.11 ^{15}N NMR Data for Nitrido and Imido Complexes

Complex	Solvent	δ (ppm)	J(NH) (Hz)	Ref.
Ta(NPh)Cl$_3$(thf)$_2$	thf	−12.9[a]		118
Ta(NPh)Cl$_3$(PEt$_3$)$_2$	thf	−28.9[a]		118
Ta(NPh)Cl(dmpe)$_2$	C$_6$H$_5$Cl	−78.9[a]		118
Ta(NPh)Cl(PMe$_3$)$_4$	thf	−77.9[a]		118
Mo(N)Cl(dppe)$_2$	thf	+166.8		148
Mo(N)Br(dppe)$_2$	thf	+190.6		148
Mo(N)(dtc)$_3$	CH$_2$Cl$_2$	+40.0		155
[Mo(NH)Cl(dppe)$_2$]Cl	CH$_2$Cl$_2$	+33.3	72	148
[Mo(NH)Br(dppe)$_2$]Br	thf	+10.6	72	148
[Mo(NH)(OMe)(dppe)$_2$]BPh$_4$	CH$_2$Cl$_2$	−58.6	75	148
[Mo(NMe)Cl(dppe)$_2$]I	CH$_2$Cl$_2$	−24.4		148
[W(NH)Br(dppe)$_2$]Br	CH$_2$Cl$_2$	−25.2	75	148
W(NMe)F$_4$(NCMe)	MeCN	+18.8[b,c]		152
W(NMe)F$_4$(MeCO$_2$Et)	MeCO$_2$Et	+9.8[b,c]		152
W(NMe)F$_4$[(MeO)$_2$SO]	(MeO)$_2$SO	+10.1[b,c]		152
W(N^{t}Bu)$_2$(NH^{t}Bu)$_2$	C$_6$D$_{12}$	+3.7[b,d]		156
Re(N)Cl$_2$(P^{n}PrPh$_2$)$_2$	CH$_2$Cl$_2$	+85.8		155
Re(N)Cl$_2$(PMe$_2$Ph)$_3$	CH$_2$Cl$_2$	+68.2		155
Re(N)Cl(dppe)$_2$]Cl	CH$_2$Cl$_2$	+67.1		155

[a] Original data referenced to NH$_3$ corrected to nitromethane scale by adding 381.9 ppm.
[b] Original data referenced to aqueous nitrate corrected to nitromethane scale by subtracting 6.0 ppm.
[c] ^{14}N double resonance measurement.
[d] Natural abundance ^{15}N measurement; amide nitrogen resonance −218.9 ppm.

was varied along the series OMe$^-$ > F$^-$ > P(OMe)$_3$ > EtOAc > (MeO)$_2$SO > MeCN.

(It is worth noting that high symmetry is also generally required to observe ^{14}N coupling in the NMR of a transition metal to which a nitrogen atom is bonded [152,153]. For example, the ^{51}V NMR of (^{t}BuO)$_3$V=N(tolyl) appears as a 1:1:1 triplet, J(VN) = 111 Hz. Yet, upon replacement of even one t-butoxy ligand with chloride, only a broad unresolved peak is observed [154].)

All reported ^{15}N NMR studies on transition metal complexes have used isotopically enriched samples. The chemical shift data for imido and nitrido derivatives in Table 4.11 are referenced to nitromethane as standard as suggested by Mason [148]. In all examples to date, the nitrogen atom is more shielded in imido complexes (-80 to $+10$ ppm) than in nitrido derivatives ($+40$ to $+190$ ppm). Also included in Table 4.11 are the $^1H-^{15}N$ coupling constants of several imido (NH) complexes. This coupling can have diagnostic value. For example, the observation of a doublet $J(^{15}NH) = 68.5$ Hz in the ^{15}N NMR spectrum of trans-[Mo(OMe)(NH)(dppe)$_2$]BPh$_4$ permitted unequivocal identification of the location of the proton in this species [148].

The ^{15}N NMR spectra have also been reported for hydrazido(2−) compounds [157,158]. Mason and coworkers have tabulated the spectral data for 26 hydrazido complexes [148]. They conclude that for tungsten complexes the β-nitrogen resonance falls in the range of -214 to -238 ppm while that for the metal-bound α nitrogen lies at -58 to -78 ppm. The corresponding ranges for molybdenum analogs are -239 to -253 ppm and -78 to -98 ppm. In addition $^{15}N-^{15}N$ coupling in the range 9–12 Hz is frequently observed. $J(^{15}NH)$ for the β NH$_2$ group is 80–95 Hz.

In one case the ^{15}N NMR of a hydrazido(4−) derivative has been reported [111]. The spectrum of [Ta(CH^{t}Bu)(PMe$_3$)$_2$Cl]$_2$(μ-N$_2$) is a sharp singlet at $+32$ ppm (relative to nitromethane).

TABLE 4.12 ^1H Chemical Shifts for Some Alkylimido Complexes

Complex	d electrons	δ (ppm)	Reference
Cp*Ta(NMe)Me$_2$	0	3.97	159
[W$_2$(NMe)$_2$F$_9$]$^-$	0	4.82	152
[W(NMe)(μ-OMe)(OMe)$_3$]$_2$	0	4.86	160
[W(NMe)F$_5$]$^-$	0	5.50	152
W(NMe)F$_4$(MeCN)	0	5.53	152
Ta(NEt)(NEt$_2$)$_3$	0	4.04	161
W(NEt)$_2$(NEt$_2$)$_2$	0	4.22	162
[W(NEt)F$_5$]$^-$	0	5.80	152
W(NEt)(O)Cl(NHEt)	0	7.3	163
Re(NMe)Cl$_3$(PEtPh$_2$)$_2$	2	0.2	164
Os(NMe)Me$_4$	2	0.27	165
Re(NMe)Cl$_3$(AsMe$_2$Ph)$_2$	2	0.7	164
Os(NMe)(CH$_2$SiMe$_3$)$_4$	2	1.59	165
[Re(NMe)(dtc)$_2$(PMe$_2$Ph)]BF$_4$	2	1.77	166
Re(NMe)(dtc)$_3$	2	2.20	121
Re(NMe)(dtc)$_2$Cl	2	2.21	121
Re$_2$(NMe)$_2$O(dtc)$_4$	2	2.38, 2.44	121
Mo(NEt)Cl$_2$(PhCON$_2$Ph)(PMe$_2$Ph)	2?	3.9, 4.6	167

4.8 ^{1}H NMR SPECTROSCOPY

The α protons in d^0 alkylimido complexes are deshielded significantly. For the examples in Table 4.12 this resonance lies 2–4 ppm downfield of the α resonance in trialkyl amines (2.42 ppm for Et$_3$N; 2.12 ppm for Me$_3$N). The situation is more complicated for the d^2 alkylimido compounds of Table 4.12. For some compounds the α-proton resonance is actually located upfield of the α-proton resonance in amines. The origin of this effect is not known.

The position of the α-proton resonance in alkylidene complexes is sensitive to electronic and structural factors. Schrock has noted that in typical "bent" alkylidene ligands this resonance occurs at large chemical shifts [168]. For the examples in Table 4.13 with <M—C—C less than 150°, this resonance appears at δ10–14. On the other hand in distorted alkylidene structures of the type observed in 14-electron systems, the α proton is significantly more shielded [168]. In the complexes of Table 4.13 with <M—C—C greater than 150°, this resonance lies in the range δ−2 to +7.

TABLE 4.13 NMR Data for Selected Alkylidene Complexes[a]

Complex	M–C$_\alpha$–C$_\beta$ Angle	δ(^{1}H)	δ(^{13}C)	J$_{CH}$	Ref.
Cp$_2$Ta(CH$_2$)Me	126.5	10.22	228	132	169,170
Cp$_2$Ta(CHPh)(CH$_2$Ph)	135.2	10.86	246	127	171
W(CHPh)Cl$_2$(PhCCPh)(PMe$_3$)$_2$	137	11.8	283.6	126	172
W(CH^{t}Bu)(O)(PEt$_3$)Cl$_2$	140.6	9.87	295	115	173,174
W(CH^{t}Bu)(O)(PMe$_3$)$_2$Cl$_2$	141.1	11.89	319	123	173,175
Re(C^{t}Bu)(CH^{t}Bu)(py)$_2$I$_2$	150.3	14.06	307.1	123	176
Cp$_2$Ta(CH^{t}Bu)Cl	150.4		273	121	177,178
W(C^{t}Bu)(CH^{t}Bu)(CH$_2$^{t}Bu)(dmpe)	150.4		256	84	179,180
Ta(CH^{t}Bu)$_2$(mes)(PMe$_3$)$_2$	154.0			104	134,181
[Ta(CH^{t}Bu)(CH$_2$^{t}Bu)(PMe$_3$)$_2$]$_2$(μ-N$_2$)	159.4[b]	6.6	270.5	88	111,182
[Ta(CH^{t}Bu)Cl$_3$(PMe$_3$)$_2$]$_2$	161.2	5.30	276.0	101	134,183
W(CHPh)(CO)(PMe$_3$)$_2$Cl$_2$	164.6	−1.21	221.4	82	196
CpTa(CH^{t}Bu)Cl$_2$	165.0	3.62	246	84	133
Cp*Ta(CHPh)(CH$_2$Ph)$_2$	166.0	5.80	220	82	41
W(CH^{t}Bu)(CO)(PMe$_3$)$_2$Cl$_2$	168.7		240	73	184
Ta(CH^{t}Bu)$_2$(mes)(PMe$_3$)$_2$	168.9			91	134,181
Cp*Ta(CH^{t}Bu)(C$_2$H$_4$)(PMe$_3$)	170.0			74	134

[a] Chemical shifts versus internal standard SiMe$_4$; coupling constants in Hz.
[b] Average value.

4.9 ^{13}C NMR SPECTROSCOPY

^{13}C chemical shift data has been reported for a number of *t*-butylimido complexes of d^0 transition metals [159,118,185]. It has been suggested that the difference between the chemical shifts for the α and β carbon atoms, Δδ, is an appropriate probe of the

TABLE 4.14 ^{13}C Chemical Shift Data for Some *tert*-Butylimido Complexes[a]

Metal	Complex	δ, C(α)	δ, C(β)	Δδ	Reference
V	V(N^{t}Bu)(O^{t}Bu)$_3$	76.8	35.4	41	186
Nb	Nb(N^{t}Bu)(dtc)$_3$	66.7	30.8	36	185
	Nb(N^{t}Bu)(NMe$_2$)$_3$	68.6	33.5	35	185
Ta	Ta(N^{t}Bu)(dtc)$_3$	63.8	33.0	31	185
	Ta(N^{t}Bu)(NMe$_2$)$_3$	66.7	34.6	32	185
	Ta(N^{t}Bu)Cl$_3$(THF)$_2$	66.3	32.0	34	118
	Cp*Ta(N^{t}Bu)Me$_2$	64.1[b]	33.8[b]	30	159
	Ta(N^{t}Bu)Cl(PMe$_3$)$_4$	63.4	35.6	28	118
Cr	Cr(N^{t}Bu)$_2$(OSiMe$_3$)$_2$	77.8	31.3	46	185
	Cr(N^{t}Bu)(O)(OSiMe$_3$)$_2$	84.1	29.3	55	185
Mo	Mo(N^{t}Bu)Np$_3$Cl	70.7[b]	32.5[b]	38	187
	Mo(N^{t}Bu)$_2$(OSiMe$_3$)$_2$	68.8	32.2	37	185
	[Mo(N^{t}Bu)(μ-N^{t}Bu)Me$_2$]$_2$	67.2	32.2	35	185
W	W(N^{t}Bu)$_2$(OSiPh$_3$)$_2$	66.6	33.3	33	185
	W(N^{t}Bu)(NH^{t}Bu)$_2$	66.1	33.9	32	185
	[W(N^{t}Bu)(μ-N^{t}Bu)Me$_2$]$_2$	66.3	32.5	34	185
Re	Re(N^{t}Bu)$_2$Cl$_3$	78.5[c]	28.1[c]	50	176
	Re(N^{t}Bu)$_2$Me$_3$	70.0[b]	30.1[b]	40	176
	Cp*Re(N^{t}Bu)Cl$_2$	76.0[b]	31.4[b]	45	188
	Re(N^{t}Bu)$_3$(OSiMe$_3$)	67.3	31.9	35	185
Os	Os(N^{t}Bu)O$_3$	82.7	27.5	55[d]	185
	Os(N^{t}Bu)$_2$O$_2$	75.1	29.6	46	185
	Os(N^{t}Bu)$_3$O	71.4	30.1	41	185

[a] Chemical shifts versus internal standard SiMe$_4$ in toluene-d$_8$ except as otherwise indicated.
[b] Spectrum obtained in benzene-d$_6$.
[c] Spectrum obtained in CDCl$_3$.
[d] A similar Δδ of 56 in CDCl$_3$ is reported in ref. 189 which includes a discussion of s-character.

electronic structure of the ligand [185]. Inspection of the data in Table 4.14 leads to several salient conclusions: (1) In series of complexes where the imido ligand has a similar bond order (e.g., the monoimido complexes), the value of $\Delta\delta$ is sensitive to the identity of the metal in the complex. (2) Although there is some dependence on the ancillary ligands that are present, their effect is much smaller than that of the nature of the metal atom. (3) As the number of imido ligands in a complex increases (and hence, as their π-bond order decreases) the value of $\Delta\delta$ falls. This is seen most clearly in the series $(^tBuN)_nOsO_{(4-n)}$; the magnitude of $\Delta\delta$ decreases $55 > 46 > 41$ for $n = 1, 2, 3$. (4) It can also be noted that reducing the metal to the d^2 oxidation state results in a significant decrease in $\Delta\delta$. For example, in the tantalum(III) derivatives $Ta(N^tBu)(PMe_3)_4Cl$ and $Ta(N^tBu)(dmpe)_2Cl$ the $\Delta\delta$ values are 28 and 27 ppm, respectively [118]. A similar effect has been observed for tungsten alkylimido compounds [190].

Conventional wisdom has it that ^{13}C chemical shifts in purely organic molecules arise principally from diamagnetic inductive effects. It is interesting to note for series of closely related compounds in Table 4.14, the magnitude of $\Delta\delta$ can be correlated with the optical electronegativity of the metal atom. As an example, in the series $(R_3SiO)_2M(N^tBu)_2$ where M = W, Mo, Cr, the optical electronegativity increases $1.5 < 1.75 < 2.3$ [93] and $\Delta\delta$ likewise increases $33 < 37 < 46$. The $\Delta\delta$ values have some predictive value for the reaction chemistry of imido complexes. For example,

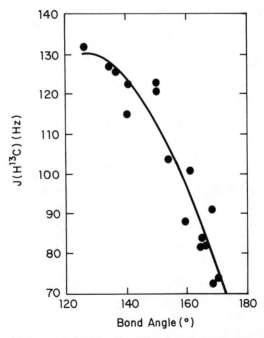

Figure 4.2 Relationship between the M—C—C bond angle in alkylidene complexes and the $^1H-^{13}C$ coupling constant for the α carbon atom.

imido ligands where $\Delta\delta < 50$ ppm react metathetically with benzaldehyde to afford benzylidene t-butylamine and the corresponding oxo complex; imido ligands where $\Delta\delta > 50$ do not undergo this reaction [190].

$$M=N^tBu + PhCH=O \rightarrow M=O + PhCH=N^tBu$$

The α-carbon resonance in alkylidene complexes is observed in the region 220–320

TABLE 4.15 ^{13}C NMR Data for Alkylidyne Complexes[a]

Complex	solvent	δ (ppm)	J(CH)	Reference
Cp*Ta(CPh)(PMe$_3$)$_2$Cl	C$_6$D$_6$	348		192
Cp*Ta(C^{t}Bu)(H)(dmpe)	C$_7$D$_8$	306.4		136
Cr(CMe)Br(CO)$_4$	CH$_2$Cl$_2$	338.2		193
Cr(CPh)Br(CO)$_4$	CH$_2$Cl$_2$	318.2		193
Mo(CPh)Br(CO)$_4$	CH$_2$Cl$_2$	284.3		193
Mo(CH)(O^{t}Bu)$_3$(quin)[b]	C$_6$D$_6$	267.2	146	194
Mo(C^{t}Bu)(CH$_2$^{t}Bu)$_3$	C$_6$D$_6$	323.8		195
Mo(C^{t}Bu)Br$_3$(dme)	C$_6$D$_6$	350.6		195
Mo(CPh)Br$_3$(PMe$_3$)$_3$	CDCl$_3$	387		196
W(CPh)Br(CO)$_4$	CH$_2$Cl$_2$	271.3		193
W(CH)Cl(PMe$_3$)$_4$	C$_6$D$_6$	250	134	197
W(CH)(O^{t}Bu)$_3$(quin)[b]	C$_6$D$_6$	247.1	147	198
W(CMe)Me(PMe$_3$)$_4$	C$_6$D$_6$	251.0		199
W(CPh)(O^{t}Bu)$_3$	C$_6$D$_6$	257		200
CpW(C^{t}Bu)Cl$_2$	C$_6$D$_6$	328.3		201
W(C^{t}Bu)H(dmpe)$_2$	C$_6$D$_6$	281.6		197
W(C^{t}Bu)(CH^{t}Bu)(CH$_2$^{t}Bu)(dmpe)	C$_6$D$_6$	296		179
W(C^{t}Bu)(O$_2$CMe)$_3$	C$_6$D$_6$	286.5		202
[CpMn(CPh)(CO)$_2$]BCl$_4$	CH$_2$Cl$_2$	357.8		193
Re(C^{t}Bu)(CH^{t}Bu)(CH$_2$^{t}Bu)$_2$	C$_6$D$_6$	295.1		176
Re(C^{t}Bu)(CH^{t}Bu)(O^{t}Bu)$_2$	C$_6$D$_6$	287.4		176
[Os(CC$_6$H$_4$Me)(CO)$_2$(PPh$_3$)$_2$]ClO$_4$	CDCl$_3$	331.0		203
[Os(CAr)Cl$_2$(PPh$_3$)$_2$(CNPh)]ClO$_4$[c]	—[d]	303.4		204

[a] Chemical shifts versus internal SiMe$_4$; coupling constants in Hz. Additional data are tabulated in refs. 193, 198, and 205.
[b] quin = quinoline = 1-azabicyclo[2.2.2]octane.
[c] Ar = p-dimethylaminophenyl.
[d] Not reported.

ppm downfield of tetramethylsilane. Schrock has noted that the $J(H^{13}C)$ coupling constant of this carbon atom is unusually small in "distorted" alkylidene ligands where the M—C—C angle differs greatly from the typical sp^2 value of 120° [168]. The general validity of this assertion can be seen from the data of Table 4.13. In Figure 4.2, we show that 16 of these data can be fitted to a simple polynomial expression with a correlation coefficient $r^2 = 0.90$. In constructing Figure 4.2 we have excluded the datum for a single complex $W(C^tBu)(CH^tBu)(CH_2{}^tBu)(dmpe)$ for which the $J(H^{13}C)$ (84 Hz) deviates markedly both from the other data and from the coupling constant for the closely related complex $W(C^tBu)(CH^tBu)(CH_2{}^tBu)(PMe_3)_2$ (113 Hz) [179]. Perhaps the low $J(H^{13}C)$ for the dmpe complex reflects some unrecognized, unique feature in its structure, or conceivably this complex has a greatly different structure in solution versus in the solid state.

The α-carbon resonance for alkylidyne complexes occurs in the 200- to 400-ppm range [191]. In Table 4.15 we have included several carbyne complexes that fall in essentially the same range. Normally in similar coordination environments the alkylidyne resonance lies downfield of the alkylidene resonance, but this is not always the case. For example, in $Re(C^tBu)(CH^tBu)py_2I_2$ the alkylidene α carbon resonates at δ 307.1 and the alkylidyne α carbon at δ 299.5 [176]. Until recently, no explanation had been offered for the origins of the chemical shifts in alkylidene or alkylidyne complexes. Indeed, Evans and Norton have described some serious pitfalls in rationalizing the chemical shifts of carbon atoms bound to transition metals [206]. However, Fenske has provided a preliminary account of efforts to predict the ^{13}C chemical shift in carbene complexes using Xα-scattered wave calculations [207]. While preliminary results for a limited set of compounds appeared promising, details of this work have yet to appear.

^{13}C Chemical shift data have been reported for several bridging alkylidyne complexes of niobium, tantalum, and tungsten, all containing the μ-trimethylsilylmethylidyne ligand. In each case, the α-carbon resonance falls in the 350–410 ppm range [208].

4.10 NMR OF THE METAL ATOM

A few comments on this topic are in order. One major body of work is that of Enemark and coworkers on the ^{95}Mo NMR of molybdenum–oxo complexes. These studies cover a broad range of oxidation states and ligand types and have recently been reviewed comprehensively [209]. Maatta has studied the ^{51}V NMR of a series of arylimido vanadium(V) complexes. He finds that the ^{51}V chemical shifts, which span a range of 1700 ppm, correlate nicely with λ_{max} of their electronic spectrum. This correlation is explained in terms of a dominating paramagnetic shielding contribution. Extended Hückel calculations on a series of model complexes appear to support this proposal [210].

REFERENCES

1. Hope, E.G.; Jones, P.J.; Levason, W.; Ogden, J.S.; Tajik, M.; Turff, J.W. *J. Chem. Soc., Dalton Trans.* **1985**, 529-533.
2. Levason, W.; Ogden, J.S.; Rest, A.J. *J. Chem. Soc., Dalton Trans.* **1980**, 419-422.
3. Beattie, I.R.; Blayden, H.E.; Crocombe, R.A.; Jones, P.J.; Ogden, J.S. *J. Raman Spect.* **1976**, *4*, 313-322.
4. Beattie, I.R.; Crocombe, R.A.; Ogden, J.S. *J. Chem. Soc., Dalton Trans.* **1977**, 1481-1489.
5. Barraclough, C.G.; Lewis, J.; Nyholm, R.S. *J. Chem. Soc.* **1959**, 3552-3555.
6. Griffith, W.P. *Coord. Chem. Rev.* **1972**, *8*, 369-396.
7. Srinivasan, K.; Kochi, J.K. *Inorg. Chem.* **1985**, *24*, 4671-4679.
8. Groves, J.T.; Kruper, W.J. *J. Am. Chem. Soc.* **1979**, *101*, 7613-7615.
9. Newton, W.E.; McDonald, J.W. *J. Less-Common Met.* **1977**, *54*, 51-62.
10. Sharpless, K.B.; Townsend, J.M.; Williams, D.R. *J. Am. Chem. Soc.* **1972**, *94*, 295-296.
11. Miller, K.F.; Wentworth, R.A.D. *Inorg. Chem.* **1979**, *18*, 984-988.
12. Wieghardt, K.; Woeste, M.; Roy, P.S.; Chaudhuri, P. *J. Am. Chem. Soc.* **1985**, *107*, 8276-8277.
13. Willner, I.; Otvos, J.W.; Calvin, M. *J. Chem. Soc., Chem. Comm.* **1980**, 964-965.
14. Collin, R.J.; Griffith, W.P.; Pawson, D. *J. Mol. Struct.* **1973**, *19*, 531-544.
15. Mayer, J.M.; Thorn, D.L.; Tulip, T.H. *J. Am. Chem. Soc.* **1985**, *107*, 7454-7462.
16. Bajdor, K.; Nakamoto, K. *J. Am. Chem. Soc.* **1984**, *106*, 3045-3046.
17. Moyer, B.A.; Meyer, T.J. *Inorg. Chem.* **1981**, *20*, 436-444.
18. Groves, J.T.; Quinn, R. *Inorg. Chem.* **1984**, *23*, 3844-3846.
19. Chisholm, M.H.; Folting, K.; Huffman, J.C.; Kirkpatick, C.C. *Inorg. Chem.* **1984**, *23*, 1021-1037.
20. Pinchas, S.; Samuel, D.; Petreanu, E. *J. Inorg. Nucl. Chem.* **1967**, *29*, 335-339.
21. Schmidt, K.H.; Flemming, V.; Müller, A. *Spectrochim. Acta, Part A* **1975**, *31*, 1913-1919.
22. Groves, J.T.; Takahashi, T.; Butler, W.M. *Inorg. Chem.* **1983**, *22*, 884-887.
23. Hill, C.L.; Hollander, F.J. *J. Am. Chem. Soc.* **1982**, *104*, 7318-7319.
24. Rummel, S.; Hermann, M.; Schmidt, D. *Z. Chem.* **1985**, *25*, 152-153.
25. Wright, M.J.; Griffith, W.P. *Trans. Met. Chem.* **1982**, *1*, 53-58.
26. Griffith, W.P.; Pawson, D. *J. Chem. Soc., Chem. Comm.* **1973**, 418-419.
27. Lewis, J.; Wilkinson, G. *J. Inorg. Nucl. Chem.* **1958**, *6*, 12-13.
28. Griffith, W.P.; Pawson, D. *J. Chem. Soc., Dalton Trans.* **1973**, 1315-1320.
29. Pasquali, M.; Marchetti, F.; Floriani, C.; Merlino, S. *J. Chem. Soc., Dalton Trans.* **1977**, 139-144.
30. Miller, F.A.; Cousins, L.R. *J. Chem. Phys.* **1957**, *26*, 329-331.

31. Scherfise, K.D.; Dehnicke, K. *Z. Anorg. Allg. Chem.* **1986**, *538*, 119-122.
32. Samsel, E.G.; Srinivasan, K.; Kochi, J.K. *J. Am. Chem. Soc.* **1985**, *107*, 7606-7617.
33. Richard, P.; Guilard, R. *Nouv. J. Chim.* **1985**, *9*, 119-124.
34. LaPointe, R.E.; Wolczanski, P.T.; Mitchell, J.F. *J. Am. Chem. Soc.* **1986**, *108*, 6382-6384.
35. Deutscher, R.L.; Kepert, D.L. *Inorg. Chim. Acta* **1970**, *4*, 645-650.
36. Griffith, W.P. *Coord. Chem. Rev.* **1970**, *5*, 459-517.
37. Arshankow, V.S.I.; Poznjak, A.L. *Z. Anorg. Allg. Chem.* **1981**, *481*, 201-206.
38. Kress, J.R.M.; Russell, M.J.M.; Wesolek, M.G.; Osborn, J.A. *J. Chem. Soc., Chem. Comm.* **1980**, 431-432.
39. Levason, W.; Narayanaswamy, R.; Ogden, J.S.; Rest, A.J.; Turff, J.W. *J. Chem. Soc., Dalton Trans.* **1981**, 2501-2507.
40. Chatt, J.; Dilworth, J.R. *J. Chem. Soc., Chem. Comm.* **1974**, 517-518.
41. Messerle, L.W.; Jennische, P.; Schrock, R.R.; Stucky, G. *J. Am. Chem. Soc.* **1980**, *102*, 6744-6752.
42. Herberhold, M.; Kniesel, H.; Haumaier, L. *J. Organomet. Chem.* **1986**, *301*, 355-367.
43. Bevan, P.C.; Chatt, J.; Dilworth, J.R.; Henderson, R.A.; Leigh, G.J. *J. Chem. Soc., Dalton Trans.* **1982**, 821-824.
44. Kolitsch, W.; Dehnicke, K. *Z. Naturforsch.* **1970**, *25B*, 1080-1082.
45. Bortolini, O.; Meunier, B. *J. Chem. Soc., Chem. Comm.* **1983**, 1364-1366.
46. Campochiaro, C.; Hofmann, Jr., J.A.; Bocian, D.F. *Inorg. Chem.* **1985**, *24*, 449-450.
47. Davison, A.; Jones, A.G.; Abrams, M.J. *Inorg. Chem.* **1981**, *20*, 4300-4302.
48. Cotton, F.A.; Davison, A.; Day, V.W.; Gage, L.D.; Trop, H.S. *Inorg. Chem.* **1979**, *18*, 3024-3029.
49. Baldas, J.; Bonnyman, J.; Williams, G.A. *Inorg. Chem.* **1986**, *25*, 150-153.
50. Abram, U.; Spies, H.; Görner, W.; Kirmse, R.; Stach, J. *Inorg. Chim. Acta* **1985**, *109*, L9-L11.
51. Lis, T.; Trzebiatowska, B.J. *Acta Cryst.* **1977**, *B33*, 1248-1250.
52. Lock, C.J.L.; Wilkinson, G. *J. Chem. Soc.* **1964**, 2281-2285.
53. Kafitz, W.; Weller, F.; Dehnicke, K. *Z. Anorg. Allg. Chem.* **1982**, *490*, 175-181.
54. Pawson, D.; Griffith, W.P. *J. Chem. Soc., Dalton Trans.* **1975**, 417-423.
55. Bartlett, N.; Jha, N.K. *J. Chem. Soc. A* **1968**, 536-543.
56. Alves, A.S.; Moore, D.S.; Andersen, R.A.; Wilkinson, G. *Polyhedron* **1982**, *1*, 83-87.
57. Buchler, J.W.; Dreher, C.; Lay, K.L. *Z. Naturforsch.* **1982**, *37B*, 1155-1162.
58. Belmonte, P.A.; Own, Z.-Y. *J. Am. Chem. Soc.* **1984**, *106*, 7493-7496.
59. Chatt, J.; Rowe, G.A. *J. Chem. Soc.* **1962**, 4019-4033.
60. Dehnicke, K.; Strähle, J. *Angew. Chem., Int. Ed. Engl.* **1981**, *20*, 413-426.

61. Farmer, R.L.; Urbach, F.L. *Inorg. Chem.* **1974**, *13*, 587-592.
62. Moyer, B.A.; Meyer, T.J. *J. Am. Chem. Soc.* **1978**, *100*, 3601-3603.
63. Aoyagi, K.; Yukawa, Y.; Shimuzu, K.; Mukaida, M.; Takeuchi, T.; Kakihana, H. *Bull. Chem. Soc. Jpn.* **1986**, *59*, 1493-1499.
64. Marmion, M.E.; Takeuchi, K.J. *J. Am. Chem. Soc.* **1986**, *108*, 510-511.
65. Hopkins, M.D.; Miskowski, V.M.; Gray, H.B. *J. Am. Chem. Soc.* **1986**, *108*, 6908-6911.
66. Che, C.-M.; Cheng, W.-K. *J. Chem. Soc., Chem. Comm.* **1986**, 1519-1521.
67. Silavwe, N.D.; Chiang, M.Y.; Tyler, D.R. *Inorg. Chem.* **1985**, *24*, 4219-4221.
68. Cotton, F.A. "Chemical Applications of Group Theory" 2nd edition, Wiley-Interscience, New York, 1971.
69. Griffith, W.P. *J. Chem. Soc. A* **1969**, 211-218.
70. Kastner, M.E.; Lindsay, M.J.; Clarke, M.J. *Inorg. Chem.* **1982**, *21*, 2037-2040.
71. Griffith, W.P.; Rossetti, R. *J. Chem. Soc., Dalton Trans.* **1972**, 1449-1453.
72. Galas, A.M.R.; Hursthouse, M.B.; Behrman, E.J.; Midden, W.R.; Green, G.; Griffith, W.P. *Trans. Met. Chem.* **1981**, *6*, 194-195.
73. Dobson, J.C.; Takeuchi, K.J.; Pipes, D.W.; Geselowitz, D.A.; Meyer, T.J. *Inorg. Chem.* **1986**, *25*, 2357-2365.
74. Ryan, R.R.; Mastin, S.H.; Reisfeld, M.J. *Acta Cryst.* **1971**, *B27*, 1270-1274.
75. Nuber, B.; Weiss, J.; Wieghardt, K. *Z. Naturforsch.* **1978**, *33B*, 265-267.
76. Ward, B.G.; Stafford, F.E. *Inorg. Chem.* **1968**, *7*, 2569-2573.
77. Butcher, R.J.; Penfold, B.R.; Sinn, E. *J. Chem. Soc., Dalton Trans.* **1979**, 668-675.
78. Schrauzer, G.N.; Hughes, L.A.; Strampach, N.; Ross, F.; Ross, D.; Schlemper, E.O. *Organometallics* **1983**, *2*, 481-485.
79. Mertis, K.; Wilkinson, G. *J. Chem. Soc., Dalton Trans.* **1976**, 1488-1492.
80. Stammreich, H.; Sala, O.; Kawai, K. *Spectrochim. Acta* **1961**, *17*, 226-232.
81. Knox, J.R.; Prout, C.K. *Acta Cryst.* **1969**, *B25*, 2281-2285.
82. Klemperer, W.G.; Mainz, V.V.; Wang, R.-C.; Shum, W. *Inorg. Chem.* **1985**, *24*, 1968-1970.
83. Griffith, W.P.; Wickins, T.D. *J. Chem. Soc. A* **1968**, 400-404.
84. Reisfeld, M.J.; Asprey, L.B.; Matwiyoff, N.A. *Spectrochim. Acta* **1971**, *27A*, 765-772.
85. Binenboym, J.; El-Gad, U.; Selig, H. *Inorg. Chem.* **1974**, *13*, 319-321.
86. Miller, F.A.; Carlson, G.L. *Spectrochim. Acta* **1960**, *16*, 1148-1154.
87. Beattie, I.R.; Jones, P.J. *Inorg. Chem.* **1979**, *18*, 2318-2319.
88. Klahn-Oliva, A.H.; Sutton, D. *Organometallics* **1984**, *3*, 1313-1314.
89. Wieghardt, K.; Pomp, C.; Nuber, B.; Weiss, J. *Inorg. Chem.* **1986**, *25*, 1659-1661.
90. Woodward, L.A.; Creighton, J.A.; Taylor, K.A. *Trans. Faraday Soc.* **1960**, *56*, 1267-1272.

91. Dolgoplosk, B.A.; Oreshkin, I.A.; Makovetsky, K.L.; Tinyakova, E.I.; Ostrovskaya, I.Ya.; Kershenbaum, I.L.; Chernenko, G.M. *J. Organomet. Chem.* **1977**, *128*, 339–344.
92. Herrmann, W.A.; Serrano, R.; Bock, H. *Angew. Chem., Int. Ed. Engl.* **1984**, *23*, 383–385.
93. Müller, A.; Diemann, E. in "MTP International Review of Science, Inorganic Chemistry, Series 2" vol. 5, Sharp, D.W.A., ed. Butterworths, London, 1974, pp. 71–110.
94. Cotton, F.A.; Wing, R.M. *Inorg. Chem.* **1965**, *4*, 867–873.
95. Chatt, J.; Heaton, B.T. *J. Chem. Soc. A* **1971**, 705–707.
96. Goedken, V.L.; Ladd, J.A. *J. Chem. Soc., Chem. Comm.* **1981**, 910–911.
97. Frank, K.-P.; Strähle, J.; Weidlein, J. *Z. Naturforsch.* **1980**, *35B*, 300–306.
98. Godemeyer, T.; Berg, A.; Gross, H.D.; Müller, U.; Dehnicke, K. *Z. Naturforsch.* **1985**, *40B*, 999–1004.
99. Schick, G.A.; Bocian, D.F. *J. Am. Chem. Soc.* **1983**, *105*, 1830–1838.
100. Crisanti, M.A.; Spiro, T.G.; English, D.R.; Hendrickson, D.N.; Suslick, K.S. *Inorg. Chem.* **1984**, *23*, 3897–3901.
101. Hofmann, Jr., J.A.; Bocian, D.F. *J. Phys. Chem.* **1984**, *88*, 1472–1479.
102. Nefedov, V.I.; Porai-Koshits, M.A.; Zakharova, I.A.; Dyatkina, M.E. *Dokl. Akad. Nauk SSSR* **1972**, *202*, 605–607.
103. Mattes, R.; Moumen, M.; Pernoll, I. *Z. Naturforsch.* **1975**, *30B*, 210–214.
104. Cleare, M.J.; Griffith, W.P. *J. Chem. Soc. A* **1970**, 1117–1125.
105. Dehnicke, K.; Strähle, J. *Z. Anorg. Allg. Chem.* **1965**, *339*, 171–181.
106. Strähle, J.; Weiher, U.; Dehnicke, K. *Z. Naturforsch.* **1978**, *33B*, 1347–1351.
107. Krüger, N.; Dehnicke, K. *Z. Naturforsch.* **1979**, *34B*, 1343–1344.
108. Dehnicke, K.; Liese, W.; Köhler, P. *Z. Naturforsch.* **1977**, *32B*, 1487.
109. Liese, W.; Dehnicke, K.; Walker, I.; Strähle, J. *Z. Naturforsch.* **1979**, *34B*, 693–696.
110. Hewkin, D.J.; Griffith, W.P. *J. Chem. Soc. A* **1966**, 472–475.
111. Rocklage, S.M.; Turner, H.W.; Fellmann, J.D.; Schrock, R.R. *Organometallics* **1982**, *1*, 703–707.
112. Clark, G.R.; Edmonds, N.R.; Pauptit, R.A.; Roper, W.R.; Waters, J.M.; Wright, A.H. *J. Organomet. Chem.* **1983**, *244*, C57–C60.
113. Fischer, E.O.; Dao, N.Q.; Wagner, W.R. *Angew. Chem., Int. Ed. Engl.* **1978**, *17*, 50–51.
114. Chatt, J.; Garforth, J.D.; Johnson, N.P.; Rowe, G.A. *J. Chem. Soc.* **1964**, 1012–1020.
115. Chatt, J.; Falk, C.D.; Leigh, G.J.; Paske, R.J. *J. Chem. Soc. A* **1969**, 2288–2293.
116. Clifford, A.F.; Kobayashi, C.S. *Inorg. Synth.* **1960**, *6*, 204–208.
117. Osborne, J.H.; Trogler, W.C. *Inorg. Chem.* **1985**, *24*, 3098–3099.
118. Rocklage, S.M.; Schrock, R.R. *J. Am. Chem. Soc.* **1982**, *104*, 3077–3081.
119. Chisholm, M.H.; Folting, K.; Huffman, J.C.; Ratermann, A.L. *Inorg. Chem.* **1982**, *21*, 978–982.

120. Tan, L.S.; Goeden, G.V.; Haymore, B.L. *Inorg. Chem.* **1983**, *22*, 1744-1750.
121. Goeden, G.V.; Hagmore, B.L. *Inorg. Chem.* **1983**, *22*, 157-167.
122. Edelblut, A.W.; Wentworth, R.A.D. *Inorg. Chem.* **1980**, *19*, 1110-1117.
123. Conley, R.T. "Infrared Spectroscopy" 2nd edition. Allyn and Bacon, Boston, 1972.
124. Krebs, B.; Müller, A. *J. Inorg. Nucl. Chem.* **1968**, *30*, 463-466.
125. Kistenmacher, T.J.; Marzilli, L.G.; Rossi, M. *Bioinorg. Chem.* **1976**, *6*, 347-364.
126. Nugent, W.A.; Harlow, R.L.; McKinney, R.J. *J. Am. Chem. Soc.* **1979**, *101*, 7265-7268.
127. Chong, A.O.; Oshima, K.; Sharpless, K.B. *J. Am. Chem. Soc.* **1977**, *99*, 3420-3426.
128. Pastuszak, R.; L'Haridon, P.; Marchand, R.; Laurent, Y. *Acta Cryst.* **1982**, *B38*, 1427-1430.
129. Seip, H.M.; Stølevik, R. *Acta Chem. Scand.* **1966**, *20*, 385-394.
130. Chatt, J.; Pearman, A.J.; Richards, R.L. *J. Chem. Soc., Dalton Trans.* **1978**, 1766-1776.
131. Chatt, J.; Heath, G.A.; Richards, R.L. *J. Chem. Soc., Dalton Trans.* **1974**, 2074-2082.
132. Rupprecht, G.A.; Messerle, L.W.; Fellmann, J.D.; Schrock, R.R. *J. Am. Chem. Soc.* **1980**, *102*, 6236-6244.
133. Wood, C.D.; McLain, S.J.; Schrock, R.R. *J. Am. Chem. Soc.* **1979**, *101*, 3210-3222.
134. Schultz, A.J.; Brown, R.K.; Williams, J.M.; Schrock, R.R. *J. Am. Chem. Soc.* **1981**, *103*, 169-176.
135. Churchill, M.R.; Wasserman, H.J.; Turner, H.W.; Schrock, R.R. *J. Am. Chem. Soc.* **1982**, *104*, 1710-1716.
136. Fellmann, J.D.; Turner, H.W.; Schrock, R.R. *J. Am. Chem. Soc.* **1980**, *102*, 6608-6609.
137. Rocklage, S.M.; Fellmann, J.D.; Rupprecht, G.A.; Messerle, L.W.; Schrock, R.R. *J. Am. Chem. Soc.* **1981**, *103*, 1440-1447.
138. Figgis, B.N.; Kidd, R.G.; Nyholm, R.S. *Proc. Roy. Soc., Ser. A* **1962**, *269*, 469-480.
139. Klemperer, W.G. *Angew. Chem., Int. Ed. Engl.* **1978**, *17*, 246-254.
140. Chisholm, M.H.; Folting, K.; Huffman, J.C.; Kirkpatrick, C.C.; Ratermann, A.L. *J. Am. Chem. Soc.* **1981**, *103*, 1305-1306.
141. Broze, M.; Luz, Z. *J. Phys. Chem.* **1969**, *73*, 1600-1602.
142. Jackson, J.A.; Taube, H. *J. Phys. Chem.* **1965**, *69*, 1844-1849.
143. Devore, D.D.; Maatta, E.A. *Inorg. Chem.* **1985**, *24*, 2846-2849.
144. English, A.D.; Jesson, J.P.; Klemperer, W.G.; Mamouneas, T.; Messerle, L.; Shum, W.; Tramontano, A. *J. Am. Chem. Soc.* **1975**, *97*, 4785-4786.
145. Filowitz, M.; Klemperer, W.G.; Messerle, L.; Shum, W. *J. Am. Chem. Soc.* **1976**, *98*, 2345-2346.
146. Kidd, R.G. *Can. J. Chem.* **1967**, *45*, 605-608.
147. Filowitz, M.; Ho, R.K.C.; Klemperer, W.G.; Shum, W. *Inorg. Chem.* **1979**, *18*, 93-103.
148. Donovan-Mtunzi, S.; Richards, R.L.; Mason, J. *J. Chem. Soc., Dalton Trans.* **1984**, 1329-1332.
149. Freeman, M.A.; Schultz, F.A.; Reilley, C.N. *Inorg. Chem.* **1982**, *21*, 567-576.

150. Figgis, B.N.; Kidd, R.G.; Nyholm, R.S. *Can. J. Chem.* **1965**, *43*, 145–153.
151. Mason, J. *Chem. Rev.* **1981**, *81*, 205.
152. Chambers, O.R.; Harman, M.E.; Rycroft, D.S.; Sharp, D.W.A.; Winfield, J.M. *J. Chem. Res. (M)* **1977**, 1849–1876.
153. Nugent, W.A.; Harlow, R.L. *J. Chem. Soc., Chem. Comm.* **1979**, 342–343.
154. Maatta, E.A. *Inorg. Chem.* **1984**, *23*, 2560–2561.
155. Dilworth, J.R.; Donovan-Mtunzi, S.; Kan, C.T.; Richards, R.L.; Mason, J. *Inorg. Chim. Acta* **1981**, *53*, L161–L162.
156. Lattman, M.; Arduengo, A.J.; Davidson, F. unpublished results.
157. Anderson, S.N.; Fakley, M.E.; Richards, R.L.; Chatt, J. *J. Chem. Soc., Dalton Trans.* **1981**, 1973–1980.
158. Chatt, J.; Fakley, M.E.; Hitchcock, P.B.; Richards, R.L.; Luong-Thi, N.T. *J. Chem. Soc., Dalton Trans.* **1982**, 345–352.
159. Mayer, J.M.; Curtis, C.J.; Bercaw, J.E. *J. Am. Chem. Soc.* **1983**, *105*, 2651–2660.
160. Nielson, A.J.; Waters, J.M.; Bradley, D.C. *Polyhedron* **1985**, *2*, 285–297.
161. Bradley, D.C.; Gitlitz, M.H. *J. Chem. Soc. A* **1969**, 980–984.
162. Bradley, D.C.; Chisholm, M.H.; Extine, M.W. *Inorg. Chem.* **1977**, *16*, 1791–1794.
163. Kuznetsova, A.A.; Podzolko, Yu.G.; Buslaev, Yu.A. *Russ. J. Inorg. Chem.* **1969**, *14*, 393–396.
164. Chatt, J.; Dilworth, J.R.; Leigh, G.J. *J. Chem. Soc. A* **1970**, 2239–2243.
165. Shapley, P.A.B.; Own, Z.-Y.; Huffman, J.C. *Organometallics* **1986**, *5*, 1269–1271.
166. Bishop, M.W.; Chatt, J.; Dilworth, J.R.; Neaves, B.D.; Dahlstrom, P.; Hyde, J.; Zubieta, J. *J. Organomet. Chem.* **1981**, *213*, 109–124.
167. Bishop, M.W.; Chatt, J.; Dilworth, J.R.; Hursthouse, M.B.; Jayaweera, S.A.A.; Quick, A. *J. Chem. Soc., Dalton Trans.* **1979**, 914–920.
168. Schrock, R.R. *Acc. Chem. Res.* **1979**, *12*, 98–104.
169. Guggenberger, L.J.; Schrock, R.R. *J. Am. Chem. Soc.* **1975**, *97*, 6578–6579.
170. Churchill, M.R.; Missert, J.R.; Youngs, W.J. *Inorg. Chem.* **1981**, *20*, 3388–3391.
171. Schrock, R.R.; Messerle, L.W.; Clayton, C.D.; Guggenberger, L.J. *J. Am. Chem. Soc.* **1978**, *100*, 3793–3800.
172. Mayr, A.; Lee, K.S.; Kjelsberg, M.A.; Van Engen, D. *J. Am. Chem. Soc.* **1986**, *108*, 6079–6080.
173. Wengrovius, J.H.; Schrock, R.R. *Organometallics* **1982**, *1*, 148–155.
174. Wengrovius, J.H.; Schrock, R.R.; Churchill, M.R.; Missert, J.R.; Youngs, W.J. *J. Am. Chem. Soc.* **1980**, *102*, 4515–4516.
175. Churchill, M.R.; Rheingold, A.L.; Youngs, W.J.; Schrock, R.R.; Wengrovius, J.H. *J. Organomet. Chem.* **1981**, *204*, C17–C20.
176. Edwards, D.S.; Biondi, L.V.; Ziller, J.W.; Churchill, M.R.; Schrock, R.R. *Organometallics* **1983**, *2*, 1505–1513.
177. Schrock, R.R. *J. Am. Chem. Soc.* **1975**, *97*, 6577–6578.

178. Churchill, M.R.; Hollander, F.J. *Inorg. Chem.* **1978**, *17*, 1957–1962.
179. Clark, D.N.; Schrock, R.R. *J. Am. Chem. Soc.* **1978**, *100*, 6774–6776.
180. Churchill, M.R.; Youngs, W.J. *Inorg. Chem.* **1979**, *18*, 2454–2458.
181. Churchill, M.R.; Youngs, W.J. *Inorg. Chem.* **1979**, *18*, 1930–1935.
182. Churchill, M.R.; Wasserman, H.J. *Inorg. Chem.* **1981**, *20*, 2899–2904.
183. Schultz, A.J.; Williams, J.M.; Schrock, R.R.; Rupprecht, G.A.; Fellmann, J.D. *J. Am. Chem. Soc.* **1979**, *101*, 1593–1595.
184. Wengrovius, J.H.; Schrock, R.R.; Churchill, M.R.; Wasserman, H.J. *J. Am. Chem. Soc.* **1982**, *104*, 1739–1740.
185. Nugent, W.A.; McKinney, R.J.; Kasowski, R.V.; Van-Catledge, F.A. *Inorg. Chim. Acta* **1982**, *65*, L91–L93.
186. Preuss, F.; Becker, H. *Z. Naturforsch.* **1986**, *41B*, 185–190.
187. Ehrenfeld, D.; Kress, J.; Moore, B.D.; Osborn, J.A.; Schoettel, G. *J. Chem. Soc., Chem. Comm.* **1987**, 129–131.
188. Herrmann, W.A.; Herdtweck, E.; Floeel, M.; Kulpe, J.; Kuesthardt, U.; Okuda, J. *Polyhedron* **1987**, *6*, 1165–1182.
189. Axelson, D.E.; Walker, I.M.; Oliver, A.J.; Holloway, C.E. *Spectrosc. Lett.* **1973**, *6*, 475–481.
190. Ashcroft, B.R.; Clark, G.R.; Nielson, A.J.; Rickard, C.E.F. *Polyhedron* **1986**, *5*, 2081–2091.
191. Schrock, R.R. *Acc. Chem. Res.* **1986**, *19*, 342–348.
192. McLain, S.J.; Wood, C.D.; Messerle, L.W.; Schrock, R.R.; Hollander, F.J.; Youngs, W.J.; Churchill, M.R. *J. Am. Chem. Soc.* **1978**, *100*, 5962–5964.
193. Fischer, E.O.; Schubert, U. *J. Organomet. Chem.* **1975**, *100*, 59–81.
194. Strutz, H.; Schrock, R.R. *Organometallics* **1984**, *3*, 1600–1601.
195. McCullough, L.G.; Schrock, R.R.; Dewan, J.C.; Murdzek, J.C. *J. Am. Chem. Soc.* **1985**, *107*, 5987–5998.
196. Mayr, A.; Asaro, M.F.; Kjelsberg, M.A.; Lee, S.L.; Van Engon, D. *Organometallics* **1987**, *6*, 432–434.
197. Holmes, S.J.; Clark, D.N.; Turner, H.W.; Schrock, R.R. *J. Am. Chem. Soc.* **1982**, *104*, 6322–6329.
198. Listemann, M.L.; Schrock, R.R. *Organometallics* **1985**, *4*, 74–83.
199. Chiu, K.W.; Jones, R.A.; Wilkinson, G.; Galas, A.M.R.; Hursthouse, M.B.; Malik, K.M.A. *J. Chem. Soc., Dalton Trans.* **1981**, 1204–1211.
200. Wengrovius, J.H.; Sancho, J.; Schrock, R.R. *J. Am. Chem. Soc.* **1981**, *103*, 3932–3934.
201. Churchill, M.R.; Ziller, J.W.; McCullough, L.; Pedersen, S.F.; Schrock, R.R. *Organometallics* **1983**, *2*, 1046–1048.
202. Schrock, R.R.; Murdzek, J.S.; Freudenberger, J.H.; Churchill, M.R.; Ziller, J.W. *Organometallics* **1986**, *5*, 25–33.
203. Roper, W.R.; Waters, J.M.; Wright, L.J.; VanMeurs, F. *J. Organomet. Chem.* **1980**, *201*, C27–C30.
204. Gallop, M.A.; Roper, W.R. *Adv. Organomet. Chem.* **1986**, *25*, 121–198.
205. Schrock, R.R.; Clark, D.N.; Sancho, J.; Wengrovius, J.H.; Rocklage, S.M.; Pedersen, S.F. *Organometallics* **1982**, *1*, 1645–1651.

206. Evans, J.; Norton, J.R. *Inorg. Chem.* **1974**, *13*, 3042-3043.
207. Fenske, R.F. in Shapiro, B.L., ed. "Organometallic Compounds: Synthesis, Structure and Theory" Texas A&M University Press, College Station, Texas, 1983, pp. 305-333.
208. Andersen, R.A.; Galyer, A.L.; Wilkinson, G. *Ang. Chem. Int. Ed. Engl.* **1976**, *15*, 609. Fanwick, P.E.; Ogilvy, A.E.; Rothwell, I.P. *Organometallics* **1987**, *6*, 73-80.
209. Minelli, M.; Enemark, J.H.; Brownlee, R.T.C.; O'Connor, M.J.; Wedd, A.G. *Coord. Chem. Rev.* **1985**, *68*, 169-278.
210. Devore, D.D.; Lichtenhan, J.D.; Takusagawa, F.; Maatta, E.A. *J. Am. Chem. Soc.* **1987**, *109*, 7408-7416.

CHAPTER 5

STRUCTURAL STUDIES

The development of the chemistry of metal–ligand multiple bonds has depended critically on structural information, available primarily through X-ray crystallography. A close analogy can be made to the chemistry of metal–metal multiple bonding [1]. Structural studies have been the primary evidence for metal–ligand multiple bonding, beginning with Fischer's discoveries of carbene and carbyne complexes [2,3]. The presence of a multiple bond is inferred from the connectivity of the molecule and particularly from a short metal–ligand distance. In addition, a linear M—C—R or M—N—R linkage in an alkylidyne or imido complex is evidence for a metal–ligand triple bond.

This chapter is divided into two parts. The first half is a description of major structural features, including important bond lengths and angles. It begins with a discussion of the distribution of compounds with metal–ligand multiple bonds as a function of metal ion and oxidation state. The second half of the chapter is a set of comprehensive tables of structures, covering metals from titanium to osmium. A complete introduction to the tables is presented in Section 5.4. There is a great deal of data presented in the tables, much more than can be discussed in the first half of the chapter. Readers are encouraged to scan the tables to derive their own perspective and to consult the original literature on subtle structural issues.

The discussion in this chapter depends heavily on the theoretical framework presented in Chapter 2. In a complex with a metal–ligand multiple bond, the electronic structure has a tremendous influence on the molecular structure. The electronic structure affects the coordination geometry, the relative disposition of ligands, the metal–ligand bond length(s), and bond lengths and angles within the multiply bonded ligand. The fact that almost all of the structures in the tables fit neatly into the bonding pictures described in Chapter 2 is perhaps the best evidence for this theoretical framework.

5.1 DISTRIBUTION OF COMPOUNDS WITH METAL–LIGAND MULTIPLE BONDS

The tables of structures at the end of the chapter give a reasonable (although crude) estimate of the distribution of compounds with metal–ligand multiple bonds. Figure 5.1 illustrates the number of structures of complexes with a metal–ligand multiple bond as a function of the metal. While only a fraction of compounds are structurally characterized, this distribution is probably representative of the overall distribution of com-

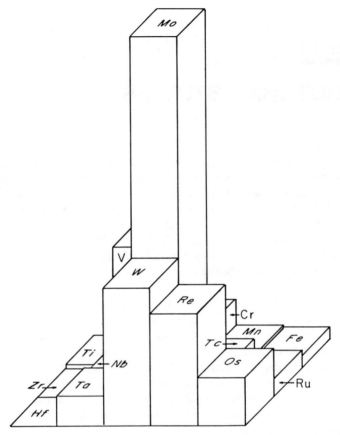

Figure 5.1 Distribution of structurally characterized compounds with metal–ligand multiple bonds as a function of the position of the metal in the periodic table.

pounds, due to the emphasis placed on structural studies in this area. There are, however, a few inherent biases in using a distribution based on structures: For instance, studies of osmium tetroxide chemistry have focused on reactivity while molybdenum complexes have been used as structural enzyme models. Technetium is presumably underrepresented because of its radioactivity.

The distribution of compounds is remarkably peaked (Fig. 5.1). The large majority of these compounds occur for metals in groups V, VI, and VII, with a number of examples for iron, ruthenium, and osmium, and a few examples in group IV. (A few alkylidene and a very few alkylidyne and imido complexes of the cobalt and nickel groups are known, a remarkably small number considering the synthetic efforts toward these molecules in recent years.) The distribution is concentrated along a diagonal from vanadium to rhenium. For example, listed in the tables are 375 molybdenum, 125 vanadium, and only 4 titanium structures. The actual distribution of compounds may be even more peaked than a distribution based on structures, because unusual compounds are more likely to be structurally characterized.

The distribution of compounds as a function of d electron count (oxidation state) is also striking. All of the titanium, niobium, and tantalum structures with multiple bonds are d^0 (although there are a few d^2 Nb and Ta imido, alkylidene, and alkylidyne species that have not been characterized structurally). Vanadium readily forms multiple bonds in both d^0 and d^1 configurations, and chromium, molybdenum, and tungsten form strong multiple bonds even at d^2. There are very few examples of terminal oxo, imido, nitrido, and alkylidyne complexes with more than two d electrons (15 out of 815 structures, see Sections 2.3.1, 2.3.3.2, and 2.5.1); these compounds all involve the later transition metals rhenium, iron, ruthenium, and osmium. Thus multiple bonding is found in higher d electron configurations on moving from titanium to osmium. Complexes with more than one multiply bonded ligand show a similar trend: Vanadium forms dioxo complexes only in the +5 (d^0) oxidation state, while rhenium and osmium often form complexes with two d electrons and two multiply bonded ligands. There are only two structures of d^1 complexes with more than one multiply bonded ligand (MnO_4^{2-} [4] and ReO_2R_2 [5]); instead, d^1 species prefer dimeric structures with both terminal and bridging ligands, such as the common oxo structures **A** and **B**.

A   **B**

It is clear that multiple bonds are very common for certain metals in specific oxidation states whereas they are rare for other metals or electronic configurations. This pattern is presumably due to a strong thermodynamic bias for or against metal–ligand multiple bonding, depending on the metal and its configuration. When thermodynamically favored, the formation of multiple bonds is difficult to avoid. For instance, when working with high-oxidation-state vanadium compounds, terminal oxo complexes are readily formed from even traces of air or water. On the other hand, when multiple bonding is not favored, as in the chemistry of zirconium or platinum, it is difficult if not impossible to make complexes with multiple bonds. These thermodynamic preferences for or against multiple bonding are discussed in more detail in Section 2.2.3.

5.2 COORDINATION GEOMETRIES

The majority of compounds with metal–ligand multiple bonds are six coordinate and adopt distorted octahedral structures (see Section 5.3.3). Octahedral coordination geometry is the most common structure for compounds of the transition metals, with or without multiply bonded ligands, due to the excellent overlap possible for both σ and π bonding.

The next most common coordination number is 5, and these compounds are found predominantly in square pyramidal structures with the multiply bonded ligand at the apex. Thus all reported [M(O)Cl$_4$]$^{n-}$ ($n = 0,1,2$) molecules or ions have a C_{4v} geometry in the gas phase or in the solid state [6–17]; in contrast, the main group analog S(O)F$_4$ has a C_{2v} structure based on a trigonal bipyramid [17]. The preference for square pyramidal coordination can be attributed both to better metal–ligand multiple bond overlap and to the large trans influence of many of these ligands (Section 5.3.3.1). The observed square pyramidal structures are closely related to the typical octahedral geometry by loss of the trans ligand and slight increase in the O–M–L angle (Section 5.3.3.2). There are a few examples of trigonal bipyramidal structures. In the case of W(NPh)(NMe$_2$)$_4$ [18], an orbital explanation has been given for a coordination geometry intermediate between square pyramidal and trigonal bipyramidal. Seven-coordinate complexes tend to be pentagonal bipyramidal, with the multiple bond in the apical position [19].

Tetrahedral structures are found primarily with a d^0 configuration, the most common examples being the tetraoxo anions and their derivatives (MnO$_4^-$, CrO$_2$Cl$_2$, Os(N)O$_3^-$, etc.) Recently tetrahedral complexes have been isolated in d^1, d^2, and d^4 configurations: Re(O)$_2$(mesityl)$_2$, Os(O)$_2$(mesityl)$_2$ [5] (mesityl = C$_6$H$_2$Me$_3$), Os(NR)$_2$(NR$_2$)$_2$ [20], and Re(O)I(MeC≡CMe)$_2$ [21]). The last appears to adopt this structure for electronic reasons (Section 2.3.3.2), whereas the first three examples are four coordinate because of the steric bulk of the ligands.

5.3 METRICAL DATA

5.3.1 Metal–Ligand Distances

The presence of a metal–ligand multiple bond is often inferred simply from a short metal–ligand distance. "Short" is defined by comparison with a typical metal–carbon, –nitrogen, or –oxygen single bond, estimated from the sum of the covalent radii, roughly 1.8–2.1 Å for bonds involving Ti—Mn, and ≳2.0 Å for Nb–Os [22]. Metal–ligand multiple bond distances in Tables 5.2–5.4 range from 1.55 Å for the shortest vanadium and chromium oxo complexes to over 2.0 Å for a few metal–carbon double bonds.

Multiple bond lengths in general vary in the order nitrido < oxo < imido < alkylidyne < alkylidene, although the observed ranges overlap to some extent. This is illustrated in Figure 5.2 for molybdenum compounds (molybdenum has been chosen because of the large number and wide variety of structures). The bond-length ordering does not simply correlate with the covalent radii of the ligating atoms, since metal–oxygen bonds lie in between nitrido and imido distances despite the larger radius of nitrogen [22]. Although all three ligands formally make triple bonds (Section 2.1.1), M—O and M—NR bonds are presumably longer than their nitrido analog because these bonds have one dative component. As a crude estimate, metal–oxo triple bond distances are 0.03–0.05 Å longer than metal–nitrido bonds, and metal–imido triple bonds are roughly the same amount longer than metal–oxo distances. Larger differences are observed between metal–ligand double-bond and triple-bond distances. For instance, the remarkable tungsten complex **C** provides a direct comparison among

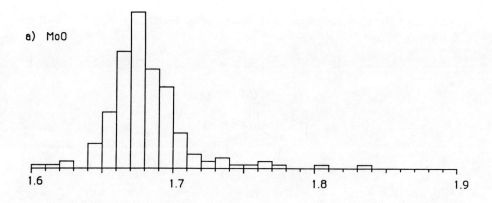
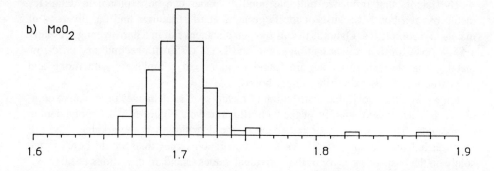
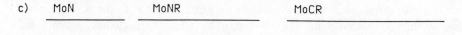
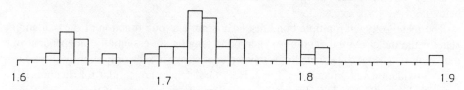

Figure 5.2 Distribution of molybdenum multiple bond distances in (*a*) monooxo (*b*) *cis*-dioxo, and (*c*) nitrido, imido, and alkylidene complexes. The vertical scale in (*c*) is half that in (*a*) and (*b*). Only bond lengths with small reported standard deviations are included: less than 0.015 Å for (*a*) and (*b*) and less than 0.020 Å for (*c*).

tungsten–alkylidyne, –alkylidene, and –alkyl bond lengths: 1.785(8), 1.942(9), 2.258(8) Å [23].

C

The factors that influence multiple bond distances can be explored in detail for metal–oxygen bonds because of the large number of structures and the diversity of metal–oxo complexes. Overall, metal–oxygen distances fall in a narrow range, usually 1.55–1.65 Å for first-row transition metals and 1.62–1.75 Å for second- and third-row metals. The shortest distances are found in monooxo complexes, with dioxo and polyoxo compounds exhibiting longer bonds.

For a particular metal, the distribution of metal–oxygen distances in monooxo complexes is narrow and sharply peaked, as illustrated in Figure 5.3 for molybdenum compounds. The smooth, gaussian shape of the distribution is remarkable. The full width at half-maximum (~0.05 Å) is not that much larger than would be expected solely on the basis of the error of the individual values (standard deviations ≤0.015 Å). A wide variety of compounds are included in this distribution, with oxidation states from four to six, coordination numbers from four to seven, and ancillary ligands from halides to Schiff bases to phosphines to cyclopentadienyls. For comparison, molybdenum–chloride distances in similar complexes vary by a couple of tenths of Ångstroms, often within the same molecule. The only type of ligand that seems to have a significant influence on the metal–oxygen distance is another multiply bonded group (see below).

The metal–oxygen multiple bond distance is not a strong function of the oxidation state of the metal. Molybdenum–oxygen bond lengths are essentially independent of oxidation state (Fig. 5.3), vanadium–oxygen bonds appear to be only slightly shorter in the +5 oxidation state than the +4, and Re—O distances are similar for rhenium(III) through rhenium(VII). This argues against a large ionic component to metal–oxygen bonding since ionic radii are a strong function of oxidation state.

The bond distance does appear, however, to be related to the metal–ligand bond order. As described in Chapter 2, monooxo complexes are best thought of as having triple bonds, *cis*-dioxo species a bond order of 2.5, and *trans*-dioxo and facial-trioxo complexes a bond order of 2. In both vanadium and molybdenum compounds, M—O distances in *cis*-dioxo complexes are on average 0.03–0.04 Å longer than in monooxo compounds (Fig. 5.2). A similar difference is observed between monooxo and facial-trioxo complexes of rhenium. Bond lengths in *trans*-dioxo (d^2) species, however, are as much as 0.1 Å longer than in monooxo compounds. The reasons for the unusually long bonds in *trans*-dioxo compounds are not understood.

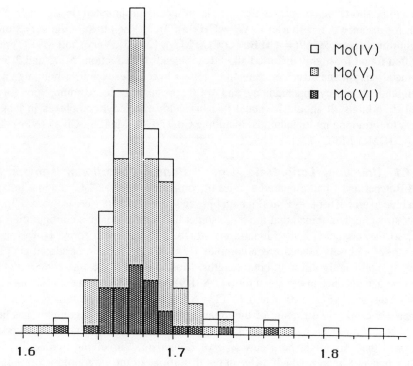

Figure 5.3 Distribution of molybdenum–oxygen distance (Å) for monooxo complexes in the oxidation states +4, +5, and +6. Only bond lengths with standard deviations less than 0.015 Å are included.

There is not yet enough data to determine whether these trends are valid for other multiply bonded ligands. Based on a limited number of structures, molybdenum–imido distances seem to be independent of oxidation state, but "Schrock-type" tungsten–alkylidyne complexes (d^0, hard ligands) appear to have shorter W—C distances than "Fischer-type" carbyne complexes (d^2, carbonyl ligands). The multiple bond distance in imido (NR), alkylidyne (CR), and alkylidene (CR_2) complexes can be influenced by the nature of the R group (cf. Section 2.1.3). The difference between alkyl and phenyl substituents (e.g., ≡C^{t}Bu versus ≡CPh) is small, but significantly longer bonds are often observed for ligands with a strongly conjugating substituent, such as hydrazido (NNR_2) or Fischer carbene ligands [e.g. $C(NR_2)R$] which contain an NR_2 group. These data and theoretical studies suggest substantial delocalization in these ligands, with concomitant reduction in the metal–ligand bond order. Hydrazido, aminoalkylidyne, and diazoalkane ligands are probably best thought of as vinylidene analogs, with M—N or M—C double bonds.

Alkylidene ligands have a much wider range of bond lengths than the ligands that form triple bonds, presumably because metal–carbon double bonds are weaker and because there is only one π interaction. Bond lengths from 1.83(2) to 2.14(2) Å are found for W═CR_2 groups with R = H, alkyl, or aryl. The metal–alkylidene distance seems to be shorter in complexes that are electronically unsaturated (less than 18 electron): for instance 0.1 Å shorter in W(=CH^{t}Bu)(O)Cl$_2$(PEt$_3$) than in the very similar six-coordinate complex W(=CH^{t}Bu)(O)Cl$_2$(PMe$_3$)$_2$ [24,25]. Very short M—C bonds are often found for highly distorted alkylidene ligands (see Sections 5.3.2 and 2.5.3). The metal–alkylidene distance seems to be longer in complexes with π-bonding ancillary ligands such as cyclopentadienyl and/or CO, presumably due to competition for the metal d_π orbitals. It should be noted that biscyclopentadienyl complexes in general appear to form long multiple bonds, including Cp$_2$Ta(CH$_2$)Me [26], Cp$^*_2$V(NPh) [27], and (C$_5$H$_4$Me)$_2$Mo(O) [28].

5.3.1.1 Unusually Long Metal–Ligand Bonds; Distortional Isomers.

In 1970 Butcher and Chatt prepared a series of compounds Mo(O)Cl$_2$L$_3$ ranging in color from blue to green; for L = PMe$_2$Ph both blue and green isomers were isolated [29]. An X-ray structure determined that the blue isomer had a meridonal-cis configuration (**D**) [30], so they assigned a meridonal-trans structure to the green form. However the structure of a closely related green complex (L = PEt$_2$Ph) is also meridonal-cis [31], but with significantly different bond lengths. Most striking is the very long molybdenum–oxygen distance in the green form (1.801(9) Å), well outside the normal range of Mo=O distances (Fig. 5.3), and 0.127(13) Å longer than the bond length in the blue isomer. (Recently the structure of the green isomer of the original complex has been described and it too has a long Mo—O distance, 1.80(2) Å [32].) Chatt and coworkers suggested that this might be a new type of isomerism, "involving two equilibrium arrangements of ligands which differ in the distortions of the ... coordination polyhedron of the metal" [33]. They called this distortional isomerism.

$$\begin{array}{c} \text{O} \\ \text{Cl} \diagdown \| \diagup \text{PR}_3 \\ \text{Mo} \\ \text{PR}_3 \diagup | \diagdown \text{PR}_3 \\ \text{Cl} \end{array}$$

D

To our knowledge this is the only suggestion that molecules of any kind have two stable structures with substantially different bond distances. Therefore, if distortional isomerism is a real phenomenon and not an artifact of some kind, it has significant implications beyond metal–ligand multiple bonding. Since Chatt's proposal there have been a number of other structures reported with metal–ligand multiple bond distances more than 0.1 Å longer than isomeric or very similar species (Table 5.1). The long bond distances are very unusual: as noted in Section 5.3.1, metal–oxo and nitrido bonds are commonly found in a narrow range. It is remarkable that the few examples of long multiple bonds occur not in unusual coordination geometries or with unusual ancillary ligands, but rather as "distortional isomers" of molecules with normal bond lengths. In one case, the blue and green forms of [W(O)Cl$_2${c-(NMeCH$_2$CH$_2$)$_3$}]PF$_6$ (**E**), the two isomers are stable in solution. This example indicates that distortional

TABLE 5.1 Distortional Isomers and Related Species

Complex	Form	M–O (Å)	Reference
$Mo(O)Cl_2(PEt_2Ph)_3$	green	1.801(9)	31
$Mo(O)Cl_2(PMe_2Ph)_3$	green	1.80(2)	32
$Mo(O)Cl_2(PMe_2Ph)_3$	blue	1.676(7)	30
$Re(N)Cl_2(PEt_2Ph)_3$	yellow	1.788(11)	34
$Re(N)Cl_2(PMe_2Ph)_3$	yellow	1.660(8)	35
$[Mo(O)(CN)_4(H_2O)](PPh_4)_2$	green	1.72(2)	36
$[Mo(O)(CN)_4(H_2O)](AsPh_4)_2$	blue	1.60(2)	36
$\{Mo(O)Cl(HBpz_3)\}_2(\mu\text{-}O)$	C_i form	1.779(6)	37
$\{Mo(O)Cl(HBpz_3)\}_2(\mu\text{-}O)$	C_2 form	1.671(4)	37
$[Mo(O)(OH)(dppe)_2]BF_4$		1.833(5)	38
$[Mo(O)Cl(dppe)_2]^+$		1.708(12)	39
$[Mo(O)Br_4]PPh_4$		1.726(14)	40
$[Mo(O)Cl_4]AsPh_4$		1.610(10)	11
$Mo(O)_2(ONCH_2CH_2CH_2CH_2CH_2)_2$		1.879(5), 1.701(5)	41
$Mo(O)_2(ONEt_2)_2$		1.714(2), 1.713(2)	42
$[W(O)Cl_2\{c\text{-}(NMeCH_2CH_2)_3\}]PF_6$	blue	1.719(18)	43
$[W(O)Cl_2\{c\text{-}(NMeCH_2CH_2)_3\}]PF_6$	green	1.893(20)	43
$[Ru(O)Cl\{c\text{-}(NMe)_4C_{10}H_{20}\}]ClO_4$		1.765(7)	44
$[Ru(O)Cl(py)_4]ClO_4$		1.862(8)	45
$[Nb(O)Cl_5](AsPh_4)_2$		1.967(6)	46
$[Nb(O)F_5]^{2-}$		1.75(2)	47
$[Mo(NNH_2)(triphos)(PPh_3)]Cl$[a]	M–N: 1.778(18)[a]	∢MNN: 178.1(15)[a]	48
$[Mo(NNH_2)(triphos)(PPh_3)]Cl$[a]	1.694(12)[a]	172.2(11)[a]	48

[a] Values reported are the molybdenum–nitrogen distance (Å) and the Mo–N–N angle (°).

isomerism is not solely a solid-state phenomenon, due to crystal packing forces or disorder. Further studies, both theoretical and experimental, clearly are warranted [49].

E

5.3.2 Bond Angles within the Multiply Bonded Ligand

The M—N—R or M—C—R bond angle in imido, hydrazido, alkylidyne, or vinylidene complexes indicates, in a simple valence bond picture, the hybridization of the central carbon or nitrogen atom. An angle of 180° implies sp hybridization while bending suggests a component of sp^2 hybridization and a lone pair at the nitrogen or carbon. All alkylidyne ligands are thought to be sp hybridized because of the reluctance of carbon to make less than four bonds (Section 2.1.2). All but two of the 51 structures of terminal alkylidyne ligands have bond angles between 171° and 180° (Fig. 5.4), the two exceptions being the imprecise structure of W($\equiv$CPh)(CO)$_4$I (162(4)°) [50] and the structure of Os($\equiv$CC$_6$H$_4$Me)Cl(CO)(PPh$_3$)$_2$ (164(2)°) [51]. The large angle in the structure of the latter complex may be related to its being the only study of a formally d^4 alkylidyne complex, all others being d^2 or d^0 when the alkylidyne ligand is taken to be CR^{3-}.

In this simple picture, imido ligands are more likely to be bent because nitrogen is much more willing to have a lone pair of electrons. However, M—N—R angles in imido compounds are found over only a slightly wider range than M—C—R angles, usually from 161° to 180° (Fig. 5.4). This suggests that almost all imido complexes have an essentially sp-hybridized nitrogen and close to an M–N triple bond.

Imido ligands are predicted to be bent (following the bonding discussion of Section 2.1.1) when the metal does not have two π-symmetry orbitals to bind to the nitrogen. This case occurs most often in octahedral complexes with two multiply bonded ligands sharing the three d_π symmetry orbitals. The classic, and at the moment, the *only* example of a strongly bent imido ligand ($<$Mo—N—C = 139.4(4)°) is found in the bisimido complex Mo(NPh)$_2$(S$_2$CNEt$_2$)$_2$ [52]. The bent imido ligand seems to have a Mo–N double bond of 1.789(4) Å while the other imido group forms a typical M–N triple bond (Mo—N = 1.754(4) Å, $<$ Mo—N—C = 169.4(4)°), thus accounting for the three molybdenum d_π orbitals. In contrast, the other bisimido and oxo–imido complexes—including three species isoelectronic with Mo(NPh)$_2$(S$_2$CNEt$_2$)$_2$—all have M—N—C bond angles greater than 153°. This implies a sharing of π-bonding among the ligands and indicates that there is a continuum between the double and triple bond

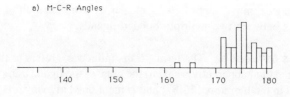

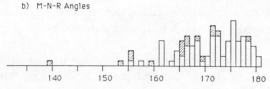

Figure 5.4 Distribution of (a) M—C—R angles in alkylidene complexes and (b) M—N—R angles in imido complexes. In (b), the shaded areas represent compounds with another multiply bonded ligand in addition to the imido group.

descriptions. The M—N—R angle appears to give a crude estimate of the bond order: Of the seven structurally characterized imido ligands with bond angles less than 161°, five are in compounds with another multiply bonded ligand. In the eight bisimido structures, the imido ligand with the smaller M—N—R angle has the longer metal–nitrogen distance where these differences are statistically significant.

Alkylidene ligands in which the two substituents (R) are the same, CR_2, usually have roughly equal M—C—R angles between 120° and 135°, close to the 120° angle of sp^2 hybridization. Ligands with one alkyl and one hydrogen substituent are usually unsymmetrical, with small M—C—H and larger M—C—R angles, the latter as high as 170°. The complexes with very large M—C—R angles (>151°) are invariably electron deficient (less than 18 electron) [53,54]; the reason for this distortion of the alkylidene ligand is thought to be an interaction between the α-hydrogen atom and the metal that increases the electron count (Section 2.5.3). For instance, the Ta—H distance of 2.042(5) Å and the Ta—C—H angle of 78.1(3)° in Ta(=CH^{t}Bu)Cp*(C$_2$H$_4$)(PMe$_3$) clearly indicate Ta—H bonding [55]. This interaction has been described as an agostic hydrogen atom or as hyperconjugation, the latter picture providing a simple rationalization for the short M—C bonds observed in highly distorted compounds. Since the C—H bond is weakened by the interaction, the distortion can also be observed spectroscopically by a reduced $^1J_{CH}$ coupling constant (Section 4.9, Fig. 4.2) and a low C—H stretching frequency (Section 4.5).

5.3.3 Distortions from Idealized Coordination Geometries

Complexes with a multiply bonded ligand, especially those with a metal–ligand triple bond, usually have structures that are sigificantly distorted from an idealized coordination geometry. Three related types of distortion are discussed in this section: structural trans influences, movement of cis ligands away from the multiply bonded group, and large angles between two multiply bonded ligands.

5.3.3.1 Trans Influence. The term "trans influence" refers to the ability of one ligand to lengthen (and apparently weaken) the bond to the ligand trans to it. It has been much discussed in the literature [56,57] and is mentioned in many reports of structures containing multiply bonded ligands because it can be dramatic. In the structure of $[Os(N)Cl_5]K_2$, for example, the four Os—Cl bonds cis to the nitride ligand are 2.362 ± 0.005 Å in length while the trans Os—Cl distance is 2.605(4) Å [58]. Lengthening of the trans ligand is almost invariably accompanied by a bending of the cis ligands away from the multiple bond (Section 5.3.3.2); in the osmium case the N—Os—Cl $_{cis}$ angles are 96.2 ± 1.3°.

Both steric and electronic explanations of the trans effect have been given by a number of workers. Bright and Ibers suggested in 1969 [58] that the structures of molecules with multiply bonded ligands "are largely determined by intramolecular packing," that is by the steric repulsion of neighboring ligands [see also ref. 60]. Because of the very short multiple bond distance, an undistorted structure would place the cis ligands very close to the multiply bonded ligand. Therefore the cis ligands bend away, pushing the trans ligand away from the metal. Viewed another way, this picture involves a fairly regular octahedron of ligands with the metal atom not located at the center but shifted toward the multiply bonded ligand, which generates both the trans influence and the angular distortions (Section 5.3.3.2). This is exactly the way the structures of polyoxoanions and solid-state metal oxides have been described: a close-packed array of oxide ions forms octahedral holes in which the metal may (ReO_3) or may not (MoO_3) be disposed symmetrically [61,62].

The steric argument, while quite powerful, does not appear to explain the detailed nature of the trans influence, such as the dependence of the distortion on d electron count (see below) or the lack of a strong dependence on the type of cis ligands. A number of electronic explanations for trans influences have been presented, both for the multiple bonds discussed here and for singly bonded ligands. These are based on overlap arguments that the metal orbitals will be more involved with the stronger metal–ligand bond, thus reducing the overlap and weakening the bond to the trans ligand [56,63,64]. While a review of these models is beyond the scope of this volume, it is clear that the general features of the observed distortions can be accounted for on the basis of electronic as well as steric effects.

The trans influence in a specific complex can be hard to quantify because of difficulties in determining the "unperturbed" bond length of the trans ligand [56]. However, if a complex contains the same ligand both cis and trans to a multiple bond (e. g. , Cl^- in the osmium nitride case above), then the difference in bond lengths is usually taken as a measure of the trans influence. The lengthening varies from very small to a few tenths of Ångstroms, frequently as much as 10% of the metal–ligand distance. In the extreme, square pyramidal structures can be thought of as the result of an infinite trans influence of the apical multiple bond. Although the number of useful structures is small, Shustorovich and coworkers have suggested [56] that the order of trans influence is nitrido > oxo > imido (>nitrosyl > carbonyl). Consistent with this order, nitrido complexes are frequently square pyramidal while imido compounds are usually octahedral. The alkylidyne ligand also appears to have a substantial trans influence [65–70], possibly as much as the oxo group. In Re($\equiv$C^{t}Bu)(=CH^{t}Bu)I$_2$(py)$_2$ the neopentylidyne ligand has a larger trans influence than the neopentylidene [65]. Nugent [71] and Maata [72] have noted that the trans influence in imido complexes is dependent on the oxidation state and electron count, being largest (~0.2Å) for 16-electron d^0 complexes and negligibly small for 18-electron d^2 species. The trans influence of the oxo ligand may also be affected by oxidation state (although there is considerable scatter in the data), but oxo and nitrido ligands always seem to exert some influence.

5.3.3.2 Angular Distortions.
Complexes with a single multiply bonded ligand almost always exhibit a $> 90°$ angle between the multiple bond and the cis ligands. This parameter, the average multiple bond metal–cis-ligand angle, seems to be related to the trans influence, as suggested by the steric argument above. Large angles have also been explained on electronic grounds: Bending away of the cis ligands causes a rehybridization of the metal and increased multiple bond overlap [73]. The cis angle is a useful parameter because it is directly determined, unlike the trans influence which requires an assumed "unperturbed" bond length. (However, the average cis angle can be misleading in complexes with very different types of ligands.) The large majority of octahedral monooxo and nitrido compounds (122 out of 143) have average cis angles between 95° and 104°. The five lowest values (88.3–90.8°) are for the three ruthenium structures with an unfavorable d^4 configuration and for molybdenum–oxo and –nitrido structures with exceptionally long multiple bonds (see Section 5.3.1.1). The values for imido complexes vary from 92° to 99° with a bias toward lower values for d^2 complexes, consistent with their smaller trans influence.

The conformations of compounds with more than one multiply bonded ligand are determined primarily by the electron count (Section 2.3.1). All d^0 complexes have a cis arrangement in the case of two multiply bonded ligands, and facial if there are three ligands. In contrast, octahedral d^2 complexes have a trans structure, with only one structurally characterized exception (**F**, [74], see also [75]).

F

The angle between cis multiply bonded ligands in d^0 complexes is almost always found in the 102–112° range, for both five- and six-coordinate structures. Dioxo compounds are by far the most common examples of these structures, but the pattern also holds fairly well for imido, alkylidyne, and alkylidene complexes. This angle is larger than the 90° expected in an octahedral structure but smaller than the 120° expected for trigonal bipyramidal coordination. This and recent theoretical work [76,77 see also 78] suggest that the observed angle is the preferred geometry of the ML_2 fragment, rather than simply being due to repulsion between the multiply bonded ligands (a VSEPR-type argument [79]). A comparison of the structures of the gas-phase tetrahedral molecules CrO_2Cl_2 and SO_2Cl_2 (<O—Cr—O = 105 ± 4, <O—S—O = 120 ± 5 [80]) illustrates this point. It has also been noted [81] that the O—M—O angle in trans osmium–dioxo–alkoxide complexes often deviates from 180°: In more than half of the reported structures this angle falls in the range of 161-167°.

5.4 STRUCTURAL TABLES

Tables 5.2–5.4 contain data from structural studies of almost a thousand complexes with metal–oxygen, metal–nitrogen, and metal–carbon multiple bonds. Complexes of metals in the cobalt, nickel, and copper triads have not been included; to our knowledge, there are at this time no structures of complexes of these metals with terminal oxo, imido, nitrido, or alkylidyne ligands. An attempt has been made to make the tables complete through the first half of 1987, searching the literature through the *Cambridge Crystallographic Database* and *Chemical Abstracts*. To limit the size of the tables, coverage has been restricted to structures that clearly contain a multiple bond, erring on the side of exclusion. Thus the table with metal–carbon multiple bonds (Table 5.4) does not include highly conjugated organic ligands or Fischer carbene compounds. The metal–oxygen table (Table 5.2) contains only structures with terminal oxo ligands; polyoxoanions and solid-state metal oxides are not included. Disordered structures and those with large (or unreported) standard deviations have often been omitted.

The order of the compounds in each table is by metal (Ti, V, Nb, Ta, Cr, Mo, ...) and by ligand type. Compounds with metal–nitrogen bonds are ordered imido, hydrazido, nitrido; similarly, alkylidene complexes are followed by vinylidenes, and subsequently by alkylidynes. Compounds are listed by oxidation state (starting with d^0 species) and then by coordination number. The large number of molybdenum(V)–oxo compounds are further separated into monomeric species, then $(MoO)_2(\mu\text{-}O)$, $(MoO)_2(\mu\text{-}O)_2$, and finally other dimeric complexes. Multiply bonded ligands with similar substituents are grouped together: for example, d^2 six-coordinate rhenium aryl–imido complexes are separate from alkyl–imido species. Within these categories, compounds are grouped by the nature of the ancillary ligands, and "simpler" compounds are given before more complex ones.

The data in the tables are presented in six columns, starting with the formula of the compound and the metal–ligand multiple bond distance. For alkylidyne, vinylidene, imido, and hydrazido complexes, the angle within the multiply bonded ligand is given, either M—C—R or M—N—R. For alkylidene ligands of the type M=C(H)R the M—C—R angle is given; for M=CR$_2$ complexes the average of the two angles is listed. The listings for metal–oxo and –nitrido complexes give the average oxo/nitrido–metal–cis-ligand angle when the value is available and when it is meaningful. In structures with more than one multiply bonding ligand (in particular dioxo complexes) the angle between the two ligands is listed. Almost all of the structures have been determined by single crystal X-ray crystallography (the exceptions are noted) and the values in parentheses are estimated standard deviations. Occasionally chemically equivalent distances and/or angles that are within three standard deviations have been averaged to save space. The last three columns give the d electron count, the coordination number, and the reference. The coordination number is defined in the usual fashion, with the following special cases: cyclopentadienyl groups are counted as tridentate ligands, peroxide (O_2^{2-}) and the related groups are counted as bidentate ligands, and olefins and acetylenes are counted as monodentate groups. Only metal–metal bonds without bridging ligands are counted.

TABLE 5.2 Metal–Oxo Structures

Complex	M=O (Å)	∠ O–M–L	d e−	CN	Ref.
[Ti(O)Cl$_4$][Et$_4$N]$_2$	1.79	102	0	5	6
Ti(O)(Et$_8$porphyrin)	1.613(5)	105.2±1.1	0	5	82
Ti(O)(α,γ-Me$_2$-α,γ-H$_2$-Et$_8$porphyrin)	1.619(4)	105.9±0.9	0	5	83
Ti(O)(phthalocyaninato)	1.650(4)	107.6±1.6	0	5	84
	1.626(7)	107.9±1.0			
V(O)Cl$_3$[a]	1.56±4[a]	108±2[a]	0	4	80
[V(O)Br$_4$]PPh$_4$	1.552(9)	103.1(1)	0	5	7
V(O)Cl$_3$(NCMe)	1.552(2)		0	5	85
V(O)Cl$_3$(NCPh)	1.557(5)		0	5	86
V(O)Cl$_3$(NCCH$_2$Ph)	1.592(6)		0	5	86
{V(O)(O$_2$NC$_{13}$H$_{18}$)}$_2$(μ–O)•0.5 dioxane	1.55,1.56(1)		0	5&6	88
{V(O)F(μ–F)$_2$}$_\infty$	1.57(1)		0	6	89
[V(O)F$_5$](enH$_2$)	1.54(1)	98.2±2.8	0	6	90
[V(O)F$_4$(H$_2$O)]·0.5(enH$_2$)	1.577(3)	99.1±1.1	0	6	91
V(O)(salen)(ClO$_4$) (trans)	1.579(3)	99.0±5.2	0	6	92
[V(O)(H$_2$O)(salen)]$_2$[Cu$_2$Cl$_4$] (trans)	1.590(5)		0	6	93
V(O)(OCHMe$_2$)(8-hydroxyquinolinato)$_2$	1.600(2)		0	6	94
{V(O)(quinolato)$_2$}$_2$(μ–O)	1.587(6)		0	6	95
[{V(O)(C$_2$O$_4$)}$_4$(μ–O)$_4$(μ–H$_2$O)$_2$]K$_4$	1.62(1)	100.6±4.0	0	6	96
V(O)Cp*Cl$_2$	1.576(8)		0	6	97
{V(O)(ONEt$_2$)$_2$}$_2$(μ–O)	1.586,1.599(3)		0	6	98
{V(O)F(bipy)}$_2$(μ–O)$_2$	1.618(8)		0	6	99

TABLE 5.2 Metal–Oxo Structures *(Cont.)*

Complex	M=O (Å)	< O-M-L	d e-	CN	Ref.
[V(O)(O$_2$)$_2$(C$_2$O$_4$)]K$_3$	1.620(3)		0	7	100
[V(O)(O$_2$)$_2$(bipy)](NH$_4$)	1.619(3)		0	7	101
V(O)(O$_2$)(2-O$_2$Cpy)(bipy)	1.604(5)		0	7	102
V(O)(O$_2$)(2-O$_2$Cpy)(H$_2$O)$_2$	1.583(2)		0	7	103
V(O)(O$_2$)[2,6-(O$_2$C)$_2$py](H$_2$O)	1.579(2)		0	7	104
V(O)(O$_2$)(O$_3$NC$_4$H$_5$O)	1.587(3)		0	7	105
V(O)(H$_2$O)(OO^{t}Bu)[2,6-(O$_2$C)$_2$py]	1.574(3)	98.0±6.8	0	7	106
V(O)(ONH$_2$)[2,6-(O$_2$C)$_2$py]	1.587(3)		0	7	107
V(O)(S$_2$CNEt$_2$)$_3$	1.65(2)		0	7	108
V(O)(NO$_3$)$_3$(MeCN)	1.55(2)	97.5±3.5	0	7	109
[{V(O)N(CH$_2$CO$_2$)$_3$}$_2$(μ-O)](NH$_4$)$_3$	1.607(6)		0.5	6	110
[{V(O)(O$_2$NC$_{11}$H$_{12}$O$_2$)}$_2$(μ-O)]Na	1.618(4)		0.5	6	111
[{V(O)(O$_2$N$_2$C$_{10}$H$_{10}$O$_2$)}$_2$(μ-O)]H	1.592(5)		0.5	6	112
V(O)(acac)$_2$	1.56(1)	106.2±2.0	1	5	113
V(O)(MeCOCHCOPh)$_2$	1.612(10)	105.8±1.8	1	5	114
[V(O)(catacholato)$_2$]K$_2$	1.616(4)		1	5	115
{V(O)(acac)}$_2$(μ-OMe)$_2$	1.587(3)	108.0±2.0	1	5	116
V(O)(S$_2$CNEt$_2$)$_2$	1.591(4)	108.3±1.9	1	5	117
[V(O)(SCH$_2$CH$_2$S)$_2$]Na(NMe$_4$)	1.625(2)	106.3±4.0	1	5	118
[V(O)(SCH$_2$CH$_2$S)(S$_2$VS$_2$)](Et$_4$N)$_2$Na	1.626(3)		1	5	119
V(O)(2-Me-8-quinolato)$_2$	1.600(8)	~tbp	1	5	120
V(O)(OC$_6$H$_4$CHNC$_6$H$_4$Cl)$_2$	1.615(8)	~tbp	1	5	121
V(O)Cl$_2$(NMe$_3$)$_2$	1.59(2)	tbp	1	5	122
V(O)Cl$_2$(OPPh$_3$)$_2$	1.584(5)	104.6±3.1	1	5	123
V(O)Cl$_2$[OC(NMe$_2$)$_2$]$_2$	1.61(3)	105.2±0.6	1	5	124
V(O)(acacen)	1.585(7)	106.8±3.7	1	5	125
[V(O)(salen)Na]BPh$_4$	1.585(3)	107.8±7.4	1	5	126
[V(O)(Ph$_2$COCO$_2$)$_2$][Et$_4$N]Na	1.584(11)	108.9±5.5	1	5	127
[{V(O)}$_2$(tartrate)$_2$]Na$_4$	1.619(7)	106.4±3.4	1	5	128
[{V(O)}$_2$(tartrate)$_2$][Et$_4$N]$_4$	1.599(3)	106.2±2.1	1	5	129
[{V(O)}$_2$(Me-tartrate)$_2$]Na$_4$	1.613(4)	106.4±1.7	1	5	130
[{V(O)}$_2$(Me$_2$-tartrate)$_2$]Na$_4$	1.623(3)	101.5±2.6	1	5	131
V(O)(O$_4$C$_{16}$H$_{22}$N$_2$)	1.600(5)	107.0±1.4	1	5	135
V(O)(O$_4$C$_{16}$H$_{20}$N$_2$Cu)	1.598(9)	107.2±1.5	1	5	135
V(O)(O$_4$C$_{17}$H$_{24}$N$_2$)	1.590(3)	106.4±0.7	1	5	135

TABLE 5.2 Metal–Oxo Structures *(Cont.)*

Complex		M=O (Å)	< O–M–L	d e–	CN	Ref.
V(0)(O$_4$C$_{18}$H$_{13}$N$_2$O$_2$Cu)		1.576(9)	103.5±3.1	1	5	132
V(0)(O$_4$C$_{26}$H$_{26}$N$_2$)		1.625(4)	105.6±1.3	1	5	134
V(0)(O$_4$C$_{26}$H$_{24}$N$_2$Ni)		1.591(4)	108.1±0.6	1	5	133
V(0)(N$_4$C$_{22}$H$_{18}$)		1.61(1)	108.2±2.6	1	5	136
V(0)(phthalocyaninato)		1.580(3)		1	5	137
V(0)(Ph$_4$porphyrin)		1.625(16)	104.9(1)	1	5	138
V(0)(etioporphyrin)		1.599(6)	103.8±1.1	1	5	138
{V(0)(etioporphyrin)}$_2${C$_6$H$_4$(OH)$_2$}^{b}		1.614(9)	104.2±1.3	1	5	138
V(0)(deoxophylloerythro-etioporphyrin)		1.62(1)	103.5±3.5	1	5	139
[V(0)(H$_2$O)$_5$]SO$_4$		1.591(5)	98.0±1.7	1	6	140
[V(0)(OSMe$_2$)$_5$](ClO$_4$)$_2$		1.591(9)	97.7±1.4	1	6	141
[V(0)F(C$_2$O$_4$)$_2$]Na$_3$		1.606(3)		1	6	142
[{V(0)(H$_2$O)F$_2$}$_2$(μ-F$_2$)](Et$_4$N)$_2$		1.594(4)	100.4±1.7	1	6	143
V(0)F$_2$(H$_2$O)(phen)		1.623(4)		1	6	144
V(0)(H$_2$O)$_2$[2,6-(O$_2$C)$_2$py]		1.591(11)		1	6	145
{V(0)(acac)$_2$}$_2$(dioxane) (trans)		1.62	107.2	1	6	146
[V(0)(H$_2$O)(O$_2$CCH$_2$CO$_2$)$_2$]$^{2-}$ (trans)		1.589(4)	98.7±0.4	1	6	147
[V(0)(H$_2$O)(C$_2$O$_4$)$_2$](NH$_4$)$_2$ (cis)		1.594(3)	98.8±6.7	1	6	148
V(0)(H$_2$O)(O$_2$NC$_{11}$H$_{12}$O$_2$) (cis)		1.602(3)		1	6	111
V(0)(H$_2$O)(N$_2$O$_2$C$_{10}$H$_{10}$O$_2$) (cis)		1.60(1)	100.9±6.3	1	6	149
V(0)(H$_2$O)(O$_4$macrocycle...) (cis)		1.589(9)		1	6	150
V(0)(bipy)(O$_2$CCH$_2$NCH$_2$CO$_2$)		1.596(2)	100.3±2.0	1	6	151
V(0)(O$_3$N$_2$C$_{17}$H$_{15}$O)		1.606(1)		1	6	152
V(0)(N$_3$O$_2$C$_{22}$H$_{27}$O$_2$)		1.617(3)		1	6	153
V(0)(O$_3$N$_2$-chelate)		1.605(8)		1	6	154
[{V(0)(H$_2$O)F}$_2$(μ-F)$_2$](NMe$_4$)$_2$		1.607(5)		1	6	155
[{V(0)[c-N$_3$(CH$_2$)$_6$]}$_2$(μ-OH)$_2$]Br$_2$		1.603(5)		1	6	156
{V(0)}$_2$(μ-O$_2$CCF$_3$)$_8$(thf)$_6$(H$_2$O)$_2$Na$_4$		1.591(2)		1	6	157
{V(0)(O$_2$C$_6$H$_2^t$Bu$_2$)$_2$}$_2$		1.581(4)		1	6	158
{V(0)}$_2$V(μ-PhCO$_2$)$_6$(μ$_3$-O)thf		1.575(6)	98.1±2.2	1	6	159
{V(0)Cl$_2$}$_2$(μ-OC$_6$H$_7$N)$_3$		1.586(2)	100.7±4.1	1	6	160

TABLE 5.2 Metal–Oxo Structures *(Cont.)*

Complex	M=O (Å)		< O–M–O	d e−	CN	Ref.
VO_4^{3-} c	1.71 c			0	4	161
$[V(O)_2(\mu-O)K]_\infty$	1.66(2)		107(1)	0	4	162
$[V(O)_2(\mu-O)Na]_\infty$	1.631(2),	1.653(2)	109.1(1)	0	4	162
$[V(O)_2Cl_2]PPh_3Me$	1.604(6),	1.559(6)	107.1(3)	0	4	163
$[V(O)_2(H_2O)(\mu-O)K]_\infty$	1.63(2),	1.67(2)	106(1)	0	5	162
$[V(O)_2F(C_2O_4)](NH_4)_2$	1.614(4),	1.617(4)	107.5(2)	0	5	162
$[V(O)_2(2,6-(O_2C)_2py)]Cs$	1.610(6),	1.615(6)	109.9(3)	0	5	164
$V(O)_2(2-O_2Cpy)(hmpa)$	1.600(5),	1.606(5)	108.7(3)	0	5	165
$\{[V(O)_2(O_3PMe)]K\}_\infty$	1.593(2)		106.8(2)	0	5	166
$[V(O)_2(NO_2C_{12}H_9N_2)]NH_4$	1.609(3),	1.654(3)	107.5(2)	0	5	167
$[V(O)_2(NHC(SMe)NNCHC_6H_4O)]NH_4$	1.625(5),	1.659(5)	107.2(8)	0	5	168
$V(O)_2[NHC(SMe)NHNCHC_6H_4O]$	1.606(3),	1.629(4)	108.6(4)	0	5	169
$[V(O)_2(ON_2C_{11}H_7NO)]PPh_4$ (~tbp)	1.615(2)		109.3(1)	0	5	170
$[V(O)_2(ON_2C_9H_5NOS)]PPh_4$	1.613(2),	1.622(3)	109.1(1)	0	5	171
$[V(O)_2(ON_2C_9H_9NS)]NH_4$	1.625(5),	1.659(5)	107.2(8)	0	5	172
$[V(O)_2(ON_2C_9H_9NOS)]NH_4$	1.579(13),	1.676(13)	106.4	0	5	173
$\{V(O)_2(O_2C_8H_{10}NO)\}_2$	1.606(4),	1.669(4)	110.7(2)	0	5	174
$\{[V(O)_2F_2(\mu-F)]K_2\}_\infty$	1.636(2)		102.8(2)	0	6	175
$V(O)_2(acac)(phen)$	1.613(4),	1.670(5)	105.5(2)	0	6	176
$[\{V(O)_2F_2\}_2(\mu-F)_2]K_4$	1.636(2)		102.8(2)	0	6	162
$[V(O)_2(C_2O_4)_2](NH_4)_3$	1.635(2),	1.648(2)	103.8(1)	0	6	177
$[V(O)_2(C_2O_4)_2]K_3$	1.628(2),	1.639(2)	104.4(1)	0	6	178
$[V(O)_2(8-quinolato)_2]Na$	1.628(2),	1.647(4)	105.7(2)	0	6	179
$[V(O)_2(8-quinolato)_2]Bu_4N$	1.62(1)		106.5(5)	0	6	179
$[V(O)_2(8-quinolato)_2]Bu_4N$	1.64(1)		105.4(5)	0	6	180
$[V(O)_2(edta)]Na_3$	1.639(2),	1.657(1)	107.0(1)	0	6	181
$[V(O)_2(edtaH_2)]NH_4$	1.623(2),	1.657(2)	107.1(1)	0	6	182
$[V(O)_2(O_2NC_{11}H_{12}O_2)]Li$	1.619(2)		105.7(1)	0	6	111

Complex	M=O (Å)	< O–M–L	d e−	CN	Ref.
$[Nb(O)Cl_4]^-[NbCl_4(diars)_2]^+$	1.70(2)	99.0±0.3	0	5 d	8
$[Nb(O)(NCS)_5](AsPh_4)_2$	1.70(4)		0	6	183
$[Nb(O)F_5]^{2-}[N_2H_6]^{2+}$	1.75(2)	101 ±1	0	6	47
$[Nb(O)Cl_5](AsPh_4)_2$	1.967(6)	93.4±3.5	0	6	46

TABLE 5.2 Metal–Oxo Structures *(Cont.)*

Complex	M=O (Å)	<O–M–L	d e-	CN	Ref.
[Nb(O)Cl$_4$(MeCN)]PPh$_3$Me	1.688(2)	97.7±1.7	0	6	184
[Nb(O)Cl$_4$(H$_2$O)]PPh$_4$	1.74(1)	98.3(1)	0	6	185
[Nb(O)Cl$_4$(thf)]$^-$[CpNbCl(MeCN)$_4$]$^+$ (tr)	1.665(11)	97.8(4)	0	6	186
[Nb(O)Cl$_4$(OPCl$_2$O)](PPh$_4$)$_2$	1.77(1)	95.6±1.1	0	6	185
Nb(O)Cl$_3$(MeCN)$_2$ (mer)	1.68(2)	97.5±4.6	0	6	187
Nb(O)Cl$_3$(hmpa)$_2$ (mer)	1.692(5)	96.1±0.9	0	6	188
Nb(O)Cl$_3$[CF$_3$C(O)CHC(O)(c-C$_4$H$_3$S)]	1.704(3)		0	6	189
Nb(O)Cl$_2$(OEt)bipy	1.710(35)		0	6	183
Nb(O)F(Ph$_4$porphyrin)	1.749(3)		0	6	190
Nb(O)(O$_2$CMe)(Ph$_4$porphyrin)	1.720(6)		0	7	191
[Nb(O)(C$_2$O$_4$)$_3$](NH$_4$)$_3$	1.710(10)		0	7	192
Nb(O)(S$_2$CNEt$_2$)$_3$	1.74(1)		0	7	108
Nb(O)Cp$_2$[C$_7$H$_5$(CF$_3$)$_2$]	1.63(3)		0	8	193
Ta(O)[N(CHMe$_2$)$_2$]$_3$	1.725(7)		0	4	194
Cr(O)(O$_2$)$_2$(py)	1.576(18)		0	6	195
Cr(O)(O$_2$)$_2$(bipy)	1.57(2)		0	7	196
Cr(O)(O$_2$)$_2$(phen)	1.56(2)		0	7	197
[Cr(O)Cl$_4$]AsPh$_4$	1.519(12)	104.5(1)	1	5	9
[Cr(O)(O$_2$CCOMeEt)$_2$]K	1.554(14)	108.0±8.2	1	5	198
[Cr(O)(Me$_2$salen)]CF$_3$SO$_3$	1.545(2)	106.4±4.3	1	5	199
[Cr(O)(Me$_2$salen)(ONC$_5$H$_5$)]CF$_3$SO$_3$	1.554(4)	97.7±3.9	1	6	199
[Cr(O)Cp*](μ-O)$_2$	1.594(3)		1	6	200
Cr(O)(tol$_4$porphyrin)	1.572(6)	103.3±2.7	2	5	201
Cr(O)(Ph$_4$porphyrin)	1.62(2)		2	5	202

Complex	M=O (Å)	<O–M–O	d e-	CN	Ref.
CrO$_4^{2-}$ c	1.65^c		0	4	161
[CrO$_4$]K$_2$	1.636 – 1.651(3)		0	4	203
Cr(O)$_2$(OC$_{15}$H$_{25}$)$_2$	1.57(1)	107.2(8)	0	4	204
CrO$_2$Cl$_2$ ^{a}	1.57±3^a	105±4^a	0	4	80
Cr(O)$_2$(OSiPh$_3$)$_2$	1.514(13), 1.568(12)	108.9(7)	0	4	205

163

TABLE 5.2 Metal–Oxo Structures *(Cont.)*

Complex	M=O (Å)	∠ O–M–L	d e−	CN	Ref.
Mo(O)(μ_3-S)$_3$(CuPPh$_3$)$_3$(μ_3-Cl)	1.769(10)		0	4	206
[Mo(O)(μ_3-S)$_3$(CuCl)$_3$]$^{2-}$	1.693(4)	111.5±1.0	0	4	207
Mo(O)F$_4$[a]	1.650(3)[a]	103.8(2)[a]	0	5	208
Mo(O)Cl$_4$[a]	1.658(2)[a]	102.8(3)[a]	0	5	10
Mo(O)[OC(CF$_3$)$_3$]$_4$	1.60(3)		0	5	209
Mo(O)(S$_2$)(S$_2$FeCl$_2$)](PPh$_4$)$_2$	1.671(5)	107.6±1.5	0–1	5	210
[Mo(O)F$_5$]K	1.66(2)	93.8±3.3	0	6	211
Mo(O)(O^{i}Pr)$_3$(μ-O^{i}Pr)$_2$Mo...	1.691(2)	98.4±4.8	0	6	212
{Mo(O)(O$_2$C$_2$Me$_4$)(OC$_2$Me$_4$OH)}$_2$(μ-O)	1.67(1)	99.5±2.0	0	6	213
{Mo(O)(O$_2$C$_6$H$_2$^{t}Bu$_2$)$_2$}$_2$	1.668(3)	103.4±5.1	0	6	214
{Mo(O)(OEt)[μ-(OCH$_2$)$_3$CMe]}$_2$	1.689(3)	100.3±8.5	0	6	215
[{Mo(O)(O$_2$)$_2$(H$_2$O)}$_2$(μ-O)](pyH)$_2$	1.674(7)	100.9±1.5	0	7	216
Mo(O)(O$_2$)$_2$(H$_2$O)$_2$	1.647(5)		0	7	217
[Mo(O)(O$_2$)$_2$(C$_2$O$_4$)]K$_2$	1.678(14)	100.7±6.1	0	7	218
Mo(O)(O$_2$)$_2$(MeCHOHCONMe$_2$)	1.671(5)	101.4±7.0	0	7	219
Mo(O)(O$_2$)$_2$(O$_2$C$_6$H$_6$O$_5$)]K$_2$	1.655(8)		0	7	220
[{Mo(O)(O$_2$)$_2$(μ-OOH)}$_2$](pyH)$_2$	1.669(6)	102.3±3.5	0	7	216
Mo(O)(O$_2$)$_2$(hmpa)(H$_2$O)	1.622(5)	100.1±5.7	0	7	221
Mo(O)(O$_2$)$_2$(hmpa)(py)	1.658(12)	101.2±5.4	0	7	221
Mo(O)(O$_2$)$_2$(bipy)	1.682(4)	101.2±8.8	0	7	222
[Mo(O)(O$_2$)$_2${2-(O$_2$C)py)}]$^-$	1.685(2)	100.8±8.1	0	7	223
[{Mo(O)(S$_2$)$_2$}$_2$(μ-S)](Et$_4$N)$_2$	1.676(6)	102.1(10)	0	7	224
[Mo(O)(S$_2$)$_2$(C$_2$O$_3$S)]Cs$_2$	1.69(3)	97.1±1.2	0	7	225
Mo(O)(O$_2$)(ONPhCPhO)$_2$	1.733(2)		0	7	226
Mo(O)(O$_2$)Cl[2-(O$_2$C)py](hmpa)	1.663(3)		0	7	227
[Mo(O)(O$_2$)F{2,6-(O$_2$C)$_2$py}]NH$_4$	1.661(5)	97.0±5.6	0	7	228
[{Mo(O)(O$_2$)[2,6-(O$_2$C)$_2$py]}$_2$(μ-F)]Et$_4$N	1.659(3)	99.1±5.5	0	7	229
[Mo(O)(O$_2$)F$_4$]K$_2$	1.64(2)		0	7	230
[Mo(O)(O$_2$)(H$_2$O){2,6-(O$_2$C)$_2$py}]$^-$	1.670(5)	98.4±6.3	0	7	223
{Mo(O)(O$_2$)(MeOH)(μ-OCH$_2$CHMeO)}$_2$	1.665(2)		0	7	231
Mo(O)(ONPh)[2,6-(O$_2$C)$_2$py](hmpa)	1.669(4)		0	7	232
Mo(O)(ONC$_5$H$_{10}$)$_2$(O$_2$C$_6$H$_4$)	1.688(6)	95.1±7.7	0	7	233
Mo(O)(ONHMe)$_2$(ONMeCSNH)	1.677(8)		0	7	234
Mo(O)(ONHMe)$_2$[ONMeC(S)S]	1.689(6)	96.2±9.6	0	7	235
[Mo(O)(ONHMe)(ONMeCMeNH)$_2$]ClO$_4$	1.677(10)	96.5±13	0	7	236

TABLE 5.2 Metal–Oxo Structures *(Cont.)*

Complex	M=O (Å)	< O–M–L	d e–	CN	Ref.
Mo(O)(ONHMe)(ONMeCSNH)(ONMeCSNH$_2$)	1.691(3)		0	7	234
[{Mo(O)(ONEt$_2$)$_2$(C$_2$O$_4$)}$_2$(μ–O)$_2$]$^{2-}$	1.714(4)		0	7	237
Mo(O)(ONMe$_2$)(ONCOPh)(ONHCOPh)	1.685(3)	98.6±7.4	0	7	238
[Mo(O)(S$_2$CNEt$_2$)$_3$]$^+$	1.684(6)	97.9±9.0	0	7	239
Mo(O)Cl$_2$(S$_2$CNEt$_2$)$_2$	1.701(4)	94.6±2.4	0	7	240
Mo(O)Br$_2$(S$_2$CNEt$_2$)$_2$	1.656(5)	94.7±1.5	0	7	240
Mo(O)(S$_2$CNMe$_2$)$_2$(OCPhNNCOPh)	1.685(10)	95.3±7.7	0–2	7	241
{[Mo(O)Cl(O^{i}Pr)]$_2$(μ–O)(μ–O^{i}Pr)}$_2$	1.625(16)	98.2±9.2	0.5	6	242
[Mo(O)Cl$_4$]AsPh$_4$	1.610(10)	105.2(1)	1	5	11
[Mo(O)Br$_4$]PPh$_4$	1.726(14)	102.0(1)	1	5	40
Mo(O)Cl$_3$(SPPh$_3$)	1.647(3)	106.1±5.2	1	5	243
[Mo(O)(SPh)$_4$]AsPh$_4$	1.669(9)	109.9±1.4	1	5	244
[Mo(O)(SCH$_2$CH$_2$CH$_2$S)$_2$]PPh$_4$	1.667(5)		1	5	245
[Mo(O)(S$_2$C$_6$H$_4$)$_2$]PPh$_4$	1.668(3)	108.1(1)	1	5	246
Mo(O)(S$_2$CNMe$_2$)[μ–NNC(S)SEt]$_2$MoS$_2$CNMe$_2$	1.672(8)		1	5	247
[Mo(O)Br$_5$](NH$_4$)$_2$	1.86(3)		1	6	248
[Mo(O)Cl$_4$(thf)]	1.655(7)	97.3±0.5	1	6	249
[Mo(O)Cl$_4$(H$_2$O)]$^-$	1.657(9)	97.4±1.5	1	6	250
[Mo(O)Cl$_4$(H$_2$O)]AsPh$_4$	1.672(15)	99.0(9)	1	6	251
[Mo(O)Br$_4$(H$_2$O)]Et$_4$N	1.65(1)	97.7±0.5	1	6	252
[Mo(O)I$_4$(H$_2$O)]Et$_4$N	1.65(1)	97.8±0.3	1	6	252
Mo(O)Cl$_3$(OPPh$_3$)$_2$ (cis, mer)	1.662(13)	95.6±2.0	1	6	253
Mo(O)Cl$_3$(hmpa)$_2$ (cis, mer)	1.669(5)	95.9±0.8	1	6	254
[Mo(O)Cl$_3$(O$_2$PCl$_2$)]$_2$(AsPh$_4$)$_2$	1.662(4)	96.6±2.1	1	6	255
[Mo(O)Cl$_3$(8–hydroxyquinolato)]$^-$	1.673(6)		1	6	256
Mo(O)Cl$_3$(NOC$_{13}$H$_{11}$N$_2$O)	1.70(1)	96.4±7.4	1	6	257
Mo(O)(SPh)$_2$[HB(Me$_2$pz)$_3$]	1.676(4)	96.3±6.4	1	6	258
[Mo(O)Cl$_2$(OC$_6$H$_4$CHNC$_6$H$_4$O)]PPh$_4$	1.673(3)	99.1±4.9	1	6	259
Mo(O)(Ph$_4$porphyrin)Cl (trans)	1.714(3)	100.6(1)	1	6	260
[Mo(O)(salen)(MeOH)]Br (trans)	1.666(8)	100.6±4.4	1	6	261
Mo(O)Cl(8–mercaptoquinolinato)$_2$	1.716(4)	98.0±8.8	1	6	262
{Mo(O)[OCMeCHCMeNNC(OMe)S]}$_2$(μ–O)	1.664(8)	107.3±4.4	1	5	263
{Mo(O)(SC$_2$H$_4$OC$_2$H$_4$S)}$_2$(μ–O)	1.738(10)		1	5	264
{Mo(O)(SC$_2$H$_4$SC$_2$H$_4$S)}$_2$(μ–O)	1.642(9)		1	5	264
{Mo(O)[NMe(C$_2$H$_4$S)$_2$]}$_2$(μ–O) (tbp)	1.667(8)		1	5	265

TABLE 5.2 Metal–Oxo Structures *(Cont.)*

Complex	M=O (Å)	< O-M-L	d e-	CN	Ref.
{Mo(O)Cl$_2$(MeOC$_2$H$_4$OMe)}$_2$(μ-O)	1.655(3)	97.5±7.0	1	6	266
{Mo(O)Cl$_2$py$_2$}$_2$(μ-O)	1.675(4)	94.1±6.4	1	6	267
{Mo(O)Cl(HBpz$_3$)}$_2$(μ-O) (>isomers)	1.671(4)	97.7±7.2	1	6	37
{Mo(O)Cl(HBpz$_3$)}$_2$(μ-O)	1.779(6)	96.5±6.0	1	6	37
{Mo(O)(N$_2$S$_2$C$_8$H$_{18}$)}$_2$(μ-O)	1.665(7)	98.7±9.3	1	6	268
{Mo(O)(S$_2$CN^{i}Pr$_2$)$_2$}$_2$(μ-O)	1.671(12)	99.9±11	1	6	269
{Mo(O)I(S$_2$CNEt$_2$)(thf)}$_2$(μ-O)	1.664(4)	99.7±3.1	1	6	270
{Mo(O)(2-Spy)$_2$}$_2$(μ-O)	1.673(4)	101.4±6.0	1	6	271
{Mo(O)(S$_2$CPh)$_2$}$_2$(μ-O)	1.675(10)	101.5±9.6	1	6	272
{Mo(O)(S$_2$COEt)$_2$}$_2$(μ-O)	1.647(29)		1	6	273
{Mo(O)[S$_2$P(OEt)$_2$]$_2$}$_2$(μ-O)	1.647(14)	99.0±8.5	1	6	274
{Mo(O)CpI}$_2$(μ-O)	1.684(6)		1	6	275
{Mo(O)Cp(SPh)}$_2$(μ-O)	1.688(9)		1	6	276
{Mo(O)(Ph$_4$porphyrin)}$_2$(μ-O)	1.707(3)	92.3±1.5	1	6	277
[{Mo(O)Cl$_2$}$_2$(μ-O)$_2$](AsPh$_4$)$_2$	1.69(1)	106.7±4.9	1	5	278
{Mo(O)[NH$_2$CH(CO$_2$Et)CH$_2$S]}$_2$(μ-O)$_2$	1.663(19)	106.9±11	1	5	279
{Mo(O)(SPh)$_2$}$_2$(μ-O)$_2$	1.672(8)	109.2±9.0	1	5	280
{Mo(O)(S$_2$CNEt$_2$)}$_2$(μ-O)$_2$	1.679(2)		1	5	281
Mo(O)(S$_2$CNEt$_2$)(μ-NNCSOMe)$_2$Mo(S$_2$CNEt$_2$)$_2$	1.69(1)		1?	5	263
{Mo(O)[S$_2$P(OEt)$_2$]}$_2$(μ-O)(μ-NH)	1.678(3)		1	5	282
{Mo(O)}$_2$(μ-O)$_2$(μ-S$_2$C$_9$H$_{22}$N$_2$)$_2$	1.676(10)	109.6±5.7	1	5	268
[{Mo(O)(S$_2$C$_4$O$_2$)}$_2$(μ-O)(μ-S)](^{n}Bu$_4$N)$_2$	1.675(7)		1	5	283
{Mo(O)(S$_2$CN^{n}Pr$_2$)}$_2$(μ-O)(μ-S)	1.665(1)	108.4±2.2	1	5	284
{Mo(O)(S$_2$CNH$_2$)}$_2$(μ-S)$_2$	1.677(3)	106.9±3.4	1	5	285
{Mo(O)(S$_2$CN^{n}Bu$_2$)}$_2$(μ-S)$_2$	1.655(10)	107.7±5.3	1	5	286
[{Mo(O)(S$_2$)}$_2$(μ-S)$_2$](Me$_4$N)$_2$	1.675(11)	109.7±3.7	1	5	287
[{Mo(O)(S$_2$)}$_2$(μ-S)$_2$](Et$_4$N)$_2$	1.683(3)	99.3±1.5	1	5	288
[Mo(O)(S$_2$)(μ-S)$_2$Mo(O)(S$_2$SO$_2$)](PPh$_4$)$_2$	1.690(11)		1	5	289
{Mo(O)[SSC(CO$_2$Me)C(CO$_2$Me)]}$_2$(μ-S)$_2$	1.676(6)		1	5	290
[{Mo(O)[S$_2$CC(CN)$_2$]}$_2$(μ-S)$_2$]$^{2-}$	1.664(7)	107.5±2.7	1	5	291
[{Mo(O)}$_2$(μ-S)$_2$(As$_4$S$_{12}$)](PPh$_4$)$_2$	1.703(14)		1	5	292
{Mo(O)(NSC$_6$H$_7$N)}$_2$(μ-O)(μ-S)(MeOH)	1.664, 1.670(6)		1	5&6	293
{Mo(O)[S$_2$P(O^{i}Pr)$_2$]}$_2$(μ-S)$_2$	1.71(1)	106.7±5.2	1	5	294
{Mo(O)[S$_2$P(O^{i}Pr)$_2$]}$_2$(μ-O)(μ-S)	1.643(8)	107.0±6.5	1	5	295
{Mo(O)[S$_2$P(O^{i}Pr)$_2$]}$_2$(μ-O)(μ-S)(N$_2$C$_4$H$_4$)	1.671(11)	106.1±2.7	1	5	296
	1.678(15)	103.2±3.0	1	6	

TABLE 5.2 Metal–Oxo Structures *(Cont.)*

Complex	M=O (Å)	< O–M–L	d e-	CN	Ref.
{Mo(O)[S$_2$P(O^{i}Pr)$_2$]}$_2$(μ–O)(μ–S)(μ–py)	1.672(11)	103.6±3.0	1	6	297
{Mo(O)[S$_2$P(O^{i}Pr)$_2$]}$_2$(μ–O)(μ–S)(N$_2$C$_4$H$_4$)	1.656(16)	101.4±3.5	1	6	297
Mo(O)Mo(NTol)(μ–S)$_2${S$_2$P(OMe)$_2$}$_2$	1.675(7)	105.9±3.5	1	5	298
{Mo(O)Mo(NPh)(μ–S)$_2$[S$_2$P(OEt)$_2$]$_2$}$_2$	1.684(7)	99.8±5.2	1	6	298
{Mo(O)(μ–O)}$_2$(μ–O^{i}Pr)$_4$Mo...	1.672(2)	106.9±3.2	1	5	212
[{Mo(O)Cl$_2$(H$_2$O)}$_2$(μ–O)$_2$](pyH)$_2$	1.665(6)	99.6±7.6	1	6	299
{Mo(O)(μ–O)}$_2$(CO$_3$)(OH)$_4$[MoCO(PMe$_3$)$_3$]$_2$	1.672(4)		1	6	300
{Mo(O)(HBpz$_3$)(μ–O)$_2$Mo(O)(MeOH)$_2$}$_2$	1.682, 1.678(3)		1	6	37
{Mo(O)bipy[OP(O)H$_2$]}$_2$(μ–O)$_2$	1.680(12)		1	6	301
{Mo(O)Cl(bipy)}$_2$(μ–O)$_2$	1.607, 1.651(7)		1	6	302
{Mo(O)py}$_4$(μ–O)$_4$(OCHMe$_2$)$_4$	1.682, 1.697(3)		1	6	303
[{Mo(O)}$_2$(μ–O)$_2$(μ–N$_2$O$_4$C$_{11}$H$_{14}$O$_4$)]Na$_2$	1.68(2)	99.7±13	1	6	304
[{Mo(O)[c–(NHCH$_2$CH$_2$)$_3$]}$_2$(μ–O)$_2$](SCN)$_2$	1.695(3)		1	6	305
[{Mo(O)[c–(NHCH$_2$CH$_2$)$_3$]}$_2$(μ–O)$_2$]I$_2$	1.696(5)		1	6	305
[{Mo(O)[c–(NHCH$_2$CH$_2$)$_3$]}$_2$(μ–S)$_2$](PF$_6$)$_2$	1.754(7)	97.1±5.8	1	6	306
{Mo(O)(2–Spy)(py)}$_2$(μ–O)$_2$	1.682(3)		1	6	307
[{Mo(O)(SCH$_2$CH(NH$_2$)CO$_2$)}$_2$(μ–O)$_2$]Na$_2$	1.693(6)		1	6	308
[{Mo(O)(C$_2$O$_4$)}$_2$(μ–O)(μ–S)]$_2$(μ–C$_2$O$_4$)K$_6$	1.685(5)	100.1±3.8	1	6	309
[{Mo(O)}$_2$(μ–O)(μ–S)(N$_2$O$_4$C$_{11}$H$_{14}$O$_4$)]Na$_2$	1.684(6)	99.1±11	1	6	304
{Mo(O)(N$_2$OC$_6$H$_5$NO)}$_2$(μ–S)$_2$	1.71(2)	96.0±10	1	6	310
[{Mo(O)(NCS)$_3$}$_2$(μ–O)(μ–S)](pyH)$_4$	1.669(6)	97.1±7.6	1	6	311
[{Mo(O)(C$_2$O$_4$)}$_2$(μ–O)$_2$(μ–S)$_2$]$_2$(C$_2$O$_4$)Cs$_6$	1.69(1)		1	6	312
[{Mo(O)(C$_2$O$_4$)(H$_2$O)}$_2$(μ–S)$_2$]Cs$_2$	1.697(6)		1	6	313
[{Mo(O)Cl$_2$(H$_2$O)}$_2$(μ–S)$_2$]Cs$_2$	1.69(1)		1	6	312
{Mo(O)(N$_2$OC$_6$H$_8$NO)}$_2$(μ–S)$_2$	1.676(7)	96.9±9.5	1	6	314
[{Mo(O)}$_2$(μ–S)$_2$(μ–N$_2$O$_4$C$_{11}$H$_{14}$O$_4$)]Na$_2$	1.705(10)	98.9±9.7	1	6	304
[{Mo(O)}$_2$(μ–S)$_2$(edta)]Cs$_2$	1.683(6)	98.3±7.3	1	6	315
[{Mo(O)(NCS)$_3$}$_2$(μ–S)$_2$](pyH)$_4$	1.683(6)	96.7±7.0	1	6	311
{Mo(O)Cp}$_2$(μ–O)$_2$	1.699(4)		1	6	316
{Mo(O)Cp*}$_2$(μ–O)$_2$	1.692(4)		1	6	317
{Mo(O)Cp}$_2$(μ–O)(μ–NTol)	1.708(2)		1	6	318
{Mo(O)Cp}$_2$(μ–O)(μ–NCOOEt)	1.694(5)		1	6	319
{Mo(O)Cp}$_2$(μ–NCOOEt)$_2$	1.695(3)		1	6	319
{Mo(O)Cp}$_2$(μ–S)$_2$	1.679(8)		1	6	320
{Mo(O)Cp*}$_2$(μ–S)$_2$	1.685(4)		1	6	321
{Mo(O)Cl$_2$(CF$_3$COCHCOCF$_3$)}$_2$(μ–Cl)$_2$	1.648(9)	96.3±5.0	1	6	322

TABLE 5.2 Metal–Oxo Structures *(Cont.)*

Complex	M=O (Å)	∠ O–M–L	d e⁻	CN	Ref.
{Mo(O)}$_4$(μ_3–O)$_2$Cl$_2$(O$_2$CPh)$_6$	1.646,1.678(4)		1	6	323
[{Mo(O)(NCS)$_2$}$_2$(μ–O)(μ–O$_2$CMe)](pyH)$_3$	1.683(6)		1	6	324
{Mo(O)(NOC$_9$H$_6$)}$_2$(μ–O)(μ–OC$_2$H$_4$S)	1.693(6)		1	6	325
[{Mo(O)(OC$_2$H$_4$S)}$_2$(μ–O)(μ–OC$_2$H$_4$S)]$^{2-}$	1.699(6)	101.1±13	1	6	326
{Mo(O)(S$_2$CNEt$_2$)}$_2$(μ–O)(μ–SPh)$_2$	1.673(5)	100.3±7.3	1	6	327
{Mo(O)(S$_2$CNEt$_2$)}$_2$(μ–S)(μ–SC$_2$H$_4$O)	1.686(7)	100.1±11	1	6	328
[{Mo(O)(STol)$_2$}$_2$(μ–STol)$_2$(μ–OMe)]Et$_4$N	1.671(20)		1	6	329
[{Mo(O)(S$_2$CNEt$_2$)}$_2$(μ–SPh)$_2$(μ–Cl)]$^+$	1.656(6)	100.1±4.5	1	6	330
[{Mo(O)}$_2$(SC$_2$H$_4$O)$_2$Cl$_3$]Et$_4$N	1.676(4)	98.7±9.5	1	6	331
[{Mo(O)}$_2$(SC$_2$H$_4$O)$_3$Cl]Et$_4$N	1.687(9)	99.1±11	1	6	331
[{Mo(O)}$_2$(SC$_2$H$_4$O)$_3$(SC$_2$H$_4$OH)]Pr$_3$NH	1.692(4)	99.9±9.7	1	6	331
[{Mo(O)}$_2$(SC$_3$H$_6$S)$_3$(μ–N$_3$)]PPh$_4$	1.665(10)		1	6	245
Mo(O)(S$_2$CNEt$_2$)(μ–NNCOPh)$_2$Mo(S$_2$CNEt$_2$)	1.66(1)	109.1±5.1	1–2?	5	332
Mo(O)(S$_2$CNEt$_2$)(μ–NNCSPh)$_2$Mo(S$_2$CNEt$_2$)	1.67(1)	108.7±2.9	1–2?	5	332
Mo(O)(Tol$_4$porphyrin)	1.656(6)	107.6±2.1	2	5	333
Mo(O)(phthalocyaninato)	1.668(6)	109.3±1.6	2	5	334
Mo(O)(S$_2$CNEt$_2$)(μ–NNCPhO)$_2$Mo(S$_2$CNEt$_2$)	1.660(6)		2	5	335
[Mo(O)(S$_4$)$_2$](Et$_4$N)$_2$	1.685(7)	108.7±1.9	2	5	336
[Mo(O)(S$_2$C$_2$O$_2$)$_2$]Cs$_2$	1.660(8)	107.2±1.0	2	5	225
Mo(O)(S$_2$CN^{i}Pr$_2$)$_2$	1.664(8)	110.0±1.7	2	5	269
Mo(O)(S$_2$CS^{i}Pr)(η^3–S$_2$CS^{i}Pr)	1.66(1)		2	5	337
Mo(O)(S$_3$CPh)(η^3–S$_2$CPh)	1.673(2)		2	5	272
Mo(O)[S$_2$C(PMe$_3$)O^{i}Pr](η^3–S$_2$CO^{i}Pr)	1.668(3)		2	5	338
[Mo(O)(S$_2$C$_6$H$_4$)$_2$](NEt$_4$)$_2$	1.699(6)	108.2(1)	2	5	246
Mo(O)(SCH$_2$CH$_2$PPh$_2$)$_2$	1.733(9)	110.7±9.7	2	5	339
[Mo(O)[μ–S)$_2$Mo(S)$_2$}$_2$](PPh$_4$)$_2$	1.746(9)	105.0±2.4	2?	5	340
Mo(O)(acac)$_2$(PMe$_3$)	1.676(5)	97.7±7.7	2	6	341
[Mo(O)(CN)$_5$](PPh$_4$)$_3$	1.705(4)	100.2±2.4	2	6	36
[Mo(O)(CN)$_4$(H$_2$O)][Pt(en)$_2$]	1.668(5)	99.1±1.1	2	6	342
[Mo(O)(CN)$_4$(H$_2$O)](AsPh$_4$)$_2$ (blue)^e	1.60(2)	100 ±4	2	6	36
[Mo(O)(CN)$_4$(H$_2$O)](PPh$_4$)$_2$ (green)^e	1.72(2)	100 ±4	2	6	36
[Mo(O)(CN)$_4$(OH)]$^{3-}$	1.698(7)	92.4(4)	2	6	342
[Mo(O)(CN)$_4$(N$_3$)]Cs$_2$Na	1.70(1)	99.6±0.9	2	6	343
[Mo(O)(CN)$_3$(phen)]Na	1.659(7)	99.0±4.8	2	6	344
[Mo(O)Cl(CNMe)$_4$](I$_3$)	1.636(37)	98.9±4.3	2	6	345

TABLE 5.2 Metal–Oxo Structures *(Cont.)*

Complex	M=O (Å)	∠ O–M–L	d e-	CN	Ref.
[Mo(O)Cl$_3$(p-ClC$_6$H$_4$CONNPh)]$^-$	1.620(14)		2	6	39
[Mo(O)Cl(dppe)$_2$]$^+$	1.708(12)		2	6	39
[Mo(O)(OH)(dppe)$_2$]BF$_4$ (trans)	1.833(5)	90.8±5.8	2	6	38
[Mo(O)(S$_4$C$_{12}$H$_{24}$)(μO)Mo(S$_4$C$_{12}$H$_{24}$)OEt]$^{3+}$	1.764(14)	97.9±1.1	2	6	346
[Mo(O)(SH)(S$_4$C$_{12}$H$_{24}$)]CF$_3$SO$_3$	1.667(3)	91.7±0.4	2	6	347
Mo(O)Cl$_2$(PMe$_2$Ph)$_3$ (mer,cis, blue)^e	1.676(7)	95.7±10	2	6	30
Mo(O)Cl$_2$(PMe$_2$Ph)$_3$ (mer,cis, green)^e	1.80(2)		2	6	32
Mo(O)Cl$_2$(PEt$_2$Ph)$_3$ (mer,cis, green)^e	1.801(9)	92.9±6.0	2	6	31
Mo(O)Cl$_2$(PMePh$_2$)$_3$ (mer,cis, green)	1.667(4)	95.4±8.9	2	6	49
Mo(O)(NCO)$_2$(PEt$_2$Ph)$_3$ (mer,cis)	1.684(8)	94.8±12	2	6	49
Mo(O)[HB(Me$_2$pz)$_3$](S$_2$CNEt$_2$)	1.669(3)	97.1±3.4	2	6	348
Mo(O)[S$_2$C$_2$(CN)$_2$](dppe)(OCMe$_2$)	1.682(2)	99.8±6.0	2	6	349
Mo(O)(S$_2$CNMe$_2$)$_2$[C$_2$(COTol)$_2$]	1.686(2)	99.8±8.9	2	6	350
Mo(O)(S$_2$CN^{n}Pr$_2$)$_2$[C$_2$(CN)$_4$]	1.682(4)		2	6	351
Mo(O)(C$_2$Ph$_2$)(fulvalene)Ru(CO)$_2$	1.698(1)		2–3	6	352
Mo(O)Cp(CF$_3$C≡CCF$_3$)(SC$_6$F$_5$)	1.678(5)		2	6	353
Mo(O)(η^5-C$_5$H$_4$Me)$_2$	1.721(2)		2	7	28
Mo(O)Cp(μ$_3$-N)[MoCp(CO)$_2$]$_2$	1.708(4)		2?		354

Complex	M=O (Å)	∠ O–M–O	d e-	CN	Ref.
[Mo(O)$_4$]$^{2-}$ ^{c}	1.76 – 1.77^c		0	4	161
[Mo(O)(S)$_3$]Cs$_2$	1.785(10)	109.9±1.3	0	4	356
[{Mo(O)$_3$}$_2$(μ-O)](^{n}Bu$_4$N)$_2$	1.716(6)		0	4	357
[Mo(O)$_3$(OSiPh$_3$)]^{n}Bu$_4$N	1.690(4)–1.711(4)	108.9–110.4(2)	0	4	358
[{Mo(O)$_3$}$_2$(edta)]$^{4-}$	1.731(6)–1.749(6)	104.5–107.5(4)	0	6	359
[Mo(O)$_3$(O$_2$NC$_6$H$_6$O$_4$)]K$_3$	1.736(5)	102.1–107.1(2)	0	6	360
Mo(O)$_3$(H$_2$NC$_2$H$_4$NHC$_2$H$_4$NH$_2$)	1.735(6)–1.739(6)	105.3–106.8(3)	0	6	361
[{Mo(O)$_2$(μ-S)$_2$}$_2$Fe$_2$S$_2$](PPh$_4$)$_4$	1.736(15),1.821(14)		0	4	362
{Mo(O)$_2$(μ-O)$_2$Mo(C$_5$H$_4$Me)$_2$}$_2$	1.72(1), 1.72(1)	106.1(7)	0	5	363
{Mo(O)$_2$(NOC$_{10}$H$_8$N)}$_2$(μ-O)	1.687(3), 1.699(3)		0	5	364
{Mo(O)$_2$(μ-O$_2$C$_{14}$H$_8$)}$_2$(μ-O)	1.681(5), 1.691(5)	105.1(2)	0	6	365
[{Mo(O)$_2$(μ-O$_2$C$_6$H$_4$)}$_2$(μ-O)]Ba	1.722(8), 1.727(10)	104.6(7)	0	6	366
[{Mo(O)$_2$(C$_2$O$_4$)(H$_2$O)}$_2$(μ-O)]K$_2$	1.680(19),1.700(20)	106.5(10)	0	6	367

TABLE 5.2 Metal–Oxo Structures *(Cont.)*

Complex	M=O (Å)	< O–M–O	d e⁻	CN	Ref.
[{Mo(O)$_2$}$_2$(μ–O)(O$_6$C$_6$H$_{11}$)]NH$_4$	1.68 – 1.72(1)	107.2(9)	0	6	368
[{Mo(O)$_2$}$_2$(μ–O)(O$_4$C$_6$H$_{11}$O$_2$)]Na	1.693(3), 1.727(3)	103.0, 104.2(1)	0	6	369
{Mo(O)$_2$}$_2$(μ–O)(μ–O$_4$C$_5$H$_6$O)	1.71(1), 1.71(1)		0	6	370
{Mo(O)$_2$}$_2$(μ–O)(μ–ONMeCMeNH)$_2$	1.68(2), 1.70(2)	103.5(8)	0	6	237
{Mo(O)$_2$[HB(Me$_2$pz)$_3$]}$_2$(μ–O)	1.696(2), 1.701(2)	102.7(1)	0	6	371
[{Mo(O)$_2$[N$_3$(CH$_2$)$_6$]}$_2$(μ–O)]$^{2+}$	1.694(7), 1.696(7)	104.7(4)	0	6	372
{Mo(O)$_2$(μ–OSC$_6$H$_4$)}$_2$(μ–O)	1.699(2), 1.712(2)		0	6	373
{Mo(O)$_2$(N$_2$SC$_6$H$_{15}$)}$_2$(μ–O)	1.685(8), 1.737(7)	108.9(4)	0	6	374
{Mo(O)$_2$(N$_2$SC$_8$H$_{19}$)}$_2$(μ–O)	1.712(6), 1.715(6)	105.9, 108.9(3)	0	6	375
Mo(O)$_2$[C(P^{n}Bu$_3$)(Mes)](Mes)	1.68(2), 1.69(2)	114(1)	0	4	376
Mo(O)$_2$(Mes)$_2$(CH$_2$P^{n}Bu$_3$)	1.688(5), 1.706(5)	113.5(3)	0	5	377
Mo(O)$_2$[2,6-(OCH^{t}Bu)$_2$py]	1.695(5), 1.701(5)	109.8(3)	0	5	378
Mo(O)$_2$[(SCPh$_2$CH$_2$)$_2$C$_5$H$_3$N]	1.691(6), 1.696(6)	110.5(2)	0	5	379
Mo(O)$_2$(OC$_2$H$_4$OC$_2$H$_4$O)	1.63(4), 1.73(2)	101(3)	0	5–6	380
Mo(O)$_2$(OC$_2$H$_4$OH)$_2$	1.714(10), 1.733(11)	106.1±1.5	0	6	381
Mo(O)$_2$(OCHMeCHMeOH)$_2$	1.662(7)	105.2(5)	0	6	382
Mo(O)$_2$(acac)$_2$	1.64(2), 1.72(2)	107.7±3.4	0	6	383
Mo(O)$_2$(PhCOCHCOPh)$_2$	1.695(8), 1.697(8)	104.8(4)	0	6	384
[Mo(O)$_2$(O$_2$C$_6$H$_4$)$_2$]K$_2$	1.77(2)	101(1)	0	6	385
[Mo(O)$_2$(O$_2$C$_4$H$_4$O$_3$)$_2$]Cs$_2$	1.708(9)	104.5(6)	0	6	386
[Mo(O)$_2$(O$_2$C$_4$H$_4$O$_4$)$_2$](Me$_4$N)$_2$	1.666(8), 1.699(10)	102.3(4)	0	6	387
{Mo(O)$_2$(OCH$_2$CMe$_2$CH$_2$O)(H$_2$O)}$_2$	1.685(3), 1.711(4)	104.3(2)	0	6	388
{Mo(O)$_2$(OSM$_2$)(OEt)}$_2$(μ–OEt)$_2$	1.703(3), 1.708(3)	102.9(1)	0	6	216
{Mo(O)$_2$}$_4$(OEt)$_2$[(OCH$_2$)$_3$CMe]$_2$	1.686(5), 1.689(5)	106.2(3)	0	6	389
{Mo(O)$_2$[μ–(OCH$_2$)$_2$CMeNH$_2$]}$_2$	1.708(8), 1.715(8)	105.1(4)	0	6	216
[Mo(O)$_2$(ONCOPh)$_2$]Cs$_2$	1.705(7)–1.750(10)	103.8, 101.5(4)	0	6	239
Mo(O)$_2$(OCMeNPhO)$_2$	1.692(3), 1.704(3)	105.2(2)	0	6	390
Mo(O)$_2$(O$_2$C$_{10}$H$_{12}$NO)$_2$	1.703(2), 1.703(2)	103.9(1)	0	6	390
[Mo(O)$_2$F$_4$]K$_2$	1.68(2), 1.73(2)	95.1(10)	0	6	391
[{Mo(O)$_2$F$_2$(μ–F)}$_2$]$^{2-}$	1.667(7), 1.690(7)	102.6(4)	0	6	240
Mo(O)$_2$Cl$_2$(H$_2$O)$_2$	1.701(8)	103.0(5)	0	6	392
Mo(O)$_2$Cl$_2$(MeOCH$_2$CH$_2$OMe)	1.667(11), 1.673(10)	105.0(5)	0	6	393
Mo(O)$_2$Cl$_2$(dmf)$_2$	1.68(1)	102.2(7)	0	6	394
Mo(O)$_2$Cl$_2$[OCNEt$_2$CH$_2$PS(OEt)$_2$]$_2$	1.702(4)	101.7(3)	0	6	395
Mo(O)$_2$Cl$_2$(hmpa)$_2$	1.71(1)	101(1)	0	6	396
Mo(O)$_2$Cl$_2$(OPPh$_3$)$_2$	1.673(1), 1.695(1)	103.2(1)	0	6	382

TABLE 5.2 Metal–Oxo Structures *(Cont.)*

Complex	M=O (Å)	< O–M–O	d e–	CN	Ref.
$Mo(O)_2Br_2(OPPh_3)_2$	1.69(1), 1.73(1)	103.2(5)	0	6	382
$Mo(O)_2Cl_2(O_2C_{12}H_{26}NP)$	1.677(2), 1.687(2)	102.8(1)	0	6	397
$Mo(O)_2Cl_2(O_2C_{14}H_8)$	1.671(3), 1.671(3)	104.8(2)	0	6	398
$Mo(O)_2(O^iPr)_2(bipy)$	1.689(7), 1.723(6)	108.0(3)	0	6	213
$Mo(O)_2Cl_2(phen)$	1.695(3)	106.8(2)	0	6	399
$Mo(O)_2Br_2(bipy)$	1.643(17), 1.826(18)	103.3(9)	0	6	400
$Mo(O)_2Me_2(bipy)$	1.707(2), 1.708(2)	110.2(1)	0	6	401
$Mo(O)_2Et_2(bipy)$	1.695(6), 1.709(5)	110.4(3)	0	6	5
$Mo(O)_2(CH_2CMe_3)_2(bipy)$	1.706(3), 1.709(3)	110.0(2)	0	6	402
$Mo(O)_2(CH_2Ph)_2(bipy)$	1.68(1), 1.71(1)	110.2(6)	0	6	403
$Mo(O)_2(ONHMe)_2$	1.700(8), 1.728(8)	113.7(3)	0	6	404
$Mo(O)_2(ONEt_2)_2$	1.713(2), 1.714(2)	116.6(1)	0	6	42
$Mo(O)_2(ONC_5H_{10})_2$	1.701(5), 1.879(5)	119.3(3)	0	6	41
$Mo(O)_2(O_3NC_8H_6N_2O_4)$	1.680(5), 1.689(5)	106.6(2)	0	6	405
$[\{Mo(O)_2\}_2(\mu\text{-}O)(O_2C_6H_3N_4)_2]Na_2$	1.688(7), 1.709(6)	105.2(3)	0	6	406
$Mo(O)_2(NO_2C_{33}H_{27})(dmso)$	1.702(4), 1.708(4)	105.4(2)	0	6	379
$\{Mo(O)_2[(OCH_2)_2py]\}_\infty$	1.710(3), 1.719(3)	106.0(2)	0	6	407
$Mo(O)_2[(OC_6H_4CHN)_2CH_2CHMe]$	1.709(3), 1.710(3)	103.2(1)	0	6	408
$Mo(O)_2[(OC_6H_4CHN)_2(CH_2)_3]$	1.701(4), 1.714(4)	103.3(2)	0	6	409
$Mo(O)_2(NOC_9H_6)_2$	1.71(3)	104(2)	0	6	410
$Mo(O)_2(NCS)_2(hmpa)_2$	1.68(1)	101.2(5)	0	6	411
$Mo(O)_2(OC_6H_4CHNMe)_2$	1.682(8), 1.688(7)	107.7(4)	0	6	412
$Mo(O)_2(Tol_4porphyrin)^f$	1.709(9), 1.744(9)	95.1(4)^f	0	6^f	413
$Mo(O)_2(SOC_9H_{10}N)_2$	1.706(5), 1.713(5)	103.7(2)	0	6	414
$Mo(O)_2[(SCH_2)_2py](OSC_4H_8)$	1.694(3), 1.723(3)	106.0(2)	0	6	407
$Mo(O)_2(N_2O_3C_{24}H_{34})$	1.705(10), 1.713(10)	109.2(6)	0	6	415
$Mo(O)_2Cl(\eta^3\text{-}Mo_5O_{18}TiCp)$	1.679(10), 1.696(10)		0	6	416
$Mo(O)_2[SCH_2CH(NH_2)COOMe]_2$	1.714(4)	108.1(3)	0	6	417
$Mo(O)_2[(SC_2H_4NMe)_2C_2H_4)_2]$	1.677(9)–1.750(10)	107.8, 109.1(7)	0	6	418
$Mo(O)_2[C_3H_6(NMeC_2H_4S)_2]$	1.704(6)	107.8(3)	0	6	418
$Mo(O)_2[C_2H_4(NHC_6H_4S)_2]$	1.708(3), 1.715(3)	109.9(1)	0	6	418
$Mo(O)_2(SC_2H_4NHC_2H_4SC_2H_4S)$	1.718(3), 1.721(3)	108.8(1)	0	6	418
$Mo(O)_2[NMe_2C_2H_4N(C_2H_4S)_2]$	1.699(2), 1.705(2)	107.9(1)	0	6	419
$Mo(O)_2[(SC_2H_4)_2NC_2H_4SMe)]$	1.694, 1.695	108.6	0	6	419
$Mo(O)_2(mercaptoquinolato)_2$	1.694(6), 1.712(7)	106.3(3)	0	6	263
$Mo(O)_2(SCMe_2CH_2NHMe)_2^g$	1.711(5), 1.723(5)	122.2(3)	0	6^g	420

TABLE 5.2 Metal–Oxo Structures *(Cont.)*

Complex	M=O (Å)	< O–M–O	d e–	CN	Ref.
Mo(O)$_2$(SCMe$_2$CMe$_2$NHMe)$_2$ [g]	1.727(1), 1.731(1)	120.7(1)	0	6[g]	420
Mo(O)$_2$(SCMe$_2$CH$_2$NMe$_2$)$_2$ [g]	1.720(3)	122.0(2)	0	6[g]	420
Mo(O)$_2$(NSC$_6$H$_{12}$O$_2$)$_2$	1.707(3), 1.717(4)	107.3(2)	0	6	421
Mo(O)$_2$(S$_2$CNEt$_2$)$_2$	1.703(2)	105.8(1)	0	6	422
Mo(O)$_2$(S$_2$CN^{i}Pr$_2$)$_2$	1.695(5), 1.696(5)	105.7(1)	0	6	270
Mo(O)$_2$(S$_4$C$_{14}$H$_{12}$)	1.71(2), 1.72(2)	111.1(1)	0	6	423
Mo(O)(S)(ONC$_5$H$_{10}$)$_2$	1.711(4), 1.734(4)	116.2,117.5(1)	0	6	424
Mo(O)(Se)(ONC$_5$H$_{10}$)$_2$	1.719(4)	115.3(1)	0	6	425
Mo(O)(NH)Cl$_2$(OPPh$_2$Et)$_2$	1.66(1)	101.8(6)	0	6	426
[Mo(O)(NNMe$_2$)(SPh)$_3$]PPh$_4$	1.705(8)	105.2(5)	0	5	427
Mo(O)(NNMe$_2$)(NOC$_9$H$_6$)$_2$	1.671(9)	100.8(4)	0	6	428
Mo(O)(NNMe$_2$)(S$_2$CNMe$_2$)$_2$	1.708(6)	104.4(3)	0	6	429
Mo(O)(NNPh$_2$)(N$_2$S$_2$C$_8$H$_{18}$)	1.696(2)	105.9(1)	0	6	430
[Mo(O)$_2$(CN)$_4$]NaK$_3$	1.834(9)	180	2	6	431

Complex	M=O (Å)	< O–M–L	d e–	CN	Ref.
W(O)(μ_3–S)$_3$(CuPPh$_3$)$_3$(μ_3–Cl)	1.752(14)		0	4	432
{W(O)(μ_3–S)$_3$(CuPPh$_3$)$_2$}$_2$	1.696(8)	108.8±1.3	0	4	433
{W(O)(μ_3–S)$_3$(CuPTol$_3$)$_2$}$_2$	1.70(2)	109.0±0.9	0	4	434
W(O)F$_4$ [a]	1.666(4)[a]	104.8(3)[a]	0	5	435
W(O)Cl$_4$ [a]	1.684(4)[a]	102.6(5)[a]	0	5	12
W(O)Br$_4$ [a]	1.684(3)[a]	102.8(6)[a]	0	5	436
{W(O)(CH$_2$CMe$_3$)$_3$}$_2$(μ–O)	1.689(13)	94.0(3)	0	5	437
W(O)(OC$_5$H$_4$FeCp)(C$_5$H$_4$FeCp)$_3$ (tbp)	1.705(5)	92.3±1.5	0	5	438
W(O)(O^{t}Bu)$_4$(thf)	1.77(>3)	99.2(2)	0	6	439
{W(O)F$_4$}$_2$(μ–F)	1.57(3)	99 ±2	0	6	440
W(O)Cp(CH$_2$SiMe$_3$)$_3$	1.664(8)		0	6	441
[W(O)(O$_2$)F$_4$]$^{2-}$	1.74(3)		0	7	442
[W(O)(O$_2$)$_2$(C$_2$O$_4$)]K$_2$	1.716(7)	100.9±6.7	0	7	443
W(O)Cl$_4$(Me$_2$AsC$_6$H$_4$AsMe)$_2$	1.89(4)		0	7	444
{W(O)Cl$_2$(H$_2$O)(μ–O)}$_4$(NHMe$_3$)$_2$	1.711(9)	98.4±2.7	0.5	6	445
[W(O)Cl$_4$]PPh$_4$	1.676(7)	105.3(2)	1	5	13
[W(O)(CH$_2$C$_6$H$_4$CH$_2$)$_2$]$_2$Mg(thf)$_4$	1.71(1)		1	5	446

TABLE 5.2 Metal–Oxo Structures *(Cont.)*

Complex	M=O (Å)	< O-M-L	d e-	CN	Ref.
[{W(O)Cl$_2$}$_2$(μ-S)$_2$](AsPh$_4$)$_2$	1.699, 1.755(15)		1	5	447
[W(O)Cl$_2${c-(NMeCH$_2$CH$_2$)$_3$}]PF$_6$ [e]	1.719(18)	97.0±5.9	1	6	43
[W(O)Cl$_2${c-(NMeCH$_2$CH$_2$)$_3$}]PF$_6$ [e]	1.893(20)	96.7±5.8	1	6	43
[{W(O)Cl$_2$}$_2$(μ-Cl)(μ-SCHMeEt)$_2$]AsPh$_4$	1.69(2)		1	6	448
[{W(O)Cl$_2$}$_2$(μ-Cl)(μ-SPh)$_2$]PPh$_4$	1.667(8)		1	6	449
{W(O)(SEt)(PMe$_2$Ph)}$_2$(μ-S)(μ-SEt)$_2$	1.676(9)		1	6	450
[W(O)(η^2-WS$_4$)$_2$(dmf)](PPh$_4$)$_2$	1.631(15)		2?	5	341
[W(O)F(CN)$_4$]K$_3$	1.770(9)	94.8±1.8	2	6	451
[W(O)(CN)$_3$(2-O$_2$Cpy)](AsPh$_4$)$_2$	1.676(9)	99.0±6.8	2	6	452
W(O)Cl$_2$(PMe$_3$)$_3$	1.67(1)	96.7±1.9	2	6	453
W(O)Cl$_2$(C$_2$H$_4$)(PMePh$_2$)$_2$	1.714(6)	95.5±4.9	2	6	454
W(O)Cl$_2$(CO)(PMePh$_2$)$_2$	1.689(6)	97.2±7.6	2	6	455
W(O)Cp(Ph)(PhC≡CPh)	1.69(2)		2	6	456
W(O)Cp(CH$_2$COOEt)(HC≡CPh)	1.714(2)		2	6	457
W(O)Cp(μ_3-PhC≡CPh)[CpW(CO)$_2$]Fe(CO)$_3$	1.726(7)		3?		458

Complex	M=O (Å)	< O-M-O	d e-	CN	Ref.
[WO$_4$]$^{2-}$ [c]	1.78 [c]		0	4	161
[W(O)(S)$_3$]K$_3$Cl	1.763(12)	109.8±0.6	0	4	459
[W(O)$_2$(OC$_2$Me$_4$O)(CH$_2$CMe$_3$)]$^-$	1.712(5), 1.720(5)	110.4(3)	0	5	460
W(O)(CH^{t}Bu)Cl$_2$(PEt$_3$)	1.661(11)	106.7(6)	0	5	24
W(O)(CH^{t}Bu)Cl$_2$(PMe$_3$)$_2$	1.697(15)	102.0(9)	0	6	25
W(O)[C(CH$_2$)$_5$]Cl$_2$(PMePh$_2$)$_2$	1.708(8)	94.9(4)	0	6	461
W(O)$_2$Cl$_2$(OPPh$_3$)$_2$ (cis)	1.702(9), 1.706(8)	102.2(4)	0	6	462
W(O)$_2$Cl$_2$(acac)$^-$	1.722(10), 1.736(10)		0	6	463
W(O)$_2$Cp(CH$_2$SiMe$_3$)	1.716(5), 1.723(5)	107.8(3)	0	6	442
MnO$_4^-$ [c]	1.61 [c]		0	4	161
[MnO$_4$]K	1.600–1.612(5)	109.4±0.6	0	4	4
Mn(O)$_3$F [a]	1.586{5} [a]		0	4	464
[MnO$_4$]K$_2$	1.633–1.660(5)	109.5±0.6	1	4	4

TABLE 5.2 Metal–Oxo Structures *(Cont.)*

Complex		M=O (Å)	∢ O–M–L	d e–	CN	Ref.
{Tc(O)F$_3$(μ–F)}$_3$		1.66(3)	103(3)	1	6	465
[Tc(O)Cl$_4$][(Ph$_3$P)$_2$N]		1.610(4)	106.8±4.7	2	5	14
Tc(O)Cl(OC$_6$H$_4$NCHC$_6$H$_4$O)		1.634(7)	109.2±2.9	2	5	466
Tc(O)(N$_4$O$_2$C$_{13}$H$_{25}$)		1.679(3)	109.8±1.2	2	5	467
[Tc(O)(NHC$_6$H$_4$S)$_2$]Bu$_4$N		1.73(2)		2	5	468
[Tc(O)(SCH$_2$CH$_2$O)$_2$]AsPh$_4$		1.662(5)	110.0±2.0	2	5	469
[Tc(O)(SCH$_2$CH$_2$S)$_2$]AsPh$_4$		1.64(1)	109.5±3.2	2	5	470
[Tc(O){SC(O)CH$_2$S}$_2$]Bu$_4$N		1.672(8)	100.0±0.3	2	5	471
[Tc(O){SCH(CO$_2$Me)CH(CO$_2$Me)S}$_2$]AsPh$_4$		1.672(6)	110.1±2.5	2	5	472
[Tc(O){Se$_2$C=C(CN)$_2$}$_2$]Et$_4$N		1.67(2)	110.9±2.8	2	5	473
[Tc(O)(H$_2$O)(acac$_2$en)]$^+$	(trans)	1.648(2)	100.8±1.5	2	6	474
Tc(O)Cl(sal$_2$en)	(trans)	1.626(11)	97.2±4.1	2	6	474
Tc(O)Cl(N$_2$O$_2$C$_{17}$H$_{16}$)	(trans)	1.66(1)		2	6	475
Tc(O)Cl(2–Me–8–quinolinato)$_2$	(cis)	1.649(3)		2	6	476
Tc(O)Cl(OC$_6$H$_4$NPh)$_2$	(cis)	1.67(1)		2	6	477
Tc(O)(O$_2$NC$_{13}$H$_{15}$O$_4$)(O$_2$C$_7$H$_5$)		1.656(9)	102.4±7.2	2	6	478
Tc(O)Cl$_2$(HBpz$_3$)		1.656(3)		2	6	479
Tc(O)(NSO$_2$C$_5$H$_9$)(NSO$_2$C$_5$H$_{10}$)		1.657(4)		2	6	480
Tc(O)(OEt)Br$_2$(4–NO$_2$py)$_2$	(all trans)	1.684(6)		2	6	481
{Tc(O)(N$_2$O$_2$C$_{17}$H$_{16}$)}$_2$(μ–O)	(trans)	1.69(1)		2	6	475

Complex	M=O (Å)	∢ O–M–O	d e–	CN	Ref.
[TcO$_4$]K	1.711(3)		0	4	482
[Tc(O)$_4$]NH$_4$	1.702(2)		0	4	483
[Tc(O)$_2$(4–^{t}Bupy)$_4$]$^+$	1.737(6), 1.748(6)	180	2	6	484
[Tc(O)$_2$(imidazole)$_4$]Cl	1.71(2)	180	2	6	485
[Tc(O)$_2$(en)$_2$]Cl	1.741(1), 1.752(1)	178.6(3)	2	6	486
[Tc(O)$_2${c–(NH)$_4$(CH$_2$)$_{10}$}]$^+$	1.749(3), 1.754(3)	180	2	6	487

TABLE 5.2 Metal–Oxo Structures *(Cont.)*

Complex	M=O (Å)	∠ O–M–L	d e-	CN	Ref.
Re(O)F$_5$[a]	1.642(20)[a]	93.1(2)[a]	0	6	15
Re(O)(C$_6$H$_2$Me$_3$)$_4$	1.679(8)	105.7±2.9	1	5	5
{Re(O)(CH$_2$SiMe$_3$)$_3$}$_2$(μ-O)	1.671(5)	109.2±6.9	1	5	331
{Re(O)Me$_3$}$_2$(μ-O)	1.742(18)		1	5	331
{Re(O)(μ-O)(CH$_2$CMe$_2$Ph)$_2$}$_2$	1.64(1)		1	5	488
Re(O)F$_4$[a]	1.609(8)[a]	108.8(11)	1	5	15
Re(O)Cl$_4$[a]	1.663(5)[a]	105.5(8)[a]	1	5	15
Re(O)Cl$_4$(H$_2$O)	1.63(2)	98 ±4	1	6	489
[Re(O)(OCHMe$_2$)$_5$]Li	1.707(9)	96.5±5.5	1	6	490
{Re(O)(OMe)$_2$}$_2$(μ-O)(μ-OMe)$_2$	1.697(12)		1	6	490
Re(O)(R)$_2$(μ-O)Re(PMe$_3$)$_4$Re(O)$_2$R R=CH$_2$SiMe$_3$	1.71(2)		2	4	491
{Re(O)Me$_4$}$_2$Mg(thf)$_4$	1.694(8)	114.1±3.5	2	5	492
{Re(O)(CH$_2$SiMe$_3$)$_4$}$_2$Mg(thf)$_2$	1.742(7)	111.8±3.8	2	5	492
[Re(O)Cl$_4$]AsPh$_4$	1.627(24)	100.5(5)	2	5	16
[Re(O)(SPh)$_4$]AsPh$_4$	1.686(9)	108.2±2.1	2	5	493
[Re(O)(S$_2$C$_2$O$_2$)$_2$]Cs	1.674(7)	108.8(av)	2	5	494
[Re(O)Cl$_4$(H$_2$O)]$^-$ (trans)	1.660(9)	97.7±2.9	2	6	495
[Re(O)Br$_4$(H$_2$O)]$^-$ (trans)	1.71(4)	98.2(8)	2	6	496
[Re(O)Br$_4$(NCMe)]AsPh$_4$ (trans)	1.73(6)	98 ±6	2	6	497
Re(O)Cl$_3$(PEt$_2$Ph)$_2$ (trans,mer)	1.660(9)	96.1±5.2	2	6	498
Re(O)Cl$_3$(PEt$_2$Ph)(OPEt$_2$Ph) (mer)	1.672(9)		2	6	499
Re(O)Br$_3$(PEt$_2$Ph)(OPEt$_2$Ph) (mer)	1.663(7)		2	6	499
Re(O)Cl$_3$(dppe) (fac)	1.680(5)		2	6	499
Re(O)Br$_3$(dppe) (fac)	1.679(8)		2	6	499
Re(O)Cl$_3$(CN^{t}Bu)$_2$ (cis,fac)	1.671(2)		2	6	500
Re(O)Cl$_3$(dmf)(PPh$_3$) (mer)	1.664(4)		2	6	499
Re(O)Cl$_3$(H$_2$O)[SC(NH$_2$)$_2$] (mer)	1.713(19)	98.9±4.1	2	6	501
[Re(O)Cl$_2$(H$_2$O){SC(NH$_2$)$_2$}$_2$]Cl	1.654(10)	98.3±1.8	2	6	502
Re(O)Cl$_2$(PPh$_3$)[N(CMe$_2$)NCPhO]	1.685(8)		2	6	503
Re(O)Cl(OC$_6$H$_4$CHNMe)$_2$	1.680(4)		2	6	504
Re(O)Cl$_2$(acac)PPh$_3$	1.69(1)		2	6	505
Re(O)Cl$_2$(OC$_6$H$_4$CHNMe)PPh$_3$ (cis Cl$_2$)	1.660(8)		2	6	506
Re(O)Cl$_2$(OC$_6$H$_4$CHNMe)PPh$_3$ (trans Cl$_2$)	1.701(5)	94.7±3.3	2	6	506
Re(O)(OEt)Cl$_2$(py)$_2$	1.684(7)	91.9±5.5	2	6	507

TABLE 5.2 Metal–Oxo Structures *(Cont.)*

Complex	M=O (Å)	< O–M–L	d e−	CN	Ref.
$Re(O)(OEt)Cl_2(PPh_3)_2$	1.76(1)		2	6	499
$Re(O)(OEt)I_2(PPh_3)_2$	1.715(9)	93.0±3.5	2	6	508
$Re(O)(OMe)I_2(PPh_3)_2$	1.698(5)	92.7±4.5	2	6	508
$\{Re(O)Cl_2(py)_2\}_2(\mu-O)$	1.715, 1.764(16)		2	6	509
$\{Re(O)Cl_2(en)\}_2(\mu-O)$	1.67(5)		2	6	510
$[\{Re(O)(acacen)(\mu-O)\}_2Re(acacen)]^+$	1.683(5)	97.9±3.7	2	6	511
$Re(O)Cp^*Cl_2$	1.700(4)		2	6	512
$Re(O)Cp^*I_2$	1.687(3)		2	6	512
$Re(O)Cp^*(Me)_2$	1.681(7)		2	6	512
$Re(O)Cp^*(CH_2Ph)_2$	1.686(2)		2	6	512
$Re(O)Cp^*(OCH_2CH_2O)$	1.696(2)		2	6	513
$Re(O)Cp^*[O_2C=CPh_2]$	1.674(5)		2	6	514
$Re(O)Cp^*[OC(O)CPh_2O]$	1.68(1)		2	6	515
$Re(O)Cp^*(\mu-O)_2ReCp^*Cl_2$	1.681(3)		2	6	512
$Re(O)Cp^*(\mu-O)_2ReCp^*(OReO_3)_2$	1.72(1)		2	6	514
$Re(O)Cp^*[(CMe=CMe)_2O]$	1.716(8)		2	6	516
$Re(O)(MeCCMe)(\mu O)(\mu MeCCMe)Re(MeCCMe)_2$	1.64(2)		?	4	517
$Re(O)I(MeC\equiv CMe)_2$	1.697(3)	111.3±3.5	4	4	518
$[Re(O)(MeC\equiv CMe)_2py]SbF_6$	1.692(3)	112.2±5.3	4	4	519
$[Re(O)(MeC\equiv CMe)_2bipy]SbF_6$	1.692(3)		4	5	519
$\{Re(O)MeC\equiv CMe)_2\}_2$	1.693(11)	112.0±3.0	5	4	517

Complex	M=O (Å)	< O–M–O	d e−	CN	Ref.
ReO_4^-	1.721(4)		0	4	161
$Re(O)_3F^a$	1.692\{3\}[a]		0	4	464
$Re(O)_3Cl^a$	1.702\{4\}[a]		0	4	464
$Re(O)_3(OSiMe_3)$	1.55(8) − 1.71(8)	100 − 118(4)	0	4	520
Re_2O_7	1.68(3) − 1.74(3)	99.6, 109.1(16)	0	4&6	521
$Re(O)_3(H_2O)_2(\mu-O)Re(O)_3$	1.73(2) − 1.77(3)	109.3, 104.2(12)	0	4&6	522
$Re(O)_3py(\mu-O)Re(O)_3py_2$ (fac)	1.695(8)−1.706(9)	113.4−103.6(5)	0	5	523
	1.710(8)−1.724(9)	103.3, 104.7(4)	0	6	
$[Re(O)_3Cl_2(H_2O)]^-$ (fac)	1.691(16)−1.786(16)	102.6−109.4(9)	0	6	524
$Re(O)_3Cl(py)_2$ (fac)	1.708(9), 1.718(15)	103.5, 107.1(4)	0	6	525

TABLE 5.2 Metal–Oxo Structures *(Cont.)*

Complex	M=O (Å)	< O–M–O	d e–	CN	Ref.
[Re(O)$_3${c-(NHCH$_2$CH$_2$)$_3$}]Cl	1.756(5)	102.7(4)	0	6	526
{Re(O)$_3$}$_2$(μ–OH)$_2$(μ–dioxane)	1.697(6)–1.715(5)	102.7–105.7(2)	0	6	527
Re(O)$_3$(C$_5$Me$_4$Et)	1.664(4)–1.716(4)	104.7–106.7(2)	0	6	512
Re(O)$_2$(C$_6$H$_2$Me$_3$)$_2$	1.688(5)	121.5(4)	1	4	5
Re(O)$_2$(R)ReL$_4$(μ–O)Re(O)R$_2$ R=CH$_2$SiMe$_3$, L=PMe$_3$	1.61(3), 1.64(3)	108(1)	2	4	491
Re(O)$_2$I(PPh$_3$)$_2$	1.742(11)	138.7(6)	2	5	508
[Re(O)$_2$py$_4$]Cl	1.745(12), 1.782(13)	174.5(4)	2	6	528
[Re(O)$_2$(4-Mepy)$_4$]ReO$_4$	1.75(2), 1.74(2)	180	2	6	529
[Re(O)$_2$(en)$_2$]Cl	1.761(7), 1.769(7)	179.3(3)	2	6	528
[Re(O)$_2$(en)$_2$]NO$_3$	1.766(9)	180	2	6	530
[Re(O)$_2$(CN)$_4$]K$_3$	1.773(8)	180	2	6	531

Complex	M=O (Å)	< O–M–L	d e–	CN	Ref.
Fe(O)[(^{t}BuCONHC$_6$H$_4$)$_4$porphyrin](thf)	1.604(19)?		4	6	532
Fe(O)(Ph$_4$porphyrin)(N–Me–imidazole)[h]	1.64–1.66[h]		4	6	533
[Fe(O)(Mesityl$_4$porphyrin)(MeOH)]$^{+}$[h]	1.62–1.66[h]		4	6	533
[Ru(O)(MeCN){c-(NMe)$_4$C$_{10}$H$_{20}$}](PF$_6$)$_2$	1.765(5)	88.3±2.0	4	6	534
[Ru(O)Cl{c-(NMe)$_4$C$_{10}$H$_{20}$}]ClO$_4$ (trans)	1.765(7)	90.8±2.1	4	6	44
[Ru(O)NCO{c-(NMe)$_4$C$_{10}$H$_{20}$}]ClO$_4$ (trans)	1.765(5)	91.7(2)	4	6	534
[Ru(O)N$_3${c-(NMe)$_4$C$_{10}$H$_{20}$}]ClO$_4$ (trans)	1.765(5)		4	6	534
[Ru(O)Cl(py)$_4$]ClO$_4$ (trans)	1.862(8)	90.1±1.5	4	6	45

Complex	M=O (Å)	< O–M–O	d e–	CN	Ref.
Ru(O)$_4$[a]	1.705(3)[a]		0	4	535
[Ru(O)$_3$(OH)$_2$]Ba (tbp)	1.751, 1.759	118.4, 120.8	2	5	536
[Ru(O)$_2${c-(NMe)$_4$(CH$_2$)$_{11}$}]$^{2+}$	1.718(5)	180	2	6	537
[Ru(O)$_2${c-(NMe)$_4$(CH$_2$)$_{12}$}]$^{2+}$	1.705(7)	180	2	6	537
Ru(O)$_2$[OC(O)Me]$_2$py$_2$	1.726(1)	180	2	6	538

TABLE 5.2 Metal–Oxo Structures *(Cont.)*

Complex	M=O (Å)	< O–M–L	d e-	CN	Ref.
$Os(O)F_4$[a]	1.624(25)[a]		2	5	17
$Os(O)Cl_4$[a]	1.663(5)[a]	108.3(2)[a]	2	5	17
$Os(O)(O_2C_2H_4)_2$	1.670(12)	110.1±2.7	2	5	539
$Os(O)(O_2C_2Me_4)_2$	1.62		2	5	540
$[Os(O)(O_2C_2Me_4)]_2(\mu-O)_2$	1.675(7)	111.1±3.7	2	5	541
$\{Os(O)(OCMe_2CH_2N^tBu)\}_2(\mu-O)_2$	1.67		2	5	542
$[Os(O)(O_2C_6H_{10})(NC_7H_{13})]_2(\mu-O)_2$	1.73		2	6	543

Complex	M=O (Å)	< O–M–O	d e-	CN	Ref.
$Os(O)_4$	1.684(7), 1.710(7)	106.7–110.7(4)	0	4	161
$Os(O)_4$[a]	1.711(3)[a]		0	4	544
$Os(O)_4$(quinuclidine)	1.697 – 1.722		0	5	545
$[Os(O)_4Cl]PPh_4$	1.711(6)–1.722(5)	101.3–117.2(3)	0	5	546
$[Os(O)_3N]Cs$	1.739(8), 1.741(14)	109.8, 111.4(5)	0	4	547
$Os(O)_3$(N-adamantyl)	1.710(3)–1.720(4)	109.0–109.7(2)	0	4	548
$Os(O)_2(N^tBu)_2$	1.744(6)	109.9(4)	0	4	548
$\{Os(O)_3(NCMe_2CH_2CMe_3)\}_2$(DABCO)	1.71(1)		0	5	549
$Os(O)_2(C_6H_2Me_3)_2$	1.690(7), 1.700(7)	136.1(3)	2	4	5
$[Os(O)_2(en)_2](HSO_4)_2$	1.74(1)	180	2	6	550
$[Os(O)_2Cl_4]K_2$	1.750(22)	180	2	6	551
$[Os(O)_2(OH)_4]K_2$	1.77	180	2	6	552
$Os(O)_2(NH_2CH_2COO)_2$	1.731(3)	180	2	6	553
$Os(O)_2(OH)_2$(phen)	1.737(4), 1.747(4)	166.2(2)	2	6	554
$Os(O)_2(py)_2(O_2adenine)$	1.78(4)	164(3)	2	6	555
$Os(O)_2(py)_2(O_2thymine)$	1.82, 1.87(3)	162(1)	2	6	556
$Os(O)_2(py)_2[O_2(1-Me-thymine)]$	1.732(7), 1.753(8)	164.0(5)	2	6	81
$Os(O)_2(py)_2(O_2benzanthracene)$	1.71, 1.73	"trans"	2	6	557
$[Os(O)_2(py)_2]_2(O_4c-octadiene)$	1.72	163.5(7)	2	6	558
$[Os(O)_2(py)_2]_2(\mu-O)_2$	1.73(2), 1.75(2)	163.5(8)	2	6	554
$Os(O)_2(4-^tBupy)_2(\mu-O)(\mu-OC..)$	1.711(8), 1.726(8)	161.2(4)	2	6	559
$Os(O)_2(\eta^2-O_2CMe)(\eta^1-O_2CMe)_2$	1.700(7), 1.722(8)	125.2(3)	2	6	74

[a]Structure determined by electron diffraction or microwave spectroscopy; measurement errors given inside {} brackets are 99% confidence limits.

TABLE 5.2 Metal–Oxo Structures *(Cont.)*

[b]There is a hydrogen bonded between the vanadyl oxygen and the phenol.
[c]Several structure determinations exist for these ions; the metal-oxygen distance given is a roughly average value.
[d]There is a weak axial interaction (Nb–Cl = 3.011Å), so this complex could be considered to be six coordinate.
[e]Distortional isomers; data for both structures are reported.
[f]Structure is not octahedral, rather "strongly distorted trigonal prism."
[g]Structure is not octahedral, rather "skew-trapezoidal bipyramidal."
[h]Structrural data determined by EXAFS.

TABLE 5.3 Imido, Hydrazido, and Nitrido Structures

Complex	M–N (Å)	< M–N–R[a]	d e-	CN	Ref.
V(N^{t}Bu)(SSiMe$_3$)$_3$	1.622(2)	170.7(4)	0	4	560
{V(N^{t}Bu)(S^{t}Bu)$_2$}$_2$(μ–O)	1.628(3)	161.8(3)	0	4	560
V(N–adamantyl)(OSiMe$_3$)$_3$	1.614(2)	175.8(2)	0	4	561
V(NSiMe$_3$)Cl$_3$	1.59(1)	177.5(7)	0	4	562
V[N{Pt(Me)(PEt$_3$)$_2$}](OSiMe$_3$)$_3$	1.600(7)	168.5(4)	0	4	355
V(N^{t}Bu)Cl$_2$(N^{t}BuSiMe$_2$NH^{t}Bu)	1.636(2)	161.5(2)	0	5	563
{V(N^{t}Bu)}$_3$Cl$_2$(μ_2–NPh)$_3$(μ_3–PhNCONH^{t}Bu)	1.608–1.648(9)		0	5	564
{V(NCl)Cl$_2$}$_2$(μ–Cl)$_2$	1.642(9)	175.2(3)	0	5	565
{V(N^{t}Bu)(NH^{t}Bu)(NH$_2$^{t}Bu)Cl}$_2$(μ–Cl)$_2$	1.624(3)	164.2(3)	0	6	566
V(NCl)Cl$_3$(bipy)	1.688(6)		0	6	567
V(NCl)Cl$_3$(SbCl$_5$)$_2$	1.655(3)		0	6	567
V(NSiMe$_3$)Cp$_2$	1.665(10)	178.0(6)	1	7	568
V(NPh)Cp*$_2$	1.730(5)	178.2(6)	1	7	27
V(NC$_6$H$_3$Me$_2$)Cp*$_2$	1.707(6)	179.7(5)	1	7	569
V[NN(SiMe$_3$)$_2$]Cp$_2$	1.666(6)	180	1	5–7	570
{V(μ–N)Cl$_2$py$_2$}$_\infty$ (V≡N→V)	1.571(7)	180	0	6	355
{Nb(NPh)Cl$_2$(SMe$_2$)}$_2$(μ–Cl)$_2$	1.733(7)	175.1(1)	0	6	571
[Cl$_4$(MeCN)Nb(NCMeMeCN)NbCl$_4$(NCMe)]$^{2-}$	1.752(6)		0	6	572
{Nb(NPPh$_3$)Cl$_3$}$_2$(μ–Cl)$_2$	1.776(8)	171.1(6)	0	6	573
Nb(NTol)(S$_2$CNEt$_2$)$_3$	1.783(3)	167.4(3)	0	7	574
[Nb$_2$(μ–N)Br$_{10}$](NH$_4$)$_3$	1.845(2)	180	0	6	575

TABLE 5.3 Imido, Hydrazido, and Nitrido Structures *(Cont.)*

Complex	M–N (Å)	∠ M–N–R[a]	d e–	CN	Ref.
Ta(N^{t}Bu)(NMe$_2$)$_3$	1.77(2)	180	0	4	576
Ta(N^{t}Bu)[N(SiMe$_3$)$_2$]$_2$Cl	1.763(6)	165.8(6)	0	4	577
Ta(NAr)(NArC$_2$H$_4$Ph)(OAr)$_2$ (Ar=C$_6$H$_3$Me$_2$)	1.776(8)	172.0(6)	0	4	578
{Ta(NSiMe$_3$)[N(SiMe$_3$)$_2$]Br}$_2$(μ-Br)$_2$	1.761(9)	166.1(6)	0	5	579
(μ-N$_2$){Ta(=CH^{t}Bu)(CH$_2^t$Bu)(PMe$_3$)$_2$}$_2$	1.840(8)	171.9(7)	0	5	580
{Ta(N^{t}Bu)Cl$_2$(NH$_2^t$Bu)}$_2$(μ-OMe)$_2$	1.70(2)	167(2)	0	6	581
{Ta(N^{t}Bu)(NH^{t}Bu)Cl(NH$_2^t$Bu)}$_2$(μ-Cl)$_2$	1.61(3)	169(2)	0	6	582
Ta(NPh)Cl$_3$(PEt$_3$)(thf)	1.765(5)	178.9(4)	0	6	583
{Ta(NPh)Cl$_2$(SMe$_2$)}$_2$(μ-Cl)$_2$	1.747(8)	176.4(8)	0	6	584
Cl$_3$(thf)$_2$Ta(NCMeCMeN)TaCl$_3$(thf)$_2$	1.747(1)	178.7(9)	0	6	585
(μ-N$_2$){TaCl$_3$[P(CH$_2$Ph)$_3$](thf)}$_2$	1.796(5)	178.9(4)	0	6	586
{Ta(NPPh$_3$)Cl$_3$}$_2$(μ-Cl)$_2$	1.801(8)	176.8(7)	0	6	587
[Ta$_2$(μ-N)Br$_{10}$](NH$_4$)$_3$	1.849(2)	180	0	6	588
[Ta$_2$(μ-N)I$_{10}$](NH$_4$)$_3$	1.847(6)	180	0	6	575
Cr(N^{t}Bu)$_2$(C$_6$H$_2$Me$_3$)$_2$	1.623(5)	159.6(3)	0	4	589
{Cr(NSiMe$_3$)Cp}$_2$(μ-NSiMe$_3$)$_2$	1.65(1)	161(2)	1	6	590
Cr(N)(Ph$_4$porphyrin)	1.565(6)	101.9±2.4[a]	1	5	591
{Mo(N^{t}Bu)Me$_2$}$_2$(μ-N^{t}Bu)$_2$	1.730(2)	167.4(1)	0	5	592
{Mo(NTol)(O^{t}Bu)$_2$}$_2$(μ-NTol)$_2$	1.751(7)	175.2(7)	0	5	593
Mo(NTol)Cl$_4$(thf)	1.717(3)	174.6(3)	0	6	594
Mo(NC$_2$Cl$_5$)Cl$_4$(OPCl$_3$)	1.692(9)	171.8(9)	0	6	595
Mo(NPh)$_2$(S$_2$CNEt$_2$)$_2$	1.754(4)	169.4(4)	0	6	52
∠ N–Mo–N = 103.5(2)	1.789(4)	139.4(4)			
Mo(NH)(O)Cl$_2$(PEtPh$_2$)$_2$	1.70(1)	157(10)	0	6	596
Mo(NPh)Cl$_2$(S$_2$CNEt$_2$)$_2$	1.734(4)	166.8(3)	0	7	597
[Mo(NCPh$_3$)(S$_2$CNMe$_2$)$_3$]BF$_4$	1.731(2)	175.1(5)	0	7	598
[Mo(NSO$_2$Ph)(S$_2$CNMe$_2$)$_3$]PF$_6$	1.70(2)	161.3(13)	0	7	598
Mo(NTol)Cl$_3$(PEtPh$_2$)$_2$	1.725(6)	176.7(5)	1	6	72
{Mo(NTol)(S$_2$CN^{i}Bu$_2$)}$_2$(μ-S)$_2$	1.720(8)	170.0(7)	1	5	599
{Mo(NTol)(μ-S)$_2$Mo(O)}[S$_2$P(OMe)$_2$]$_2$	1.724(9)	167.4(8)	1	5	299
{Mo(NTol)}$_2${MoO}$_2$(μ_3-S)$_4$[S$_2$P(OMe)$_2$]$_4$	1.716(8)	166.9(9)	1	6	299
Mo(NTol)Cl$_2$[N(Tol)C(O)Ph](PMe$_2$Ph)	1.726(9)	177	1	6	600
Mo(NTol)[S$_2$P(OEt)$_2$]$_3$	1.732(4)	168.4(4)	1	6	601

TABLE 5.3 Imido, Hydrazido, and Nitrido Structures *(Cont.)*

Complex	M–N (Å)	∠ M–N–R[a]	d e–	CN	Ref.
{Mo(NTol)(μ_3–S)[S_2P(OEt)$_2$]}$_4$	1.700(12)– 1.745(12)	164.7(10)– 175.0(11)	1	6	602
{Mo(NTol)(μ_3–S)(S_2CN^{i}Bu$_2$)}$_4$	1.72(2)	157–173(1)	1	6	603
{Mo(NTol)[S_2P(OEt)$_2$]}$_2$(S)(HS)(O_2CCF$_3$)	1.74(1)	174(1)	1	6	604
{Mo(NTol)[S_2P(OEt)$_2$]}$_2$(S)(MeS)(O_2CCF$_3$)	1.726(6)	172.2–6.4(4)	1	6	604
[Mo(NH)Br(dppe)$_2$]Br	1.73(2)		2	6	605
Mo(NTol)Cl$_2$(PMe$_3$)$_3$	1.739(2)	175.4(2)	2	6	606
Mo(NPh)Cp(μ–CNPh)Mo(CO)$_2$Cp	1.754(2)	166.3(3)	2	6?	607
Mo(NEt)Cp*[μ–C(O)N(Et)NN]MoCp*(CO)$_2$	1.726(9)	174.6(8)	2?	6	608
Mo(NC$_6$H$_4^t$Bu)Cp[C(O)N(Tol)NN]MoCp(CO)$_2$	1.75(1)		2?	6	609
Mo(NNCPh$_2$)(O^{t}Bu)$_4$ (tbp)	1.797(3)	168.2(3)	0	5	610
[Mo(NNMe$_2$)(O)(SPh)$_3$]PPh$_4$	1.821(9)	152.5(10)	0	5	427
[Mo(NNMe$_2$)$_2$Cl(PPh$_3$)$_2$]BPh$_4$ (~tbp)	1.752(5) 1.763(4)	168.2(7) 163.8(4)	0	5	611
Mo(NNMe$_2$)$_2$(PPh$_3$)(μ–S)$_2$MoS$_2$	1.78–1.85(3)	165–178(3)	0	5	612
[Mo(NNMe$_2$)$_2$(bipy)$_2$](BF$_4$)$_2$	1.795(10)	170.9±1.7	0	6	611
Mo(NNMePh)$_2$(S_2CNMe$_2$)$_2$ (cis)	1.709(9)	172.6(8)	0	6	613
[Mo(NNMePh)(NHNMePh)(S_2CNMe$_2$)$_2$]BPh$_4$	1.75(1)		0	6	614
Mo(NNPh$_2$)$_2$(S_2CNMe$_2$)$_2$ (cis)	1.709(8)	169.9(8)	0	6	613
Mo(NNMe$_2$)(O)(NOC$_9$H$_6$ (cis)	1.800(9)	155.5(9)	0	6	428
Mo(NNPh$_2$)(O)[(SCH$_2$CH$_2$NMeCH$_2$)$_2$] (cis)	1.778(3)	172.9(2)	0	6	615
Mo(NNMe$_2$)(O)(S_2CNMe$_2$)$_2$	1.799(8)	168.0(7)	0	6	429
Mo(NNCO$_2$Me)(NHNHCO$_2$Me)(S_2CNMe$_2$)$_2$	1.74(1)	179.6(15)	0	6	614
Mo[NNC(O)Ph)]Cl(NHNCOPh)(PMePh$_2$)$_2$	1.782(9)	175(1)	0?	6	616
[Mo(NNEtPh){S_2CN(CH$_2$)$_5$}$_3$]BPh$_4$	1.715(16)	170(2)	0	7	617
Mo(NNCTol$_2$)(=CTol$_2$)Cp[MoCp(CO)$_3$]	1.741(10)	174.7(9)	1	6	587
{Mo(NNCPh$_2$)}$_2$(O^{i}Pr)$_3$(py)(μ–O^{i}Pr)$_3$	1.770(9)	155.3–164.4(8)	1	6	610
[Mo(NNH$_2$)(NOC$_9$H$_6$)(PMe$_2$Ph)$_3$]I	1.743(4)	172.3(5)	2	6	618
[Mo(NNH$_2$)F(dppe)$_2$]BF$_4$	1.762(12)	176.4(13)	2	6	619
[Mo(NNH$_2$)(triphos)(PPh$_3$)]Cl (>isomers)	1.694(12)	172.2(11)	2	6	48
[Mo(NNH$_2$)(triphos)(PPh$_3$)]Cl	1.778(18)	178.1(15)	2	6	48
[Mo(NNHC$_8$H$_{17}$)I(dppe)$_2$]I	1.801(11)	174(1)	2	6	620
[Mo(NNPh)Mo(NNHPh)($S_2C_2H_4$)$_3$($S_2C_2H_5$)]$^{2-}$	1.740(7) 1.793(6)	162.2(6) 161.1(7)	2 2	6 6	621
{Mo(NNPh)$_2$(acac)}$_2$(μ–OMe)$_2$	1.826(14)	175.9(12)	2	6	622

TABLE 5.3 Imido, Hydrazido, and Nitrido Structures *(Cont.)*

Complex		M–N (Å)	∠ M–N–R[a]	d e–	CN	Ref.
[{Mo(NNPh)$_2$(μ-OMe)(μ-O$_2$MoO$_2$)}$_2$](Bu$_4$N)$_2$		1.823(9)		2	6	623
[Mo(NNMePh)(η^2-NHNMePh)(S$_2$CNMe$_2$)$_2$]Ph$_4$B		1.752(10)	169.6(7)	2	7	624
{Mo[NNC(O)SMe][NH$_2$NC(O)SMe]}$_2$(μ-OMe)$_2$		1.749(25)	169(2)	2	7	248
Mo[NNC(S)SMe][NH$_2$NC(S)SMe](Me$_2$NO)		1.763(8)	177.8(2)	2	7	248
Mo[NNC(O)SMe][NH$_2$NC(O)SMe](S$_2$CNEt$_2$)$_2$		1.764(6)	176.1(5)	2	7	625
Mo[NNC(S)OMe][NH$_2$NC(S)OMe](S$_2$CNEt$_2$)$_2$		1.72(3)	174(2)	2	7	264
Mo[NNC(S)SEt][NH$_2$NC(S)SEt](S$_2$CNMe$_2$)$_2$		1.776(9)	178.3(8)	2	7	248
Mo(NNCTol$_2$)(CTol$_2$)Cp[MoCp(CO)$_3$]		1.741(10)	174.7(9)	3?	6	626
[Mo(NNMe$_2$)$_2${S$_2$MoS$_2$}$_2$](PPh$_4$)$_2$		2.13(1)	165.9(16)	4?	6	627
{Mo(μ-N)(O^{t}Bu)$_3$}$_\infty$	(Mo≡N→Mo)	1.661(4)	180	0	5	628
{Mo(μ-N)(O^{i}Pr)$_3$}$_\infty$	(Mo≡N→Mo)	1.579(26)	180	0	5	628
[Mo(N)Cl$_4$]PPh$_4$		1.637(4)	103.1±0.2[a]	0	5	629
[Mo(N)Br$_4$]PPh$_4$		1.628(15)	103.0(1)[a]	0	5	630
[Mo(N)(N$_3$)$_4$]AsPh$_4$		1.630(6)	99.5±3.5[a]	0	5	631
Mo(N)(N$_3$)$_3$(py)		1.634(3)	100.2±6.2[a]	0	5	632
Mo(N)(N$_3$)$_3$(bipy)		1.642(5)	98.5±7.4[a]	0	6	633
Mo(N)(N$_3$)$_2$[HB(Me$_2$pz)$_3$]		1.642(4)	98.2±1.6[a]	0	6	634
{Mo(μ-N)Cl$_3$}$_4$	(Mo≡N–Mo)	1.638(11)	167.3(6)	0	6	635
		1.672(10)	178.1(4)			
{Mo(μ-N)Cl$_3$(^{n}Bu$_2$O)}$_4$	(Mo≡N–Mo)	1.642(6)	173.0(4)	0	6	636
		1.665(6)	175.9(4)			
{Mo(μ-N)Cl$_3$(OPCl$_3$)}$_4$	(Mo≡N–Mo)	1.659(5)	167.2(3)	0	6	637
		1.661(5)	176.2(3)			
{Mo(μ-N)[S$_2$P(OMe)$_2$]$_2$}	(Mo=N=Mo)	1.867(12)		1	6	638
Mo(N)(N$_3$)$_2$Cl(terpy)		1.662(7)	95.1±10[a]	0	7	639
Mo(N)(S$_2$CNEt$_2$)$_3$		1.641(9)	97.2±6.0[a]	0	7	640
{Mo(μ-N)(S$_2$CNEt$_2$)$_3$}$_2$Mo(S$_2$CNEt$_2$)$_3$		1.65(1)		0	7	641
[Mo(N)Cl$_4$](PMePh$_3$)$_2$		1.634(6)	100.1±0.7[a]	1	5	642
Mo(N)(N$_3$)(dppe)$_2$		1.79(2)	90.2±6.9[a]	2	6	605
W(NC$_6$H$_3$Me$_2$)(CSiMe$_3$)(CCSiMe$_3$)(O^{i}Pr)$_4$W		1.763(6)		0?	4	643
W(N^{t}Bu)(NH^{t}Bu)$_2$(O$_2$C$_2$Ph$_4$)		1.753(4)	161.0(4)	0	5	644
W(N^{t}Bu)$_2$(NH$_2$^{t}Bu)[O$_2$C$_2$(CF$_3$)$_4$]		1.742(4)	165.3(3)	0	5	644
{W(N^{t}Bu)Me$_2$}$_2$(μ-N^{t}Bu)$_2$		1.736(5)	168.3(4)	0	5	645
W(N^{t}Bu)Me[N(^{t}Bu)CMe$_2$][N(^{t}Bu)CMe=CMe$_2$]		1.757(12)	165.0(11)	0	5	646
W(NPh)(NMe$_2$)$_4$	(tbp)	1.758(5)	180	0	5	18

TABLE 5.3 Imido, Hydrazido, and Nitrido Structures *(Cont.)*

Complex	M–N (Å)	∢ M–N–R[a]	d e–	CN	Ref.
[W(N^{i}Pr)Cl$_5$]NEt$_4$	1.763(16)	174.4(16)	0	6	647
{W(N^{i}Pr)Cl$_3$}$_2$(μ–Cl)$_2$	1.697(12)	171.5(9)	0	6	647
{W(N^{t}Bu)Cl$_3$}$_2$(μ–Cl)$_2$	1.704(5)	172.9(5)	0	6	648
{W(N^{t}Bu)Cl$_2$(NH$_2$^{t}Bu)}$_2$(μ–NPh)$_2$	1.729(4)		0	6	649
W(N^{t}Bu)$_2$(bipy)Cl$_2$	1.779(9)	165.8(8)	0	6	650
W(NC$_6$H$_{11}$)[N(C$_6$H$_{11}$)C(O)C(CO)^{t}Bu]Cl$_3$	1.667(11)	177.6(11)	0	6	651
W(NPh)Cl(O^{t}Bu)$_3$(NH$_2$^{t}Bu)	1.71(2)	171(1)	0	6	652
{W(NPh)(OMe)$_3$}$_2$(μ–OMe)$_2$	1.61(4)	174(2)	0	6	652
{W(NPh)Me$_2$(PMe$_3$)}$_3$(μ–O)$_3$	1.758(11)		0	6	653
{W(NPh)(O^{t}Bu)$_2$}$_2$(μ–O)(μ–O^{t}Bu)$_2$	1.738(5)	164.8(5)	0	6	654
{W(NPh)Me$_3$}$_2$(μ–N$_2$H$_4$)(μ–N$_2$H$_2$)	1.728(13)		0	6	655
[W(NC$_2$Cl$_5$)Cl$_5$]AsPh$_4$	1.684(9)	168.0(10)	0	6	656
{W(NC$_2$Cl$_5$)Cl$_3$}$_2$(μ–Cl)$_2$	1.71(2)	177(2)	0	6	657
[W(NCBr$_3$)Br$_5$]PPh$_3$Me	1.75(2)	178(1)	0	6	658
[{W(NCPh$_3$)Cl$_4$}$_2$(μ–F)$_2$W(N)Cl$_2$]CPh$_3$	1.67(3)	179(2)	0	6	659
W(NSO$_2$NH$_2$)Cl$_4$(NCMe)	1.730(12)	168.3(7)	0	6	660
W(NSO$_2$Ph)$_2$Cl$_2$(NCMe)$_2$	1.765(5)	171.5(3)	0	6	661
	1.787(5)	166.5(3)			
W(NPh)Cl$_3$(PMe$_3$)$_2$	1.731(6)	175.8(6)	1	6	662
W(NPh)Cl$_3$(PPh$_3$)$_2$	1.742(8)	172.3(7)	1	6	663
W(NPh)Cl$_2$(PMe$_3$)$_3$	1.755(3)	179.5(3)	2	6	663
[W(NH)Br(dppe)$_2$]Br	1.69(4)		2	6	664
(μ–N$_2$){WCp*Me$_3$}$_2$	1.753(18)	168.6(16)	0	7	665
(μ–N$_2$){WCl$_2$(PhCCPh)(dme)}$_2$	1.756(11)	176.0(10)	2	6	666
W[NN(H)BPh$_3$](H)(Cl)(Br)(PMe$_2$Ph)$_3$	1.781(5)	170.0(5)	0	7	667
[W(NNHTol)Cp$_2$H]PF$_6$	1.837(7)	146.4(5)	0	8	668
W(NNH$_2$)Cl$_3$(PMe$_2$Ph)$_2$	1.752(10)	178.7(9)	1	6	669
W(NNHPh)Cl$_3$(PMe$_2$Ph)$_2$	1.738(5)	172.5(6)	1	6	670
[W(NNH$_2$)(NOC$_9$H$_6$)(PMe$_2$Ph)$_3$]I	1.753(10)	174.7(9)	2	6	618
[W(NNH$_2$)Cl(dppe)$_2$]BPh$_4$	1.73(1)	171(1)	2	6	671
[W(NNHMe)Br(dppe)$_2$]Br	1.768(14)	174(1)	2	6	672
[W(NNCMe$_2$)Br(dppe)$_2$]Br	1.724(12)	171.3(7)	2	6	673
[W{NNC(Me)CH$_2$COMe}F(dppe)$_2$]BF$_4$	1.770(17)	173.8(15)	2	6	674
[W{NNCH(CH$_2$)$_3$OH}Br(dppe)$_2$]PF$_6$	1.772(13)	172.6(12)	2	6	673
[W(NNHCMe=CHCOMe)(acac)(PMe$_2$Ph)$_3$]Br	1.793(13)	176.4(16)	2	6	675
W(NNHSiMe$_3$)I$_2$(PMe$_2$Ph)$_3$	1.777(17)	171.4(16)	2	6	676

TABLE 5.3 Imido, Hydrazido, and Nitrido Structures *(Cont.)*

Complex		M–N (Å)	∠ M–N–R[a]	d e–	CN	Ref.
[W{NNC(NMeCH$_2$)$_2$}Br(dppe)$_2$]PF$_6$		1.776(7)	163.0(5)	2	6	677
[W{NN(COCH$_2$)$_2$}F(dppe)$_2$]BF$_4$		1.75(2)	174(2)	2	6	678
[W(NNCCl$_2$)Br(dppe)$_2$]PF$_6$		1.75(2)	169(2)	2	6	679
[W{NNC(Cl)C(CN)$_2$}Br(dppe)$_2$]		1.783(5)	171.5(5)	2	6	679
W(NNSiMe$_3$)I(PMe$_2$Ph)$_4$		1.815(17)	167.9(19)	4	6	676
W[NC(Br)C(CN)C(CN)$_2$]Br(dppe)$_2$		1.777(6)	171.4(5)	4?	6	680
{W(μ-N)(O^{t}Bu)$_3$}$_\infty$	(W≡N→W)	1.740(15)	180	0	5	681
[W(N)Cl$_2${(μ-F)W(NCPh$_3$)Cl$_4$}$_2$]CPh$_3$		1.66(2)	tbp	0	5	659
{W(μ-N)Cl$_3$}$_4$(N$_3$H)$_2$		1.69(3)	162,177(1)	0	6	682
{W(μ-N)Cl$_3$(POCl$_3$)}$_4$	(W≡N–W)	1.556(2)–	153.0(3)–	0	6	683
		1.691(2)	177.9(2)			
[W$_2$(μ-N)Cl$_{10}$]PPh$_4$	(W=N=W)	1.79,1.88(2)	173(1)	0	6	684
[W$_3$(μ-N)$_2$Cl$_{14}$](PPh$_4$)$_2$	(W=N=W)	1.81,1.86(1)	175.6(8)	0	6	685
[W$_2$(μ-N)Cl$_{10}$](AsPh$_4$)$_2$		1.71(7)	180	0?	6	686
Mn(N)(Ph$_4$porphyrin)		1.515(3)		2	5	687
Mn(N)(Me$_2$Et$_8$porphyrin)		1.512(2)		2	5	688
[Tc(N)Cl$_4$]AsPh$_4$		1.581(5)	103.34(3)[a]	1	5	689
[Tc(N)Br$_4$]AsPh$_4$		1.596(6)	103.04(2)[a]	1	5	690
Tc(N)(S$_2$CNEt$_2$)$_2$		1.604(6)	108.1±1.1[a]	2	5	691
Tc(N)(SNC$_9$H$_6$)$_2$		1.623(4)	105.6±6.6[a]	2	5	692
Tc(N)(NCS)$_2$(NCMe)(PPh$_3$)$_2$		1.629(4)	96.3±4.9[a]	2	6	693
{Re(N^{t}Bu)$_2$(OSiMe$_3$)}$_2$(O)(OSiMe$_3$)(ReO$_4$)		1.70(1)	153–168(1)	0	6	694
Re(NF)F$_5$		1.711(11)	177.5(11)	0	6	695
Re(NCl)F$_5$		1.73(2)	176.7(14)	0	6	695
Re(NCl)F$_4$(μ-F)Re(N)F$_4$		1.64(2)	177(2)	0	6	696
Re(NC$_2$Cl$_5$)Cl$_4$(OPCl$_3$)		1.69(1)	172(1)	1	6	697
[Re(NMe)Cl(NH$_2$Me)$_4$](ClO$_4$)$_2$		1.694(11)	179.8(18)	2	6	698
Re(NMe)Cl$_3$(PEtPh$_2$)$_2$		1.685(11)	173.4(10)	2	6	699
Re(NPh)Cl$_3$(PPh$_3$)$_2$		1.726(6)	172.6(6)	2	6	700
Re(NC$_6$H$_4$OMe)Cl$_3$(PEt$_2$Ph)$_2$		1.709(4)	175.8(1)	2	6	701
Re[NC$_6$H$_4$C(O)Me]Cl$_3$(PEt$_2$Ph)$_2$		1.690(5)	171.8(4)	2	6	701
Re(NTol)Cl$_2$(MeNCHC$_6$H$_4$O)(PPh$_3$)		1.71(1)	171(1)	2	6	702
Re(NTol)Cl$_2$(PhNCHC$_6$H$_4$O)(PPh$_3$)		1.75(1)	168(1)	2	6	702
Re(NTol)(OEt)(S$_2$CNMe$_2$)$_2$		1.744(5)	155.5(5)	2	6	703

TABLE 5.3 Imido, Hydrazido, and Nitrido Structures *(Cont.)*

Complex	M–N (Å)	∠ M–N–R[a]	d e–	CN	Ref.
Re(NBCl$_3$)Cl$_2$(PMe$_2$Ph)$_3$	1.728(7)	176.5(6)	2	6	704
[Re(NNHPh)Cl$_2$(PMe$_2$Ph)$_2$(NH$_3$)]Br	1.750(12)	172(1)	2	6	705
Re[NNHC(O)Ph][NHNHC(O)Ph]Cl$_2$(PPh$_3$)$_2$	1.730(7)	174.7(7)	2	6	614
Re(NNHPh)(NNPh)(PPh$_3$)$_2$	1.922(11)	131.2(10)	2	6	706
	1.793(11)	172.4(10)			
Re[NN(Me)C$_6$H$_4$OMe]Cp(CO)$_2$	1.937(7)	138.1(6)	4	6	707
Re(NNC$_6$H$_4$Cl)(OSN$_2$C$_8$H$_{18}$O$_3$)PPh$_3$	1.75(2)	158.8(17)	4	6	708
Re[NNC(O)C$_6$H$_4$Cl](S$_2$N$_2$C$_8$H$_{18}$)PPh$_3$	1.76(2)	165.1(12)	4	6	708
Re[NNC(O)Ph](S$_4$C$_7$H$_{14}$)PPh$_3$	1.776(12)	167.5(12)	4	6	708
Re(NNCO$_2$Me)(S$_4$C$_7$H$_{14}$)PPh$_3$	1.770(11)	167.0(10)	4	6	708
Re(NPPh$_3$)Cl$_3$(NO)(OPPh$_3$)	1.855(8)	138.5(5)	4	6	709
{Re(μ–N)Cl$_4$}$_\infty$ (Re≡N→Re)	1.58(4)	174(2)	0	6	710
Re(N)F$_4$(μ–F)Re(NCl)F$_4$	1.59(2)	98.9±5.9[a]	0	6	696
{[Re(μ–N)(NC)$_4$]K$_2$·H$_2$O}$_\infty$	1.53	180	0	6	711
[Re(N)Cl$_4$]AsPh$_4$	1.619(10)	103.49(6)[a]	1	5	712
[Re(N)Br$_4$]AsPh$_4$	1.620(14)	103.0[a]	1	5	713
[Re(N)(NCS)$_5$](AsPh$_4$)$_2$	1.657(12)	96.0±0.8[a]	1	6	714
Re(N)(S$_2$CNEt$_2$)$_2$	1.656(8)	107.7±1.4[a]	2	5	715
Re(N)Cl$_2$(PPh$_3$)$_2$	1.602(9)	98.9±11[a]	2	5	716
Re(N)Cl$_2$(PMe$_2$Ph)$_3$	1.660(8)	95.2±8.6[a]	2	6	35
Re(N)Cl$_2$(PEt$_2$Ph)$_3$	1.788(11)	93.9±5.3[a]	2	6	34
Fe(NNC$_9$H$_{18}$)[(ClC$_6$H$_4$)$_4$porphyrin]	1.809(4)	180	6?	5	717
{Fe(Ph$_4$porphyrin)}$_2$(μ–N)	1.6605(7)	180	4.5	5	718
Ru(NPEt$_2$Ph)Cl$_3$(PEt$_2$Ph)$_2$	1.841(3)	174.9(3)	4	6	719
[Ru(N)Cl$_4$]AsPh$_4$	1.570(7)	104.58(4)[a]	2	5	720
[Ru(N)Br$_4$]AsPh$_4$	1.580(11)	104.25(3)[a]	2	5	721
[Ru$_2$(μ–N)Cl$_8$(H$_2$O)$_2$]K$_3$	1.720(4)	180	4	6	722
[Ru$_2$(μ–N)(en)$_5$]Cl$_5$	1.742(1)	174.6(4)	4	6	723
Os(N–adamantyl)(O)$_3$	1.697(4)	171.4(4)	0	4	548
Os(N^{t}Bu)$_2$(O)$_2$	1.710(8)	178.9(9)	0	4	548
	1.719(8)	155.1(8)			
{Os(NCMe$_2$CH$_2$CMe$_3$)O$_3$}$_2$(DABCO)	1.73(1)		0	5	549
Os(N^{t}Bu)$_2$[N^{t}Bu(CHCO$_2$Me)$_2$NSO$_2$C$_6$H$_2$^{i}Pr$_3$]	1.648(4)	172.3(3)	2	4	20
	1.719(3)	165.4(4)			

TABLE 5.3 Imido, Hydrazido, and Nitrido Structures *(Cont.)*

Complex	M–N (Å)	∠ M–N–R[a]	d e–	CN	Ref.
$Os(NMe)(CH_2SiMe_3)_4$	1.686(5)	163.6(6)	2	5	724
$[Os(N)O_3]Cs$	1.676(15)		0	4	547
$[Os(N)Cl_4]AsPh_4$	1.604(10)	104.53(3)[a]	2	5	725
$[Os(N)Br_4]AsPh_4$	1.583(15)	104.29(3)[a]	2	5	726
$[Os(N)I_4]AsPh_4$	1.626(17)	103.73(2)[a]	2	5	727
$[Os(N)(CH_2SiMe_3)_4]NBu_4$	1.631(8)	107.7±1.1[a]	2	5	724
$[Os(N)(N_2O_2C_{20}H_{12}O_2)]^-$	1.640(7)	106.3±2.6[a]	2	5	728
$[Os(N)Cl_5]K_2$	1.614(13)	96.2±1.3[a]	2	6	58
$Os_2(\mu\text{-}N)(S_2CNMe_2)_5$	1.76(3)	164.6(18)	4	6	729
$U(NPh)(C_5H_4Me)_3$	2.019(6)		f^1	10	730
$U[NC(Me)CHPMePh_2]Cp_3$	2.06(1)	163(1)	f^2	10	731

[a] Values marked by an [a] are average nitrogen–metal–cis ligand angles.

TABLE 5.4 Structures with Metal–Carbon Multiple Bonds

Complex	M–C (Å)	∠ M–C–R	d e–	CN	Ref.
$Nb_2(\mu\text{-}CSiMe_3)_2(CH_2SiMe_3)_4$	1.954, 1.995(9)		0	4	732
$Ta(=CHSiMe_3)(CH_2SiMe_3)(OC_6H_3{}^tBu_2)_2$	1.888(29)		0	4	733
$Ta(=CH^tBu)_2(C_6H_2Me_3)(PMe_3)_2$	1.932(7)	168.9(6)	0	5	734
∠ C–Ta–C = 109.0(3)	1.955(7)	154.0(6)			
$\{Ta(=CH^tBu)(CH_2{}^tBu)(PMe_3)_2\}_2N_2$	1.935(9)	159.4(7)	0	5	580
$\{Ta(=CH^tBu)Cl_2(\mu\text{-}Cl)(PMe_3)\}_2$	1.898(2)[a]	161.2(1)[a]	0	6	55
$Ta(=CH^tBu)Cp^*(C_2H_4)PMe_3$	1.946(3)[a]	170.0(2)[a]	0	6	55
$Ta(=CHPh)Cp^*(CH_2Ph)_2$	1.883(14)	166.0(10)	0	6	735
$Ta(=CHPh)Cp_2(CH_2Ph)$	2.07(1)	135.2(7)	0	8	736
$Ta(=CH^tBu)Cp_2Cl$	2.030(6)	150.4(5)	0	8	737
$Ta(=CH_2)Cp_2(Me)$	2.026(10)		0	8	26
$Ta(=CPhCPhCMeN^tBu)Cp(Me)$	1.98(1)		0	6	738
$Ta(\equiv C^tBu)(H)(ClAlMe_3)(dmpe)_2$	1.850(5)	178.7(4)	0	7	739

TABLE 5.4 Metal–Carbon Multiple Bonds *(Cont.)*

Complex	M–C (Å)	∠ M–C–R	d e−	CN	Ref.
Ta(≡CPh)Cp*Cl(PMe$_3$)$_2$	1.849(8)	171.8(6)	0	7	740
Ta(≡C^{t}Bu·LiN$_2$C$_6$H$_{14}$)(CH$_2$^{t}Bu)$_3$	1.76(2)	165(1)	0	4	741
(CH$_2$SiMe$_3$)$_2$Ta(μ-CSiMe$_3$)$_2$Ta(OC$_6$H$_3$^{t}Bu$_2$)$_2$	2.000(6)	122,134(.6)	0	4	742
Ta$_2$(μ-CSiMe$_3$)$_2$(CH$_2$SiMe$_3$)$_3$(OC$_6$H$_3$Ph$_2$)	1.93(2)–1.98(1)	131.8±13	0	4	742
{Ta(μ-C^{t}Bu)Cl$_2$(dme)}$_2$(μ-Cl)$_2$Zn	1.79,1.86(2)	159,156(1)	0	6	743
Cr(=CPh$_2$)Cp(CO)(NO)	1.912(8)	123.9±1.3	4	6	744
Cr(=CCPh=CPh)(CO)$_5$	2.05(1)		4	6	745
Cr[=C=C(Ph)NMe$_2$](CO)$_5$	2.015(15)		4	6	746
Cr(≡CMe)Cl(CO)$_4$	1.710		2	6	747
Cr(≡CMe)I(CO)$_4$	1.69(1)		2	6	50
Cr(≡CMe)Br(CO)$_3$PMe$_3$	1.68(3)	177(3)	2	6	50
[Cr≡CMe)(CO)$_4$(PMe$_3$)]BCl$_4$	1.67	175	2	6	748
Cr(≡C-menthyl)Br(CO)$_4$	1.67(2)	174(1)	2	6	749
Cr[≡C(c-C$_3$H$_5$)](CO)$_4$(μ-Br)Cr(CO)$_5$	1.714(13)	177.7(19)	2	6	750
Cr(≡CPh)Cl(CO)$_4$[a]	1.728(4)[a]	180	2	6	751
Cr(≡CPh)Br(CO)$_4$	1.68(3)	180	2	6	752
Cr(≡CPh)Br(CO)$_2$(CN^{t}Bu)$_2$	1.76(3)	171(2)	2	6	753
Cr(≡CPh)Br(CO)$_2$[P(OPh)$_3$]$_2$	1.68(1)	173.5(10)	2	6	753
Cr(≡CC$_6$H$_4$CF$_3$)Br(CO)$_4$	1.68(2)	171(3)	2	6	754
Cr(≡CC$_5$H$_4$FeCp)Br(CO)$_4$	1.71(2)	175.1(15)	2	6	755
Cr(≡C-cyclopentenyl)I(CO)$_4$	1.65(2)	176.0(15)	2	6	756
Cp(CO)$_2$Cr≡CCPh=CPhC≡Cr(CO)$_2$Cp	1.707(2)	173.9(2)	2	6	757
[Cr(≡CNEt$_2$)(CO)$_5$]BF$_4$	1.797(9)	175.8(8)	2	6	758
[Cr(≡CNEt$_2$)(CO)$_4$(PPh$_3$)]BF$_4$	1.757(11)	175.2(17)	2	6	758
Cr(≡CNEt$_2$)Br(CO)$_4$	1.720(10)	177.2	2	6	758
Cr(≡CNEt$_2$)(SnPh$_3$)(CO)$_4$	1.744(12)	177.0(12)	2	6	758
Cr(≡CNEt$_2$)Br(CO)$_3$(PPh$_3$)	1.752(12)	175.8(8)	2	6	758
Cr(≡CN^{i}Pr$_2$)Cl(CO)$_4$	1.747(5)	177.4(4)	2	6	759
Mo(=CTol$_2$)Cp(NNCTol$_2$)MoCp(CO)$_3$	1.98(1)		1	6	760
{Mo(=CHSiMe$_3$)Br(PMe$_3$)$_2$}$_2$	1.949(5)	129.8(3)	3	5	761
Mo(=C^{t}BuCH=C^{t}BuCC^{t}Bu)[OCH(CF$_3$)$_2$]$_2$(py)	1.989(3)		2	5	762
Mo[=C^{t}Bu(C$_6$H$_2$^{t}Bu$_3$)(O$_2$CMe)$_2$	1.945(3)		2	7	762
Mo(=C=CHPh)Cp(Br)[P(OMe)$_3$]$_2$	1.917(5)		2	7	763

TABLE 5.4 Metal–Carbon Multiple Bonds *(Cont.)*

Complex	M–C (Å)	∢ M–C–R	d e–	CN	Ref.
Mo[=C=C(CN)$_2$]Cp(Cl)[P(OMe)$_3$]$_2$	1.833(6)	166.6(4)	2	7	764
[Mo(≡CCH$_2$^{t}Bu)Cp(H){P(OMe)$_3$}$_2$]BF$_4$	1.798(2)		0	7	765
Mo(≡CPh)Cl{P(OMe)$_3$}$_4$	1.793(8)		2	6	766
Mo(≡CPh)(CO)$_4$Re(CO)$_5$	1.835(25)	180	2	6	767
Mo(≡CCl)(Bpz$_4$)(CO)$_2$	1.894(9)		2	6	768
Mo(≡CSiMe$_3$)Br(dppe)$_2$	1.819(12)	174.2(7)	2	6	761
{Mo(≡CNEt$_2$)(CO)$_3$(μ-I)}$_2$	1.795(18)	172.0(14)	2	6	70
{Mo(≡CNEt$_2$)(CO)$_3$(μ-NCO)}$_2$	1.816(16)	174.2(12)	2	6	70
Mo(≡CSC$_6$H$_4$NO$_2$)[HB(Me$_2$pz)$_3$](CO)$_2$	1.801(4)	179.5(2)	2	6	769
W(=CH^{t}Bu)(≡C^{t}Bu)(CH$_2$^{t}Bu)(dmpe) (spy)	1.942(9)	150.4(8)	0	5	23
W(=CH^{t}Bu)(O)Cl$_2$(PEt$_3$) (tbp)	1.882(14)	140.6(11)	0	5	24
W(=CH^{t}Bu)(O)Cl$_2$(PMe$_3$)$_2$	1.986(21)	142.4(19)	0	6	25
W[=C(CH$_2$)$_5$]Cl$_2$(PMePh$_2$)$_2$	1.980(12)	126.4±5.9	0	6	461
[W(=CH$_2$)I(PMe$_3$)$_4$]CF$_3$SO$_3$	1.83(2)		2	6	770
W(=CH^{t}Bu)Cl$_2$(CO)(PMe$_3$)$_2$	1.859(4)	168.7(3)	2	6	771
W(=CHPh)Cl$_2$(CO)(PMe$_3$)$_2$	1.860(7)	164.6(6)	2	6	772
W(=CHPh)Cl$_2$(PhCCPh)(PMe$_3$)$_2$	1.97(2)	137(1)	2	6	773
W(=CPhCH=CHMe)Br$_2$(CO)$_2$(4-Mepy)	1.98(2)		2	7	774
W(=CHTol)Cp(I)(CO)$_2$	2.05(2)	139(1)	2	7	775
W(=CHPh)Cp$_2$	2.05(2)	133(2)	2	7	776
W(=CHTol)Cp(CO)$_2$(SnPh$_3$)	2.032(7)	137.8(5)	2	7	777
W(=CPh$_2$)(CO)$_5$	2.14(2)	124.0±3.3	4	6	778
W(=CPh$_2$)(CO)$_4$(trans-cyclooctene)	2.067(4)		4	6	779
W(=C=CHCO$_2$Me)(CO)$_3$(dppe)	1.98(1)	173(1)	4	6	780
W$_2$(μ-CSiMe$_3$)$_2$(CH$_2$SiMe$_3$)$_4$	1.89(3)		1	4	781
W$_2$(μ-CSiMe$_3$)(CPhCPhCSiMe$_3$)(CH$_2$SiMe$_3$)$_4$	1.96, 2.00(1)		?	5	782
W$_2$(μ-CPh)$_2$(O^{t}Bu)$_4$	1.938(11)		1	4	783
W(C$_3$Et$_3$)(OC$_6$H$_3$^{i}Pr$_2$)$_3$	1.883(10)		0	5	784
W(C$_3$Et$_3$)[OCH$_3$(CF$_3$)$_2$]$_3$	1.862(17)		0	5	785
W(CPhC^{t}BuCPh)CpCl$_2$	1.943(5)		0	7	786
W(CHCHCPhN)W(O^{t}Bu)$_6$	1.980(6)	119.8(5)	1	5	787
W(C^{t}BuCC^{t}Bu)Cp(Cl)	1.924(8)		?	?	788
W(≡CPh)(O^{t}Bu)$_3$	1.758(5)	175.8(4)	0	4	440
{W(≡CMe)(O^{t}Bu)$_2$(μ-O^{t}Bu)}$_2$	1.759(6)	179.8(6)	0	5	681
{W(≡CNMe$_2$)(O^{t}Bu)$_2$(μ-O^{t}Bu)}$_2$	1.76(2)	179(2)	0	5	789

TABLE 5.4 Metal–Carbon Multiple Bonds *(Cont.)*

Complex	M–C (Å)	∠ M–C–R	d e-	CN	Ref.
W(≡C^{t}Bu)(=CH^{t}Bu)(CH$_2$^{t}Bu)(dmpe) (spy) ∠ C≡W=C = 108.7(4)	1.785(8)	175.3(7)	0	5	23
{W(≡CEt)(O^{i}Pr)$_2$(μ-O^{i}Pr)(NHMe$_2$)}$_2$	1.772(11)		0	6	68
{W(≡C^{t}Bu)(C$_5$Me$_4$^{t}Bu)I}$_2$(μ-N$_2$H$_2$)	1.769(8)	171.9(7)	0	6	790
W(≡C^{t}Bu)Cl$_2$(PHPh)(PEt$_3$)$_2$	1.808(6)	174.0(4)	0	6	66
W(≡C^{t}Bu)Cl$_3$(PMe$_3$)$_3$	1.793(6)	178.6(4)	0	7	791
{W(≡CPMe$_3$)Cl(μ-Cl)(PMe$_3$)$_2$}$_2$[AlCl$_4$]$_2$	1.833(30)	174.0(19)	1	6	792
W(≡CAl$_2$Me$_4$Cl)Me(PMe$_3$)$_2$(CH$_2$CH$_2$)	1.813(5)		2	5	793
W(≡CH)Cl(PMe$_3$)$_4$	~1.84		2	6	794
W(≡CH·AlMe$_{1.8}$Cl$_{1.2}$)Cl(PMe$_3$)$_3$	1.807(6)	163.8(44)	2	6	794
W(≡CMe)Me(PMe$_3$)$_4$	1.891(25)		2	6	67
W(≡CMe)Br(CO)$_4$	1.82(4)	178(4)	2	6	795
W(≡CMe)I(CO)$_4$	1.77(4)	180	2	6	795
W(≡CPh)I(CO)$_4$	1.90(5)	162(4)	2	6	50
W(≡CPh)(CO)$_4$[Co(CO)$_4$] (trans)	1.821(12)	175.6(1)	2	6	796
(CO)$_4$BrW≡CC$_6$H$_4$C≡WBr(CO)$_4$	1.89(3)	172(3)	2	6	797
W[≡CPhCr(CO)$_3$]Cl(CO)$_4$	1.84(3)	173.9	2	6	798
W(≡CSiPh$_3$)Cp(CO)$_2$	1.81(2)	176(1)	2	6	799
W(≡CTol)Cp(CO)$_2$	1.82(2)	172(2)	2	6	800
W(≡CTol)(Bpz$_4$)(CO)$_2$	1.821(7)		2	6	801
W(≡CPh)Br(CO)$_2$(py)$_2$ (all trans)	1.84(2)	173.8(6)	2	6	802
W(≡CPh)Cl(CO)(py)$_2$(maleic anhydride)	1.801(6)		2	6	69
[W(≡CNEt$_2$)(CO)$_5$]SbCl$_6$	1.85(3)	172(3)	2	6	803
{W(≡CNEt$_2$)(CO)$_3$(μ-N$_3$)}$_2$	1.748(37)	176.1(28)	2	6	70
{W(≡CNEt$_2$)(CO)$_3$(μ-SPh)}$_2$	1.816(25)	171.6(20)	2	6	70
W(≡CSPh)Cp(CO)PPh$_3$	1.807(10)	174.2(6)	2	6	804
Mn(=CMe$_2$)Cp(CO)$_2$	1.793(5)	125.3±0.9	4	6	805
Mn(=CPh$_2$)Cp(CO)$_2$	1.885(2)	123.7±7.0	4	6	744
Mn(=CC$_{14}$H$_{12}$)Cp(CO)$_2$	1.853(5)		4	6	806
Mn[=C(Ph)C(O)Ph]Cp(CO)$_2$	1.88(2)		4	6	807
Mn(=C=CH$_2$)Cp(CO)$_2$	1.79(2)	176(2)	4	6	808
Mn(=C=CHPh)Cp(CO)$_2$	1.68(2)	174(2)	4	6	809
Mn[=C=C=C(C$_6$H$_{11}$)$_2$]Cp(CO)$_2$	1.806(6)	177.9(5)	4	6	810
[Mn(≡CCH=CPh$_2$)Cp(CO)$_2$]BF$_4$	1.665(5)	174.6(4)	2	6	811

TABLE 5.4 Metal–Carbon Multiple Bonds *(Cont.)*

Complex	M–C (Å)	∢ M–C–R	d e–	CN	Ref.
Re(=CH^{t}Bu)(≡C^{t}Bu)I$_2$(py)$_2$ ∢ C–Re–C = 98.1(4)	1.873(9)	150.3(7)	0	6	65
[Re(=CH$_2$)Cp*(NO){P(OPh)$_3$}]PF$_6$	1.898(18)	128±10	4	6	812
[Re(=CHPh)Cp(NO)PPh$_3$]PF$_6$	1.949(6)	136.2(5)	4	6	813
Re(=CHSiPh$_3$)Cp(CO)$_2$	1.92(2)	135(1)	4	6	814
Re(=C=CHPh)Cl(dppe)$_2$	2.046(8)	166.6(12)	4	6	815
Re(≡CSiMe$_3$)(CH$_2$SiMe$_3$)$_3$Cl	1.726(11)	178.8(6)	0	5	816
Re(≡C^{t}Bu)(=CH^{t}Bu)I$_2$(py)$_2$	1.742(9)	174.8(7)	0	6	65
[Re(≡CNH$_2$)Cl(dppe)$_2$]BF$_4$	1.802(4)	171.9(3)	2	6	817
[Re(≡CNHMe)Cl(dppe)$_2$]BF$_4$	1.798(30)	175.2(18)	2	6	818
Fe(=CCPh=CPh)(CO)$_2$(μ–S)$_2$Fe$_2$(CO)$_6$	1.901(7)		4	6	819
(Ph$_4$porphyrin)Fe=C=Fe(Ph$_4$porphyrin)	1.675	180	4	5	820
[Fe{=C(CH)$_6$}Cp(CO)$_2$]PF$_6$	1.979(3)	118.8±0.2	4	6	821
[Fe(=CC$_{10}$H$_8$)Cp(CO)$_2$]PF$_6$	1.996(2)	118.6±0.9	4	6	821
[Cp(dppe)Fe=C=CMeCMe=C=Fe(dppe)Cp]$^{2+}$	1.756(9)	172.5±2.5	4	6	822
[Fe{=C=CMeC(S)SMe}Cp(dppe)]I	1.74(2)	176(1)	4	6	823
Fe(=C=CC$_9$H$_{18}$)(CO)$_2$(μ–S$_2$C$_{11}$H$_{18}$)FeL$_3$	1.79(1)	177(1)	5	5	824
[Fe(≡CN^{i}Pr$_2$)(CO)$_3$(PPh$_3$)]BCl$_4$ (tbp)	1.734(6)		4	6	825
[Ru(=C=CHMe)Cp(PMe$_3$)$_2$]PF$_6$	1.845(7)	180(2)	4	6	826
[Ru(=C=C=CPh$_2$)Cp(PMe$_3$)$_2$]PF$_6$	1.884(5)	175.9(5)	4	6	827
[CpL$_2$Ru(=C=CHC$_6$H$_{11}$C=)RuCpL$_2$](PF$_6$)$_2$	1.83(2)	166(1)	4	6	828
" Ru=CR$_2$ (L=PPh$_3$)	2.30(2)	124±1	4	6	
[Ru{=C=C(Br)C$_6$H$_4$Br}Cp(PPh$_3$)$_2$]$^+$	1.85(1)	169.4(14)	4	6	829
Os(=CH$_2$)Cl[η^2–C(O)(o-Tol)](PPh$_3$)$_2$	1.856(12)		4	6	830
Os(=CHPh)Cl$_2$(CO)(PPh$_3$)$_2$	1.94(1)	140(1)	4	6	831
Os(=CH$_2$)Cl(NO)(PPh$_3$)$_2$	1.92(1)		6	5	832
[Os(=C=CH^{t}Bu)Cp*(CO)(PPh$_3$)]BF$_4$	1.879(6)	175.0(5)	4	6	833
Os(=C=CC$_6$H$_8$)(CO)$_2$(PPh$_3$)$_2$	1.90(1)	169(3)	6	6	834
Os(≡CC$_6$H$_4$NMe$_2$)(NCS)Cl$_2$(PPh$_3$)$_2$	1.75(1)		2	6	835
[Os(≡CC$_6$H$_4$NMe$_2$)Cl$_2$(CNTol)(PPh$_3$)$_2$]ClO$_4$	1.78(1)		2	6	835
Os(≡CTol)Cl(CO)(PPh$_3$)$_2$	1.78(2)	165(2)	4	5	51
Os(≡CTeMe)(CO)$_2$(PPh$_3$)$_2$	1.841(16)		4	6	831
U(=CHPMe$_2$Ph)Cp$_3$	2.29(3)	142(1)	f^2	10	836

[a]Structure determined by neutron diffraction.

REFERENCES

1. Cotton, F.A.; Walton, R.A. "Multiple Bonds Between Metal Atoms" J. Wiley & Sons, New York, 1982.
2. Fischer, E.O.; Maasböl, A. Angew. Chem., Int. Ed. Engl. **1964**, *3*, 580-581.
3. Fischer, E.O.; Kreis, G.; Kreiter, C.G.; Müller, J.; Huttner, G.; Lorenz, H. Angew. Chem., Int. Ed. Engl. **1973**, *12*, 564-565.
4. Palenik, G.J. Inorg. Chem. **1967**, *6*, 507-511.
5. Stavropoulos, P.; Edwards, P.G.; Behling, T.; Wilkinson, G.; Motevalli, M.; Hursthouse, M.B. J. Chem. Soc., Dalton Trans. **1987**, 169.
6. Haase, W.; Hoppe, H. Acta Cryst. **1968**, *B24*, 282-283.
7. Müller, U.; Shihada, A.-F.; Dehnicke, K. Z. Naturforsch. **1982**, *37B*, 699.
8. Dewan, J.C.; Kepert, D.L.; Raston, C.L.; White, A.H. J. Chem. Soc., Dalton Trans. **1975**, 2031.
9. Gahan, B.; Garner, C.D.; Hill, L.H.; Mabbs, F.E.; Hargrave, K.D.; McPhail, A.T. J. Chem. Soc., Dalton Trans. **1977**, 1726.
10. Iijima, K.; Shibata, S. Bull. Chem. Soc. Jpn. **1975**, *48*, 666-668.
11. Garner, C.D.; Hill, L.H.; Mabbs, F.E.; McFadden, D.L.; McPhail, A.T. J. Chem. Soc., Dalton Trans. **1977**, 853.
12. Iijima, K.; Shibata, S. Chem. Lett. **1972**, 1033.
13. Fenske, D.; Stahl, K.; Hey, E.; Dehnicke, K. Z. Naturforsch. **1984**, *39B*, 850.
14. Cotton, F.A.; Davison, A.; Day, V.W.; Gage, L.D.; Trop, H.S. Inorg. Chem. **1979**, *18*, 3024-3029.
15. Hagen, K.; Hobson, R. J.; Rice, D. A.; Turp, N. J. Mol. Struct. **1985**, *128*, 33 and references therein.
16. Lis, T.; Jezowska-Trzebiatowska, B. Acta Cryst. **1977**, *B33*, 1248.
17. Hagen, K.; Hobson, R. J.; Holwill, C. J.; Rice, D. A. Inorg. Chem. **1986**, *25*, 3659-3661 and references therein.
18. Berg, D.M.; Sharp, P.R. Inorg. Chem. **1987**, *26*, 2959-2962.
19. Melník, M.; Sharrock, P. Coord. Chem. Rev. **1985**, *65*, 49-85.
20. Hentges, S.G.; Sharpless, K.B.; Tulip, T.H. unpublished results.
21. Mayer, J.M.; Thorn, D.L.; Tulip, T.H. J. Am. Chem. Soc. **1985**, *107*, 7454-7462.
22. Pauling, L. "The Nature of the Chemical Bond" 3rd edition, Cornell University Press, Ithaca, New York, 1960, pp. 224ff.
23. Churchill, M.R.; Youngs, W.J. Inorg. Chem. **1979**, *18*, 2454-2458.
24. Churchill, M.R.; Missert, J.R.; Youngs, W.J. Inorg. Chem. **1981**, *20*, 3388-3391.
25. Churchill, M.R.; Rheingold, A.L. Inorg. Chem. **1982**, *21*, 1357-1359.
26. Guggenberger, L.J.; Schrock, R.R. J. Am. Chem. Soc. **1975**, *97*, 6578-6579.
27. Gambarotta, S.; Chiesi-Villa, A.; Guastini, C. J. Organomet. Chem. **1984**, *270*, C49-C52.
28. Silavwe, N.D.; Chiang, M.Y.; Tyler, D.R. Inorg. Chem. **1985**, *24*, 4219-4221.
29. Butcher, A.V.; Chatt, J. J. Chem. Soc. A **1970**, 2652-2656

30. Manojlović-Muir, L. *J. Chem. Soc. A* **1971**, 2796-2800.
31. Manojlović-Muir, L.; Muir, K.W. *J. Chem. Soc., Dalton Trans.* **1972**, 686.
32. Haymore, B.L.; Goddard, W.A., III; Allison, J.N. *Proc. Int. Conf. Coord. Chem., 23rd.* **1984**, 535.
33. Chatt, J.; Manojlovic-Muir, L.; Muir, K.W. *J. Chem. Soc., Chem. Comm.* **1971**, 655-656.
34. Corfield, P.W.R.; Doedens, R.J.; Ibers, J.A. *Inorg. Chem.* **1967**, *6*, 197-204.
35. Forsellini, E.; Casellato, U.; Graziani, R.; Magon, L. *Acta Cryst.* **1982**, *B38*, 3081-3083.
36. Wieghardt, K.; Backes-Dahmann, G.; Holzbach, W.; Swiridoff, W.J.; Weiss, J. *Z. Anorg. Allg. Chem.* **1983**, *499*, 44.
37. Lincoln, S.; Koch, S.A. *Inorg. Chem.* **1986**, *25*, 1594-1602.
38. Churchill, M.R.; Rotella, F.J. *Inorg. Chem.* **1978**, *17*, 668.
39. Bishop, M.W.; Chatt, J.; Dilworth, J.R.; Hursthouse, M.B.; Motevalli, M. *J. Chem. Soc., Dalton Trans.* **1979**, 1603.
40. Schumacher, C.; Weller, F.; Dehnicke, K. *Z. Anorg. Allg. Chem.* **1982**, *495*, 135.
41. Wieghardt, K.; Hahn, M.; Weiss, J.; Swiridoff, W. *Z. Anorg. Allg. Chem.* **1982**, *492*, 164.
42. Saussine, L.; Mimoun, H.; Mitschler, A.; Fisher, J. *Nouv. J. Chim.* **1980**, *4*, 235.
43. Wieghardt, K.; Backes-Dahmann, G.; Nuber, B.; Weiss, J. *Angew. Chem., Int. Ed. Engl.* **1985**, *24*, 777-778.
44. Che, C.-M.; Wong, K.-Y.; Mak, T.C.W. *J. Chem. Soc., Chem. Comm.* **1985**, 988.
45. Aoyagi, K.; Yukawa, Y.; Shimuzu, K.; Mukaida, M.; Takeuchi, T.; Kakihana, H. *Bull. Chem. Soc. Jpn.* **1986**, *59*, 1493-1499.
46. Müller, U.; Lorenz, I. *Z. Anorg. Allg. Chem.* **1980**, *463*, 110.
47. Gorbunov, Yu. E.; Pakhomov, V.I.; Kuznetsov, V.G.; Kovaleva, E.S. *Zh. Strukh. Khim.* **1972**, *13*, 165-166.
48. Gebreyes, K.; Zubieta, J.; George, T.A.; Koczon, L.M.; Tisdale, R.C. *Inorg. Chem.* **1986**, *25*, 405-407.
49. Cotton, F.A.; Diebold, M.P; Roth, W.J. *Inorg. Chem.* **1987**, *26*, 2848-2852.
50. Huttner, G.; Lorenz, H.; Gartzke, W. *Angew. Chem.* **1974**, *86*, 667.
51. Clark, G.R.; Marsden, K.; Roper, W.R.; Wright, L.J. *J. Am. Chem. Soc.* **1980**, *102*, 6570-6571. Clark, G.R.; Cochrane, C.M.; Marsden, K.; Roper, W.R.; Wright, L.J. *J. Organomet. Chem.* **1986**, *315*, 211-230.
52. Haymore, B.L.; Maatta, E.A.; Wentworth, R.A.D. *J. Am. Chem. Soc.* **1979**, *101*, 2063-2068.
53. Schrock, R.R. *Acc. Chem. Res.* **1979**, *12*, 98-104.
54. Schrock, R.R. *Acc. Chem. Res.* **1986**, *19*, 342-348.
55. Schultz, A.J.; Brown, R.K.; Williams, J.M.; Schrock, R.R. *J. Am. Chem. Soc.* **1981**, *103*, 169-176.
56. Shustorovich, E.M.; Porai-Koshits, M.A.; Buslaev, Yu.A. *Coord. Chem. Rev.* **1975**, *17*, 1-98.
57. Appleton, T.G.; Clark, H.C.; Manzer, L.E. *Coord. Chem. Rev.* **1973**, *10*, 335-422. Hartley, F.R. *Chem. Soc. Rev.* **1973**, *2*, 163. Yatsimiskii, K.B. *Pure Appl. Chem.* **1974**, *38*, 341.
58. Bright, D.; Ibers, J.A. *Inorg. Chem.* **1969**, *8*, 709-716.

59. Anhaus, J.; Siddiqi, Z.A.; Schimkowiak, J.; Roesky, H.W.; Lueken, H. *Z. Naturforsch.* **1984**, *39B*, 1722-1728.
60. Yamanouchi, K.; Enemark, J.H. "Proceedings of the Climax Third International Conference on the Chemistry and Uses of Molybdenum" Barry, H.F.; Mitchell, P.C.H. eds.; Climax Molybdenum: Ann Arbor MI **1979**, pp. 24-27.
61. Pope, M.T. "Heteropoly and Isopoly Oxometallates" Springer-Verlag, New York, 1983, pp. 18-30.
62. Wells, A.F. "Structural Inorganic Chemistry" 5th edition, Oxford, New York, **1984**, pp. 327, 560ff.
63. McWeeny, R.; Mason, R.; Towl, A.D.C. *Disc. Faraday Soc.* **1969**, *47*, 20. Mason, R.; Towl, A.D.C. *J. Chem. Soc. A* **1970**, 1601.
64. Burdett, J.K.; Albright, T.A. *Inorg. Chem.* **1979**, *18*, 2112-2120.
65. Edwards, D.S.; Biondi, L.V.; Ziller, J.W.; Churchill, M.R.; Schrock, R.R. *Organometallics* **1983**, *2*, 1505-1513.
66. Rocklage, S.M.; Schrock, R.R.; Churchill, M.R.; Wasserman, H.J. *Organometallics* **1982**, *1*, 1332-1338.
67. Chiu, K.W.; Jones, R.A.; Wilkinson, G.; Galas, A.M.R.; Hursthouse, M.B.; Malik, K.M.A. *J. Chem. Soc., Dalton Trans.* **1981**, 1204-1211.
68. Chisholm, M.H.; Conroy, B.K.; Huffman, J.C. *Organometallics* **1986**, *5*, 2384-2386.
69. Mayr, A.; Dorries, A.M.; McDermott, G.A.; Geib, S.J.; Rheingold, A.L. *J. Am. Chem. Soc.* **1985**, *107*, 7775-7776.
70. Fischer, E.O.; Wittmann, D.; Himmelreich, D.; Cai, R.; Ackermann, K.; Neugebauer, D. *Chem. Ber.* **1982**, *115*, 3152.
71. Nugent, W.A.; Haymore, B.L. *Coord. Chem. Rev.* **1980**, *31*, 123-175.
72. Chou, C.Y.; Huffman, J.C.; Maatta, E.A. *Inorg. Chem.* **1986**, *25*, 822-826.
73. DuBois, D.L.; Hoffmann, R. *Nouv. J. Chim.* **1977**, *1*, 479-492.
74. Behling, T.; Capparelli, M.V.; Skapski, A.C.; Wilkinson, G. *Polyhedron* **1982**, *1*, 840-841.
75. Dobson, J.C.; Takeuchi, K.J.; Pipes, D.W.; Geselowitz, D.A.; Meyer, T.J. *Inorg. Chem.* **1986**, *25*, 2357-2365.
76. Tatsumi, K.; Hoffmann, R. *Inorg. Chem.* **1980**, *19*, 2656-2658.
77. Brower, D.C.; Templeton, J.L.; Mingos, D.M.P. *J. Am. Chem. Soc.* **1987**, *109*, 5203-5208.
78. Kubacek, P.; Hoffmann, R. *J. Am. Chem. Soc.* **1981**, *103*, 4320-4332.
79. Burdett, J.K. "Molecular Shapes" J. Wiley & Sons, New York, 1980, p. 38ff and references therein.
80. Palmer, K.J. *J. Am. Chem. Soc.* **1938**, *60*, 2360-2369.
81. Kistenmacher, T.J.; Marzilli, L.G.; Rossi, M. *Bioinorg. Chem.* **1976**, *6*, 347-364.
82. Guilard, R.; Latour, J.M.; Lecomte, C.; Marchon, J.-C.; Protas, J.; Ripoll, D. *Inorg. Chem.* **1978**, *17*, 1228-1237.
83. Dwyer, P.N.; Puppe, L.; Buchler, J.W.; Scheidt, W.R. *Inorg. Chem.* **1975**, *14*, 1782-1785.
84. Hiller, W.; Strähle, J.; Kobel, W.; Hanack, M. *Z. Kristallogr.* **1982**, *159*, 173.
85. Daran, J.C.; Jeannin, Y.; Constant, G.; Morancho, R. *Acta Cryst.* **1975**, *B31*, 1833.
86. Gourdon, A.; Jeannin, Y. *Acta Cryst.* **1980**, *B36*, 304.

88. Casellato, U.; Vigato, P.A.; Graziani, R.; Vidali, M.; Milani, F.; Musiani, M.M. *Inorg. Chim. Acta* **1982**, *61*, 121.
89. Edwards, A.J.; Taylor, P. *J. Chem. Soc., Chem. Comm.* **1970**, 1474-1475.
90. Rieskamp, H.; Mattes, R. *Z. Naturforsch.* **1976**, *31B*, 1453.
91. Rieskamp, H.; Mattes, R. *Z. Naturforsch.* **1976**, *31B*, 541.
92. Bonadies, J.A.; Butler, W.M.; Pecoraro, V.L.; Carrano, C.J. *Inorg. Chem.* **1987**, *26*, 1218-1222.
93. Banci, L.; Bencini, A.; Dei, A.; Gatteschi, D. *Inorg. Chim. Acta* **1984**, *84*, L11.
94. Scheidt, W.R. *Inorg. Chem.* **1973**, *12*, 1758-1761.
95. Yamada, S.; Katayama, C.; Tanaka, J.; Tanaka, M. *Inorg. Chem.* **1984**, *23*, 253.
96. Rieskamp, H.; Gietz, P.; Mattes, R. *Chem. Ber.* **1976**, *109*, 2090-2096.
97. Bottomley, F.; Darkwa, J.; Sutin, L.; White, P.S. *Organometallics* **1986**, *5*, 2165-2171.
98. Saussine, L.; Mimoun, H.; Mitschler, A.; Fisher, J. *Nouv. J. Chim.* **1980**, *4*, 235.
99. Edwards, A.J.; Slim, D.R.; Guerchais, J.E.; Sala-Pala, J. *J. Chem. Soc., Dalton Trans.* **1977**, 984-986.
100. Campbell, N.J.; Capparelli, M.V.; Griffith, W.P.; Skapski, A.C. *Inorg. Chim. Acta* **1983**, *77*, L215.
101. Szentivanyi, H.; Stomberg, R. *Acta Chem. Scand.* **1983**, *A37*, 553.
102. Szentivanyi, H.; Stomberg, R. *Acta Chem. Scand.* **1983**, *A37*, 709.
103. Mimoun, H.; Saussine, L.; Daire, E.; Postel, M.; Fischer, J.; Weiss, R. *J. Am. Chem. Soc.* **1983**, *105*, 3101-3110.
104. Drew, R.E.; Einstein, F.W.B. *Inorg. Chem.* **1973**, *12*, 829.
105. Djordjevic, C.; Craig, S.A.; Sinn, E. *Inorg. Chem.* **1985**, *24*, 1281.
106. Mimoun, H.; Chaumette, P.; Mignard, M.; Saussine, L.; Fischer, J.; Weiss, R. *Nouv. J. Chim.* **1983**, *7*, 467.
107. Nuber, B.; Weiss, J. *Acta Cryst.* **1981**, *B37*, 947.
108. Dewan, J.C.; Kepert, D.L.; Raston, C.L.; Taylor, D.; White, A.H. *J. Chem. Soc., Dalton Trans.* **1973**, 2082-2086.
109. Einstein, F.W.B.; Enwall, E.; Morris, D.M.; Sutton, D. *Inorg. Chem.* **1971**, *10*, 678.
110. Nishizawa, M.; Hirotsu, K.; Ooi, S.; Saito, K. *J. Chem. Soc., Chem. Comm.* **1979**, 707.
111. Kojima, A.; Okazaki, K.; Ooi, S.; Saito, K. *Inorg. Chem.* **1983**, *22*, 1168.
112. Launay, J.-P.; Jeannin, Y.; Daoudi, M. *Inorg. Chem.* **1985**, *24*, 1052.
113. Dodge, R.P.; Templeton, D.H.; Zalkin, A. *J. Chem. Phys.* **1961**, *35*, 55-67.
114. Hon, P.-K.; Belford, R.L.; Pfluger, C.E. *J. Chem. Phys.* **1965**, *43*, 1323-1333.
115. Cooper, S.R.; Koh, Y.B.; Raymond, K.N. *J. Am. Chem. Soc.* **1982**, *104*, 5092.
116. Musiani, M.M.; Milani, F.; Graziani, R.; Vidali, M.; Casellato, U.; Vigato, P.A. *Inorg. Chim. Acta* **1982**, *61*, 115.
117. Henrick, K.; Raston, C.L.; White, A.H. *J. Chem. Soc., Dalton Trans.* **1976**, 26.

118. Money, J.K.; Huffman, J.C.; Cristou, G. *Inorg. Chem.* **1985**, *24*, 3297.
119. Money, J.K.; Nicholson, J.R.; Huffman, J.C.; Christou, G. *Inorg. Chem.* **1986**, *25*, 4072–4074.
120. Shiro, M.; Fernando, Q. *Anal. Chem.* **1971**, *43*, 1222.
121. Pasquali, M.; Marchetti, F.; Floriani, C.; Merlino, S. *J. Chem. Soc., Dalton Trans.* **1977**, 139–144.
122. Drake, J.E.; Vekris, J.; Wood, J.S. *J. Chem. Soc. A* **1968**, 1000.
123. Caira, M.R.; Gellatly, B.J. *Acta Cryst.* **1980**, *B36*, 1198.
124. Coetzer, J. *Acta Cryst.* **1970**, *B26*, 872.
125. Bruins, D.; Weaver, D.L. *Inorg. Chem.* **1970**, *9*, 130.
126. Pasquali, M.; Marchetti, F.; Floriani, C.; Cesari, M. *Inorg. Chem.* **1980**, *19*, 1198.
127. Chasteen, N.D.; Belford, R.L.; Paul, I.C. *Inorg. Chem.* **1969**, *8*, 408–418.
128. Tapscott, R.E.; Belford, R.L.; Paul, I.C. *Inorg. Chem.* **1968**, *7*, 356.
129. Ortega, R.B.; Campana, C.F.; Tapscott, R.E. *Acta Cryst.* **1980**, *B36*, 1786.
130. Ortega, R.B.; Tapscott, R.E.; Campana, C.F. *Inorg. Chem.* **1982**, *21*, 672.
131. Hahs, S.K.; Ortega, R.B.; Tapscott, R.E.; Campana, C.F.; Morosin, B. *Inorg. Chem.* **1982**, *21*, 664.
132. Adams, H.; Bailey, N.A.; Fenton, D.E.; Leal-Gonzalez, M.S.; Phillips, C.A. *J. Chem. Soc., Dalton Trans.* **1983**, 371.
133. Kahn, O.; Galy, J.; Journaux, Y.; Jaud, J.; Morgenstern-Badarau, I. *J. Am. Chem. Soc.* **1982**, *104*, 2165.
134. Lintvedt, R.L.; Glick, M.D.; Tomlonovic, B.K.; Gavel, D.P. *Inorg. Chem.* **1976**, *15*, 1646.
135. Glick, M.D.; Lintvedt, R.L.; Gavel, D.P.; Tomlonovic, B. *Inorg. Chem.* **1976**, *15*, 1654.
136. Greenwood, A.J.; Henrick, K.; Owston, P.G.; Tasker, P.A. *J. Chem. Soc., Chem. Comm.* **1980**, 88.
137. Ziolo, R.F.; Griffiths, C.H.; Troup, J.M. *J. Chem. Soc., Dalton Trans.* **1980**, 2300.
138. Drew, M.G.B.; Mitchell, P.C.H.; Scott, C.E. *Inorg. Chim. Acta* **1984**, *82*, 63.
139. Pettersen, R.C. *Acta Cryst.* **1969**, *B25*, 2527–2539.
140. Tachez, M.; Théobald, F. *Acta Cryst.* **1980**, *B36*, 1757–1761.
141. Khodashova, T.S.; Porai-Koshits, M.A.; Sergienko, V.S.; Butman, L.A.; Uslaev, Yu.A.; Kovalev, V.V.; Kuznetsova, A.A. *Koord. Khim.* **1978**, *4*, 1909.
142. Rieskamp, H.; Mattes, R. *Z. Naturforsch.* **1976**, *31B*, 537.
143. Demsar, A.; Bukovec, P. *Croat. Chem. Acta* **1984**, *57*, 673.
144. Demsar, A.; Bukovec, P. *J. Fluorine Chem.* **1984**, *24*, 369.
145. Bersted, B.H.; Belford, R.L.; Paul, I.C. *Inorg. Chem.* **1968**, *7*, 1557.
146. Dichmann, K.S.; Hamer, G.; Nyburg, S.C.; Reynolds, W.F. *J. Chem. Soc. D* **1970**, 1295.
147. Pajunen, A.; Pajunen, S. *Acta Cryst.* **1980**, *B36*, 2425.
148. Oughtred, R.E.; Raper, E.S.; Shearer, H.M.M. *Acta Cryst.* **1976**, *B32*, 82.
149. Ooi, S.; Nishizawa, M.; Matsumoto, K.; Kuroya, H.; Saito, K. *Bull. Chem. Soc. Jpn.* **1979**, *52*, 452.

150. Bencini, A.; Benelli, C.; Dei, A.; Gatteschi, D. Inorg. Chem. **1985**, *24*, 695.
151. Ghosh, M.; Ray, S. Acta Cryst. **1983**, *C39*, 1367.
152. Pecoraro, V.; Bonadies, J.A.; Marrese, C.A.; Carrano, C.J. J. Am. Chem. Soc. **1984**, *106*, 3360.
153. Alyea, E.C.; Dee, T.D.; Ferguson, G. J. Cryst. Spectroscop. **1985**, *15*, 29.
154. Fallon, G.D.; Gatehouse, B.M. Acta Cryst. **1976**, *B32*, 71.
155. Bukovec, P.; Milicev, S.; Demsar, A.; Golic, L. J. Chem. Soc., Dalton Trans. **1981**, 186, 1802.
156. Wieghardt, K.; Bossek, U.; Volckmar, K.; Swiridoff, W.; Weiss, J. Inorg. Chem. **1984**, *23*, 1387.
157. Cotton, F.A.; Lewis, G.E.; Mott, G.N. Inorg. Chem. **1983**, *22*, 1825.
158. Cass, M.E.; Greene, D.L.; Buchanan, R.M.; Pierpont, C.G. J. Am. Chem. Soc. **1983**, *105*, 2680.
159. Cotton, F.A.; Lewis, G.E.; Mott, G.N. Inorg. Chem. **1982**, *21*, 3127.
160. Cotton, F.A.; Lewis, G.E.; Mott, G.N. Inorg. Chem. **1983**, *22*, 378.
161. Krebs, B.; Hasse, K.-D. Acta Cryst. **1976**, *B32*, 1334-1337 and references therein.
162. Rieskamp, H.; Mattes, R. Z. Anorg. Allg. Chem. **1976**, *419*, 193-199 and references therein.
163. Fenske, D.; Shihada, A.-F.; Schwab, H.; Dehnicke, K. Z. Anorg. Allg. Chem. **1980**, *471*, 140.
164. Nuber, B.; Weiss, J.; Wieghardt, K. Z. Naturforsch. **1978**, *33B*, 265-267.
165. Mimoun, H.; Saussine, L.; Daire, E.; Postel, M.; Fischer, J.; Weiss, R. J. Am. Chem. Soc. **1983**, *105*, 3101.
166. Mattes, R.; Preuss, A.; Richter, K.-L. Z. Naturforsch. **1984**, *39B*, 1331.
167. Kravtsov, V.Kh.; Belyaeva, K.F.; Byushkin, V.N.; Struchkov, Yu.T.; Bodyu, V.G.; Samus, N.M. Koord. Khim. **1981**, *7*, 1569.
168. Petrovic, A.F.; Ribar, B.; Petrovic, D.M.; Leovac, V.M.; Gerbeleu, N.V. J. Coord. Chem. **1982**, *11*, 239-245.
169. Petrovic, A.F.; Leovac, V.M.; Ribar, B.; Argay, G.; Kalman, A. Trans. Met. Chem. **1986**, *11*, 207-213.
170. Galesic, N.; Siroki, M. Acta Cryst. **1979**, *B35*, 2931.
171. Galesic, N.; Siroki, M. Acta Cryst. **1984**, *C40*, 378.
172. Ribar, B.; Kozmidis-Petrovic, A.; Leovac, V. Cryst. Struct. Commun. **1980**, *9*, 1237.
173. Leovac, V.M.; Petrovic, A.F. Trans. Met. Chem. **1983**, *8*, 337.
174. Sabirov, V.Kh.; Batsanov, A.S.; Struchkov, Yu.T.; Azizov, M.A.; Shabilalov, A.A.; Pulatov, A.S. Koord. Khim. **1984**, *10*, 275.
175. Ryan, R.R.; Mastin, S.H.; Reisfeld, M.J. Acta Cryst. **1971**, *B27*, 1270-1274.
176. Isobe, K.; Ooi, S.; Nakamura, Y.; Kawaguchi, S.; Kuroya, H. Chem. Lett. **1975**, 35.
177. Scheidt, W.R.; Tsai, C.-C.; Hoard, J.L. J. Am. Chem. Soc. **1971**, *93*, 3867-3872.
178. Drew, R.E.; Einstein, F.W.B.; Gransden, S.E. Can. J. Chem. **1974**, *52*, 2184-2189.

179. Giacomelli, A.; Floriani, C.; Duarte, A.O. de S.; Villa, A.C.; Guastini, C. *Inorg. Chem.* **1982**, *21*, 3310.
180. Jeannin, Y.; Launay, J.P.; Sedjadi, M.A.S. *J. Coord. Chem.* **1981**, *11*, 27.
181. Scheidt, W.R.; Countryman, R.; Hoard, J.L. *J. Am. Chem. Soc.* **1971**, *93*, 3878-3881.
182. Scheidt, W.R.; Collins, D.M.; Hoard, J.L. *J. Am. Chem. Soc.* **1971**, *93*, 3873-3877.
183. Kamenar, B.; Prout, C.K. *J. Chem. Soc. A* **1970**, 2379-2384.
184. Hiller, W.; Strähle, J.; Prinz, H.; Dehnicke, K. *Z. Naturforsch.* **1984**, *39B*, 107.
185. Klingelhofer, P.; Müller, U. *Z. Anorg. Allg. Chem.* **1984**, *516*, 85.
186. Aspinall, H.C.; Roberts, M.M.; Lippard, S.J. *Inorg. Chem.* **1984**, *23*, 1782.
187. Chavant, C.; Daran, J.-C.; Jeannin, Y.; Constant, G.; Morancho, R. *Acta Cryst.* **1975**, *B31*, 1828.
188. Hubert-Pfalzgraf, L.G.; Pinkerton, A.A. *Inorg. Chem.* **1977**, *16*, 1895.
189. Daran, J.-C.; Jeannin, Y.; Guerchais, J.E.; Kergoat, R. *Inorg. Chim. Acta* **1979**, *33*, 81.
190. Lecomte, C.; Protas, J.; Richard, P.; Barbe, J.M.; Guilard, R. *J. Chem. Soc., Dalton Trans.* **1982**, 247.
191. Lecomte, C.; Protas, J.; Guilard, R.; Fliniaux, B.; Fournari, P. *J. Chem. Soc., Dalton Trans.* **1979**, 1306.
192. Mathern, G.; Weiss, R. *Acta Cryst.* **1971**, *B27*, 1610.
193. Mercier, R.; Douglade, J.; Amaudrut, J.; Sala-Pala, J.; Guerchais, J.E. *J. Organomet. Chem.* **1983**, *244*, 145.
194. Bradley, D.C. private communication.
195. Stomberg, R. *Arkiv. Kem.* **1963**, *22*, 29-47.
196. Stomberg, R.; Ainalem, I.-B. *Acta Chem. Scand.* **1968**, *22*, 1439.
197. Stomberg, R. *Arkiv. Kem.* **1965**, *24*, 111.
198. Krumpolc, M.; Deboer, B.G.; Roček, J. *J. Am. Chem. Soc.* **1978**, *100*, 145.
199. Srinivasan, K.; Kochi, J.K. *Inorg. Chem.* **1985**, *24*, 4671-4679.
200. Herberhold, M.; Kniesel, H.; Haumaier, L. *J. Organomet. Chem.* **1986**, *301*, 355-367.
201. Groves, J.T.; Kruper, Jr., W.J.; Haushalter, R.C.; Butler, W.M. *Inorg. Chem.* **1982**, *21*, 1363.
202. Budge, J.R.; Gatehouse, B.M.K.; Nesbit, M.C.; West, B.O. *J. Chem. Soc., Chem. Comm.* **1981**, 370.
203. McGinnety, J.A. *Acta Cryst.* **1972**, *B28*, 2845-2852.
204. Amirthalingam, V.; Grant, D.F.; Senol, A. *Acta Cryst.* **1972**, *B28*, 1340.
205. Stensland, B.; Kierkegaard, P. *Acta Chem. Scand.* **1970**, *24*, 211.
206. Müller, A.; Bogge, H.; Tolle, H.-G.; Jostes, R.; Schimanski, U.; Dartmann, M. *Angew. Chem., Int. Ed. Engl.* **1980**, *19*, 654.
207. Clegg, W.; Garner, C.D.; Nichlson, J.R.; Raithby, P.R. *Acta Cryst.* **1983**, *C39*, 1007.
208. Iijima, K. *Bull. Chem. Soc. Jpn.* **1977**, *50*, 373-375.
209. Johnson, D.A.; Taylor, J.C.; Waugh, A.B. *J. Inorg. Nucl. Chem.* **1980**, *42*, 1271-1275.
210. Clegg, W.; Garner, C.D.; Fletcher, P.A. *Acta Cryst.* **1984**, *C40*, 754.

211. Grandjean, D.; Weiss, R. *Bull. Soc. Chim. France* **1967**, 3054–3058.
212. Chisholm, M.H.; Folting, K.; Huffman, J.C.; Kirkpatick, C.C. *Inorg. Chem.* **1984**, *23*, 1021–1037.
213. Matheson, A.J.; Penfold, B.R. *Acta Cryst.* **1979**, *B35*, 2707.
214. Buchanan, R.M.; Pierpont, C.G. *Inorg. Chem.* **1979**, *18*, 1616.
215. McKee, V.; Wilkins, C.J. *J. Chem. Soc., Dalton Trans.* **1987**, 523–528.
216. Le Carpentier, J.-M.; Mitschler, A.; Weiss, R. *Acta Cryst.* **1972**, *B28*, 1288.
217. Shoemaker, C.B.; McAfee, L.V.; Dekock, C.W.; Shoemaker, D.P. *Acta Cryst.* **1984**, *A40*, C307.
218. Stomberg, R. *Acta Chem. Scand.* **1970**, *24*, 2024.
219. Winter, W.; Mark, C.; Schurig, V. *Inorg. Chem.* **1980**, *19*, 2045.
220. Flanagan, J.; Griffith, W.P.; Skapski, A.C.; Wiggins, R.W. *Inorg. Chim. Acta* **1985**, *96*, L23.
221. Le Carpentier, J.-M.; Schlupp, R.; Weiss, R. *Acta Cryst.* **1972**, *B28*, 1278.
222. Schlemper, E.O.; Schrauzer, G.N.; Hughes, L.A. *Polyhedron* **1984**, *3*, 377–380.
223. Jacobson, S.E.; Tang, R.; Mares, F. *Inorg. Chem.* **1978**, *17*, 3055.
224. Coucouvanis, D.; Hadjikyriacou, A. *Inorg. Chem.* **1987**, *26*, 1–2.
225. Mennemann, K.; Mattes, R. *J. Chem. Res.* **1979**, *102*, 1372.
226. Tomioka, H.; Takai, K.; Oshima, K.; Nozaki, H.; Toriumi, K. *Tetrahedron Lett.* **1980**, 4843.
227. Chaumette, P.; Mimoun, H.; Saussine, L.; Fischer, J.; Mitschler, A. *J. Organomet. Chem.* **1983**, *250*, 291.
228. Edwards, A.J.; Slim, D.R.; Guerchais, J.E.; Kergoat, R. *J. Chem. Soc., Dalton Trans.* **1977**, 1966.
229. Edwards, A.J.; Slim, D.R.; Guerchais, J.E.; Kergoat, R. *J. Chem. Soc., Dalton Trans.* **1980**, 289.
230. Grandjean, D.; Weiss, R. *Bull. Soc. Chim. France* **1967**, 3044–3049.
231. Shum, W. *Inorg. Chem.* **1986**, *25*, 4329–4330.
232. Liebeskind, L.S.; Sharpless, K.B.; Wilson, R.D.; Ibers, J.A. *J. Am. Chem. Soc.* **1978**, *100*, 7061–7063.
233. Bristow, S.; Enemark, J.H.; Garner, C.D.; Minelli, M.; Morris, G.A.; Ortega, R.B. *Inorg. Chem.* **1985**, *24*, 4070–4077.
234. Wieghardt, K.; Hofer, E.; Holzbach, W.; Nuber, B.; Weiss, J. *Inorg. Chem.* **1980**, *19*, 2927.
235. Holzbach, W.; Wieghardt, K.; Weiss, J. *Z. Naturforsch.* **1981**, *36B*, 289.
236. Wieghardt, K.; Holzbach, W.; Hofer, E.; Weiss, J. *Chem. Ber.* **1981**, *114*, 2700.
237. Wieghardt, K.; Holzbach, W.; Weiss, J. *Inorg. Chem.* **1981**, *20*, 3436.
238. Wieghardt, K.; Holzbach, W.; Hofer, E.; Weiss, J. *Inorg. Chem.* **1981**, *20*, 343.
239. Dirand, J.; Ricard, L.; Weiss, R. *Trans. Met. Chem.* **1975**, *1*, 2.
240. Dirand, J.; Ricard, L.; Weiss, R. *J. Chem. Soc., Dalton Trans.* **1976**, 278–282.
241. Marabella, C.P.; Enemark, J.H.; Newton, W.E.; McDonald, J.W. *Inorg. Chem.* **1982**, *21*, 623.

242. Beaver, J.A.; Drew, M.G.B. *J. Chem. Soc., Dalton Trans.* **1973**, 1376.
243. Garner, C.D.; Howlander, N.C.; Mabbs, F.E.; Boorman, P.M.; King. T.J. *J. Chem. Soc., Dalton Trans.* **1978**, 1350.
244. Bradbury, J.R.; Mackay, M.F.; Wedd, A.G. *Aust. J. Chem.* **1978**, *31*, 2423.
245. Bishop, P.T.; Dilworth, J.R.; Hutchinson, J.; Zubieta, J.A. *J. Chem. Soc., Chem. Comm.* **1982**, 1052.
246. Boyde, S.; Ellis, S.R.; Garner, C.D.; Clegg, W. *J. Chem. Soc., Chem. Comm.* **1986**, 1541-1543.
247. Mattes, R.; Scholand, H.; Mikloweit, U.; Schrenk, V. *Z. Naturforsch.* **1987**, *42B*, 589-598.
248. Atovmyan, L.O.; D'yachenko, O.A.; Lobkovskii, E.B. *Zh. Strukh. Khim.* **1970**, *11*, 469-471.
249. Bird, P.H.; Wickramasinghe, W.A. *Can. J. Chem.* **1981**, *59*, 2879.
250. Bino, A.; Cotton, F.A. *J. Am. Chem. Soc.* **1979**, *101*, 4150.
251. Garner, C.D.; Hill, L.H.; Mabbs, F.E.; McFadden, D.L.; McPhail, A.T. *J. Chem. Soc., Dalton Trans.* **1977**, 1202.
252. Bino, A.; Cotton, F.A. *Inorg. Chem.* **1979**, *18*, 2710.
253. Garner, C.D.; Howlader, N.C.; Mabbs, F.E.; McPhail, A.T.; Onan, K.D. *J. Chem. Soc., Dalton Trans.* **1978**, 1848.
254. Khodadad, P.; Viossat, B.; Rodier, N. *Acta Cryst.* **1977**, *B33*, 1035.
255. Dorner, H.-D.; Dehnicke, K.; Fenske, D. *Z. Anorg. Allg. Chem.* **1982**, *486*, 136.
256. Yamanouchi, K.; Huneke, J.T.; Enemark, J.H.; Taylor, R.D.; Spence, J.T. *Acta Cryst.* **1979**, *B35*, 2326.
257. Domiano, P.; Musatti, A.; Nardelli, M.; Pelizzi, C.; Predieri, G. *Trans. Met. Chem.* **1980**, *5*, 172.
258. Cleland, Jr., W.E.; Barnhart, K.M.; Yamanouchi, K.; Collison, D.; Mabbs, F.E.; Ortega, R.B.; Enemark, J.H. *Inorg. Chem.* **1987**, *26*, 1017-1025.
259. Yamanouchi, K.; Yamada, S.; Enemark, J.H. *Inorg. Chim. Acta* **1984**, *85*, 129.
260. Ledon, H.; Mentzen, B. *Inorg. Chim. Acta* **1978**, *31*, L393-L394.
261. Gheller, S.F.; Bradbury, J.R.; Mackay, M.F.; Wedd, A.G. *Inorg. Chem.* **1981**, *20*, 3899.
262. Yamanouchi, K.; Enemark, J.H.; McDonald, J.W.; Newton, W.E. *J. Am. Chem. Soc.* **1977**, *99*, 3529-3531.
263. Mattes, R.; Scholand, H.; Mikloweit, U.; Schrenk, V. *Z. Naturforsch.* **1987**, *42B*, 599-604.
264. Hyde, J.R.; Zubieta, J. *Cryst. Struct. Commun.* **1982**, *11*, 929.
265. Tsao, Y.-Y.P.; Fritchie, Jr., C.J.; Levy, H.A. *J. Am. Chem. Soc* **1978**, *100*, 4089.
266. Kamenar, B.; Penavic, M.; Korpar-Colig, B.; Markovic, B. *Inorg. Chim. Acta* **1982**, *65*, L245-L247.
267. El-Essawi, M.M.; Weller, F.; Stahl, K.; Kersting, M.; Dehnicke, K. *Z. Anorg. Allg. Chem.* **1986**, *542*, 175-181.
268. Dahlstrom, P.L.; Hyde, J.R.; Vella, P.A.; Zubieta, J. *Inorg. Chem.* **1982**, *21*, 927.
269. Ricard, L.; Estienne, J.; Karagiannidis, P.; Toledano, P.; Fischer, J.; Mitschler, A.; Weiss, R. *J. Coord. Chem.* **1974**, *3*, 277-285.

270. Baird, D.M.; Rheingold, A.L.; Croll, S.D.; DiCenso, A.T. *Inorg. Chem.* **1986**, *25*, 3458-3461.
271. Cotton, F.A.; Fanwick, P.E.; Fitch, III, J.W. *Inorg. Chem.* **1978**, *17*, 3254.
272. Tatsumisago, M.; Matsubayashi, G.-E.; Tanaka, T.; Nishigaki, S.; Nakatsu, K. *J. Chem. Soc., Dalton Trans.* **1982**, 121.
273. Blake, A.B.; Cotton, F.A.; Wood, J.S. *J. Am. Chem. Soc.* **1964**, *86*, 3024-3031.
274. Knox, J.R.; Prout, C.K. *Acta Cryst.* **1969**, *B25*, 2281-2285.
275. Prout, K.; Couldwell, C. *Acta Cryst.* **1980**, *B36*, 1481-1482.
276. Benson, I.B.; Killops, S.D.; Knox, S.A.R.; Welch, A.J. *J. Chem. Soc., Chem. Comm.* **1980**, 1137.
277. Johnson, J.F.; Scheidt, W.R. *Inorg. Chem.* **1978**, *17*, 1280-7.
278. Moynihan, K.J.; Boorman, P.M.; Ball, J.M.; Patel, V.D.; Kerr, K.A. *Acta Cryst.* **1982**, *B38*, 2258.
279. Drew, M.G.B.; Kay, A. *J. Chem. Soc. A* **1971**, 1846-1850.
280. Dance, I.G.; Wedd, A.G.; Boyd, I.W. *Aust. J. Chem.* **1978**, *31*, 519.
281. Ricard, L.; Martin, C.; Wiest, R.; Weiss, R. *Inorg. Chem.* **1975**, *14*, 2300.
282. Edelblut, A.W.; Haymore, B.L.; Wentworth, R.A.D. *J. Am. Chem. Soc.* **1978**, *100*, 2250.
283. Altmeppen, D.; Mattes, R. *Acta Cryst.* **1980**, *B36*, 1942.
284. Dirand-Colin, J.; Ricard, L.; Weiss, R. *Inorg. Chim. Acta* **1976**, *18*, L21.
285. Howlader, N.C.; Haight, Jr., G.P.; Hambley, T.W.; Snow, M.R.; Lawrance, G.A. *Inorg. Chem.* **1984**, *23*, 1811.
286. Winograd, R.; Spivack, B.; Dori, Z. *Cryst. Struct. Commun.* **1976**, *5*, 373.
287. Clegg, W.; Mohan, N.; Müller, A.; Neumann, A.; Rittner, W.; Sheldrick, G.M. *Inorg. Chem.* **1980**, *19*, 2066.
288. Clegg, W.; Sheldrick, G.M.; Garner, C.D.; Christou, G. *Acta Cryst.* **1980**, *B36*, 2784.
289. Müller, A.; Reinsch-Vogell, U.; Krickemeyer, E.; Bogge, H. *Angew. Chem., Int. Ed. Engl.* **1982**, *21*, 796.
290. Halbert, T.R.; Pan, W.-H.; Stiefel, E.I. *J. Am. Chem. Soc.* **1983**, *105*, 5476.
291. Gelder, J.I.; Enemark, J.H. *Inorg. Chem.* **1976**, *15*, 1839.
292. Zank, G.A.; Rauchfuss, T.B.; Wilson, S.R.; Rheingold, A.L. *J. Am. Chem. Soc.* **1984**, *106*, 7621.
293. Goodgame, D.M.L.; Rollins, R.W.; Skapski, A.C. *Inorg. Chim. Acta* **1985**, *96*, L61.
294. Atovmyan, L.O.; Tkachev, V.V.; Shchepinov, S.A. *Koord. Khim.* **1978**, *4*, 610.
295. Drew, M.G.B.; Mitchell, P.C.H.; Read, A.R.; Colclough, T. *Acta Cryst.* **1981**, *B37*, 1758.
296. Baricelli, P.J.; Drew, M.G.B.; Mitchell, P.C.H. *Acta Cryst.* **1983**, *C39*, 843.
297. Drew, M.G.B.; Baricelli, P.J.; Mitchell, P.C.H.; Read, A.R. *J. Chem. Soc., Dalton Trans.* **1983**, 649.
298. Noble, M.E.; Folting, K.; Huffman, J.C.; Wentworth, R.A.D. *Inorg. Chem.* **1983**, *22*, 3671-3676.
299. Glowiak, T.; Sabat, M. *J. Cryst. Mol. Struct.* **1975**, *5*, 247.

300. Carmona, E.; Gonzalez, F.; Poveda, M.L.; Marin, J.M.; Atwood, J.L.; Rogers, R.D. *J. Am. Chem. Soc.* **1983**, *105*, 3365.
301. Gatehouse, B.M.; Nunn, E.K. *Acta Cryst.* **1976**, *B32*, 2627.
302. Beck, J.; Hiller, W.; Schweda, E.; Strähle, J. *Z. Naturforsch.* **1984**, *39B*, 1110.
303. Chisholm, M.H.; Huffman, J.C.; Kirkpatrick, C.C.; Leonelli, J.; Folting, K. *J. Am. Chem. Soc.* **1981**, *103*, 6093.
304. Kojima, A.; Ooi, S.; Sasaki, Y.; Suzuki, K.Z.; Saito, K.; Kuroya, H. *Bull. Chem. Soc. Jpn.* **1981**, *54*, 2457.
305. Wieghardt, K.; Hahn, M.; Swiridoff, W.; Weiss, J. *Angew. Chem., Int. Ed. Engl.* **1983**, *22*, 491.
306. Cotton, F.A.; Dori, Z.; Llusar, R.; Schwotzer, W. *Inorg. Chem.* **1986**, *25*, 3654-3658.
307. Cotton, F.A.; Ilsley, W.H. *Inorg. Chim. Acta* **1982**, *59*, 213.
308. Liu, H.; Williams, G.J.B. *Acta Cryst.* **1981**, *B37*, 2065.
309. Shibahara, T.; Ooi, S.; Kuroya, H. *Bull. Chem. Soc. Jpn.* **1982**, *55*, 3742.
310. Spivack, B.; Guughan, A.P.; Dori, Z. *J. Am. Chem. Soc.* **1971**, *93*, 5265-5266.
311. Shibahara, T.; Kuroya, H.; Matsumoto, K.; Ooi, S. *Bull. Chem. Soc. Jpn.* **1983**, *56*, 2945.
312. Mennemann, K.; Mattes, R. *J. Chem. Res.* **1979**, *100*, 1343.
313. McDonald, W.S. *Acta Cryst.* **1978**, *B34*, 2850.
314. Spivack, B.; Dori, Z. *J. Chem. Soc., Dalton Trans.* **1975**, 1077.
315. Spivack, B.; Dori, Z. *J. Chem. Soc., Dalton Trans.* **1973**, 1173-1177.
316. Couldwell, C.; Prout, K. *Acta Cryst.* **1978**, *B34*, 933-934.
317. Arzoumanian, H.; Baldy, A.; Pierrot, M.; Petrignani, J.-F. *J. Organomet. Chem.* **1985**, *294*, 327-331.
318. Alper, H.; Petrignani, J.-F.; Einstein, F.W.B.; Willis, A.C. *J. Am. Chem. Soc.* **1983**, *105*, 1701-1702.
319. Korswagen, R.; Weidenhammer, K.; Ziegler, M.L. *Acta Cryst.* **1979**, *B35*, 2554.
320. Stevenson, D.L.; Dahl, L.F. *J. Am. Chem. Soc.* **1967**, *89*, 3721-3726.
321. Dubois, M.R.; Dubois, D.L.; Vanderveer, M.C.; Haltiwanger, R.C. *Inorg. Chem.* **1981**, *20*, 3064.
322. Drew, M.G.B.; Shanton, K.J. *Acta Cryst.* **1978**, *B34*, 276.
323. Kamenar, B.; Penavic, M.; Korpar-Colig, B.; Markovic, B. *Cryst. Struct. Commun.* **1981**, *10*, 961.
324. Glowiak, T.; Sabat, M.; Sabat, H.; Rudolf, M.F. *J. Chem. Soc., Chem. Comm.* **1975**, 712.
325. Gelder, J.I.; Enemark, J.H.; Wolterman, G.; Boston, D.A.; Haight, G.P. *J. Am. Chem. Soc.* **1975**, *97*, 1616.
326. Dance, I.G.; Landers, A.E. *Inorg. Chem.* **1979**, *18*, 3487.
327. Yamanouchi, K.; Enemark, J.H. *Inorg. Chem.* **1979**, *18*, 1626-1633.
328. Huneke, J.T.; Yamanouchi, K.; Enemark, J.H. *Inorg. Chem.* **1978**, *17*, 3695.
329. Buchanan, I.; Clegg, W.; Garner, C.D.; Sheldrick, G.M. *Inorg. Chem.* **1983**, *22*, 3657.
330. Bunzey, G.; Enemark, J.H.; Gelder, J.I.; Yamanoochi, K.; Newton, W.E. *J. Less-Common Met.* **1977**, *54*, 101.
331. Boyd, I.W.; Dance, I.G.; Landers, A.E.; Wedd, A.G. *Inorg. Chem.* **1979**, *18*, 1875.

332. Bishop, M.W.; Chatt, J.; Dilworth, J.R.; Hyde, J.R.; Kim, S.; Venkatasubramanian, K.; Zubieta, J. *Inorg. Chem.* **1978**, *17*, 2917.
333. Diebold, T.; Chevrier, B.; Weiss, R. *Inorg. Chem.* **1979**, *18*, 1193.
334. Borschel, V.; Strähle, J. *Z. Naturforsch.* **1984**, *39B*, 1664.
335. Bishop, M.W.; Chatt, J.; Dilworth, J.R.; Kaufman, G.; Kim, S.; Zubieta, J. *J. Chem. Soc., Chem. Comm.* **1977**, 70.
336. Draganjac, M.; Simhon, E.; Chan, L.T.; Kanatzidis, M.; Baenziger, N.C.; Coucouvanis, D. *Inorg. Chem.* **1982**, *21*, 3321.
337. Hyde, J.; Venkatasubramanian, K.; Zubieta, J. *Inorg. Chem.* **1978**, *17*, 414.
338. Carmona, E.; Galindo, A. Gutiérrez-Puebla, E.; Monge, A.; Puetra, C. *Inorg. Chem.* **1986**, *25*, 3804-3807.
339. Chatt, J.; Dilworth, J.R.; Schmutz, J.A.; Zubieta, J.A. *J. Chem. Soc., Dalton Trans.* **1979**, 1595.
340. Müller, A.; Hellmann, W.; Romer, C.; Romer, M.; Bogge, H.; Jostes, R.; Schimanski, U. *Inorg. Chim. Acta* **1984**, *83*, L75.
341. Rogers, R.D.; Carmona, E.; Galindo, A.; Atwood, J.L.; Canada, L.G. *J. Organomet. Chem.* **1984**, *277*, 403.
342. Robinson, P.R.; Schlemper, E.O.; Murmann, R.K. *Inorg. Chem.* **1975**, *14*, 2035-2041.
343. Basson, S.S.; Leipoldt, J.G.; Potgieter, I.M.; Roodt, A. *Inorg. Chim. Acta* **1985**, *103*, 121-125.
344. Basson, S.S.; Leipoldt, J.G.; Potgieter, I.M. *Inorg. Chim. Acta* **1984**, *87*, 71.
345. Lam, C.T.; Lewis, D.L.; Lippard, S.J. *Inorg. Chem.* **1976**, *15*, 989.
346. Desimone, R.E.; Cragel, Jr., J.; Ilsley, W.H.; Glick, M.D. *J. Coord. Chem.* **1979**, *9*, 167.
347. Desimone, R.E.; Glick, M.D. *Inorg. Chem.* **1978**, *17*, 3574.
348. Young, C.G.; Roberts, S.A.; Ortega, R.B.; Enemark, J.H. *J. Am. Chem. Soc.* **1987**, *109*, 2938-2946.
349. Nicholas, K.M.; Khan, M.A. *Inorg. Chem.* **1987**, *26*, 1633-1636.
350. Newton, W.E.; McDonald, J.W.; Corbin, J.L.; Ricard, L.; Weiss, R. *Inorg. Chem.* **1980**, *19*, 1997.
351. Ricard, L.; Weiss, R. *Inorg. Nucl. Chem. Lett.* **1974**, *10*, 217-220.
352. Drage, J.S.; Tilset, M.; Vollhardt, P.C.; Weidman, T.W. *Organometallics* **1984**, *3*, 812.
353. Howard, J.A.K.; Stansfield, R.F.D.; Woodward, P. *J. Chem. Soc., Dalton Trans.* **1976**, 246.
354. Feasey, N.D.; Knox, S.A.R.; Orpen, A.G. *J. Chem. Soc., Chem. Comm.* **1982**, 75.
355. Doherty, N.M.; Critchlow, S.C. *J. Am. Chem. Soc.* **1987**, *109*, 7906-7908. Critchlow, S.C.; Lerchen, M.E.; Smith, R.C.; Doherty, N.M. *ibid.* **1988**, *110*, in press. Willing, W.; Christophersen, R.; Müller, U.; Dehnicke, K. *A. Anorg. Allg. Chem.* **1987**, in press.
356. Krebs, B.; Müller, A.; Kindler, E. *Z. Naturforsch.* **1970**, *25B*, 222.
357. Day, V.W.; Fredrich, M.F.; Klemperer, W.G.; Shum, W. *J. Am. Chem. Soc.* **1977**, *99*, 6146.
358. Klemperer, W.G.; Mainz, V.V.; Wang, R.-C.; Shum, W. *Inorg. Chem.* **1985**, *24*, 1968-1970.

359. Park, J.J.; Glick, M.D.; Hoard, J.L. *J. Am. Chem. Soc.* **1969**, *91*, 301-307.
360. Butcher, R.J.; Penfold, B.R. *J. Cryst. Mol. Struct.* **1976**, *6*, 13.
361. Cotton, F.A.; Elder, R.C. *Inorg. Chem.* **1964**, *3*, 397-401.
362. Anglin, R.J.; Kurtz, D.M., Jr.; Kim, S.; Jacobson, R.A. *Inorg. Chem.* **1987**, *26*, 1470-1472.
363. Prout, K.; Daran, J.-C. *Acta Cryst.* **1978**, *B34*, 3586.
364. Piggott, B.; Sheppard, R.N.; Williams, D.J. *Inorg. Chim. Acta* **1984**, *86*, L65.
365. Pierpont, C.G.; Buchanan, R.M. *J. Am. Chem. Soc.* **1975**, *97*, 6450.
366. Tkachev, V.V.; Atovmyan, L.O. *Koord. Khim.* **1976**, *2*, 110.
367. Cotton, F.A.; Morehouse, S.M.; Wood, J.S. *Inorg. Chem.* **1964**, *3*, 1603-1608.
368. Godfrey, J.E.; Waters, J.M. *Cryst. Struct. Commun.* **1975**, *4*, 5.
369. Hedman, B. *Acta Cryst.* **1977**, *B33*, 3077.
370. Taylor, G.E.; Waters, J.M. *Tetrahedron Lett.* **1981**, 1277.
371. Barnhart, K.M.; Enemark, J.H. *Acta Cryst.* **1984**, *C40*, 1362.
372. Wieghardt, K.; Backes-Dahmann, G.; Herrmann, W.; Weiss, J. *Angew. Chem., Int. Ed. Engl.* **1984**, *23*, 899.
373. Garner, C.D.; Nicholson, J.R.; Clegg, W. *Angew. Chem., Int. Ed. Engl.* **1984**, *23*, 972.
374. Gebreyes, K.; Shaikh, S.N.; Zubieta, J. *Acta Cryst.* **1985**, *C41*, 871.
375. Marabella, C.P.; Enemark, J.H.; Miller, K.F.; Bruce, A.E.; Pariyadath, N.; Corbin, J.L.; Stiefel, E.I. *Inorg. Chem.* **1983**, *22*, 3456.
376. Arzoumanian, H.; Baldy, A.; Lai, R.; Metzger, J.; Peh, M.-L.N.; Pierrot, M. *J. Chem. Soc., Chem. Comm.* **1985**, 1151-1152.
377. Lai, R.; LeBot, S.; Baldy, A.; Pierrot, M.; Arzoumanian, H. *J. Chem. Soc., Chem. Comm.* **1986**, 1208-1209.
378. Hawkins, J.M.; Dewan, J.C.; Sharpless, K.B. *Inorg. Chem.* **1986**, *25*, 1501-1503.
379. Berg, J.M.; Holm, R.H. *J. Am. Chem. Soc.* **1985**, *107*, 917-925.
380. Wilson, A.J.; Penfold, B.R.; Wilkins, C.J. *Acta Cryst.* **1983**, *C39*, 329-330.
381. Schroder, F.A.; Scherle, J.; Hazell, R.G. *Acta Cryst.* **1975**, *B31*, 531.
382. Butcher, R.J.; Penfold, B.R.; Sinn, E. *J. Chem. Soc., Dalton Trans.* **1979**, 668-675.
383. Kamenar, B.; Penavic, M.; Prout, C.K. *Cryst. Struct. Commun.* **1973**, *2*, 41.
384. Kojic-Prodic, B.; Ruzic-Toros, Z.; Grdenic, D.; Golic, L. *Acta Cryst.* **1974**, *B30*, 300.
385. Tkachev, V.V.; Atovmyan, L.O. *Koord. Khim.* **1975**, *1*, 845.
386. Knobler, C.B.; Wilson, A.J.; Hider, R.N.; Jensen, I.W.; Penfold, B.R.; Robinson, W.T.; Wilkins, C.J. *J. Chem. Soc., Dalton Trans.* **1983**, 1299.
387. Robinson, W.T.; Wilkins, C.J. *Trans. Met. Chem.* **1986**, *11*, 86-89.
388. Chew, C.K.; Penfold, B.R. *J. Cryst. Mol. Struct.* **1975**, *5*, 413.
389. Wilson, A.J.; Robinson, W.T.; Wilkins. C.J. *Acta Cryst.* **1983**, *C39*, 54.
390. Brewer, G.A.; Sinn, E. *Inorg. Chem.* **1981**, *20*, 1823.
391. Grandjean, D.; Weiss, R. *Bull. Soc. Chim. France* **1967**, 3049-3054.

392. Kamenar, B.; Penavic, M. *Acta Cryst.* **1976**, *B32*, 3323.
393. Kamenar, B.; Penavic, M.; Korpar-Colig, B.; Markovic, B. *Inorg. Chim. Acta* **1982**, *65*, L245-L247.
394. Florian, L.R.; Corey, E.R. *Inorg. Chem.* **1968**, *7*, 722-725.
395. McCabe, D.J.; Duesler, E.N.; Paine, R.T. *Inorg. Chem.* **1987**, *26*, 2300-2304.
396. Viossat, B.; Khodadad, P.; Rodier, N. *Acta Cryst.* **1977**, *B33*, 2523.
397. Bowen, S.M.; Duesler, E.N.; McCabe, D.J.; Paine, R.T. *Inorg. Chem.* **1985**, *24*, 1191.
398. Pierpont, C.G.; Downs, H.H. *Inorg. Chem.* **1977**, *16*, 2970.
399. Viossat, B.; Rodier, N. *Acta Cryst.* **1979**, *B35*, 2715-2718.
400. Fenn, R.H. *J. Chem. Soc. A* **1969**, 1764-1769.
401. Schrauzer, G.N.; Hughes, L.A.; Strampach, N.; Robinson, P.R.; Schlemper, E.O. *Organometallics* **1982**, *1*, 44-47.
402. Schrauzer, G.N.; Hughes, L.A.; Strampach, N.; Ross, F.; Ross, D.; Schlemper, E.O. *Organometallics* **1983**, *2*, 481-485.
403. Schrauzer, G.N.; Hughes, L.A.; Schlemper, E.O.; Ross, F.; Ross, D. *Organometallics* **1983**, *2*, 1163-1166.
404. Wieghardt, K.; Holzbach, W.; Weiss, J.; Nuber, B.; Prikner, B. *Angew. Chem., Int. Ed. Engl.* **1979**, *18*, 548.
405. Butcher, R.J.; Penfold, B.R. *J. Cryst. Mol. Struct.* **1976**, *6*, 1.
406. Burgmayer, S.J.N.; Stiefel, E.I. *J. Am. Chem. Soc.* **1986**, *108*, 8310-8311.
407. Berg, J.M.; Holm, R.H. *Inorg. Chem.* **1983**, *22*, 1768-1771.
408. Gullotti, M.; Pasini, A.; Zanderighi, G.M.; Ciani, G.; Sironi, A. *J. Chem. Soc., Dalton Trans.* **1981**, 902.
409. Villa, A.C.; Coghi, L.; Manfredotti, A.G.; Guastini, C. *Cryst. Struct. Commun.* **1974**, *3*, 551.
410. Atovmyan, L.O.; Sokolova, Yu.A. *J. Chem. Soc., Chem. Comm.* **1969**, 649.
411. Viossat, B.; Rodier, N.; Khodadad, P. *Acta Cryst.* **1979**, *B35*, 2712.
412. Tsukuma, K.; Kawaguchi, T.; Watanabe, T. *Acta Cryst.* **1975**, *B31*, 2165-2167.
413. Mentzen, B.F.; Bonnet, M.C.; Ledon, H.J. *Inorg. Chem.* **1980**, *19*, 2061.
414. Cliff, C.A.; Fallon, G.D.; Gatehouse, B.M.; Murray, K.S.; Newman, P.J. *Inorg. Chem.* **1980**, *19*, 773.
415. Subramanian, P.; Spence, J.T.; Ortega, R.; Enemark, J.H. *Inorg. Chem.* **1984**, *23*, 2564.
416. Day, V.W.; Fredrich, M.F.; Thompson, M.R.; Klemperer, W.G.; Liu, R.-S.; Shum, W. *J. Am. Chem. Soc.* **1981**, *103*, 3597.
417. Buchanan, I.; Minelli, M.; Ashby, M.T.; King, T.J.; Enemark, J.H.; Gainer, C.D. *Inorg. Chem.* **1984**, *23*, 495.
418. Bruce, A.; Corbin, J.L.; Dahlstrom, P.L.; Hyde, J.R.; Minelli, M.; Stiefel, E.I.; Spence, J.T.; Zubieta, J. *Inorg. Chem.* **1982**, *21*, 917-926.
419. Berg, J.M.; Hodgson, K.O.; Cramer, S.P.; Corbin, J.L.; Elsberry, A.; Pariyadath, N.; Stiefel, E.I. *J. Am. Chem. Soc.* **1979**, *101*, 2774-2776.
420. Berg, J.M.; Spira, D.J.; Hodgson, K.O.; Bruce, A.E.; Miller, K.F.; Corbin, J.L.; Stiefel, E.I. *Inorg. Chem.* **1984**, *23*, 3412.

421. Buchanan, I.; Garner, C.D.; Clegg, W. *J. Chem. Soc., Dalton Trans.* **1984**, 1333.
422. Berg, J.M.; Hodgson, K.O. *Inorg. Chem.* **1980**, *19*, 2180–2181.
423. Kaul, B.B.; Enemark, J.H.; Merbs, S.L.; Spence, J.T. *J. Am. Chem. Soc.* **1985**, *107*, 2885.
424. Bristow, S.; Collison, D.; Garner, C.D.; Clegg, W. *J. Chem. Soc., Dalton Trans.* **1983**, 2495.
425. Traill, P.R.; Tiekink, E.R.T.; O'Connor, M.J.; Snow, M.R.; Wedd, A.G. *Aust. J. Chem.* **1986**, *39*, 1287–1295.
426. Chatt, J.; Choukroun, R.; Dilworth, J.R.; Hyde, J.; Vella, P.; Zubieta, J. *Trans. Met. Chem.* **1979**, *4*, 59.
427. Burt, R.J.; Dilworth, J.R.; Leigh, G.J.; Zubieta, J.A. *J. Chem. Soc., Dalton Trans.* **1982**, 2295–2298.
428. Chatt, J.; Crichton, B.A.L.; Dilworth, J.R.; Dahlstrom, P.; Zubieta, J.A. *J. Chem. Soc., Dalton Trans.* **1982**, 1041–1047.
429. Bishop, M.W.; Chatt, J.; Dilworth, J.R.; Hursthouse, M.B.; Motevalli, M. *J. Chem. Soc., Dalton Trans.* **1979**, 1600–1602.
430. Dahlstrom, P.L.; Dilworth, J.R.; Shulman, P.; Zubieta, J. *Inorg. Chem.* **1982**, *21*, 933.
431. Day, V.W.; Hoard, J.L. *J. Am. Chem. Soc.* **1968**, *90*, 3374–3379.
432. Müller, A.; Kwang, T.K.; Bogge, H. *Angew. Chem., Int. Ed. Engl.* **1979**, *18*, 628.
433. Müller, A.; Bogge, H.; Hwang, T.K. *Inorg. Chim. Acta* **1980**, *39*, 71.
434. Doherty, R.; Hubbard, C.R.; Mighell, A.D.; Siedle, A.R.; Stewart, J. *Inorg. Chem.* **1979**, *18*, 2991.
435. Robiette, A.G.; Hedberg, K.; Hedberg, L. *J. Mol. Struct.* **1977**, *37*, 105.
436. Page, E.M.; Rice, D.A.; Hagen, K.; Hedberg, L.; Hedberg, K. *Inorg. Chem.* **1987**, *26*, 467–468.
437. Feinstein-Jaffe, I.; Gibson, D.; Lippard, S.J.; Schrock, R.R.; Spool, A. *J. Am. Chem. Soc.* **1984**, *106*, 6305–6310.
438. Herberhold, M.; Kniesel, H.; Haumaier, L.; Thewalt, U. *J. Organomet. Chem.* **1986**, *301*, 355–367.
439. Cotton, F.A.; Schwotzer, W.; Shamshoum, E.S. *J. Organomet. Chem.* **1985**, *296*, 55–68.
440. Hoskins, B.F.; Linden, A.; O'Donnell, T.A. *Inorg. Chem.* **1987**, *26*, 2223–2228.
441. Legzdins, P.; Rettig, S.J.; Sánchez, L. *Organometallics* **1985**, *4*, 1470–1471.
442. Ruzic-Toros, Z.; Kojic-Prodic, B.; Gabela, F.; Sljukic, M. *Acta Cryst.* **1977**, *B33*, 692.
443. Stomberg, R.; Olson, S. *Acta Chem. Scand.* **1985**, *A39*, 79.
444. Drew, M.G.B.; Mandyczewsky, R. *J. Chem. Soc., Chem. Comm.* **1970**, 292–293.
445. Jeannin, Y.; Launay, J.-P.; Livage, J.; Nel, A. *Inorg. Chem.* **1978**, *17*, 374.
446. Lappert, M.F.; Raston, C.L.; Rowbottom, G.L.; White, A.H. *J. Chem. Soc., Chem. Comm.* **1981**, 6.
447. Drew, M.G.B.; Page, E.M.; Rice, D.A. *Inorg. Chim. Acta* **1983**, *76*, L33.
448. Patel, V.D.; Boorman, P.M.; Kerr, K.A.; Moynihan, K.J. *Inorg. Chem.* **1982**, *21*, 1383.

449. Ball, J.M.; Boorman, P.M.; Moynihan, K.J.; Richardson, J.F. *Acta Cryst.* **1985**, *C41*, 47.
450. Povey, D.C.; Richards, R.L. *J. Chem. Soc., Dalton Trans.* **1984**, 2585.
451. Leipoldt, J.G.; Basson, S.S.; Roodt, A.; Potgieter, I.M. *S. Afr. J. Chem.* **1986**, *39*, 179-183.
452. Leipoldt, J.G.; Basson, S.S.; Roodt, A.; Potgieter, I.M. *Trans. Met. Chem.* **1986**, *11*, 323-326.
453. Chiu, K.W.; Lyons, D.; Wilkinson, G.; Thornton-Pett, M.; Hursthouse, M.B. *Polyhedron* **1983**, *2*, 803.
454. Su, F.-M.; Cooper, C.; Geib, S.J.; Rheingold, A.L.; Mayer, J.M. *J. Am. Chem. Soc.* **1986**, *108*, 3545-3547.
455. Bryan, J.C.; Geib, S.J.; Rheingold, A.L.; Mayer, J.M. *J. Am. Chem. Soc.* **1987**, *109*, 2826-2828.
456. Bokiy, N.G.; Gatilov, Yu.V.; Struchov, Yu.T.; Ustynyuk, N.A. *J. Organomet. Chem.* **1973**, *54*, 213-219.
457. Burkhardt, E.R.; Doney, J.J.; Bergman, R.G.; Heathcock, C.H. *J. Am. Chem. Soc.* **1987**, *109*, 2022-2039.
458. Busetto, L.; Jeffery, J.C.; Mills, R.M.; Stone, F.G.A.; Went, M.J.; Woodward, P. *J. Chem. Soc., Dalton Trans.* **1983**, 101.
459. Krebs, B.; Buss, B.; Ferwanah, A. *Z. Anorg. Allg. Chem.* **1972**, *387*, 142-153.
460. Feinstein-Jaffe, I.; Dewan, J.C.; Schrock, R.R. *Organometallics* **1985**, *4*, 1189-1193.
461. Bryan, J.C.; Mayer, J.M. *J. Am. Chem. Soc.*, **1987**, *109*, 7213-7214.
462. De Wet, J.F.; Caira, M.R.; Gellatly, B.J. *Acta Cryst.* **1978**, *B34*, 762.
463. Drew, M.G.B.; Fowles, G.W.A.; Rice, D.A.; Shanton, K.J. *J. Chem. Soc., Chem. Comm.* **1974**, 614-615.
464. Lotspeich, J.F.; Javan, A.; Englebrecht, A. *J. Chem. Phys.* **1959**, *31*, 633.
465. Edwards, A.J.; Jones, G.R.; Sills, R.J.C. *J. Chem. Soc. A* **1970**, 2521-2523.
466. Bandoli, G.; Mazzi, U.; Wilcox, B.E.; Jurisson, S.; Deutsch, E. *Inorg. Chim. Acta* **1984**, *95*, 217.
467. Fair, C.K.; Troutner, D.E.; Schlemper, E.O.; Murmann, R.K.; Hoppe, M.L. *Acta Cryst.* **1984**, *C40*, 1544.
468. Bandoli, G.; Gerber, T.I.A. *Inorg. Chim. Acta* **1987**, *126*, 205-208.
469. Jones, A.G.; DePamphilis, B.V.; Davison, A. *Inorg. Chem.* **1981**, *20*, 1617-1618.
470. Smith, J.E.; Byrne, E.F.; Cotton, F.A.; Sekatowski, J.C. *J. Am. Chem. Soc.* **1978**, *100*, 5571-5572.
471. DePamphilis, B.V.; Jones, A.G.; Davis, M.A.; Davison, A. *J. Am. Chem. Soc.* **1978**, *100*, 5570-5571.
472. Bandoli, G.; Nicolini, M.; Mazzi, U.; Spies, H.; Munze, R. *Trans. Met. Chem.* **1984**, *9*, 127.
473. Bandoli, G.; Mazzi, U.; Abram, U.; Spies, H.; Münze, R. *Polyhedron* **1987**, *6*, 1547-1550.
474. Jurisson, S.; Lindoy, L.F.; Dancey, K.P.; McPartlin, M.; Tasker, P.A.; Uppal, D.K.; Deutsch, E. *Inorg. Chem.* **1984**, *23*, 227.
475. Bandoli, G.; Nicolini, M.; Mazzi, U.; Refosco, F. *J. Chem. Soc., Dalton Trans.* **1984**, 2505.

476. Wilcox, B.E.; Heeg, M.J.; Deutsch, E. *Inorg. Chem.* **1984**, *23*, 2962-2967.
477. Bandoli, G.; Mazzi, U.; Clemente, D.A.; Roncari, E. *J. Chem. Soc., Dalton Trans.* **1982**, 2455-2459.
478. Duatti, A.; Marchi, A.; Magon, L.; Deutch, E.; Bertolasi, V.; Gilli, G. *Inorg. Chem.* **1987**, *26*, 2182-2186.
479. Thomas, R.W.; Estes, G.W.; Elder, R.C.; Deutsch, E. *J. Am. Chem. Soc.* **1979**, *101*, 4581.
480. Franklin, K.J.; Howard-Lock, H.E.; Lock, C.J.L. *Inorg. Chem.* **1982**, *21*, 1941-1946.
481. Fackler, P.H.; Kastner, M.E.; Clarke, M.J. *Inorg. Chem.* **1984**, *23*, 3968-3972.
482. Krebs, B.; Hasse, K.-D. *Acta Cryst.* **1976**, *B32*, 1334-1337 and references therein.
483. Faggiani, R.; Lock, C.J.L.; Pocé, J. *Acta Cryst.* **1980**, *B36*, 231-233.
484. Kastner, M.E.; Fackler, P.H.; Clarke, M.J.; Deutsch, E. *Inorg. Chem.* **1984**, *23*, 4683.
485. Fackler, P.H.; Lindsay, M.J.; Clarke, M.J.; Kastner, M.E. *Inorg. Chim. Acta* **1985**, *109*, 39.
486. Kastner, M.E.; Lindsay, M.J.; Clarke, M.J. *Inorg. Chem.* **1982**, *21*, 2037-2040.
487. Zuckman, S.A.; Freeman, G.M.; Troutner, D.E.; Volkert, W.A.; Holmes, R.A.; Vanderveer, D.G.; Barefield, E.K. *Inorg. Chem.* **1981**, *20*, 2386.
488. Huggins, J.M.; Whitt, D.R.; Lebioda, L. *J. Organomet. Chem.* **1986**, *312*, C15-C19.
489. Frais, P.W.; Lock, C.J.L. *Can. J. Chem.* **1972**, *50*, 1811-1818.
490. Edwards, P.G.; Wilkinson, G.; Hursthouse, M.B.; Malik, K.M.A. *J. Chem. Soc., Dalton Trans.* **1980**, 2467-2475.
491. Chiu, K.W.; Wong, W.-K.; Wilkinson, G.; Galas, A.M.R.; Hursthouse, M.B. *Polyhedron* **1982**, *1*, 31.
492. Stravropoulos, P.; Edwards, P.G.; Wilkinson, G.; Motevalli, M.; Malik, K.M.A.; Hursthouse, M.B. *J. Chem. Soc., Dalton Trans.* **1985**, 2167-2175.
493. McDonell, A.C.; Hambley, T.W.; Snow, M.R.; Wedd, A.G. *Aust. J. Chem.* **1983**, *36*, 253.
494. Mattes, R.; Weber, H. *Z. Anorg. Allg. Chem.* **1981**, *474*, 216.
495. Lis, T. *Acta Cryst.* **1979**, *B35*, 3041-3044.
496. Cotton, F.A.; Lippard, S.J. *Inorg. Chem.* **1965**, *4*, 1621-1629.
497. Cotton, F.A.; Lippard, S.J. *Inorg. Chem.* **1966**, *5*, 416-423.
498. Sergienko, V.S.; Porai-Koshits, M.A. *Koord. Khim.* **1982**, *8*, 251.
499. Sergienko, V.S.; Porai-Koshits, M.A.; Mistryukov, V.E.; Kotegov, K.V. *Koord. Khim.* **1982**, *8*, 230.
500. Bryan, J.C.; Stenkamp, R.E.; Tulip, T.H.; Mayer, J.M. *Inorg. Chem.* **1987**, *26*, 2283-2288.
501. Lis, T. *Acta Cryst.* **1977**, *B33*, 944-946.
502. Lis, T. *Acta Cryst.* **1976**, *B32*, 2707-2709.
503. Hursthouse, M.B.; Jayaweera, S.A.A.; Quick, A. *J. Chem. Soc., Dalton Trans.* **1979**, 279-282.
504. Gilli, G.; Sacerdoti, M.; Bertolasi, V.; Rossi, R. *Acta Cryst.* **1982**, *B38*, 100.
505. Lock, C.J.L.; Wan, C. *Can. J. Chem.* **1975**, *53*, 1548-1553.

506. Bertolasi, V.; Ferretti, V.; Sacerdoti, M.; Marchi, A. *Acta Cryst.* **1984**, *C40*, 971.
507. Lock, C.J.L.; Turner, G. *Can. J. Chem.* **1977**, *55*, 333-339.
508. Ciani, G.F.; D'Alfonso, G.; Romiti, P.F.; Sironi, A.; Freni, M. *Inorg. Chim. Acta* **1983**, *72*, 29-37.
509. Lock, C.J.L.; Turner, G. *Can. J. Chem.* **1978**, *56*, 179-188.
510. Glowiak, T.; Lis, T.; Jezowska-Trzebiatowska, B. *Bull. Acad. Pol. Sci., Sci. Chim.* **1972**, *20*, 199.
511. Carrondo, M.A.A.F. de C.T.; Middleton, A.R.; Skapski, A.C.; West, A.P.; Wilkinson, G. *Inorg. Chim. Acta* **1980**, *44*, L7.
512. Herrmann, W.A.; Herdtweck, E.; Flöel, M.; Kulpe, J.; Küsthardt, U.; Okuda, J. *Polyhedron* **1987**, *6*, 1165-1182.
513. Herrmann, W.A.; Marz, D.; Herdtweck, E.; Schäfer, A.; Wagner, W.; Kneuper, H.-J. *Angew. Chem., Int. Ed. Eng.* **1987**, *26*, 462-464.
514. Herrmann, W.A.; Serrano, R.; Kusthardt, U.; Ziegler, M.L.; Guggolz, E.; Zahn, T. *Angew. Chem., Int. Ed. Engl.* **1984**, *23*, 515.
515. Herrmann, W.A.; Küsthardt, U.; Ziegler, M.L.; Zahn, T. *Angew. Chem., Int. Ed. Engl.* **1985**, *24*, 860-861.
516. DeBoer, E.J.M.; DeWith, J.; Orpen, A.G. *J. Am. Chem. Soc.* **1986**, *108*, 8271-8273.
517. Valencia, E.; Santarsiero, B.D.; Geib, S.J.; Rheingold, A.L.; Mayer, J.M. *J. Am. Chem. Soc.*, **1987**, *109*, 6896-6898.
518. Fellmann, J.D.; Schrock, R.R.; Traficante, D.D. *Organometallics* **1982**, *1*, 481-484.
519. Mayer, J.M.; Tulip, T.H.; Calabrese, J.C.; Valencia, E. *J. Am. Chem. Soc.* **1987**, *109*, 157-163.
520. Sheldrick, G.M.; Sheldrick, W.S. *J. Chem. Soc. A* **1969**, 2160.
521. Krebs, B.; Müller, A.; Beyer, H.H. *Inorg. Chem.* **1969**, *8*, 436.
522. Beyer, H.; Glemser, O.; Krebs, B.; Wagner, G. *Z. Anorg. Allg. Chem.* **1970**, *376*, 87.
523. Johnson, J.W.; Brody, J.F.; Ansell, G.B.; Zentz, S. *Acta Cryst.* **1984**, *C40*, 2024.
524. Lis, T. *Acta Cryst.* **1979**, *B35*, 1230-1232.
525. Sergienko, V.S.; Khodashova, T.S.; Porai-Koshits, M.A.; Butman, L.A. *Koord. Khim.* **1977**, *3*, 1060-1068.
526. Wieghardt, K.; Pomp, C.; Nuber, B.; Weiss, J. *Inorg. Chem.* **1986**, *25*, 1659-1661.
527. Fischer, D.; Krebs, B. *Z. Anorg. Allg. Chem.* **1982**, *491*, 73.
528. Lock, C.J.L.; Turner, G. *Acta Cryst.* **1978**, *B34*, 923-927.
529. Johnson, J.W.; Brody, J.F.; Ansell, G.B.; Zentz, S. *Inorg. Chem.* **1984**, *23*, 2415-2418.
530. Lis, T.; Glowiak, T.; Jezowska-Trzebiatowska, B. *Bull. Acad. Pol. Sci., Sci. Chim.* **1975**, *23*, 417.
531. Fenn, R.H.; Graham, A.J. *J. Chem. Soc. A* **1971**, 2880-2883.
532. Schappacher, M.; Weiss, R.; Montiel-Montoya, R.; Trautwein, A.; Tabard, A. *J. Am. Chem. Soc.* **1985**, *107*, 3736-3738.
533. Penner-Hahn, J.E.; Eble, K.S.; McMurry, T.J.; Renner, M.; Balch, A.L.; Groves, J.T.; Dawson, J.H.; Hodgson, K.O. *J. Am. Chem. Soc.* **1986**, *108*, 7819-7825.
534. Che, C.-M.; Lai, T.-F.; Wong, K.-Y. *Inorg. Chem.* **1987**, *26*, 2289-2299.

535. Schäfer, L.; Seip, H.M. Acta Chem. Scand. **1967**, *21*, 737–744.
536. Nowogrocki, G.; Abraham, F.; Tréhoux, J.; Thomas, D. Acta Cryst. **1976**, *B32*, 2413.
537. Mak, T.C.W.; Che, C.-M.; Wong, K.-Y. J. Chem. Soc., Chem. Comm. **1985**, 986.
538. Lau, T.C.; Kochi, J.K. J. Chem. Soc., Chem. Comm. **1987**, 798–799.
539. Phillips, F.L.; Skapski, A.C. Acta Cryst. **1975**, *B31*, 1814–1818.
540. Atovmyan, L.O.; Sokolova, Yu.A. Zh. Strukt. Khim. **1979**, *20*, 754.
541. Phillips, F.L.; Skapski, A.C. J. Chem. Soc., Dalton Trans. **1975**, 2586–2590.
542. Griffith, W.P.; McManus, N.T.; Skapski, A.C.; Nielson, A.J. Inorg. Chim. Acta **1985**, *103*, L5–L6.
543. Cartwright, B.A.; Griffith, W.P.; Schröder, M.; Skapski, A.C. J. Chem. Soc., Chem. Comm. **1978**, 853–854.
544. Seip, H.M.; Stølevik, R. Acta Chem. Scand. **1966**, *20*, 385–394.
545. Griffith, W.P.; Skapski, A.C.; Woode, K.A.; Wright, M.J. Inorg. Chim. Acta **1978**, *31*, L413–L414.
546. Weber, R.; Dehnicke, K.; Müller, U.; Fenske, D. Z. Anorg. Allg. Chem. **1984**, *516*, 214.
547. Pastuszak, R.; L'Haridon, P.; Marchand, R.; Laurent, Y. Acta Cryst. **1982**, *B38*, 1427–1430.
548. Nugent, W.A.; Harlow, R.L.; McKinney, R.J. J. Am. Chem. Soc. **1979**, *101*, 7265–7268.
549. Griffith, W.P.; McManus, N.T.; Skapski, A.C.; White, A.D. Inorg. Chim. Acta **1985**, *105*, L11.
550. Malin, J.M.; Schlemper, E.O.; Murmann, R.K. Inorg. Chem. **1977**, *16*, 615–619.
551. Kruse, F.H. Acta Cryst. **1961**, *14*, 1035–1041.
552. Atovmyan, L.O.; Andianov, V.G.; Porai-Koshits, M.A. Zh. Strukt. Khim. **1962**, *3*, 685.
553. Roth, W.J.; Hinckley, C.C. Inorg. Chem. **1981**, *20*, 2023–2026.
554. Galas, A.M.R.; Hursthouse, M.B.; Behrman, E.J.; Midden, W.R.; Green, G.; Griffith, W.P. Trans. Met. Chem. **1981**, *6*, 194–195.
555. Conn, J.F.; Kim, J.J.; Suddath, F.L.; Blattmann, P.; Rich, A. J. Am. Chem. Soc. **1974**, *96*, 7152–7153.
556. Neidle, S.; Stuart, D.I. Biochim. Biophys. Acta **1976**, *418*, 226–231.
557. Prangé, T.; Pascard, C. Acta Cryst. **1977**, *B33*, 621–623.
558. Cartwright, B.A.; Griffith, W.P.; Schröder, M.; Skapski, A.C. Inorg. Chim. Acta **1981**, *53*, L129–L130.
559. Audett, J.D.; Collins, T.J.; Santarsiero, B.D.; Spies, G.H. J. Am. Chem. Soc. **1982**, *104*, 7352.
560. Preuss, F.; Noichl, H.; Kaub, J. Z. Naturforsch. **1986**, *41B*, 1085–1092.
561. Nugent, W.A.; Harlow, R.L. J. Chem. Soc., Chem. Comm. **1979**, 342–343.
562. Schweda, E.; Scherfise, K.D.; Dehnicke, K. Z. Anorg. Allg. Chem. **1985**, *528*, 117–124.
563. Preuss, F.; Fuchslocher, E.; Sheldrick, W.S. Z. Naturforsch. **1985**, *40B*, 1040–1044.
564. Bradley, D.C.; Hursthouse, M.B.; Jelfs, A.N. de M.; Short, R.L. Polyhedron **1983**, *2*, 849–852.
565. Strähle, J.; Bärnighausen, H. Z. Anorg. Chem. **1968**, *359*, 325–337.

566. Preuss, F.; Fuchslocher, E.; Sheldrick, W.S. Z. Naturforsch. 1985, 40B, 363-367.
567. Lorcher, K.-P.; Strähle, J.; Walker, I. Z. Anorg. Allg. Chem. 1979, 452, 123-140.
568. Wiberg, N.; Häring, H.W.; Schubert, U. Z. Naturforsch. 1980, 35B, 599-603.
569. Osborne, J.H.; Rheingold, A.L.; Trogler, W.C. J. Am. Chem. Soc. 1985, 107, 7945-7952.
570. Veith, M. Angew. Chem., Int. Ed. Engl. 1976, 15, 387-388.
571. Cotton, F.A.; Duraj, S.A.; Roth, W.J. J. Am. Chem. Soc. 1984, 106, 4749-4751.
572. Finn, P.A.; King, M.S.; Kilty, P.A.; McCarley, R.E. J. Am. Chem. Soc. 1975, 97, 220-221.
573. Bezler, H.; Strähle, J. Z. Naturforsch. 1979, 34B, 1199-1202.
574. Tan, L.S.; Goeden, G.V.; Haymore, B.L. Inorg. Chem. 1983, 22, 1744-1750.
575. Hörner, M.; Frank, K.P.; Strähle, J. Z. Naturforsch. 1986, 41B, 423-428.
576. Nugent, W.A.; Harlow, R.L. J. Chem. Soc., Chem. Comm. 1978, 579-580.
577. Bradley, D.C.; Hursthouse, M.B.; Malik, K.M.A.; Nielson, A.J.; Chota Vuru, G.B. J. Chem. Soc., Chem. Comm. 1984, 1069-1072.
578. Chamberlain, L.R.; Rothwell, I.P.; Huffman, J.C. J. Chem. Soc., Chem. Comm. 1986, 1203-1205.
579. Bradley, D.C.; Hursthouse, M.B.; Malik, K.M.A.; Nielson, A.J.; Vuru, G.B.C. J. Chem. Soc., Dalton Trans. 1984, 1069.
580. Churchill, M.R.; Wasserman, H.J. Inorg. Chem. 1981, 20, 2899-2904.
581. Bates, P.A.; Nielson, A.J.; Waters, J.M. Polyhedron 1985, 4, 1391-1401.
582. Jones, T.C.; Nielson, A.J.; Ricard, C.E.F. J. Chem. Soc., Chem. Comm. 1984, 205-206.
583. Churchill, M.R.; Wasserman, H.J. Inorg. Chem. 1982, 21, 223-226.
584. Canich, J.A.M.; Cotton, F.A.; Duraj, S.A.; Roth, W.J. Polyhedron 1986, 5, 895-898.
585. Cotton, F.A.; Hall, W.T. Inorg. Chem. 1978, 17, 3525-3528.
586. Churchill, M.R.; Wasserman, H.J. Inorg. Chem. 1982, 21, 218.
587. Bezler, H.; Strähle, J. Z. Naturforsch. 1983, 38B, 317.
588. Frank, K.-P.; Strähle, J.; Weidlein, J. Z. Naturforsch. 1980, 35B, 300-306.
589. Hursthouse, M.B.; Motevalli, M.; Sullivan, A.C.; Wilkinson, G. J. Chem. Soc., Chem. Comm. 1986, 1398-1399.
590. Wiberg, N.; Häring, H.-W.; Schubert, U. Z. Naturforsch. 1978, 33B, 1365-1369.
591. Groves, J.T.; Takahashi, T.; Butler, W.M. Inorg. Chem. 1983, 22, 884-887.
592. Nugent, W.A.; Harlow, R.L. J. Am. Chem. Soc. 1980, 102, 1759-1760.
593. Chisholm, M.H.; Folting, K.; Huffman, J.C.; Ratermann, A.L. Inorg. Chem. 1982, 21, 978-982.
594. Chou, C.Y.; Huffman, J.C.; Maatta, E.A. J. Chem. Soc., Chem. Comm. 1984, 1184-1185.
595. Dehnicke, K.; Weiher, U.; Fenske, D. Z. Anorg. Allg. Chem. 1979, 456, 71-80.

596. Chatt, J.; Choukroun, R.; Dilworth, J.R.; Hyde, J.; Vella, P.; Zubieta, J. *Trans. Met. Chem.* **1979**, *4*, 59–63.
597. Maatta, E.A.; Haymore, B.L.; Wentworth, R.A.D. *Inorg. Chem.* **1980**, *19*, 1055–1059.
598. Bishop, M.W.; Chatt, J.; Dilworth, J.R.; Neaves, B.D.; Dahlstrom, P.; Hyde, J.; Zubieta, J. *J. Organomet. Chem.* **1981**, *213*, 109–124.
599. Wall, K.L.; Folting, K.; Huffman, J.C.; Wentworth, R.A.D. *Inorg. Chem.* **1983**, *22*, 2366–2371.
600. Bishop, M.W.; Chatt, J.; Dilworth, J.R.; Hursthouse, M.B.; Jayaweera, S.A.A.; Quick, A. *J. Chem. Soc., Dalton Trans.* **1979**, 914–920.
601. Noble, M.E.; Huffman, J.C.; Wentworth, R.A.D. *Inorg. Chem.* **1982**, *21*, 2101–2103.
602. Edelbult, A.W.; Folting, K.; Huffman, J.C.; Wentworth, R.A.D. *J. Am. Chem. Soc.* **1981**, *103*, 1927.
603. Wall, K.L.; Folting, K. Huffman, J.C.; Wentworth, R.A.D. *Inorg. Chem.* **1983**, *22*, 2366.
604. Noble, M.E.; Huffman, J.C.; Wentworth, R.A.D. *Inorg. Chem.* **1983**, *22*, 1756–1760.
605. Dilworth, J.R.; Dahlstrom, P.L.; Hyde, J.R.; Zubieta, J. *Inorg. Chim. Acta* **1983**, *71*, 21–28.
606. Chou, C.Y.; Devore, D.D.; Huckett, S.C.; Maatta, E.A.; Huffman, J.C.; Takusagawa, F. *Polyhedron* **1986**, *5*, 301–304.
607. Bernal, I.; Draux, M.; Brunner, H.; Hoffmann, B.; Wachter, J. *Organometallics* **1986**, *5*, 655–660.
608. Herrmann, W.A.; Kriechbaum, G.W.; Dammel, R.; Bock, H.; Ziegler, M.L.; Pfisterer, H. *J. Organomet. Chem.* **1983**, *254*, 219–241.
609. D'Errico, J.J.; Messerle, L.; Curtis, M.D. *Inorg. Chem.* **1983**, *22*, 849.
610. Chisholm, M.H.; Folting, K.; Huffman, J.C.; Ratermann, A.L. *Inorg. Chem.* **1984**, *23*, 2303–2311.
611. Chatt, J.; Crichton, B.A.L.; Dilworth, J.R.; Dahlstrom, P.; Gutkoska, R.; Zubieta, J. *Inorg. Chem.* **1982**, *21*, 2383.
612. Dilworth, J.R.; Zubieta, J.A. *J. Chem. Soc., Chem. Comm.* **1981**, 132–133.
613. Chatt, J.; Crichton, B.A.L.; Dilworth, J.R.; Dahlstrom, P.; Gutkowska, R.; Zubieta, J. *Inorg. Chem.* **1982**, *21*, 2383–2391.
614. Dilworth, J.R.; Henderson, R.A.; Dahlstrom, P.; Nicholson, T.; Zubieta, J.A. *J. Chem. Soc., Dalton Trans.* **1987**, 529–540.
615. Dahlstrom, P.L.; Dilworth, J.R.; Shulman, P.; Zubieta, J. *Inorg. Chem.* **1982**, *21*, 933.
616. Butcher, A.V.; Chatt, J.; Dilworth, J.R.; Leigh, G.J.; Hursthouse, M.B.; Jayaweera, S.A.A.; Quick, A. *J. Chem. Soc., Dalton Trans.* **1979**, 921.
617. March, F.C.; Mason, R.; Thomas, K.M. *J. Organomet. Chem.* **1975**, *96*, C43–C45.
618. Hanson, I.R.; Hughes, D.L. *J. Chem. Soc., Dalton Trans.* **1981**, 390–399.
619. Hidai, M.; Kodama, T.; Sato, M.; Harakawa, M.; Uchida, Y. *Inorg. Chem.* **1976**, *15*, 2694–2697.
620. Day, V.W.; George, T.A.; Iske, S.D.A.; Wagner, S.D. *J. Organomet. Chem.* **1976**, *112*, C55–C58.

621. Hsieh, T.-C.; Gebreyes, K.; Zubieta, J. J. Chem. Soc., Chem. Comm. 1984, 1172-1174.
622. Carrillo, D.; Gouzerh, P.; Jeannin, Y. Nouv. J. Chim. 1985, 9, 749-755.
623. Hsieh, T.-C.; Zubieta, J.A. Polyhedron, 1986, 5, 305-314.
624. Chatt, J.; Dilworth, J.R.; Dahlstrom, P.L.; Zubieta, J. J. Chem. Soc., Chem. Comm. 1980, 786-787.
625. Mattes, R.; Scholand, H.; Mikloweit, U.; Schrenk, V. Chem. Ber. 1987, 120, 783-787.
626. Messerle, L.; Curtis, M.D. J. Am. Chem. Soc. 1982, 104, 889.
627. Dilworth, J.R.; Zubieta, J.; Hyde, J.R. J. Am. Chem. Soc. 1982, 104, 365-367.
628. Chan, D.M.T.; Chisholm, M.H.; Folting, K.; Huffmann, J.C.; Marchant, N.S. Inorg. Chem. 1986, 25, 4170-4174.
629. Müller, U., Schweda, E.; Strähle, J. Z. Naturforsch. 1983, 38B, 1299-1300.
630. Dehnicke, K.; Krüger, N.; Kujanek, R.; Weller, F. Z. Kristallogr. 1980, 153, 181-187.
631. Dehnicke, K.; Schmitte, J.; Fenske, D. Z. Naturforsch. 1980, 35B, 1070-1074.
632. Schweda, E.; Strähle, J. Z. Naturforsch. 1981, 36B, 662-665.
633. Schweda, E.; Strähle, J. Z. Naturforsch. 1980, 35B, 1146-1149.
634. Beck, J.; Sträle, J. Z. Naturforsch. 1987, 42B, 255-259.
635. Strähle, J. Z. Anorg. Allg. Chem. 1970, 375, 238-254.
636. Müller, U.; Kujanek, R.; Dehnicke, K. Z. Anorg. Allg. Chem. 1982, 495, 127-134.
637. Strähle, J.; Weiher, U.; Dehnicke, K. Z. Naturforsch. 1978, 33B, 1347-1351.
638. Noble, M.E.; Folting, K.; Huffman, J.C.; Wentworth, R.A.D. Inorg. Chem. 1982, 21, 3772-3776.
639. Beck, J.; Schweda, E.; Strähle, J. Z. Naturforsch. 1985, 40B, 1073-1076.
640. Hursthouse, M.B.; Motevalli, M. J. Chem. Soc., Dalton Trans. 1979, 1362-1366.
641. Bishop, M.W.; Chatt, J.; Dilworth, J.R.; Hursthouse, M.B.; Motevalli, M. J. Chem. Soc., Chem. Comm. 1976, 780-781.
642. Schmitte, J.; Friebel, C.; Weller, F.; Dehnicke, K. Z. Anorg. Allg. Chem. 1982, 495, 148-156.
643. Chisholm, M.H.; Heppert, J.A.; Huffman, J.C.; Streib, W.E. J. Chem. Soc., Chem. Comm. 1985, 1771-1773.
644. Chan, D.M.-T.; Fultz, W.C.; Nugent, W.A.; Roe, D.C.; Tulip, T.H. J. Am. Chem. Soc. 1985, 107, 251-253.
645. Thorn, D.L.; Nugent, W.A.; Harlow, R.L. J. Am. Chem. Soc. 1981, 103, 357-363.
646. Chiu, K.W.; Jones, R.A.; Wilkinson, G.; Galas, A.M.R.; Hursthouse, M.B. J. Chem. Soc., Dalton Trans. 1981, 2088-2097.
647. Ashcroft, B.R.; Clark, G.R.; Nielson, A.J.; Rickard, C.E.F. Polyhedron 1986, 5, 2081-2091.
648. Stahl, K; Weller, F.; Dehnicke, K.; Paetzold, P. Z. Anorg. Allg. Chem. 1986, 534, 93-99.
649. Bradley, D.C.; Errington, R.J.; Hursthouse, M.B.; Nielson, A.J.; Short, R.L. Polyhedron 1983, 2, 843-847.

650. Ashcroft, B.R.; Bradley, D.C.; Clark, G.R.; Errington, R.J.; Nielson, A.J.; Rickard, C.E.F. *J. Chem. Soc., Chem. Comm.* **1987**, 170–171.
651. Weiss, K.; Schubert, U.; Schrock, R.R. *Organometallics* **1986**, *5*, 397–398.
652. Nielson, A.J.; Waters, J.M.; Bradley, D.C. *Polyhedron* **1985**, *2*, 285–297.
653. Bradley, D.C.; Hursthouse, M.B.; Malik, K.M.A.; Nielson, A.J. *J. Chem. Soc., Chem. Comm.* **1981**, 103–104.
654. Cotton, F.A.; Shamshoum, E.S. *J. Am. Chem. Soc.* **1984**, *106*, 3222–3225.
655. Blum, L.; Williams, I.D.; Schrock, R.R. *J. Am. Chem. Soc.* **1984**, *106*, 8316–8317.
656. Weiher, U.; Dehnicke, K.; Fenske, D. *Z. Anorg. Allg. Chem.* **1979**, *457*, 105.
657. Drew, M.G.B.; Fowles, G.W.A.; Rice, D.A.; Rolfe, N. *J. Chem. Soc., Chem. Comm.* **1971**, 231–232.
658. Schmidt, I.; Willing, W.; Mueller, U.; Dehnicke, K. *Z. Anorg. Allg. Chem.* **1987**, *545*, 169–176.
659. Fenske, D.; Kujanek, R.; Dehnicke, K. *Z. Anorg. Allg. Chem.* **1983**, *507*, 51–58.
660. Witt, M.; Roesky, H.W.; Noltemeyer, M.; Sheldrick, G.M. *Z. Naturforsch.* **1987**, *42B*, 519–521.
661. Roesky, H.W.; Sundermeyer, J., Schimkowiak, J.; Jones, P.G.; Noltemeyer, M.; Schroeder, T.; Sheldrick, G.M. *Z. Naturforsch.* **1985**, *40B*, 736–739.
662. Nielson, A.J.; Waters, J.M. *Aust. J. Chem.* **1983**, *36*, 243–251.
663. Bradley, D.C.; Hursthouse, M.B.; Malik, K.M.A.; Nielson, A.J.; Short, R.L. *J. Chem. Soc., Dalton Trans.* **1983**, 2651–2666.
664. Dilworth, J.R.; Henderson, R.; Dahlstrom, P.; Hutchinson, J.; Zubieta, J. *Cryst. Struct. Commun.* **1982**, *11*, 1135–1139.
665. Churchill, M.R.; Li, Y.-J. *J. Organomet. Chem.* **1986**, *301*, 49–59.
666. Churchill, M.R.; Li, Y.J.; Theopold, K.H.; Schrock, R.R. *Inorg. Chem.* **1984**, *23*, 4472–4476.
667. Takahashi, T.; Mizobe, Y.; Sato, M.; Uchida, Y.; Hidai, M. *J. Am. Chem. Soc.* **1980**, *102*, 7461–7467.
668. Einstein, F.W.B.; Jones, T.; Hanlan, A.J.L.; Sutton, D. *Inorg. Chem.* **1982**, *21*, 2585–2589.
669. Chatt, J.; Fakley, M.E.; Hitchcock, P.B.; Richards, R.L.; Luong-Thi, N.T. *J. Chem. Soc., Dalton Trans.* **1982**, 345–352.
670. Hughes, D.L.; *Acta Cryst.* **1981**, *B37*, 557–562.
671. Heath, G.A.; Mason, R.; Thomas, K.M. *J. Am. Chem. Soc.* **1974**, *96*, 259–260.
672. Henderson, R.A. *J. Chem. Soc., Dalton Trans.* **1982**, 917–925.
673. Head, R.A.; Hitchcock, P.B. *J. Chem. Soc., Dalton Trans.* **1980**, 1150.
674. Hidai, M.; Mizobe, Y.; Sato, M.; Kodama, T.; Uchida, Y. *J. Am. Chem. Soc.* **1978**, *100*, 5740–5748.
675. Hidai, M.; Aramaki, S.; Yoshida, K.; Kodama, T.; Takahashi, T.; Uchida, Y.; Mizobe, Y. *J. Am. Chem. Soc.* **1986**, *108*, 1562–1568.
676. Hidai, M.; Komori, K.; Kodama, T.; Jin, D.-M.; Takahashi, T.; Sugiura, S.; Uchida, Y.; Mizobe, Y. *J. Organomet. Chem.* **1984**, *272*, 155–167.

677. Colquhoun, H.M.; Williams, D.J. *J. Chem. Soc., Dalton Trans.* **1984**, 1675.
678. Iwanami, K.; Mizobe, Y.; Takahashi, T.; Kodama, T.; Uchida, Y.; Hidai, M. *Bull. Chem. Soc. Jpn.* **1981**, *54*, 1773–1776.
679. Colquhoun, H.M.; King, T.J. *J. Chem. Soc., Chem. Comm.* **1980**, 879–881.
680. Colquhoun, H.M.; Crease, A.E.; Taylor, S.A.; Williams, D.J. *J. Chem. Soc., Chem. Comm.* **1982**, 736.
681. Chisholm, M.H.; Hoffman, D.M.; Huffman, J.C. *Inorg. Chem.* **1983**, *22*, 2903–2906.
682. Walker, I.; Strähle, J.; Ruschke, P.; Dehnicke, K. *Z. Anorg. Allg. Chem.* **1982**, *487*, 26–32.
683. Musterle, W.; Strähle, J.; Liebolt, W.; Dehnicke, K. *Z. Naturforsch.* **1979**, *34B*, 942–948.
684. Godemeyer, T.; Berg, A.; Gross, H.D.; Müller, U.; Dehnicke, K. *Z. Naturforsch.* **1985**, *40B*, 999–1004.
685. Godemeyer, T.; Dehnicke, K.; Fenske, D. *Z. Naturforsch.* **1985**, *40B*, 1005–1009.
686. Weller, F.; Liebelt, W.; Dehnicke, K. *Z. Anorg. Allg. Chem.* **1982**, *484*, 124–130.
687. Hill, C.L.; Hollander, F.J. *J. Am. Chem. Soc.* **1982**, *104*, 7318–7319.
688. Buchler, J.W.; Dreher, C.; Lay, K.-L.; Lee, Y.J.A.; Scheidt, W.R. *Inorg. Chem.* **1983**, *22*, 888–891.
689. Baldas, J.; Boas, J.F.; Bonnyman, J.; Williams, G.A. *J. Chem. Soc., Dalton Trans.* **1984**, 2395–2400.
690. Baldas, J.; Bonnyman, J.; Mackay, M.F.; Williams, G.A. *Aust. J. Chem.* **1984**, *37*, 751–759.
691. Baldas, J.; Bonnyman, J.; Pojer, P.M.; Williams, G.A.; Mackay, M.F. *J. Chem. Soc., Dalton Trans.* **1981**, 1798–1801.
692. Baldas, J.; Bonnyman, J.; Williams, G.A. *Inorg. Chem.* **1986**, *25*, 150–153.
693. Baldas, J.; Bonnyman, J.; Williams, G.A. *J. Chem. Soc., Dalton Trans.* **1984**, 833–837.
694. Nugent, W.A.; Harlow, R.L. *J. Chem. Soc., Chem. Comm.* **1979**, 1105–1106.
695. Fawcett, J.; Peacock, R.D.; Russell, D.R. *J. Chem. Soc., Chem. Comm.* **1982**, 958–959; *J. Chem. Soc. Dalton Trans.* **1987**, 567.
696. Kafitz, W.; Dehnicke, K.; Schweda, E.; Strähle, J. *Z. Naturforsch.* **1984**, *39B*, 1114–1117.
697. Weiher, U.; Dehnicke, K.; Fenske, D. *Z. Anorg. Allg. Chem.* **1979**, *457*, 115–122.
698. Shandles, R.S.; Murmann, R.K.; Schlemper, E.O. *Inorg. Chem.* **1974**, *13*, 1373–1377.
699. Bright, D.A.; Ibers, J.A. *Inorg. Chem.* **1969**, *8*, 703–709.
700. Forsellini, E.; Casellato, U.; Graziani, R.; Carletti, M.C.; Magon, L. *Acta Cryst.* **1984**, *C40*, 1795–1797.
701. Bright, D.; Ibers, J.A. *Inorg. Chem.* **1968**, *7*, 1099–1111.
702. Rossi, R.; Marchi, A.; Duatti, A.; Magon, L.; Casellato, U.; Graziani, R.; Polizzotti, G. *Inorg. Chim. Acta* **1984**, *90*, 121–131.
703. Goeden, G.V.; Hagmore, B.L. *Inorg. Chem.* **1983**, *22*, 157–167.
704. Dantona, R.; Schweda, E.; Strähle, J. *Z. Naturforsch.* **1984**, *39B*, 733.

705. Mason, R.; Thomas, K.M.; Zubieta, J.A.; Douglas, P.G.; Galbraith, A.R.; Shaw, B.L. *J. Am. Chem. Soc.* **1974**, *96*, 260-262.
706. Dilworth, J.R.; Harrison, S.A.; Walton, D.R.M.; Schweda, E. *Inorg. Chem.* **1985**, *24*, 2594-2595.
707. Barrientos-Penna, C.F.; Einstein, F.W.B.; Jones, T.; Sutton, D. *Inorg. Chem.* **1982**, *21*, 2578-2585.
708. Nicholson, T.; Zubieta, J. *Inorg. Chem.* **1987**, *26*, 2094-2101.
709. Mronga, N.; Weller, F.; Dehnicke, K. *Z. Anorg. Allg. Chem.* **1983**, *502*, 35.
710. Liese, W.; Dehnicke, K.; Walker, I.; Strähle, J. *Z. Naturforsch.* **1979**, *34B*, 693-696.
711. Davies, W.O.; Johnson, N.P.; Johnson, P.; Graham, A.J. *J. Chem. Soc., Chem. Comm.* **1969**, 736-737.
712. Liese, W.; Dehnicke, K.; Rogers, R.D.; Shakir, R.; Atwood, J.L. *J. Chem. Soc., Dalton Trans.* **1981**, 1061-1063.
713. Kafitz, W.; Weller, F.; Dehnicke, K. *Z. Anorg. Allg. Chem.* **1982**, *490*, 175.
714. Carrondo, M.A.A.F. de C.T.; Shakir, R.; Skapski, A.C. *J. Chem. Soc., Dalton Trans.* **1978**, 844-848.
715. Fletcher, S.R.; Skapski, A.C. *J. Chem. Soc., Dalton Trans.* **1972**, 1079-1082.
716. Doedens, R.J.; Ibers, J.A. *Inorg. Chem.* **1967**, *6*, 204-210.
717. Mahy, J.-P.; Battioni, P.; Mansuy, D.; Fisher, J.; Weiss, R.; Mispelter, J.; Morgenstern-Badarau, I.; Gans, P. *J. Am. Chem. Soc.* **1984**, *106*, 1699-1706.
718. Scheidt, W.R.; Summerville, D.A.; Cohen, I.A. *J. Am. Chem. Soc.* **1976**, *98*, 6623-6628.
719. Phillips, F.L.; Skapski, A.C. *J. Chem. Soc., Dalton Trans.* **1976**, 1448-1453.
720. Phillips, F.L.; Skapski, A.C. *Acta Cryst.* **1975**, *B31*, 2667-2670.
721. Collison, D.; Garner, C.D.; Mabbs, F.E.; King, T.J. *J. Chem. Soc., Dalton Trans.* **1981**, 1820-1824.
722. Ciechanowicz, M.; Skapski, A.C. *J. Chem. Soc. A* **1971**, 1792-1794.
723. Griffith, W.P.; McManus, N.T.; Skapski, A.C. *J. Chem. Soc., Chem. Comm.* **1984**, 434.
724. Shapley, P.A.B.; Own, Z.-Y.; Huffman, J.C. *Organometallics* **1986**, *5*, 1269-1271.
725. Phillips, F.L.; Skapski, A.C. *J. Cryst. Mol. Struct.* **1975**, *5*, 83-92.
726. Collison, D.; Garner, C.D.; Mabbs, F.E.; Salthouse, J.A.; King, T.J. *J. Chem. Soc., Dalton Trans.* **1981**, 1812-1819.
727. Phillips, F.L.; Skapski, A.C.; Withers, M.J. *Trans. Met. Chem.* **1975**, *1*, 28-32.
728. Barner, C.J.; Collins, T.J.; Mapes, B.E.; Santarsiero, B.D. *Inorg. Chem.* **1986**, *25*, 4322-4323.
729. Given, K.W.; Pignolet, L.H. *Inorg. Chem.* **1977**, *16*, 2982-2984.
730. Brennan, J.G.; Andersen, R.A. *J. Am. Chem. Soc.* **1985**, *107*, 514.
731. Cramer, R.E.; Panchanatheswaran, K.; Gilje, J.W. *J. Am. Chem. Soc.* **1984**, *106*, 1853-1854.
732. Hug, F.; Mowat, W.; Skapski, A.C.; Wilkinson, G. *J. Chem. Soc., Chem. Comm.* **1971**, 1477-1478.
733. Chamberlain, L.; Rothwell, I.P.; Huffman, J.C. *J. Am. Chem. Soc.* **1982**, *104*, 7338-7340.

734. Churchill, M.R.; Youngs, W.J. *Inorg. Chem.* **1979**, *18*, 1930–1935.
735. Messerle, L.W.; Jennische, P.; Schrock, R.R.; Stucky, G. *J. Am. Chem. Soc.* **1980**, *102*, 6744–6752.
736. Schrock, R.R.; Messerle, L.W.; Clayton, C.D.; Guggenberger, L.J. *J. Am. Chem. Soc.* **1978**, *100*, 3793–3800.
737. Churchill, M.R.; Hollander, F.J. *Inorg. Chem.* **1978**, *17*, 1957–1962.
738. Curtis, M.D.; Real, J. *J. Am. Chem. Soc.* **1986**, *108*, 4668–4669.
739. Churchill, M.R.; Wasserman, H.J.; Turner, H.W.; Schrock, R.R. *J. Am. Chem. Soc.* **1982**, *104*, 1710–1716.
740. Churchill, M.R.; Youngs, W.J. *Inorg. Chem.* **1979**, *18*, 171.
741. Guggenberger, L.J.; Schrock, R.R. *J. Am. Chem. Soc.* **1975**, *97*, 2935.
742. Fanwick, P.E.; Ogilvy, A.E.; Rothwell, I.P. *Organometallics* **1987**, *6*, 73–80.
743. Gal, A.W.; van der Heijden, H. *J. Chem. Soc., Chem. Comm.* **1983**, 420–422.
744. Herrmann, W.A.; Hubbard, J.L.; Bernal, I.; Korp, J.D.; Haymore, B.L.; Hillhouse, G.L. *Inorg. Chem.* **1984**, *23*, 2978.
745. Huttner, G.; Schelle, S.; Mills, O.S. *Angew. Chem.* **1969**, *81*, 536.
746. Fischer, E.O.; Kalder, H.-J.; Frank, A.; Kohler, F.H.; Huttner, G. *Angew. Chem., Int. Ed. Engl.* **1976**, *15*, 623.
747. Kruger, C.; Goddard, R.; Claus, K.H. *Z. Naturforsch.* **1983**, *38B*, 1431.
748. Huttner, G.; Frank, A.; Fischer, E.O. *Isr. J. Chem.* **1977**, *15*, 133–142.
749. Fontana, S.; Orama, O.; Fischer, E.O.; Schubert, U.; Kreissl, F.R. *J. Organomet. Chem.* **1978**, *149*, C57.
750. Fischer, E.O.; Tran-Huy, N.H.; Neugebauer, D. *J. Organomet. Chem.* **1982**, *229*, 169.
751. Dao, N.Q.; Neugebauer, D.; Février, H.; Fischer, E.O.; Becker, P.J.; Pannetier, J. *Nouv. J. Chim.* **1982**, *6*, 359.
752. Frank, A.; Fischer, E.O.; Huttner, G. *J. Organomet. Chem.* **1978**, *161*, C27.
753. Frank, A.; Schubert, U.; Huttner, G. *Chem. Ber.* **1977**, *110*, 3020.
754. Fischer, E.O.; Schwanzer, A.; Fischer, H.; Neugebauer, D.; Huttner, G. *Chem. Ber.* **1977**, *110*, 53.
755. Fischer, E.O.; Schluge, M.; Besenhard, J.O.; Friedrich, P.; Huttner, G.; Kreissl, F.R. *Chem. Ber.* **1978**, *111*, 3530.
756. Fischer, E.O.; Wagner, W.R.; Kreissl, F.R.; Neugebauer, D. *Chem. Ber.* **1979**, *112*, 1320.
757. Ustynyuk, N.A.; Vinogradova, V.N.; Andrianov, V.G.; Struchkov, Yu.T. *J. Organomet. Chem.* **1984**, *268*, 73.
758. Schubert, U.; Neugebauer, D.; Hofmann, P.; Schilling, B.E.R.; Fischer, H.; Motsch, A. *Chem. Ber.* **1981**, *114*, 3349–3365.
759. Fischer, H.; Motsch, A.; Markl, R.; Ackermann, K. *Organometallics* **1985**, *4*, 726.
760. Messerle, L.; Curtis, M.D. *J. Am. Chem. Soc.* **1982**, *104*, 889–891.
761. Ahmed, K.J.; Chisholm, M.H.; Huffman, J.C. *Organometallics* **1985**, *4*, 1168–1174.
762. Strutz, H.; Dewan, J.C.; Schrock, R.R. *J. Am. Chem. Soc.* **1985**, *107*, 5999–6005.

763. Beevor, R.G.; Green, M.; Orpen, A.G.; Williams, I.D. *J. Chem. Soc., Chem. Comm.* **1983**, 673.
764. Kirchner, R.M.; Ibers, J.A. *Inorg. Chem.* **1974**, *13*, 1667.
765. Green, M.; Orpen, A.G.; Williams, I.D. *J. Chem. Soc., Chem. Comm.* **1982**, 493–495.
766. Mayr, A.; Dorries, A.M.; McDermott, G.A.; Van Engen, D. *Organometallics* **1986**, *5*, 1504–1506.
767. Fischer, E.O.; Huttner, G.; Lindner, T.L.; Frank, A.; Kreissl, F.R. *Angew. Chem., Int. Ed. Engl.* **1976**, *15*, 157.
768. Desmond, T.; Lalor, F.J.; Ferguson, G.; Parvez, M. *J. Chem. Soc., Chem. Comm.* **1983**, 457.
769. Desmond, T.; Lalor, F.J.; Ferguson, G.; Parvez, M. *J. Chem. Soc., Chem. Comm.* **1984**, 75.
770. Schultz, A.J.; Williams, J.M.; Schrock, R.R.; Holmes, S.J. *Acta Cryst.* **1984**, *C40*, 590.
771. Wengrovius, J.H.; Schrock, R.R.; Churchill, M.R.; Wasserman, H.J. *J. Am. Chem. Soc.* **1982**, *104*, 1739–1740.
772. Mayr, A.; Asaro, M.F.; Kjelsberg, M.A.; Lee, S.L.; Van Engon, D. *Organometallics* **1987**, *6*, 432–434.
773. Mayr, A.; Lee, K.S.; Kjelsberg, M.A.; Van Engen, D. *J. Am. Chem. Soc.* **1986**, *108*, 6079–6080.
774. Mayr, A.; Asaro, M.F.; Glines, T.J. *J. Am. Chem. Soc.* **1987**, *109*, 2215–6.
775. Howard, J.A.K.; Jeffrey, J.L.; Laurie, J.C.V.; Moore, I.; Stone, F.G.A.; Stringer, A. *Inorg. Chim. Acta* **1985**, *100*, 23.
776. Marsella, J.A.; Folting, K.; Huffmann, J.C.; Caulton, K.G. *J. Am. Chem. Soc.* **1981**, *103*, 5596.
777. Carriedo, G.A.; Hodgson, D.; Howard, J.A.K.; Marsden, K.; Stone, F.G.A.; Went, M.J.; Woodward, P. *J. Chem. Soc., Chem. Comm.* **1982**, 1006.
778. Casey, C.P.; Burkhardt, T.J.; Bunnell, C.A.; Calabrese, J.C. *J. Am. Chem. Soc.* **1977**, *99*, 2127.
779. Angermund, K.; Grevels, F.-W.; Kruger, C.; Skibbe, V. *Angew. Chem., Int. Ed. Engl.* **1984**, *23*, 904.
780. Birdwhistell, K.R.; Nieter-Burgmayer, S.J.; Templeton, J.L. *J. Am. Chem. Soc.* **1983**, *105*, 7789.
781. Chisholm, M.H.; Cotton, F.A.; Extine, M.W.; Murillo, C.A. *Inorg. Chem.* **1978**, *17*, 696–698.
782. Chisholm, M.H.; Heppert, J.A.; Huffman, J.C. *J. Am. Chem. Soc.* **1984**, *106*, 1151.
783. Cotton, F.A.; Schwotzer, W.; Shamshoum, E.S. *Organometallics* **1984**, *3*, 1770–1771.
784. Churchill, M.R.; Ziller, J.W.; Freudenberger, J.H.; Schrock, R.R. *Organometallics* **1984**, *3*, 1554–1562.
785. Freudenberger, J.H.; Schrock, R.R.; Churchill, M.R.; Rheingold, A.L.; Ziller, J.W. *Organometallics* **1984**, *3*, 1563–1573.
786. Churchill, M.R.; Ziller, J.W.; McCullough, L.; Pedersen, S.F.; Schrock, R.R. *Organometallics* **1983**, *2*, 1046–1048.
787. Chisholm, M.H.; Hoffman, D.M.; Huffman, J.C. *J. Am. Chem. Soc.* **1984**, *106*, 6815.
788. McCullough, L.G.; Listemann, M.L.; Schrock, R.R.; Churchill, M.R.; Ziller, J.W. *J. Am. Chem. Soc.* **1983**, *105*, 6729–6730.
789. Chisholm, M.H.; Huffman, J.C.; Marchant, N.S. *J. Am. Chem. Soc.* **1983**, *105*, 6162–6163.

790. Churchill, M.R.; Li, Y.-J.; Blum, L.; Schrock, R.R. *Organometallics* **1984**, *3*, 109.
791. Churchill, M.R.; Li, Y.-J. *J. Organomet. Chem.* **1985**, *282*, 239-246.
792. Holmes, S.J.; Schrock, R.R.; Churchill, M.R.; Wasserman, H.J. *Organometallics* **1984**, *3*, 476.
793. Churchill, M.R.; Wasserman, H.J. *Inorg. Chem.* **1981**, *20*, 4119.
794. Churchill, M.R.; Rheingold, A.L.; Wasserman, H.J. *Inorg. Chem.* **1981**, *20*, 3392-3399.
795. Neugebauer, D.; Fischer, E.O.; Dao, N.Q.; Schubert, U. *J. Organomet. Chem.* **1978**, *153*, C41.
796. Fischer, E.O.; Friedrich, P.; Lindner, T.L.; Neugebauer, D.; Kreissl, F.R.; Uedelhoven, W.; Dao, N.Q.; Huttner, G. *J. Organomet. Chem.* **1983**, *247*, 239.
797. Fischer, E.O.; Roll, W.; Huy, N.H.T.; Ackermann, K. *Chem. Ber.* **1982**, *115*, 2951.
798. Fischer, E.O.; Gammel, F.J.; Neugebauer, D. *Chem. Ber.* **1980**, *113*, 1010.
799. Fischer, E.O.; Hollfelder, H.; Friedrich, P.; Kreissl, F.R.; Huttner, G. *Angew. Chem., Int. Ed. Engl.* **1977**, *16*, 401.
800. Fischer, E.O.; Lindner, T.L.; Huttner, G.; Friedrich, P.; Kreissl, F.R.; Besenhard, J.O. *Chem. Ber.* **1977**, *110*, 3397.
801. Green, M.; Howard, J.A.K.; James, A.P.; Jelfs, A.N. de M.; Nunn, C.M.; Stone, F.G.A. *J. Chem. Soc., Chem. Comm.* **1984**, 1623.
802. Cotton, F.A.; Schwotzer, W. *Inorg. Chem.* **1983**, *22*, 387.
803. Fischer, E.O.; Wittmann, D.; Himmelreich, D.; Schubert, U.; Ackermann, K. *Chem. Ber.* **1982**, *115*, 3141.
804. Greaves, W.W.; Angelici, R.J.; Helland, B.J.; Klima, R.; Jacobson, R.A. *J. Am. Chem. Soc.* **1979**, *101*, 7618.
805. Friedrich, P.; Besl, G.; Fischer, E.O.; Huttner, G. *J. Organomet. Chem.* **1977**, *139*, C68.
806. Herrmann, W.A.; Plank, J.; Kriechbaum, G.W.; Ziegler, M.L.; Pfisterer, H.; Atwood, J.L.; Rogers, R.D. *J. Organomet. Chem.* **1984**, *264*, 327.
807. Redhouse, A.D. *J. Organomet. Chem.* **1975**, *99*, C29.
808. Berke, H.; Huttner, G.; Von Seyerl, J. *J. Organomet. Chem.* **1981**, *218*, 193.
809. Aleksandrov, G.G.; Antonova, A.B.; Kolobova, N.E.; Struchkov, Yu.-T. *Koord. Khim.* **1976**, *2*, 1684.
810. Berke, H.; Huttner, G.; Von Seyerl, J. *Z. Naturforsch.* **1981**, *36B*, 1277.
811. Kolobova, N.E.; Ivanov, L.L.; Zhvanko, O.S.; Khitrova, O.M.; Batsanov, A.S.; Struchkov, Y.T. *J. Organomet. Chem.* **1984**, *262*, 39.
812. Patton, A.T.; Strouse, C.E.; Knobler, C.B.; Gladysz, J.A. *J. Am. Chem. Soc.* **1983**, *105*, 5804.
813. Kiel, W.A.; Lin, G.-Y.; Constable, A.G.; McCormick, F.B.; Strouse, C.E.; Eisenstein, O.; Gladysz, J.A. *J. Am. Chem. Soc.* **1982**, *104*, 4865.
814. Fischer, E.O.; Rustemeyer, P.; Neugebauer, D. *Z. Naturforsch.* **1980**, *35B*, 1083.
815. Pombeiro, A.J.L.; Jeffery, J.C.; Pickett, C.J.; Richards, R.L. *J. Organomet. Chem.* **1984**, *277*, C7.

816. Savage, P.D.; Wilkinson, G.; Motewalli, M.; Hursthouse, M.B. *Polyhedron* **1987**, *6*, 1599–1601.
817. Pombeiro, A.J.L.; Hughes, D.L.; Pickett, C.J.; Richards, R.L. *J. Chem. Soc., Chem. Comm.* **1986**, 246.
818. Pombeiro, A.J.L.; Carvalho, M.F.N.N.; Hitchcock, P.B.; Richards, R.L. *J. Chem. Soc., Dalton Trans.* **1981**, 1629.
819. Dettlaf, G.; Hubener, P.; Klimes, J.; Weiss, E. *J. Organomet. Chem.* **1982**, *229*, 63.
820. Goedken, V.L.; Deakin, M.R.; Bottomley, L.A. *J. Chem. Soc., Chem. Comm.* **1982**, 607.
821. Riley, P.E.; Davis, R.E.; Allison, N.T.; Jones, W.M. *J. Am. Chem. Soc.* **1980**, *102*, 2458.
822. Iyer, R.S.; Selegue, J.P. *J. Am. Chem. Soc.* **1987**, *109*, 910–911.
823. Selegue, J.P. *J. Am. Chem. Soc.* **1982**, *104*, 119.
824. Umland, H.; Behrens, U. *J. Organomet. Chem.* **1984**, *273*, C39.
825. Fischer, E.O.; Schneider, J.; Neugebauer, D. *Angew. Chem., Int. Ed. Engl.* **1984**, *23*, 820.
826. Bruce, M.I.; Wong, F.S.; Skelton, B.W.; White, A.H. *J. Chem. Soc., Dalton Trans.* **1982**, 2203.
827. Selegue, J.P. *Organometallics* **1982**, *1*, 217.
828. Selegue, J.P. *J. Am. Chem. Soc.* **1983**, *105*, 5921.
829. Bruce, M.I.; Humphrey, M.G.; Koutsantonis, G.A.; Nicholson, B.K. *J. Organomet. Chem.* **1985**, *296*, C47.
830. Bohle, D.S.; Clark, G.R.; Rickard, C.E.F.; Roper, W.R.; Shepard, W.E.B.; Wright, L.J. *J. Chem. Soc., Chem. Comm.* **1987**, 563–565.
831. Roper, W.R. *J. Organomet. Chem.* **1986**, *300*, 167–190.
832. Hill, A.F.; Roper, W.R.; Waters, J.M.; Wright, A.H. *J. Am. Chem. Soc.* **1983**, *105*, 5939–5940.
833. Pourreau, D.B.; Geoffroy, G.L.; Rheingold, A.L.; Geib, S.J. *Organometallics* **1986**, *5*, 1337–1345.
834. Roper, W.R.; Waters, J.M.; Wright, L.J.; Van Meurs, F. *J. Organomet. Chem.* **1980**, *201*, C27.
835. Clark, G.R.; Edmonds, N.R.; Pauptit, R.A.; Roper, W.R.; Waters, J.M.; Wright, A.H. *J. Organomet. Chem.* **1983**, *244*, C57–C60.
836. Cramer, R.E.; Maynard, R.B.; Paw, J.C.; Gilje, J.W. *J. Am. Chem. Soc.* **1981**, *103*, 3589–3590.

CHAPTER 6

REACTIONS OF MULTIPLY BONDED LIGANDS

Having prepared and characterized a new complex by the approaches already described, and having determined its electronic and molecular structure, what sort of chemistry should we expect from this molecule? Will an alkylidene ligand bind to Lewis acids or to Lewis bases? Are nitrides electrophiles or nucleophiles? The answers, of course, are not that simple. The reactivity of a multiply bonded ligand is a sensitive function of its chemical environment. It will depend on (1) the nature of the transition metal to which it is bound, (2) the oxidation state of the metal, and (3) the nature of the other ligands that are present. These three factors are important because they control the nature of the ligand-p to metal-d π-interaction, which can sometimes even encompass *both* the HOMO and LUMO of the complex.

Recent years have witnessed a growing recognition that the reactions of π-bonded ligands, like those of π-bonded organic molecules, must be understood in the context of frontier molecular orbital theory. Frontier molecular orbital considerations were first applied to Fischer-type carbene [1] and carbyne [2,3] complexes by Fenske and coworkers. Subsequently, Nakatsuji and coworkers have reported *ab initio* calculations on both Fischer- and Schrock-type carbenes and carbynes. They conclude that, independent of the type of complex involved,

> the atomic charges on the carbene and carbyne carbons were calculated negative. The reactivities of the metal–carbon bonds were unifiedly understood by the frontier orbital theory. For the Schrock type complex, the HOMO has a maximum coefficient on the C_{carb} atom and the LUMO has a maximum coefficient on the Nb atom. Therefore, the electrophile attacks the C_{carb} atom and the nucleophile attacks the Nb atom. For the carbyne complexes, the differences in reactivity between the cationic and neutral complexes were explained from the existence of the nearly degenerate LUMO and next LUMO in the frontier MO region of the neutral complex. They would never be explained by the charge controlled mechanism [4].

Frontier molecular orbital considerations are clearly relevant to the other classes of multiply bonded ligands as well. An especially enlightening study is that by Hoffmann and coworkers [5] on the addition of olefins to osmium tetroxide (see Section 6.3.4.2).

6.1 A SIMPLE CONCEPTUAL MODEL

If we confine ourselves to, say, d^0 complexes containing a single multiply bonded ligand, then a simple periodic trend is observed. That is, π donation increases as one proceeds upwards and to the right among the transition metals on the periodic table. The chemical implications of increasing electronegativity of the metal were rationalized in Chapter 2 by the rather simplistic molecular orbital diagram of Figure 2.2. In situation **A** the HOMO of the molecule is localized on the α atom of the ligand which is therefore subject to attack by electrophiles. The LUMO is localized on the metal which, in the absence of steric constraints, is subject to attack and coordination by nucleophiles. In contrast, in situation **B** the p atomic orbital of the α atom is involved in a covalent bond with the metal d orbital. This will diminish markedly the nucleophilicity of the α atom and the electrophilicity of the metal. Under such circumstances we expect electrophiles to attack the bond side-on. However, the LUMO (π^*) has a node between the metal and the bound atom of the ligand; consequently, nucleophilic attack on either atom of the bond is possible.

As an example of situation **A** we consider the complex $(Et_2N)_3Ta(NEt)$. On treatment with ethanol the metal–nitrogen bond is cleaved (Eq. 1) [6]. This reaction presumably involves protonation of the imido nitrogen atom at some stage. As a model for situation **B**, we consider the alkylimido osmium(VIII) species $O_3Os(NR)$. Compounds in this series are unstable when the α carbon atom bears hydrogen, apparently due to the high acidity of these protons. For example, when R is methyl the complex explodes on warming above $-40°C$ [7]. When R = t-butyl the compound is stable and the Os—N bond is relatively insensitive to hydrolysis. Even under forcing conditions (aqueous HCl), it is not the Os—N bond that is cleaved. Instead the C—N bond solvolyzes according to Eq. 2 [8]. In fact, under appropriate conditions, the covalent π system in $O_3Os(N^tBu)$ will now react with nucleophiles such as alkenes (Eq. 3) [9]. However, it has subsequently been suggested that this reaction is an electrocyclic process comparable to the Diels–Alder transformation (Section 6.3.4.2). These chemical effects of increasing π donation are summarized in Figure 6.1.

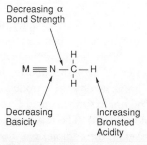

Figure 6.1 Some effects of increasing the π-acceptor capability (electronegativity) of the metal on the reactivity of a methylimido complex.

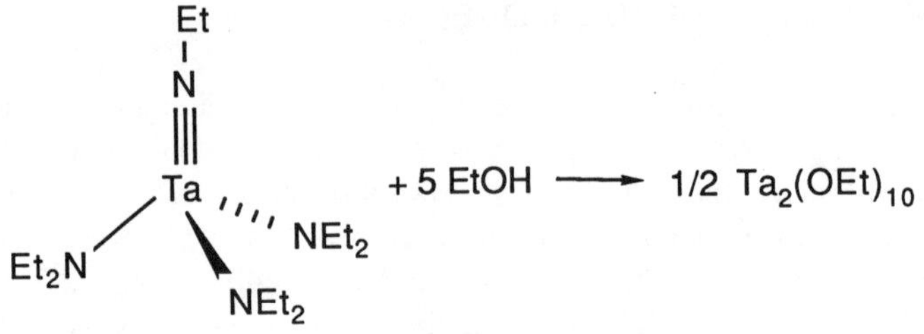

$$\underset{Et_2N}{\overset{Et}{\underset{|}{N}}}\overset{|||}{\underset{NEt_2}{Ta}}\underset{NEt_2}{\text{''''}} \quad + 5\ EtOH \longrightarrow 1/2\ Ta_2(OEt)_{10} \quad (1)$$

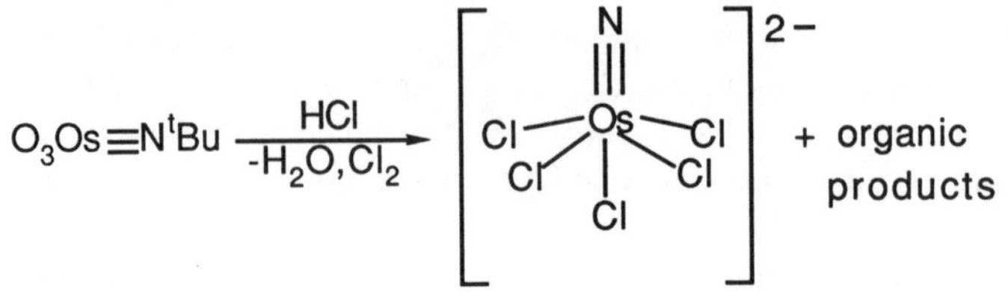

$$O_3Os\equiv N^tBu \xrightarrow[-H_2O,Cl_2]{HCl} \left[\begin{array}{c} N \\ ||| \\ Cl-Os-Cl \\ Cl\ |\ Cl \\ Cl \end{array}\right]^{2-} + \text{ organic products} \quad (2)$$

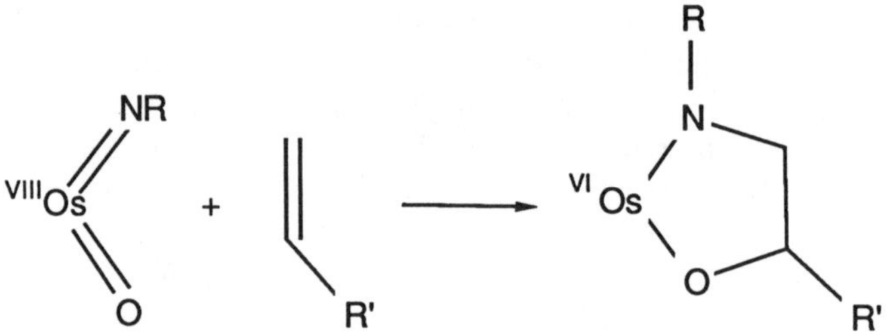

$$\overset{NR}{\underset{O}{^{VIII}Os\diagdown\diagup}} + \overset{}{\underset{R'}{||}} \longrightarrow {}^{VI}Os\diagdown\underset{O}{\overset{N-R}{\diagup\diagdown}}{R'} \quad (3)$$

A particularly nice illustration of the effect of increasing the electronegativity of the metal has been provided by Henderson. In the complexes trans-$[M(NH)X(dppe)_2]^+X^-$ on changing M from tungsten to molybdenum, the acidity of the imido proton is increased 1000-fold [10].

As we intimated in the first paragraph of this chapter, the extent of π donation in the complexes of a given metal is influenced by other factors. For example, π donation can be diminished by decreasing the oxidation state of the metal: In contrast to $O_3Os(NMe)$, the osmium(VI) methylimido derivative $OsMe_4(NMe)$ is thermally robust [11]. The degree of π donation is also diminished by the introduction of competitive π-bonding ligands. The effect of the second imido substituent [12] in $(dtc)_2Mo(NPh)_2$ is evident in its reaction with methyl bromide (Eq. 4). The reaction stops cleanly after replacement of a single imido substituent. Again, the complexes trans-$[Mo(NH)X(dppe)_2]^+$ undergo ready protonolysis to give ammonia when X = methoxide but are inert even to acidic methanol when X = halide [13].

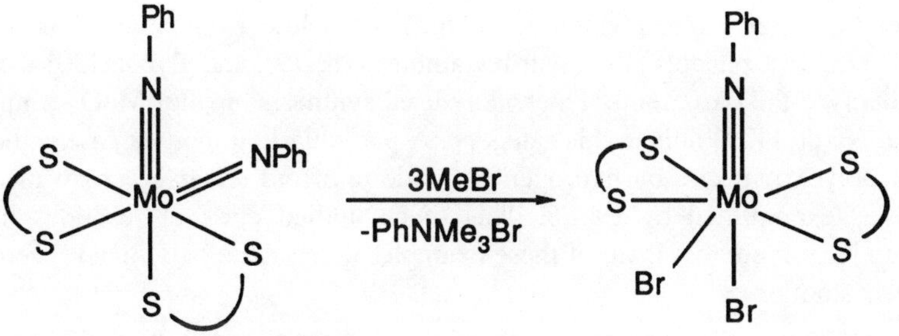

$$\text{(dtc)}_2\text{Mo(NPh)}_2 \xrightarrow[-\text{PhNMe}_3\text{Br}]{3\text{MeBr}} \text{(dtc)}_2\text{Mo(NPh)Br}_2 \quad (4)$$

S⌢S = diethyldithiocarbamate

While we have illustrated these trends with examples from imido chemistry, we shall find that they apply to other multiply bonded ligands as well. A diagram resembling Figure 2.2 was first applied to multiply bonded ligands for the case of alkylidenes by Goddard, Hoffmann, and Jemmis [14]. In the remainder of this chapter we will consider the reactions of all the ligand types with electrophiles (Section 6.2) and with nucleophiles (Section 6.3). Three special topics will be addressed separately. α-Cleavage reactions, rearrangement reactions of complex multiply bonded ligands, and the coupling of two multiply bonded ligands are discussed in Sections 6.4–6.6.

6.2 REACTIONS WITH ELECTROPHILES

6.2.1 Reaction with Bronsted Acids

Although oxo ligands are generally less prone toward protonolysis than the nitrogen and carbon bound analogs they can be protonolyzed under appropriate conditions. Again, the relevant factors are the identity of the transition metal, the oxidation state, and the presence of other ligands that lower the π bond order. For example, $\text{(dtc)}_2\text{MoO}_2$ will react with HCl with replacement of one oxo ligand (Eq. 5). The remaining (triple-bonded) oxo substituent does not react.

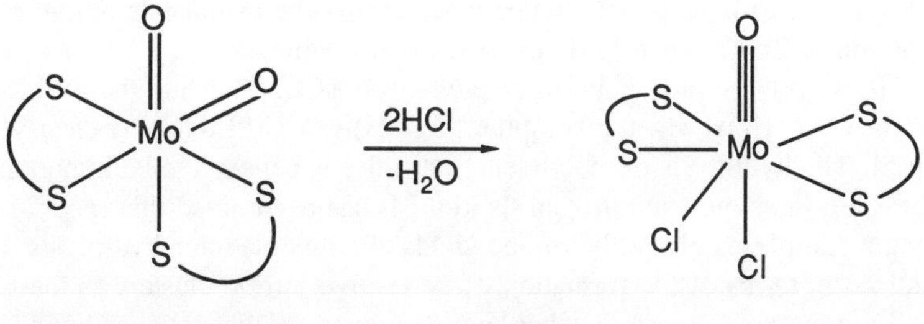

$$\text{(dtc)}_2\text{MoO}_2 \xrightarrow[-\text{H}_2\text{O}]{2\text{HCl}} \text{(dtc)}_2\text{Mo(O)Cl}_2 \quad (5)$$

S⌢S = diethyldithiocarbamate

When the metal–oxygen bond order is sufficiently low, even rather weak acids such as thiols [15, 16], phenols [17], hydroxylamines [18, 19], and glycols [20] will suffice to protonolyze the oxo group. The widely used synthesis of $(dtc)_2MoO_2$ from molybdate and Na(dtc) [21] falls in this category. A particularly important case is the hydrolysis of polyoxo species by hydrogen peroxide to afford reactive peroxo metal complexes, as exemplified by Eq. 6. This much-studied class of reactions has been reviewed [22]. Note that in all of these examples at least one oxo ligand is retained by the metal atom!

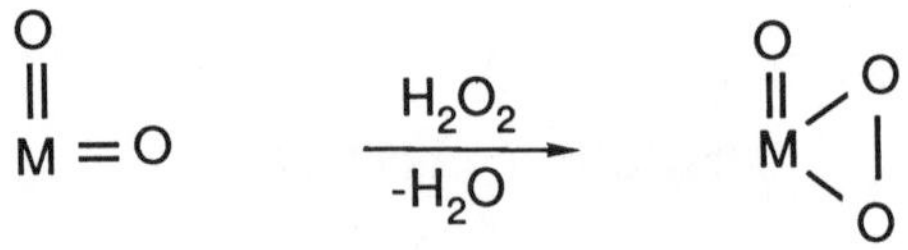

$$M(=O)_2 \xrightarrow[-H_2O]{H_2O_2} M(=O)(O_2) \qquad (6)$$

The kinetic basicity of the oxo group has been implicated in the thermal instability of alkylmolybdenum complexes when β-hydrogen is present [23]. This is exemplified by Eq. 7. However, the product hydroxo complex has not been observed and further studies would be of interest.

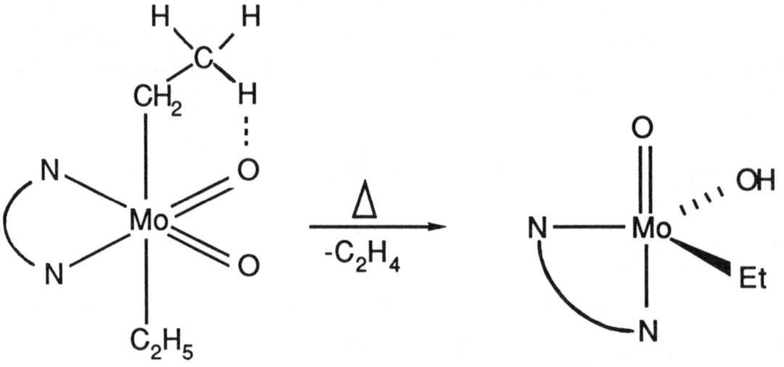

$$(N\frown N)Mo(O)_2(C_2H_5)(\cdots H-CH_2CH_3) \xrightarrow[-C_2H_4]{\Delta} (N\frown N)Mo(O)(OH)(Et) \qquad (7)$$

N⌒N = 2,2'-bipyridyl

Apparently, it is not possible to generalize about the relative thermodynamic basicity of nitrido versus oxo ligands. IR studies indicate that the osmiamate anion is preferentially protonated on oxygen [24]. Indeed, upon treatment with HCl all of the oxo ligands of [OsO$_3$N]$^-$ are protonolyzed *en route* to [OsCl$_4$N]$^-$ while the nitrido ligand remains intact [8]. However, the complex Mo(NH)(O)Cl$_2$(PEtPh$_2$)$_2$ is clearly N-protonated [25]. The hydrolysis of molybdenum nitrides is believed to be important as an elementary step in enzymatic nitrogen fixation. It has been noted that model nitrido–molybdenum complexes give only limited yields of ammonia upon hydrolysis, but this is rationalized in terms of the formation of unreactive nitrido clusters as the reaction proceeds. In support of this notion, it has been demonstrated that polymer-bound molybdenum nitrides give significantly higher yields of ammonia upon hydrolysis [26].

Organoimido complexes typically undergo facile hydrolysis reactions [27–29].

Even Os(N^{t}Bu)O$_3$, which is reportedly stable to cold dilute nitric acid [30], elsewhere is reported to be hydrolyzed to OsO$_4$ by dilute sulfuric acid [31]. Particularly when the conjugate base is not a good π donor, such as with hydrogen halides, only low bond-order polyimido complexes will be cleaved by acids (Eqs. 8 and 9) [32,33].

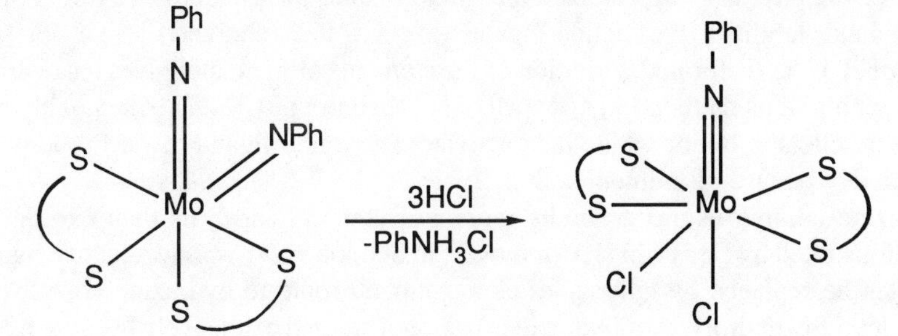

$$\text{(structure with Mo=NPh, S-donor ligands)} \xrightarrow[-\text{PhNH}_3\text{Cl}]{3\text{HCl}} \text{(Mo(NPh) with S-ligands and 2 Cl)} \tag{8}$$

$$(\text{Me}_3\text{SiO})\text{Re}(\text{N}^t\text{Bu})_3 \xrightarrow[\substack{-^t\text{BuNH}_3\text{Cl} \\ -\text{Me}_3\text{SiOH}}]{4\text{HCl}} \text{Cl}_3\text{Re}(\text{N}^t\text{Bu})_2 \tag{9}$$

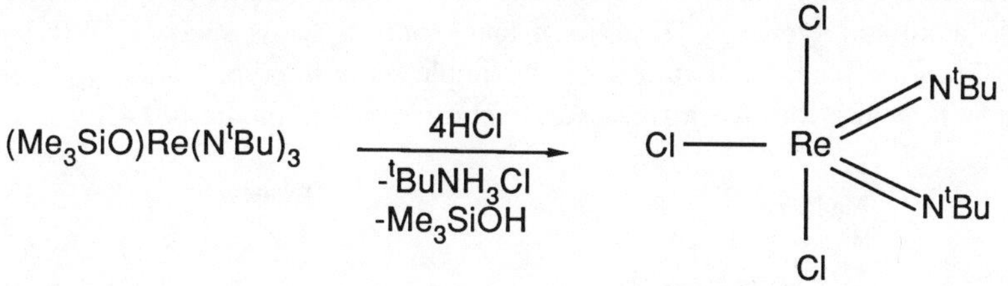

In this regard the separate reports that a tantalum imido complex (MeCN)$_2$Ta(NR)Cl$_3$ does not react with HCl in ether [34] while (Et$_2$N)$_3$Ta(NEt) is completely hydrolyzed by ethanol to Ta$_2$(OEt)$_{10}$ [35,6] seem entirely consistent. Another fascinating case in which the same imido complex reacts with different acids by different pathways has been reported by Wilkinson [36]. The rhenium(V) complex in Eq. 10 reacts with acetic acid by exclusive methyl substitution; yet with tetrafluoroboric acid N-protonation obtains.

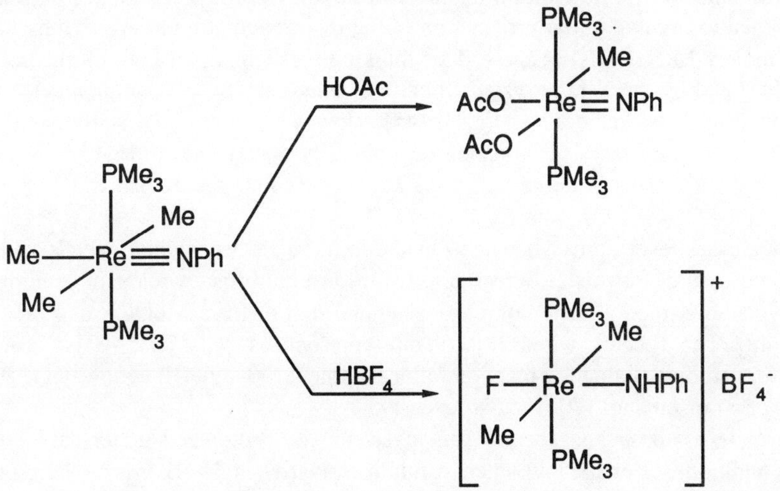

$$\text{Me}_2(\text{PMe}_3)_2\text{Re}\equiv\text{NPh} \xrightarrow{\text{HOAc}} (\text{AcO})_2(\text{PMe}_3)_2(\text{Me})\text{Re}\equiv\text{NPh} \tag{10a}$$

$$\text{Me}_2(\text{PMe}_3)_2\text{Re}\equiv\text{NPh} \xrightarrow{\text{HBF}_4} [\text{F}(\text{Me})(\text{PMe}_3)_2\text{Re}-\text{NHPh}]^+ \text{BF}_4^- \tag{10b}$$

Studies on the protonolysis of the imido complexes trans-[M(NH)X(dppe)$_2$]$^+$X$^-$, where M = Mo or W and X is halide have revealed a fascinating "domino effect" type of mechanism [10,13]. These complexes undergo ready protonolysis in basic methanol, even though the molybdenum complex has been shown to be inert to acidic methanol. The role of the base is to deprotonate the imido ligand; the greater trans effect of the nitrido ligand labilizes the halide ligand so that the relatively stable ion pair [M(N)(dppe)$_2$]$^+$X$^-$ is formed. Addition of the elements of methanol gives a new imido complex with a trans methoxo ligand, [M(NH)(OCH$_3$)(dppe)$_2$]$^+$. The methoxide moiety is an excellent π donor which then activates the imido ligand toward protonation and ultimately release of ammonia.

Hydrazido complexes are generally more resistant to hydrolysis than oxo derivatives, as illustrated by Eq. 11 [37]. (Moreover, in Section 3.2.1 we saw that oxo ligands could often be replaced by hydrazines as a synthetic route to hydrazido complexes.) Nevertheless, the hydrolysis of hydrazides has been studied extensively because hydrazine formation is often a significant side-reaction in model systems designed to mimic enzymatic ammonia synthesis [38,39]. In fact, some systems such as [W(8-quin)-(NNH$_2$)(PMe$_2$Ph)$_3$]I and the hydrazido(4−) complexes of the type [L$_2$TaX$_3$]$_2$N$_2$ upon treatment with acid afford hydrazine exclusively with no ammonia [40,41].

$$\text{(11)}$$

The mechanism of hydrazine formation from hydrazido(2−) complexes originally was believed to involve initial protonation of the α-nitrogen atom to afford a hydrazido(1−) moiety M(NHNH$_2$) [42,43]. This thinking was supported both by theoretical studies [44] and by the X-ray crystal structure of a hydrazido(1−) complex generated by protonation of hydrazido(2−) ligand [45]. However, subsequent studies indicate that, at least in some cases, protonation occurs on the metal to afford a hydridometal hydrazide [46,47]. This is suggested as the key intermediate on the pathway to hydrazine formation.

In some cases, especially when coordinated to more electronegative metals, alkylidene ligands can be surprisingly resistant toward protonolysis. A relevant example is the molybdenum–methylidene complex generated in situ from MoOCl$_3$ and methyllithium [48]. This species, which Kauffmann represents as "Cl-Mo(O)-CH$_2$," can be utilized for carbonylolefination reactions after treatment (30 min, 0°) with water (220:1 molar excess) or ethanol (70:1 molar excess).

Nevertheless, alkylidene species can invariably be protonated under sufficiently forcing conditions. Even the low-valent osmium derivative in Eq. 12 is protonated upon treatment with anhydrous HCl [49]. Furthermore, it has been recognized since the early work of Schrock that alkylidene ligands are often protonated more readily than alkyl ligands present in the same complex (Eq. 13) [50].

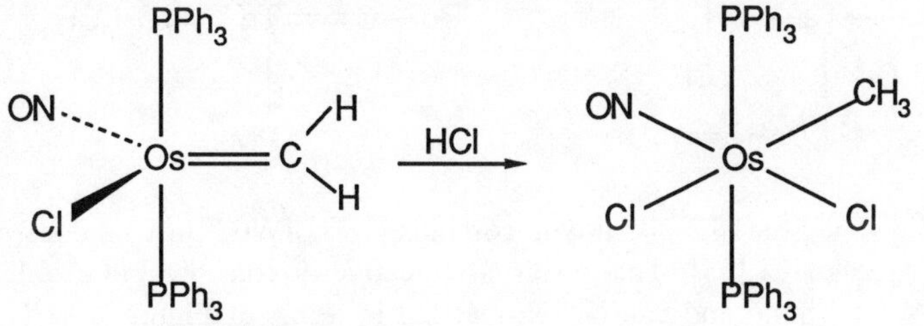

$$\text{ON}\cdots\underset{\underset{PPh_3}{|}}{\overset{\overset{PPh_3}{|}}{Os}}=C\overset{H}{\underset{H}{\diagdown}} \quad \xrightarrow{HCl} \quad \text{ON}\cdots\underset{\underset{PPh_3}{|}}{\overset{\overset{PPh_3}{|}}{Os}}\overset{CH_3}{\underset{Cl}{\diagdown}} \quad (12)$$

$$Np_3Ta=C\overset{H}{\underset{^tBu}{\diagdown}} \quad \xrightarrow[-78°C]{HCl} \quad Np_4TaCl \quad (13)$$

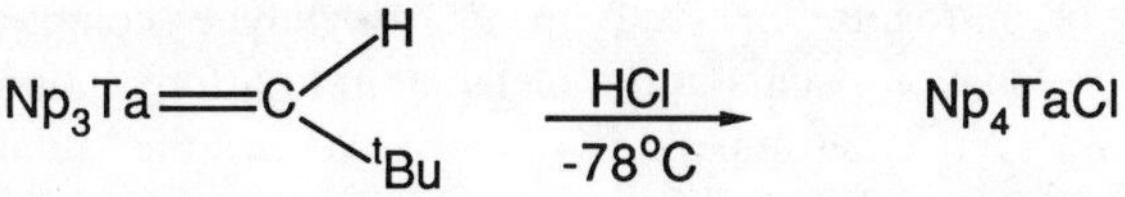

Intramolecular competition reactions provide useful insight into the protonation of alkylidene ligands. A further demonstration of the higher kinetic basicity of the alkylidene versus alkyl ligands is provided by the intramolecular competition reaction of Eq. 14 [51]. Especially noteworthy is Eq. 15 since the position of this equilibrium reflects the propensity toward α elimination versus β elimination in this system. Historically, β elimination has been regarded as the more facile of these two processes in transition metal systems. However, the equilibrium constant for Eq. 15 was shown to be ~1 [52].

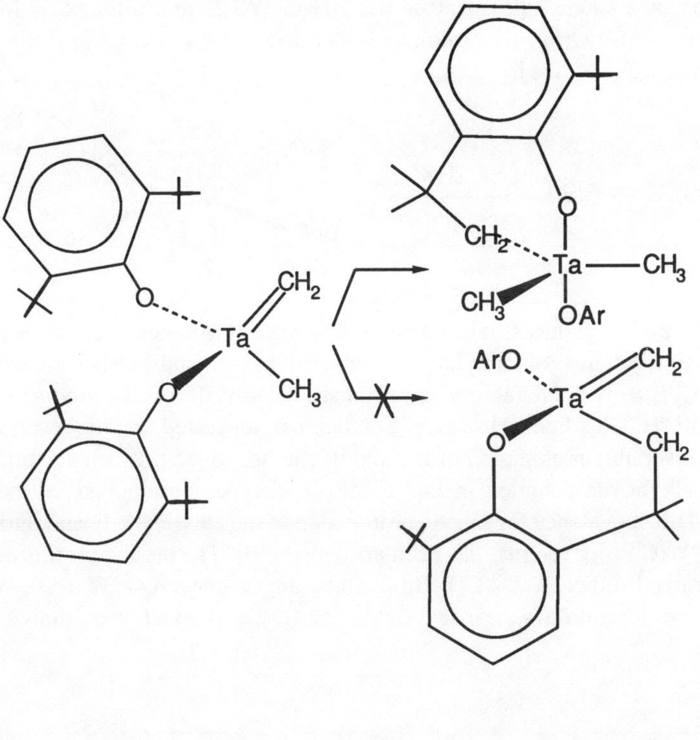

$$\text{{}^tBuCH_2-Ta\begin{smallmatrix}PMe_3\\|\\Cl\\|\\PMe_3\end{smallmatrix}\begin{smallmatrix}CH_2\\||\\CH_2\end{smallmatrix}} \rightleftharpoons {}^tBu-C\!\!=\!\!Ta\begin{smallmatrix}H\\|\\PMe_3\\|\\Cl\\|\\PMe_3\end{smallmatrix}-CH_2CH_3 \quad (15)$$

Schrock has noted that the alkylidyne complex $[NEt_4][W(C^tBu)Cl_4]$ does not react readily with excess HCl [53]. This result is indicative of considerable covalent character in the W≡C bond and can be rationalized in terms of Figure 2.2. The lower electronegativity of carbon allows covalent character with more electropositive metals than in the case of nitrogen. In fact, in d^2 alkylidyne complexes such as $(dmpe)_2W(CR)Cl$, protonation occurs at the metal atom to afford cationic alkylidyne hydride complexes, $[W(CR)H(dmpe)_2Cl]^+$ [54]. When the related complex $W(CH)(PMe_3)_4Cl$ is protonated with triflic acid, a grossly distorted "face-protonated methylidyne" results. It has been suggested that electronic factors would again favor protonation of the metal but that this is precluded by steric factors [55]. (A species suggested to be $[V=CH_2]^+$ has been reported in gas-phase studies [56]. It would seem that alternative formulation as an alkylidyne hydride should at least be considered.)

In Section 3.2.1 we noted the synthetic utility of reactions in which an alkylidyne ligand (at least formally) underwent *intramolecular* protonation by a hydroxo or amido ligand. We now note that the *intermolecular* protonation of alkylidynes is likewise a valuable route to alkylidene complexes. For example, Eq. 16 has been demonstrated for a wide range of acids (X = Cl, Br, $MeCO_2$, $PhCO_2$, PhO, C_6F_5O, p-ClC_6H_4O) [57]. Addition of a single equivalent of RCO_2H to $W(CR')(O^tBu)_3$, (R' = Me, Et) yields unusual β-hydrogen-containing alkylidene complexes of the type $W(CHR')(O^tBu)_3(O_2CR)$ [58].

$$({}^tBuO)_3W\!\equiv\!C^tBu \xrightarrow[-{}^tBuOH]{2HX} \begin{smallmatrix}{}^tBuO\\{}^tBuO\end{smallmatrix}\!\!W\!\!=\!\!C\begin{smallmatrix}H\\{}^tBu\end{smallmatrix} \text{ with } X, X \text{ axial} \quad (16)$$

As in the case of alkylidenes, it is not possible at present to generalize concerning the relative kinetic basicity of alkylidyne versus alkyl ligands. The cleavage of $Mo(C^tBu)(CH_2{}^tBu)_3$ with HCl in dimethoxyethane cleanly affords the product of alkyl cleavage, $Mo(C^tBu)Cl_3(dme)$. However, Schrock has suggested that the reaction may involve fast, reversible protonation of the alkylidyne ligand [58]. It is noteworthy that the bridging alkylidyne complex in Eq. 17 also undergoes protonolysis at the alkyl ligands [59]. Direct evidence for kinetic protonation of the alkylidyne ligand during the hydrolysis of $W(C^tBu)(CH_2{}^tBu)_3$ has been presented [60]. The product of this reaction is the oxo-bridged dimer $W_2O_3(CH_2{}^tBu)_6$ containing a linear O=W—O—W=O unit. When the reaction was carried out in D_2O the product was shown to be

$W_2O_3(CH_2{}^tBu)_4(CD_2{}^tBu)_2$. This result has been rationalized in terms of sequence of events represented by Eqs. 18 and 19.

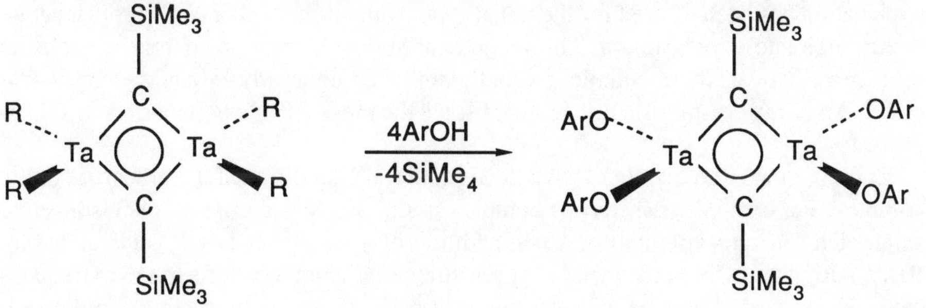

(17)

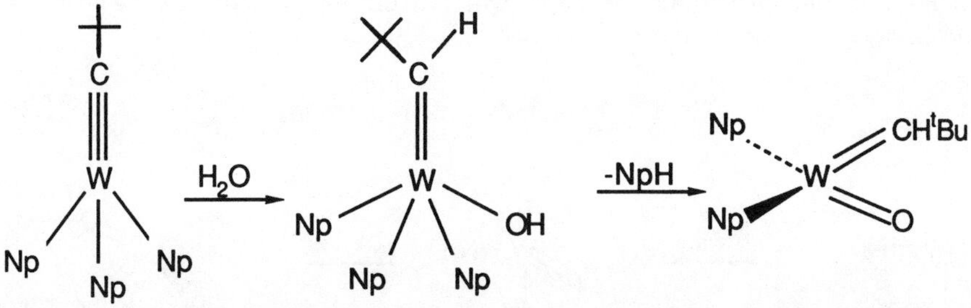

(18)

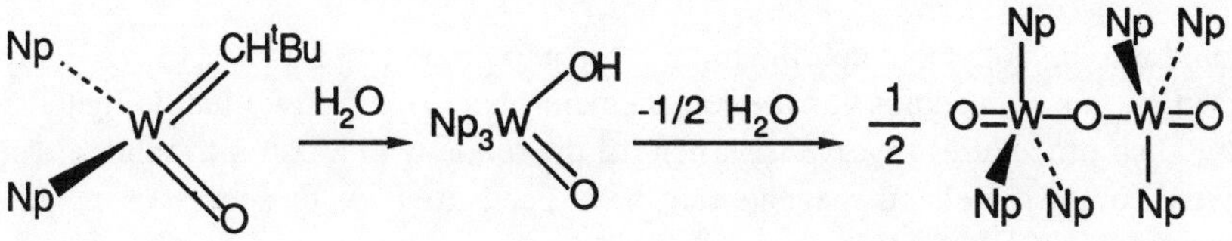

(19)

6.2.2 Reaction with Lewis Acids

Some terminal oxo ligands are sufficiently basic to coordinate with Lewis acids. Anionic oxo complexes occasionally coordinate with a main group metal counterion, even when it is not a particularly strong Lewis acid. Two structurally characterized examples have been provided by Wilkinson and coworkers [61] in the complexes $(Me_4ReO)_2Mg(thf)_4$ and $[(Me_3SiCH_2)_4ReO]_2Mg(thf)_2$. The Re—O bond lengths in these complexes were found to be lengthened by 0.05 to 0.10 Å compared with the 1.60 Å observed for example in $ReOCl_3(PEt_2Ph)_2$ [62] but do not fall outside the normal range in Table 5.2. In contrast, lengthening *is* observed when oxygen symmetrically bridges two π-acceptor rhenium atoms (in the range of 1.83 to 1.85 Å [61].)

Oxo ligands in neutral molecules can also coordinate with Lewis acids. One well-known consequence is the structure of solid $WOCl_4$ and $WOBr_4$. The presence of an open coordination site in the square-pyramidal WOX_4 monomers leads to a polymeric solid-state structure with an unsymmetrical M=O—M backbone [63]. A similar explanation has been offered for the initially puzzling properties of N,N'-propylenebis-(salicylaldimato)oxovanadium. This complex is yellow-orange, in contrast to the usual blue-green oxovanadium complexes, and exhibits an unusually low v(M—O) of 854 cm^{-1}. An X-ray crystal structure has confirmed the presence of a polymeric V=O—V chain [64].

Coordination of an oxo ligand to a Lewis acid can alter the catalytic properties of the complex. For example, the starting complex in Eq. 20 (X = Cl, Br, or OR) is inactive as an olefin metathesis catalyst. Upon addition of a variety of Lewis acids including BBr_3, $AlCl_3$, and $GaCl_3$ the metal–oxygen stretching vibration shifts to lower frequency by 30–140 cm^{-1} and the v(M—X) increases 10–25 cm^{-1}, suggesting coordination of the oxo group to the Lewis acid. (A methylimido analog is said to behave similarly.) The adducts formed in Eq. 20 are active olefin metathesis catalysts [65]. Preferential binding of the Lewis acid to oxygen is supported by molecular orbital calculations which indicate that the affinity of AlH_3 will fall off in the order oxo > chloro > alkylidene [66].

(20)

It should also be noted that coordination of the oxo group to a Lewis acid is a likely first step in certain procedures for the replacement of an oxo by two halide ligands. Typically, such procedures involve treatment of the oxide with a boron trihalide in an inert solvent [67,68]. Related reagents that have been used for this purpose include $CpTiCl_3$ and Me_3SiCl [69,70].

As first noted by Chatt and Heaton, nitrido ligands also bind Lewis acids [71,72]. Upon coordination of BCl_3, the Re–N bond length in $ReNCl_3(PMe_2Ph)_3$ increases from 1.660(8) Å to 1.728(7) Å [73,74]. The trans Re—Cl distance is ostensibly slightly shorter in the BCl_3 adduct [2.439(3) versus 2.442(2) Å], although this difference is not statistically significant. Consistent with the latter observation, v(Re—Cl) for the trans chloride ligand in a series of rhenium–nitrido complexes $ReNCl_2L_3$ increased by 10–25 cm^{-1} upon Lewis acid complexation [71]. However, despite the lengthening of the Re—N bond, the Re—N stretching frequency actually increases as a result of such coordination (see Section 4.2). As in the oxometal series, nitrido species that would be coordinatively unsaturated as monomers often adopt a solid-state structure involving unsymmetrical M=N—M bridges. Such hermaphroditic interactions can result in either a polymeric or a tetrameric structure, as previously exemplified by the compounds in Table 4.7.

Again referring to Figure 2.2, we expect the tendency of alkylidene complexes to bind to Lewis acids to be greatest on the left side of the periodic table. The classic example is the complexation of aluminum alkyls to the Tebbe–Grubbs reagents [75]. Although free titanium methylene species have been invoked as reactive intermediates, it is clear at minimum that the equilibrium in Eq. 21 lies far to the right [76]. Similarly, the tantalum methylene complex $Cp_2Ta(CH_2)(CH_3)$ reacts with $AlMe_3$ to form an isolable adduct, $Cp_2Ta(CH_2AlMe_3)(CH_3)$ [77]. In regard to the tantalum system, it was noted that aluminum alkyls are common co-catalysts in olefin metathesis and that complexation of this type might protect the alkylidene moiety both sterically and electronically. (It has subsequently been found that there are additional roles that the Lewis acids can play in olefin metathesis; see Section 7.4.1)

$$Cp_2Ti=CH_2 + AlMe_2Cl \rightleftharpoons Cp_2Ti\begin{smallmatrix}CH_2\\Cl\end{smallmatrix}Al\begin{smallmatrix}CH_3\\CH_3\end{smallmatrix} \qquad (21)$$

The osmium starting complex in Eq. 22 was suggested to contain a nucleophilic methylidene moiety on the basis of its ready reactions with reagents like HCl and $HgCl_2$. Formation of an acid–base complex in Eq. 22 is of considerable interest because both the starting complex and the adduct have been structurally characterized [49]. The Os–C bond length is not altered significantly by complexation to gold, the distances being 1.92(1) Å in the gold adduct and 1.90 Å in the uncomplexed methylidene. The most notable feature in the structure of this adduct is that the gold atom straddles the Os—C bond; the complex is the apparent product of electrophilic attack on the covalent π system. In fact, the structure is reminiscent of the "mercurinium ion" stage in the oxymercuration of olefins. Apparently the affinity of osmium for iodide is insufficient to complete the analogy by formally adding across the osmium–carbon bond.

$$\text{ON}-\text{Os}(PMe_3)_2(Cl)=CH_2 \xrightarrow[-Et_4NI]{Et_4NAuI_2} \text{ON}-\text{Os}(PMe_3)_2(Cl)(CH_2)(AuI) \qquad (22)$$

Equation 22 may remind the reader of the "face-protonation" of the tungsten(IV)–methylidyne complex $W(CH)Cl(PMe_3)_4$ by acids containing non–nucleophilic anions, which we discussed in Section 6.2.1. In fact, the same complex reacts with the aluminum reagents $AlMe_3$ and $AlMe_2Cl$ according to Eq. 23 [55]. In this case, a crystal structure [78] confirms that the formal "addition" reaction proceeds; the product is as shown in Eq. 23. Nevertheless, the W—C bond length of 1.806(6) Å remains fairly typical for a W—C triple bond and the long Al—C distance of 2.115 Å suggests that the interaction is not particularly strong. Consistent with this view, the aluminum reagent can be removed by treatment with TMEDA in the presence of PMe_3.

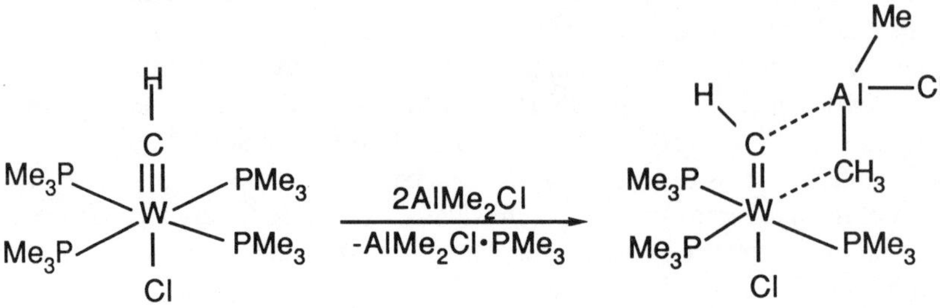

$$Me_3P\text-W(\equiv CH)(PMe_3)_3Cl + 2AlMe_2Cl \xrightarrow{-AlMe_2Cl \cdot PMe_3} \text{[bridged Al-methylidyne W complex]} \quad (23)$$

6.2.3 Reaction with Alkylating Agents and Related Electrophiles

In Section 3.3.2 we discussed the alkylation of nitrides as a synthetic entry to organoimido compounds. Here we will focus on the alkylation of oxo, imido, and alkylidene species.

Reactions of oxo ligands with alkylating agents RX typically do not yield discrete M(OR) type products. A useful exception is Eq. 24, where M is rhenium or technetium [79,80]. The oxo ligands in perrhenate and pertechnetate are not particulary nucleophilic; the silver salt is apparently required to dechlorinate the silicon, providing a very reactive electrophile. The crystal structure of trimethylsilyl perrhenate [81] shows the Re—O bond length for the siloxy oxygen [1.69(8)) Å] to be little different from other Re—O distances, which range from 1.55 to 1.71 Å.

$$AgMO_4 + Me_3SiCl \xrightarrow[-AgCl]{25°C} Me_3SiOMO_3 \quad (24)$$

Oxo ligands which are sufficiently nucleophilic undergo alkylation but the reactions are often complicated by subsequent transformations, particularly reduction of the metal or further alkylation of the initially formed M(OR) species. Both of these eventualities are illustrated by Eq.25 [70]. (Many of the classic routes to the group VI oxychlorides involve the use of a carbon electrophile to replace an oxo ligand by two chlorides. Electrophiles used in this regard have included phosgene, acetyl chloride, carbon tetrachloride, hexachloropropylene, and octachlorocyclopentene; see, for example, ref. 82. Noncarbon electrophiles such as thionyl chloride have been employed for the same purpose [83].)

$$Cp^*Re(O)_3 \xrightarrow[-3Me_3SiOSiMe_3]{6Me_3SiCl} Cp^*ReCl_4 + Cl_2 \quad (25)$$

Imido ligands can also be alkylated. Treatment of the molybenum complex in Eq. 26 with excess methyl bromide results in permethylation of one imido nitrogen atom only [12]. It has also been reported that $Cp^*_2V(NPh)$ can be methylated with methyl iodide—but the reaction is said to proceed with loss of one Cp^* ligand [84]. Intramolecular alkylation of imido ligands is also known as indicated in Eq. 27. This reaction was first developed to model chemistry that takes place on the catalyst surface during heterogeneous propylene ammoxidation (Section 7.2.1) [85]. Subsequently, it has been found that this reaction can represent a useful transformation for organic synthesis, particularly in the case where R is a triethylsilyl group [86].

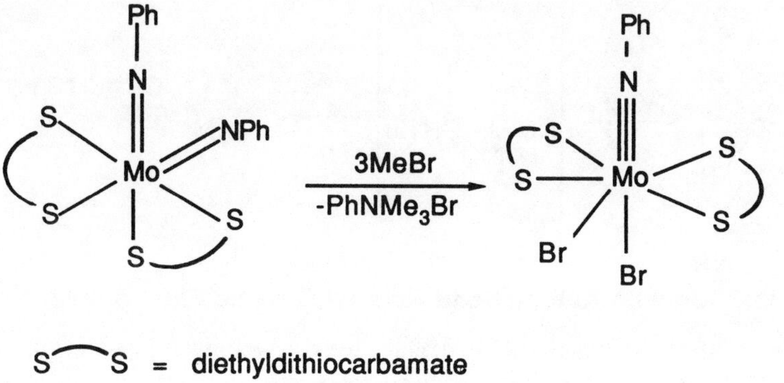

$$\text{(26)}$$

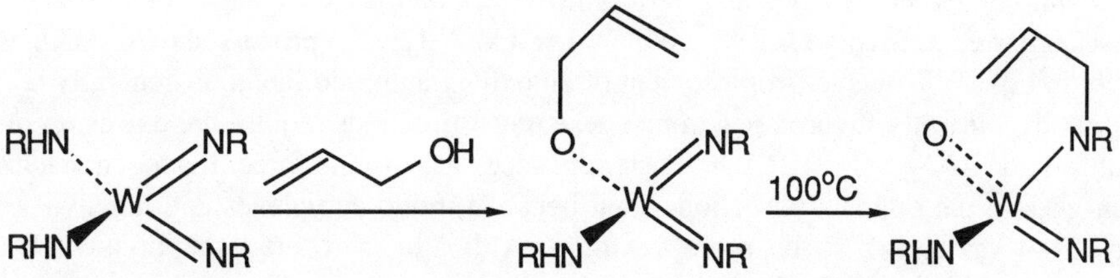

S⌒S = diethyldithiocarbamate

$$\text{(27)}$$

Studies on the alkylation of alkylidene complexes have been carried out by the Schrock group. The reaction of $Cp_2Ta(CH_2)(CH_3)$ with trimethylsilyl bromide affords the ionic complex $[Cp_2Ta(CH_2SiMe_3)(CH_3)]^+Br^-$. The reaction of this same complex with methyl iodide also apparently involves initial alkylation of the methylene ligand. However, as shown in Eq. 28, the resultant ethyltantalum intermediate undergoes a β-hydride elimination, ultimately resulting in a tantalum ethylene complex [77]. Similarly, the final products from treatment of the neopentylidene complexes in Eq. 29 with acid chlorides (M = Nb or Ta; R = Me or Ph) are the indicated enolate derivatives [50]. It again seems reasonable that the first step in Eq. 29 is acylation of the alkylidene carbon atom. Finally we note that Mayr has reported a reaction in which the first step apparently involves alkylation of a tungsten(IV)alkylidyne complex with allyl bromide [87].

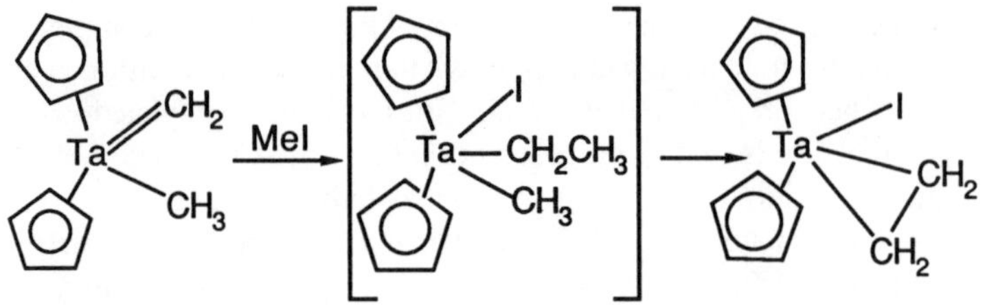

$$\text{Cp}_2\text{Ta}(=\text{CH}_2)(\text{CH}_3) \xrightarrow{\text{MeI}} [\text{Cp}_2\text{Ta}(\text{I})(\text{CH}_2\text{CH}_3)(\text{CH}_3)] \longrightarrow \text{Cp}_2\text{Ta}(\text{I})(\text{CH}_2\text{CH}_2) \quad (28)$$

$$\text{Np}_3\text{M}=\text{C}(\text{H})(^t\text{Bu}) \xrightarrow{\text{RCCl=O}} [\text{Np}_3\text{M}(\text{Cl})(\text{CH}^t\text{Bu})(\text{C}(=\text{O})\text{R})] \longrightarrow \text{Np}_3\text{M}(\text{Cl})(\text{OC}(\text{R})=\text{CH}^t\text{Bu}) \quad (29)$$

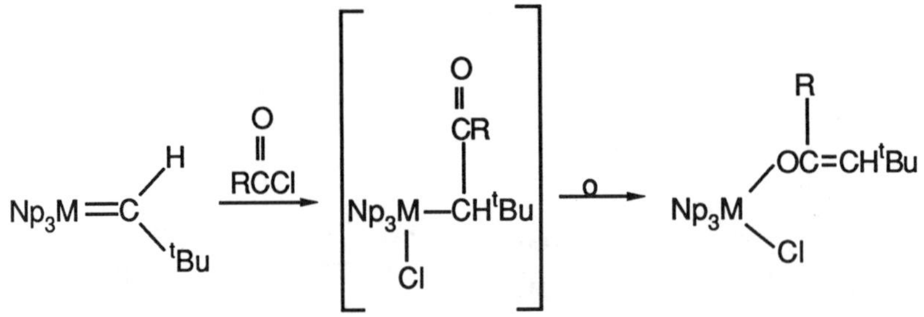

6.2.4 Reaction with Ketones and Related Organic Electrophiles

In Section 3.2.6 we saw that one particular type of carbonyl compound, the isocyanates, would react with certain metal–oxo complexes to afford the corresponding organoimido complex. This reaction (and related reactions involving sulfinylamines and phosphinimines) proceed via an electrophilic attack on the oxo ligand. The mechanism was suggested to involve the type of "Wittig-like" (2 + 2) process shown schematically in Eq. 30. However, replacement of an oxo by an imido ligand is generally not a thermodynamically favored reaction: These transformations require the extrusion of a stable oxide (CO_2, SO_2, R_3PO) to muster the necessary driving force. Indeed, a reaction that goes in the opposite direction—that between imido complexes and aldehydes in Eq. 31—was shown to be more common. Although there are exceptions such as $O_3Os(NR)$ and $(Me_3SiO)_2CrO(NR)$ which contain more electronegative metal atoms, the majority of imido complexes will react with aldehydes in this way.

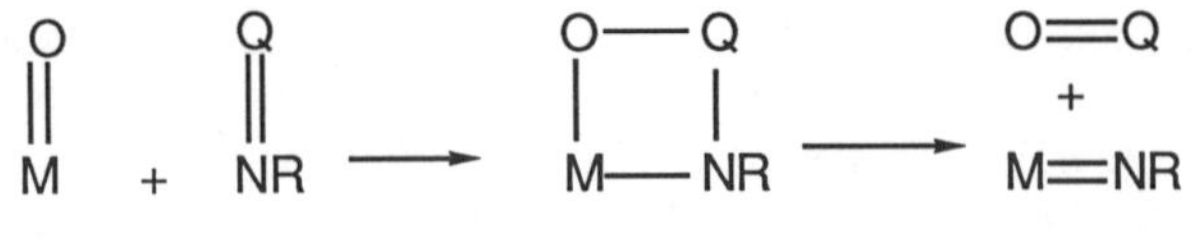

$$M=O + Q=NR \longrightarrow \underset{M-NR}{\overset{O-Q}{|\quad|}} \longrightarrow M\equiv NR + O=Q \quad (30)$$

$$Q = CO, SO, PR_3 \text{ etc}$$

$$M\equiv N\text{-}R + R'CH=O \longrightarrow M=O + R'CH=NR \quad (31)$$

It should be noted that hydrazido(2−) complexes typically react with aldehydes and ketones at the β- rather than the α-nitrogen atom, as exemplified by Eq. 32. This reaction provides a general synthetic entry to Mo– and W–diazoalkane complexes

[88,89,90]. It also reminds us that the $d_\pi - p_\pi$ bonding in a hydrazido(2−) complex results in a decrease in electron density for the α-nitrogen atom relative to the uncomplexed NH_2 group. Nevertheless, under appropriate conditions the α-nitrogen of a hydrazide can be induced to react with a carbonyl compound. This is illustrated by the reaction of the bridging hydrazido(4−) ligand in Eq. 33 with two equivalents of acetone to afford dimethylketazine [41].

$$M\equiv N-NH_2 \xrightarrow[-H_2O]{R_2C=O} M\equiv N-N=CR_2 \quad (32)$$

$$(THF)Np_3Ta=N-N=TaNp_3(THF) \xrightarrow[\substack{-\text{tantalum} \\ \text{product(s)}}]{Me_2C=O\ (xs)} \underset{60\%}{Me_2C=N-N=CMe_2} \quad (33)$$

In Section 3.2.6 we noted that alkylidene complexes readily react with organic carbonyl compounds to give olefins and metal oxo species as products. Such reactions have found little use as a synthetic route to oxometal derivatives but are invaluable as a tool for organic synthesis. It is upon the organic products of these reactions that we now focus our attention.

In a pioneering study [91], Schrock demonstrated that a tantalum neopentylidene complex would promote carbonylolefination of aldehydes and ketones in direct analogy to the phosphorus ylides of Wittig [92]. Significantly, several types of carbonyl compounds that do not react readily with Wittig reagents would undergo Wittig-type reactions with the tantalum reagent. These include esters, amides, and carbon dioxide. However, in the case of phenyl benzoate [50], the reaction takes an anomalous twist (Eq. 34).

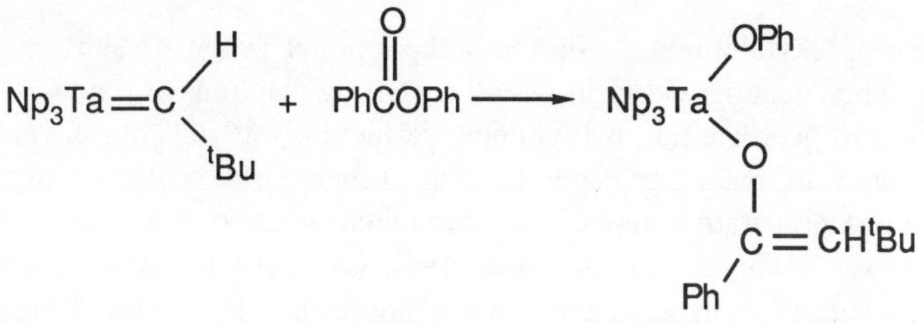

$$Np_3Ta=C\begin{smallmatrix}H\\ \\ ^tBu\end{smallmatrix} + PhCOPh \longrightarrow Np_3Ta\begin{smallmatrix}OPh\\ \\ O\\ |\\ C=CH^tBu\\ /\\ Ph\end{smallmatrix} \quad (34)$$

A practical reagent for methylenation of esters was subsequently reported by Tebbe and Parshall [75] and has been developed extensively by Grubbs and coworkers [93,94]. For applications in organic synthesis, the Tebbe–Grubbs reagent is conveniently prepared *in situ* according to Eq. 35 [95]. As shown in Eq. 36, this reagent will cleanly convert even sensitive enolizable ketones to the corresponding methylene compounds [96]. Similarly, esters are converted to the corresponding vinyl ethers, a reaction that has been applied to the synthesis of some complex and highly functionalized natural products [94]. An especially compelling demonstration of the power and selectivity of this reaction is RajanBabu's synthesis of 1-methylene sugars in Eq. 37. This approach simplifies access to many C-glycoside derivatives [97].

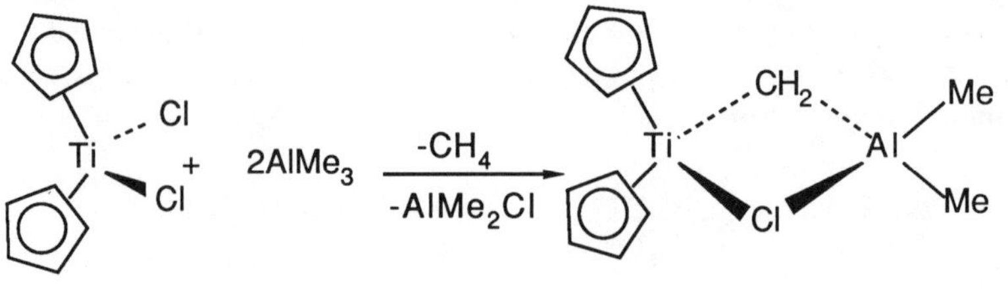

$$Cp_2TiCl_2 + 2AlMe_3 \xrightarrow[-AlMe_2Cl]{-CH_4} Cp_2Ti(\mu\text{-}CH_2)(\mu\text{-}Cl)AlMe_2 \quad (35)$$

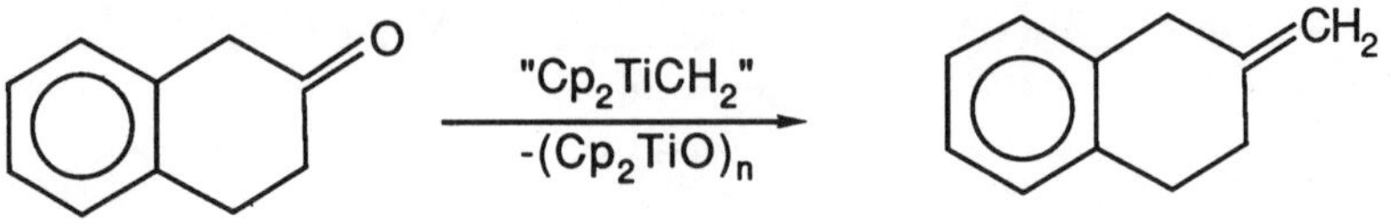

$$\text{2-tetralone} \xrightarrow[-(Cp_2TiO)_n]{"Cp_2TiCH_2"} \text{2-methylene tetralin} \quad (36)$$

$$\text{lactone sugar} \xrightarrow[-(Cp_2TiO)_n]{"Cp_2TiCH_2"} \text{1-methylene sugar (82\%)}$$

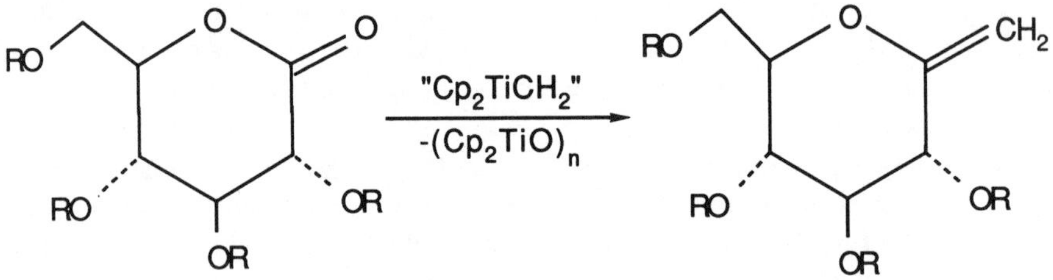

R = CH$_2$Ph   (37)

The presence of alkylaluminum chloride in the original Tebbe–Grubbs reagent can be problematic when dealing with acid-sensitive starting materials or products. In such cases, it is better to generate the methylidene via thermolysis of a titanacyclobutane. These metallacycles are readily prepared from the Tebbe–Grubbs reagent (see Section 6.3.4.3). For most preparative applications the metallacycle derived from 2-methyl-1-pentene (Eq. 38) is the reagent of choice [94]. One type of reaction where this approach has proven uniquely effective is the regioselective formation of ketone enolates from acid chlorides [98], as shown in Eq. 39. Other "protected" forms of the methylidene include the phosphine complexes Cp$_2$Ti(CH$_2$)(PR$_3$) [99] and the adduct with bis(trimethylsilyl)acetylene [100].

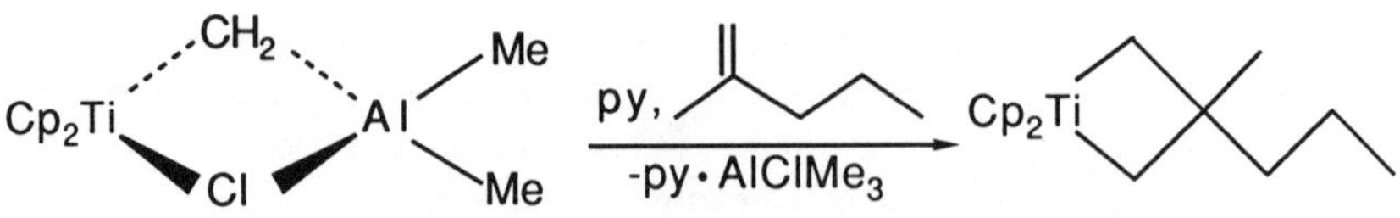

$$Cp_2Ti(\mu\text{-}CH_2)(\mu\text{-}Cl)AlMe_2 \xrightarrow[-py\cdot AlClMe_3]{py,\ 2\text{-methyl-1-pentene}} Cp_2Ti\text{(titanacyclobutane)} \quad (38)$$

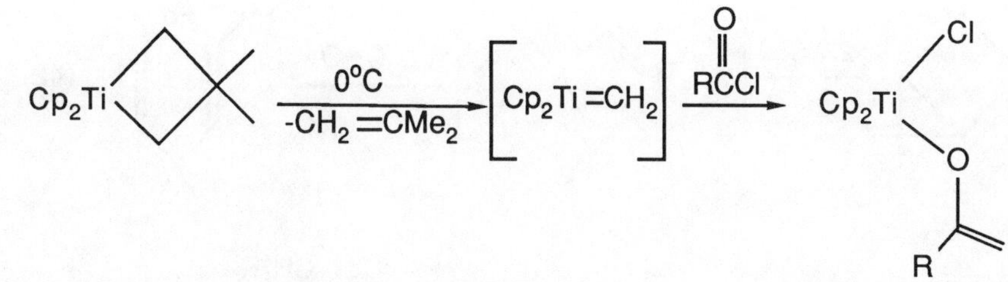

$$\text{Cp}_2\text{Ti}\underset{\text{-CH}_2=\text{CMe}_2}{\xrightarrow{0°\text{C}}} \left[\text{Cp}_2\text{Ti}=\text{CH}_2\right] \xrightarrow{\text{RCCl}} \text{Cp}_2\text{Ti}\overset{\text{Cl}}{\underset{\text{O-C(=CH}_2)\text{R}}{}} \quad (39)$$

Another fascinating (if somewhat mysterious) series of methylenating reagents has been developed by Kauffmann and coworkers. These reagents are prepared *in situ* by treating a chloride-containing tungsten or molybdenum compound with methyllithium [101–103] or trimethyl aluminum [104]. Useful transition metal components include MoCl_5, $\text{Mo(OEt)}_3\text{Cl}_2$, MoOCl_3, MoO_2Cl_2, WOCl_3, and WOCl_4. Although the resultant reagents have been represented schematically as in Eq. 40 as severely coordinatively unsaturated species, this is almost certainly an oversimplification. Remarkably, these reagents have been found to effect methylenation of ketones even in aqueous or ethanolic solvents [48]. Moreover, the presence of a hydroxyl group in the substrate as in Eq. 40 has been found to enhance the rate of the olefination reaction [105].

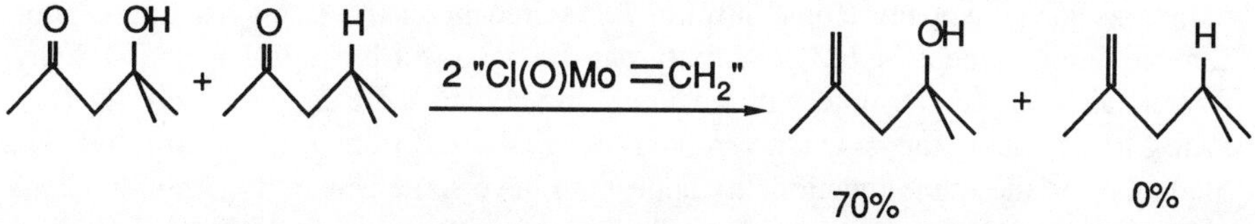

$$\text{(ketone-OH)} + \text{(ketone-H)} \xrightarrow{2 \text{ "Cl(O)Mo}=\text{CH}_2\text{"}} \text{(alkene-OH) 70\%} + \text{(alkene-H) 0\%} \quad (40)$$

It would be desirable to extend such methylenation reactions to allow the replacement of the carbonyl function with a longer-chain alkylidene moiety. To date there exists no completely satisfactory general method for accomplishing this transformation, in part because of the instability of the alkylidene reagents. In cases where the alkylidene cannot readily decompose by elimination of β hydrogen, useful carbonylolefination reactions such as Eqs. 41 and 42 have been demonstrated [106,107]. Moreover, the rather ill-defined alkylidene reagent prepared according to Eq. 43 is reported to convert ketones into the corresponding olefins [108].

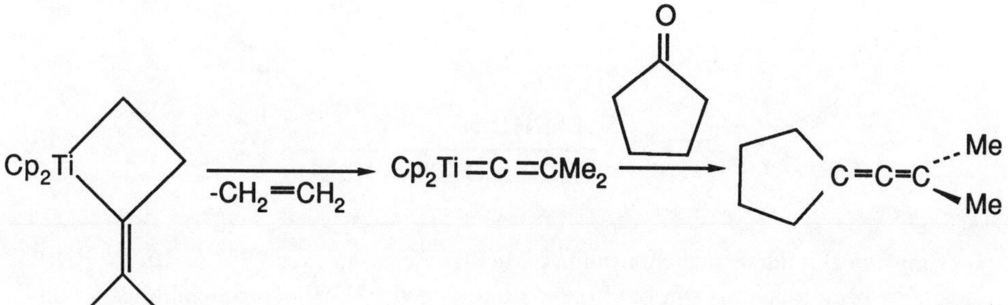

$$\text{Cp}_2\text{Ti}\xrightarrow{-\text{CH}_2=\text{CH}_2} \text{Cp}_2\text{Ti}=\text{C}=\text{CMe}_2 \xrightarrow{\text{cyclopentanone}} \text{(cyclopentylidene)C}=\text{C}=\text{C(Me)}_2 \quad (41)$$

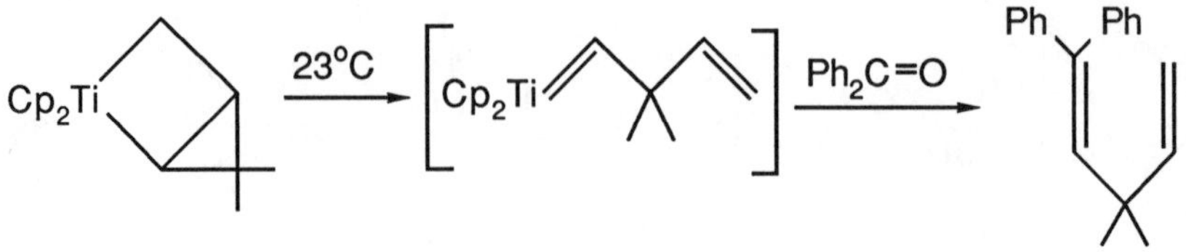

$$Cp_2Ti \diagup\!\!\!\diagup \xrightarrow{23°C} \left[Cp_2Ti = \diagdown\diagup\diagdown \right] \xrightarrow{Ph_2C=O} \begin{array}{c} Ph\ Ph \\ \diagdown\!\!=\!\!\diagup\diagdown\diagup \end{array} \quad (42)$$

$$\underset{H}{\overset{R}{>}}C=C\underset{Al^iBu_2}{\overset{H}{<}} + Cp_2TiCl_2 \longrightarrow \left[Cp_2Ti\underset{X}{\overset{CH_2R}{\underset{|}{\overset{|}{\underset{X}{\diagdown\!Al\!\diagup}}}}}X \right] \quad (43)$$

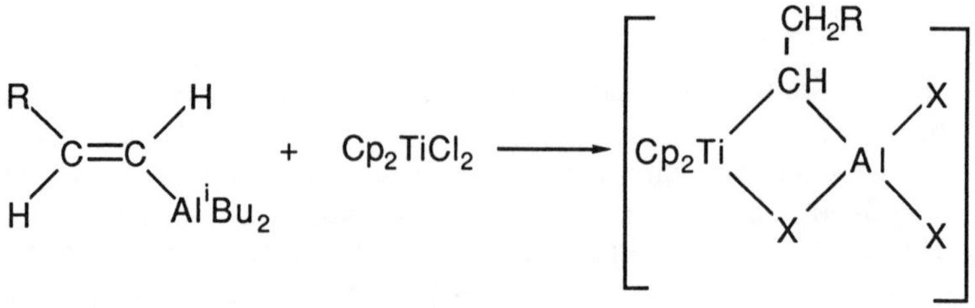

As a solution to the longer-chain alkylidene problem, Schwartz [109] has developed the use of zirconium derivatives such as that in Eq. 44 where L = phosphine or HMPA. These reagents will convert esters to vinyl ethers and ketones to olefins, although higher temperatures and longer reaction times are required when compared with titanium alkylidene reagents. The control of E/Z stereochemistry in this reaction is poor. Limited improvement in E/Z selectivity can be achieved by replacing the carbonyl functionality in the substrate with an imine moiety; in the example of Eq. 44 (L = PMe$_2$Ph), replacing the =O with =NPh lowers the Z/E from 1.0 to 0.35 [109]. The olefination of the imine functionality appears to have some generality. Another example, involving a tantalum or niobium alkylidene is shown in Eq. 45 [41].

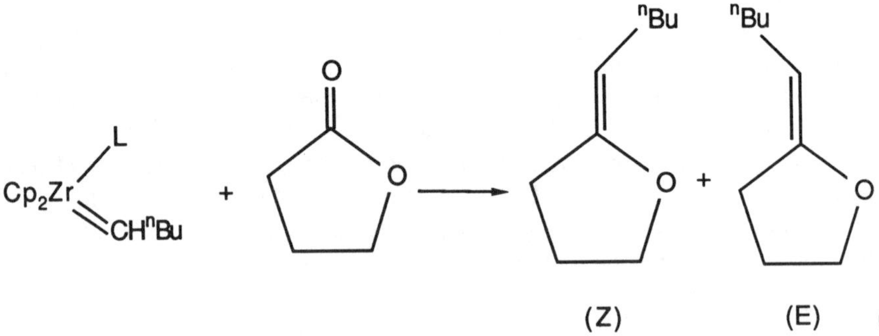

$$Cp_2Zr\underset{CH^nBu}{\overset{L}{<}} + \underset{\text{(lactone)}}{\bigcirc} \longrightarrow \underset{(Z)}{\text{vinyl ether}} + \underset{(E)}{\text{vinyl ether}} \quad (44)$$

$$\begin{array}{c} THF\diagdown\!\!\underset{X}{\overset{X}{\underset{|}{M}}}\!\!\diagup X \\ THF\diagup\!\!\underset{X}{\overset{|}{|}}\!\!\diagdown CH^tBu \end{array} \xrightarrow[-^tBuCH=CHPh]{PhCH=NR} \begin{array}{c} THF\diagdown\!\!\underset{X}{\overset{X}{\underset{|}{M}}}\!\!\diagup X \\ THF\diagup\!\!\underset{X}{\overset{|}{|}}\!\!\diagdown NR \end{array} \quad (45)$$

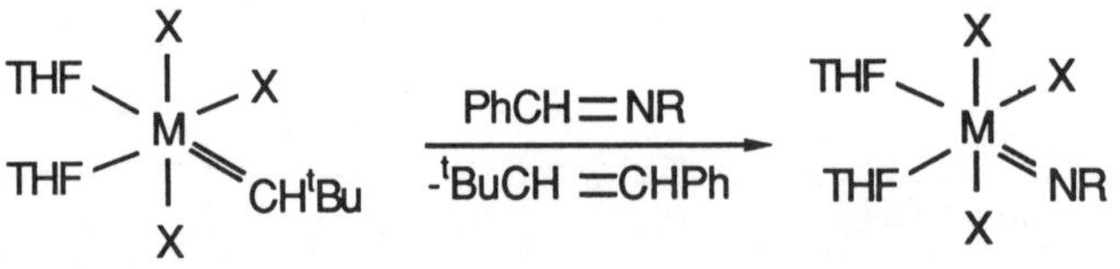

Tungsten alkylidene reagents and in particular the complexes $(^tBuCH_2O)_4W=CRR'$ have also been touted as carbonyl olefination reagents [110]. Considerable generality was demonstrated for the organic substrate, with rates falling off in the order aldehydes > ketones > formates > esters > amides. The rate is also strongly dependent on the steric bulk of the alkylidene ligand, decreasing in the order *n*-pentylidene > benzylidene > cyclopentylidene > neopentylidene. Again, control of *E/Z* stereochemistry limits the synthetic utility of the procedure.

The tungsten neopentylidyne complex in Eq. 46 (ArO = 2,6-diisopropylphenoxide) is also subject to electrophilic attack by organic carbonyl compounds. Reactions with acetone, benzaldehyde, paraformaldehyde, ethyl formate, and *N,N*-dimethylformamide all proceed according to Eq. 46 to give the corresponding oxo–vinyl complex [111]. The reaction between W(C^{t}Bu)(dme)Cl$_3$ and cyclohexyl isocyanate is suggested to commence in a similar manner (Eq. 47). However, the ketenyl complex produced in Eq. 47 further reacts with an additional equivalent of isocyanate to give a bidentate acylamido ligand [112]. It has also been shown that (ArO)$_3$W(C^{t}Bu) will react with acetonitrile to give insoluble (ArO)$_3$W(N) and ^{t}BuC≡CMe [111]. Intramolecular reactions of tungsten alkylidyne species with coordinated carbon monoxide and isonitriles are also known [113].

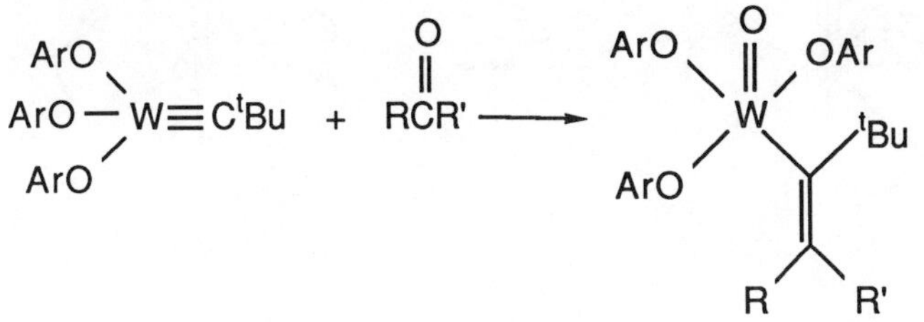

$$\text{ArO}_3\text{W}\equiv\text{C}^t\text{Bu} + \text{RCR'} \longrightarrow (\text{ArO})_3\text{W}(\text{O})(\text{C}^t\text{Bu})(\text{CRR'}) \quad (46)$$

$$\text{Cl}_3(\text{O}\cap\text{O})\text{W}\equiv\text{C}^t\text{Bu} \xrightarrow[-\text{DME}]{\text{RNCO}} [\text{Cl}_3(\text{RN})\text{W}-\text{C}(\text{O})\text{C}^t\text{Bu}] \xrightarrow{\text{RNCO}} \text{Cl}_3\text{W}(\text{NR})(\text{OC}(\text{NR})\text{C}^t\text{Bu}=\text{C}=\text{O}) \quad (47)$$

O⌒O = dimethoxyethane ; R = cyclohexyl

6.2.5 Reaction with Molecular Oxygen

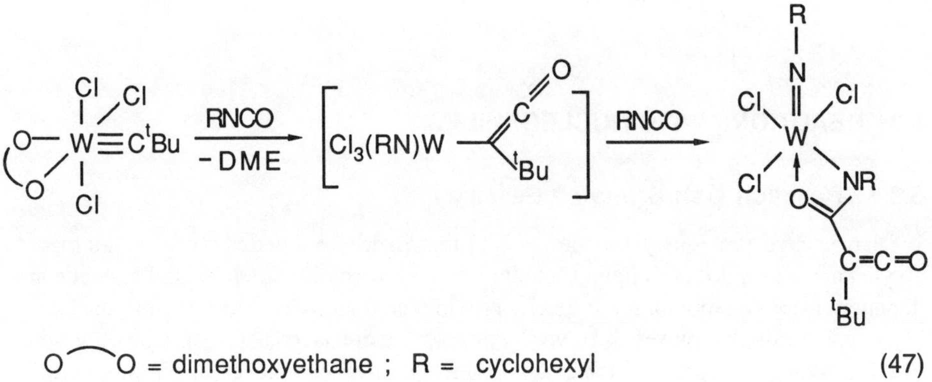

The reaction of Re(PPh$_3$)$_2$(NAr)Cl$_3$ and especially its dephosphinated analog in Eq. 48 with O$_2$ is reported to afford an arylnitroso complex as shown [114,115].

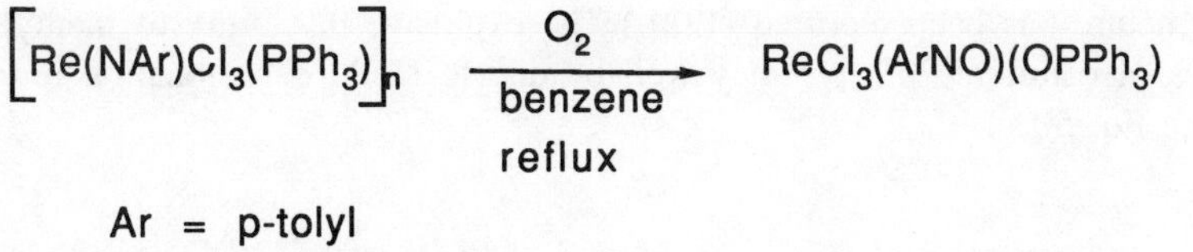

$$\left[\text{Re(NAr)Cl}_3(\text{PPh}_3)\right]_n \xrightarrow[\text{benzene reflux}]{\text{O}_2} \text{ReCl}_3(\text{ArNO})(\text{OPPh}_3) \quad (48)$$

Ar = p-tolyl

6.2.6 Reaction with Elemental Sulfur and Its Equivalents

The unusually nucleophilic nitrido complex $(dtc)_3Mo(N)$ reacts with elemental sulfur or propylene sulfide according to Eq. 49 to afford the corresponding thionitrosyl derivative [116]. Both the starting complex and the product from Eq. 49 have been structurally characterized [117]. Although the reaction of elemental sulfur itself with other nitrido complexes fails, similar transformations can be achieved by resorting to a more reactive "sulfur equivalent." Using S_2Cl_2, nitrido complexes of osmium and rhenium [116] and of technetium [118,119] can be converted into their thionitrosyl analogs. Another sulfur source that has been used for this purpose is S_4N_4 [120].

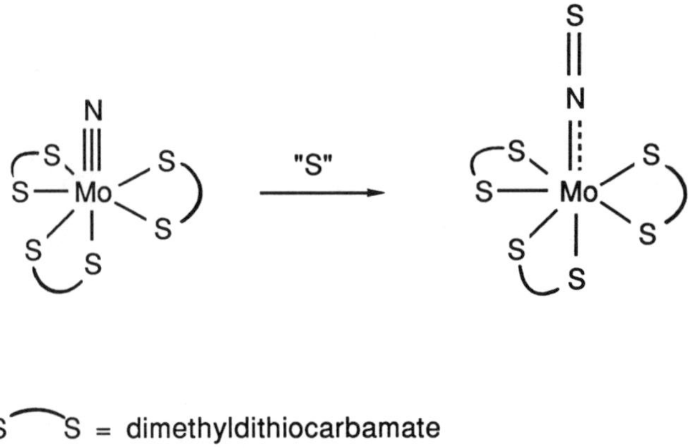

$$\text{(49)}$$

S⌒S = dimethyldithiocarbamate

6.3 REACTIONS WITH NUCLEOPHILES

6.3.1 Reaction with Bronsted Bases

We have noted previously (Section 3.2.1) that hydrogen bonded to the α atom of alkylidene and imido (NH) ligands exhibits enhanced acidity due to ligand-to-metal π donation. Deprotonation of these species provides an important synthetic entry to alkylidyne and nitrido complexes. It is worthy of note that enhanced acidity is also observed in the β-hydrogen atoms of, for example, methylimido complexes of the more electronegative transition metals. For example, treatment of the methylimido–rhenium(V) complex in Eq. 50 with pyridine results in formation of a rhenium(III)–methyleneamido complex [121]. Similar products were formed by deprotonating other ethyl- and propylimido derivatives. Although the product of Eq. 50 was not structurally characterized, the alkylidenamido moiety is suggested to have a linear structure reminiscent of a three-electron donor nitrosyl ligand. (A ruthenium alkylidenamido complex having such a structure has been reported [122].) The explosive instability of methylimido complexes of osmium(VIII) [7] may well be due to facile deprotonation processes resembling Eq. 50.

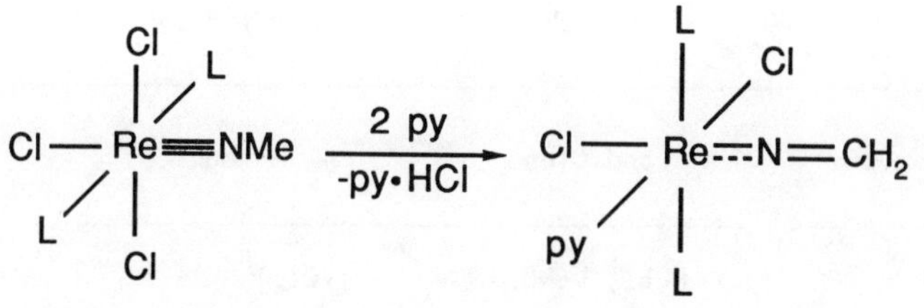

$$\text{Cl} - \underset{\underset{\text{Cl}}{|}}{\overset{\overset{\text{Cl}}{|}}{\underset{}{Re}}} \equiv \text{NMe} \quad \xrightarrow[-\text{py} \cdot \text{HCl}]{2 \text{ py}} \quad \text{Cl} - \underset{\underset{\text{L}}{|}}{\overset{\overset{\text{L}}{|}}{\underset{\text{py}}{Re}}} \cdots \text{N} = \text{CH}_2 \qquad (50)$$

L = PPh$_3$, PMePh$_2$, PEtPh$_2$

Lane and Henderson have compared the kinetic acidity of the β-hydrogen atoms in hydrazido complexes of tungsten [123] versus those of molybdenum [124]. They attribute the 10^5x slower deprotonation of the tungsten complexes to the "greater electron-releasing capability of the heavier metal".

6.3.2 Reaction with Lewis Bases

A wide range of oxometal complexes will oxidize triphenylphosphine to triphenylphosphine oxide. A number of examples where the metal-containing product has been identified are summarized in Table 6.1. This oxidation is also effected by the oxo–porphyrin systems which are commonly employed as cytochrome P-450 models. Examples include oxidation of triphenylphosphine by oxo–porphyrins of manganese [135] and iron [136] as well as by Ledon's *cis*-dioxomolybdenum(VI) tetra-*p*-tolylporphyrin complex [137].

The oxidation of triphenylphosphine by oxomolybdenum(VI) complexes as in Eq. 51 was first reported by Barral in 1972 [138]. The extensive research activity that has ensued reflects two features of this chemistry. First, because molybdenum can be reoxidized to the +6 oxidation state with dioxygen, this reaction can be carried out using a catalytic amount of the molybdenum complex. Thus, the reaction has been studied as a prototypic oxometal-mediated air oxidation. Interest is further enhanced by the presumed relevance of this transformation to the "oxo transferase" family of molybdenum-containing enzymes [139,140].

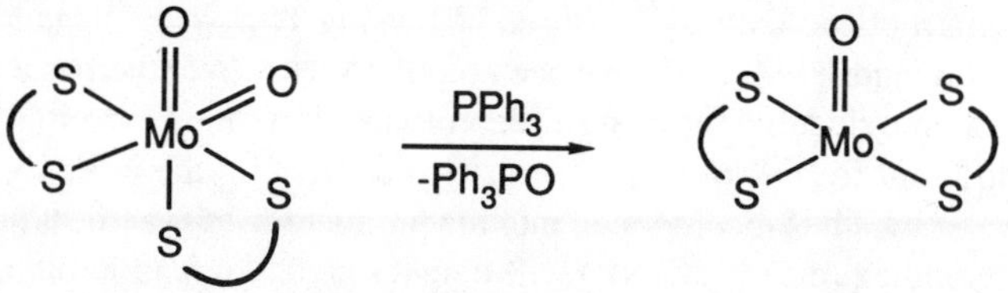

$$\underset{\text{S}}{\overset{\text{S}}{\diagdown}} \underset{\underset{\text{S}}{|}}{\overset{\overset{\text{O}}{\|}}{Mo}} \overset{\text{O}}{\diagup} \quad \xrightarrow[-\text{Ph}_3\text{PO}]{\text{PPh}_3} \quad \underset{\text{S}}{\overset{\text{S}}{\diagdown}} \overset{\overset{\text{O}}{\|}}{Mo} \underset{\text{S}}{\overset{\text{S}}{\diagup}} \qquad (51)$$

S⌒S = diethyldithiocarbamate

The scope of Eq.51 as a synthetic entry to lower-valent oxomolydenum complexes has been explored [141]. The particular suitability of chelating sulfur-containing

TABLE 6.1 Reduction of Oxo Complexes with Triorganophosphines (Selected Examples)

Starting Complex	Conditions	Product	Reference
$[VOCl_4]^-$	PPh_3, $MeNO_2$	$[VCl_6]^{3-}$	125
$[(salen)Cr(O)]^+$	PEt_3	$[(salen)Cr(OPEt_3)]^+$	126
Re_2O_7	PEt_3, py	$trans\text{-}[ReO_2(py)_4]ReO_4$	127
$Re(O)Cl_3(PPh_3)_2$	$P(OEt)_3$, C_6H_6[a]	$ReCl_3[P(OEt)_3]_3$[b]	128
$[Re(O)(SPh)_4]^-$	PPh_3, $MeCN$[a]	$Re(SPh)_3(PPh_3)(MeCN)$	129
$Cp^*Re(O)_3$	PPh_3	$[Cp^*Re(O)(\mu\text{-}O)]_2$	70
$Cp^*Re(O)_3$	PPh_3, 2-butyne	$Cp^*(O)\overline{ReCMe{=}CMeOCMe{=}CMe}$	69
$[(TMC)Ru(O)_2]^{2+}$[c]	PPh_3, $MeCN$	$[(TMC)Ru(O)(MeCN)]^{2+}$	130
OsO_4	PPh_3, $EtOH$, HCl	$Os(O)_2Cl_2(PPh_3)_2$	131
$Os(O)_2Cl_2(PPh_3)_2$	PPh_3, $EtOH$, HCl	$OsCl_4(PPh_3)_2$	131
$[Os(O)_2(bpbH_2)]^{2+}$[d]	PPh_3, $MeCN$	$trans\text{-}[(bpb)OsCl(PPh_3)]$	132
$Os(O)(N^tBu)_3$	PPh_3	$[Os(N^tBu)_3]_n$[e]	133
$[(Bu_3P)(acac)_2Co(O)]$[f]	$MeCN$	$Co(acac)_2(OPBu_3)$	134

[a] At reflux.
[b] Oxo oxygen said to be converted to Ph_3PO.
[c] TMC = 1,4,8,11-tetramethyl-1,4,8,11-tetraazacyclotetradecane.
[d] $bpbH_2$ = N,N'-bis(2'-pyridinecarboxamide)-1,2-benzene.
[e] Not isolated.
[f] Oxo complex suggested to be formed in situ by homolysis of peroxo bridged dimer.

ligands for effecting oxo transfer reactions of molybdenum has been noted [142]. In keeping with this notion, extensions of Eq. 51 using N_2S_2-type ligands [143] and cysteine-derived ligands [144,145] have been reported. However there is evidence that catalytic phosphine oxidations utilizing S-deprotonated cysteine esters as ligands are qualitatively different from Eq. 51 and, in particular, require the presence of discrete quantities of water which is incorporated into the product Ph_3PO [146]. It has also been found that the complex $(dtc)_2MoO(NR)$ will transfer an oxo ligand in preference to an imido ligand [147].

Thermodynamic studies on Eq. 51 (R = ethyl) in solvent 1,2-dichloroethane indicate that the process is substantially exoergonic with $\Delta H = -29.0 \pm 2.5$ kcal/mole^{-1} [148]. Early kinetic studies on this reaction [149,150] did not take adequate account of the potential for reversible syn-proportionation of the oxomolybenum(IV) product with starting oxomolybenum(VI) shown in Eq. 52 [151]. A general kinetic treatment taking

REACTIONS WITH NUCLEOPHILES   243

this equilibrium into account has subsequently been developed by Holm and coworkers [151]. Rate constants (in $M^{-1}\,\text{sec}^{-1}$) for the $\text{MoO}_2(\text{S}_2\text{CNEt}_2)_2/\text{PR}_3$ system were shown to increase along the series PPh_3 (0.071), PPh_2Et (0.23), PPhEt_2 (0.43), PEt_3 (0.53).

$$(\text{dtc})_2\text{MoO} \;+\; (\text{dtc})_2\text{MoO}_2 \;\rightleftharpoons\; \text{Mo}_2\text{O}_3(\text{dtc})_4 \qquad (52)$$

Another contribution of Holm and coworkers is the development of molydenum enzyme models containing sterically encumbered NS_2-type ligands [140]. These ligands inhibit the formation of μ-oxo dimers shown in Eq. 52. Reaction with triphenylphosphine therefore proceeds cleanly, as shown in Figure 6.2.

The oxidation of triphenylphosphine by ruthenium(IV) in Eq. 53 has also been the subject of a careful study [152]. Experiments using ^{18}O-labeled complex showed transfer of oxygen from Ru=O to Ph$_3$P to be quantitative within experimental error. Initial oxygen transfer is first order in each reactant with $k(26.6°C) = 1.75 \pm 0.10 \times 10^5$ $M^{-1}\text{s}^{-1}$. The initial product $[(\text{bpy})_2\text{pyRu}(O{=}\text{PPh}_3)]^{2+}$ is formed as an observable intermediate and then undergoes slow $[k(25°C) = 1.15 \pm 0.10 \times 10^{-4}\,\text{s}^{-1}]$ solvolysis. Activation parameters for Eq. 53 were reported as $\Delta H^\ddagger = 4.7 \pm 0.5$ kcal mole^{-1} and $\Delta S^\ddagger = -19 \pm 3$ eu. It was suggested that the initial two-electron reduction involves strong electronic–vibrational coupling between the electron donor and acceptor sites through Ru—O and P—O stretching vibrations.

$$\left[(\text{bipy})_2\text{pyRuO}\right]^{2+} + \text{Ph}_3\text{P} \;\xrightarrow{\text{MeCN}}\; \left[(\text{bipy})_2\text{pyRu}(\text{NCMe})\right]^{2+} + \text{Ph}_3\text{PO} \qquad (53)$$

It was pointed out by Sharpless and coworkers [153] that attack of a phosphine on a metal–oxo moiety (typically polarized $M^+{-}O^-$) was inconsistent with the then-prevalent notion of a charge-controlled process. (In a real sense, the current apprecia-

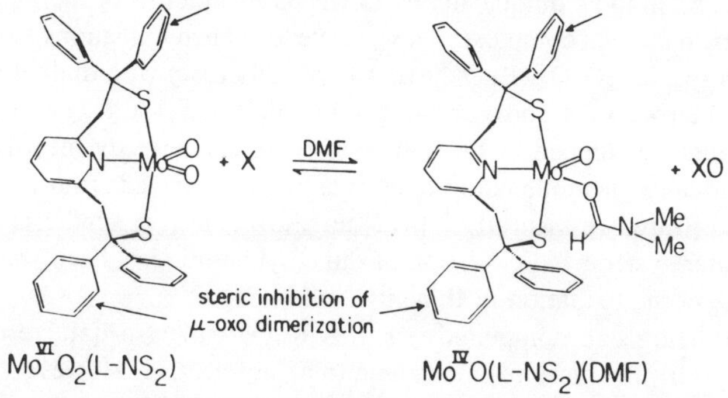

Figure 6.2 Use of bulky NS$_2$ ligands to promote clean oxo transfer uncomplicated by formation of μ-oxo dimers. (Adapted with permission from E.W. Harlan et al., *J.Am. Chem. Soc.* **1986**, *108*, 6993. Copyright 1986 American Chemical Society.)

tion of frontier orbital control in the reactions of multiply bonded ligands has grown from this proposal.) One alternative model for phosphine oxidation was suggested to involve the association of the phosphine with the metal center followed by migration of the coordinated phosphorus to an oxo group (Eq. 54). To date, mechanisms corresponding to eq. 54 have yet to be demonstrated, although there is no reason to believe that such processes cannot exist. However, it has become evident that prior coordination is not a necessary condition for phosphine oxidation. Witness reactions such as Eq. 55 in which triphenylphosphine is oxidized despite the presence of a more strongly reducing cis-triethylphosphine in the coordination sphere of the metal [154].

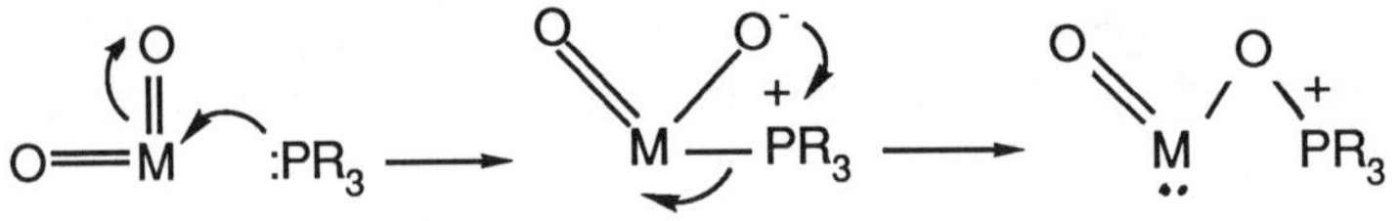

$$O=M(=O)\ :PR_3 \longrightarrow M(O)(O^-)-PR_3^+ \longrightarrow M(O)(O)PR_3^+ \qquad (54)$$

$$\left[(N\frown N)_2 Ru(O)(PEt_3) \right]^{2+} + Ph_3P \longrightarrow Ph_3PO + Ru(II)\ \text{product} \qquad (55)$$

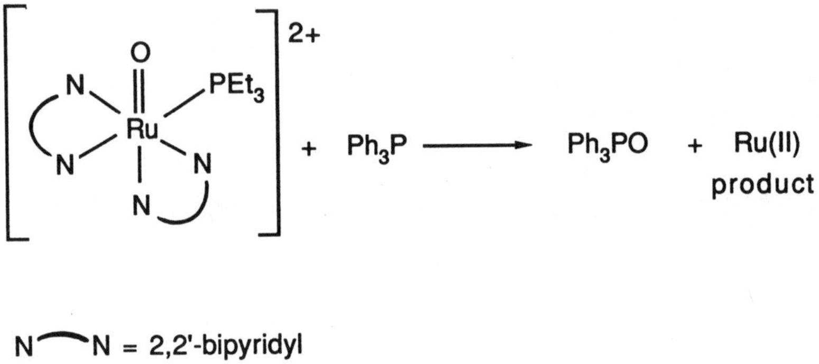

N⁀N = 2,2'-bipyridyl

Phosphines seem to be unique among Lewis bases in terms of their clean and rapid oxygenation by oxometal complexes. Nevertheless, related oxidations are known. Simple metal oxo species [OsO_4, RuO_4, MnO_4^-] typically react with dialkyl sulfides much more slowly than with the corresponding sulfoxides [155,156]. Consequently sulfide oxidation cannot be stopped at the sulfoxide stage and directly affords the sulfone. However, sterically encumbered systems such as that derived from (TPP)FeCl/PhIO [157,158] and [(bipy)$_2$(py)Ru(O)]$^{2+}$ [312] do allow selective synthesis of the sulfoxide. In the latter system, the oxidation of dimethyl sulfide has been shown to be 120 times faster than the oxidation of dimethyl sulfoxide.

Highly oxidizing oxo compounds such as RuO_4 will also oxidize amines to the amine oxide [160], but these reactions tend to be unselective. However, the oxidation of pyridine mediated by a cationic oxochromium(V) derivative is reported to afford coordinated pyridine N-oxide as shown in Eq. 56 [126].

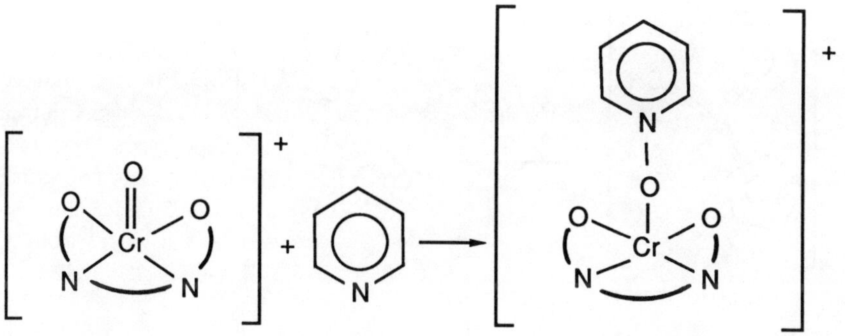

$$\left[\begin{array}{c} O \diagdown \overset{\displaystyle O}{\underset{\displaystyle \|}{Cr}} \diagup O \\ \big(\big) \\ N \diagdown \diagup N \end{array}\right]^+ + \bigcirc\!\!\!\!_N \longrightarrow \left[\begin{array}{c} \bigcirc\!\!\!\!_N \\ | \\ O \diagdown \overset{\displaystyle O}{\underset{\displaystyle |}{Cr}} \diagup O \\ \big(\big) \\ N \diagdown \diagup N \end{array}\right]^+ \quad (56)$$

O⌒N⌢N⌒O = N,N'-ethylenebis(salicylideneaminato) (2−)

Nucleophilic attack of phosphines on other multiply bonded ligands are also known. For example, the nitrido complexes in Eq. 57 (M = Ru or Os; L = Et_3As, Ph_3Sb, or bipy) react with triphenylphosphine to give the corresponding phospinimidato complexes [161]. Analogous reactions have been observed for certain nitrido complexes of molybdenum(VI) [162] and rhenium(VII) [163] and for the porphyrinato nitrides of chromium(V) and manganese(V) [164]. The complex $MoNCl_2(PPh_3)_2$ does not react with excess PPh_3 or $PMePh_2$ but apparently does form a phosphinimidate complex with the more basic PMe_2Ph [165]. A presumably related reaction is that between $ReNCl_2(PPh_3)_2$ and excess arylthiolate anion in acetonitrile. Isolable products were obtained only upon aerial oxidation of the resultant solution, whereupon $Re(NPPh_3)(SPh)_4$ was isolated [129].

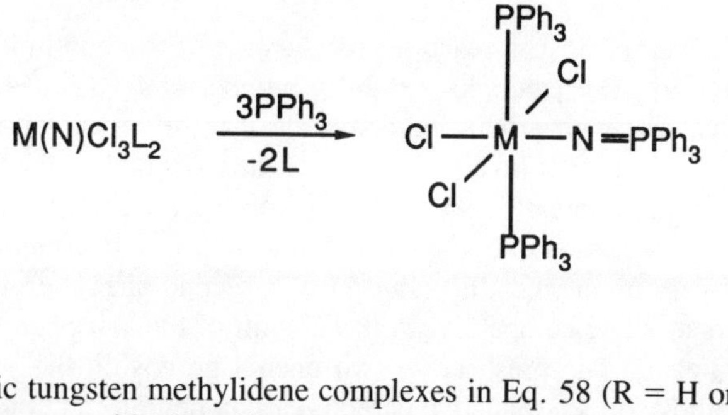

$$M(N)Cl_3L_2 \xrightarrow[-2L]{3PPh_3} \begin{array}{c} PPh_3 \\ | \quad Cl \\ Cl-M-N=PPh_3 \\ Cl \quad | \\ PPh_3 \end{array} \quad (57)$$

The cationic tungsten methylidene complexes in Eq. 58 (R = H or CH_3) have also been shown to react with phosphines as indicated [166,167]. A formally related reaction of a tungsten alkylidyne complex has also been reported [55]. However, the mechanism of Eq. 59 has not been investigated.

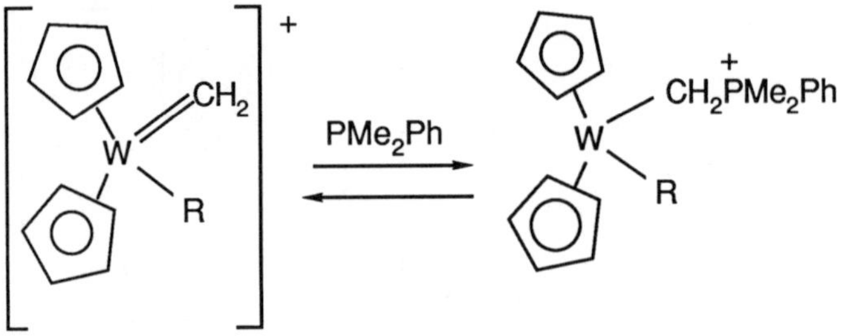

$$[\text{Cp}_2\text{W}(=\text{CH}_2)(\text{R})]^+ \xrightleftharpoons{\text{PMe}_2\text{Ph}} [\text{Cp}_2\text{W}(\text{CH}_2\text{PMe}_2\text{Ph})(\text{R})]^+ \quad (58)$$

$$\text{W(CH)(H)(PMe}_3)_3\text{Cl}_2 \xrightarrow[0^\circ\text{C}]{\text{CO}} \text{W(CH}_2\text{PMe}_3)\text{Cl}_2(\text{PMe}_3)_2(\text{CO})_2 \quad (59)$$

6.3.3 Organometallic Reagents as Nucleophiles

The formal addition of an alkyl or aryl carbanion to the α atom of an oxo, imido, or alkylidene ligand is known. In most cases there is at least indirect evidence that the reaction involves an organometallic intermediate in which the carbanionic nucleophile is σ-bonded to the transition metal. The organic ligand then migrates to the α atom, as in Eq. 60. Such reactions appear to have much in common with the most important class of molecular rearrangements in organic chemistry, including the Hoffmann, pinacol, and hydroperoxide rearrangements. Each of these reactions is best understood as a 1,2-shift to an electron-deficient atom, although Eq. 60 has not been observed directly. Consistent with this notion, the "migratory aptitude" of the phenyl group appears to be higher than that of saturated alkyl substituents in both the organic and organometallic "rearrangements."

$$\underset{|}{\overset{R}{M{=}Q}} \longrightarrow M-Q-R \quad (60)$$

Eqs. 61 and 62 exemplify such reactions for the case of oxo and imido complexes, respectively [168,169]. The examples exhibit a number of features in common: (1) Prior to hydrolysis, the main-group organometallic can be recovered as PhHgCl or PhZnOSiMe$_3$, respectively. (2) Inverse addition of the transition metal complex to the organometallic reagent suppresses the reaction involving the multiply bonded ligand and results instead in formation of high yields of biphenyl. One interpretation is that, in the presence of excess organometallic, transfer of a second phenyl group to the transition metal occurs at a rate competitive with 1,2-shift of the first phenyl group to the multiply bonded ligand. The presence of two phenyl groups on the same transition metal apparently allows facile reductive carbon–carbon bond formation. (3) Replacement of the phenyl substituent of the organometal with mesityl allows the isolation of a stable organotransition metal compound, namely (mes)$_3$VO or (mes)$_2$Cr(N^{t}Bu)$_2$ [170,171]. These observations support a stepwise mechanism as suggested by Eq. 60.

$$\text{Cl}\underset{\text{Cl}}{\overset{\text{O}}{\underset{|}{\overset{\|}{\text{V}}}}}\text{Cl} \xrightarrow[-\text{PhHgCl}]{\text{Ph}_2\text{Hg}} \left[\text{Cl}\underset{\text{Cl}}{\overset{\text{O}}{\underset{|}{\overset{\|}{\text{V}}}}}\text{Ph}\right] \longrightarrow \left[\text{VCl}_2(\text{OPh})\right]_n \quad (61)$$

$$\underset{\text{RO}}{\overset{\text{RO}}{\text{Cr}}}\underset{\text{N}^t\text{Bu}}{\overset{\text{N}^t\text{Bu}}{=}} \xrightarrow[-\text{PhZnOR}]{\text{ZnPh}_2} \left[\underset{\text{Ph}}{\overset{\text{RO}}{\text{Cr}}}\underset{\text{N}^t\text{Bu}}{\overset{\text{N}^t\text{Bu}}{=}}\right] \longrightarrow \left[\text{Cr}(\text{N}^t\text{Bu})(\text{NPh}^t\text{Bu})(\text{OR})\right]_n$$

R = SiMe$_3$ (62)

Observations on the reaction between di-n-decylmercury and CrO_2Cl_2 also support a stepwise mechanism. The reactants can be mixed at $-78°C$ and filtered to remove decylmercury chloride. Addition of bromine to the filtrate gives 1-bromodecane, suggesting the formation of an organochromium intermediate. Alternatively, warming the solution to room temperature followed by hydrolysis affords n-decyl alcohol [153].

Related reactions are also known in alkylidene chemistry. For example, Eq. 63 results in an isolable benzylidene complex [172]. The starting methylidene complex is generated *in situ* by hydrogen atom abstraction from the radical cation of $\text{Cp}_2\text{W}(\text{CH}_3)\text{Ph}$. Cooper has suggested that rearrangements like Eq. 63 should be especially facile in cationic alkylidene complexes. Indeed, evidence has been presented for a related methyl migration in a cationic iridium methylene complex which affords a stable ethyliridium derivative [159].

$$\overset{+}{\text{Cp}_2\text{W}}\underset{\text{Ph}}{\overset{\text{CH}_3}{<}} \xrightarrow[-\text{Ph}_3\text{CH}]{\text{Ph}_3\text{C}\cdot} \left[\text{Cp}_2\text{W}\underset{\text{Ph}}{\overset{\text{CH}_2}{<}}\right]^+ \xrightarrow{\text{MeCN}} \text{Cp}_2\text{W}\underset{\text{NCMe}}{\overset{\text{CH}_2\text{Ph}}{<}} \quad (63)$$

Similar 1,2-alkyl migrations apparently occur in neutral early transition metal alkylidene systems. However, such reactions are complicated by the fact that the product is an alkyl complex containing both β hydrogen and an open coordination site and thus ripe for β-hydrogen elimination. The olefin formed in Eq. 64 is exclusively the trans isomer [54]. This is consistent with the two t-butyl groups of the alkyl tungsten intermediate turning away from one another prior to elimination of a β-hydrogen atom. β-hydride eliminations also seem to occur in the products of alkyl migration of the complexes $\text{Cp}_2\text{Ta}(\text{CHMe})(\text{Me})$ and $[\text{Cp}_2\text{W}(\text{CH}_2)(\text{CH}_3)]^+$. However, in both of these cases the product olefin remains coordinated to the transition metal in the isolated product [166,60]. One practical consequence of this type of reaction pertains to olefin metathesis catalysts prepared by treating $WOCl_4$ with a main-group methyl derivative. In the absence of excess olefin the catalyst decomposes with formation of ethylene, a reaction that is proposed to proceed analogously to the above examples [173].

R = tert-butyl ; P⌒P = dmpe   (64)

Photolysis of hexamethyltungsten (VI) in neat trimethylphosphine results in formation of an ethylidyne complex *trans*-W(CMe)(Me)(PMe$_3$)$_4$. This reaction is suggested to involve migration of a methyl group to a methylidyne ligand [174]. We also note a distantly related reaction of a rhenium carbyne complex [175]. Transfer of hydride from aluminum to the carbyne carbon atom in Eq. 65 would seem unlikely to involve a 20-electron hydridorhenium complex as intermediate.

(65)

6.3.4 Reactions with the Olefinic Double Bond

6.3.4.1 Epoxidation. Olefins are epoxidized with certain high-valent oxo complexes of Cr, Mn, Fe, Ru, and Os. While it would be conceptually attractive to believe that all of these reactions proceed by a single basic mechanism, significant differences in the epoxidations promoted by these various metals have come to light in recent years. Consider the stereochemical results in Table 6.2. Most of the oxidants convert *cis*- and *trans*-stilbene (1,2-diphenylethylene) more or less stereospecifically to the *cis*- or *trans*-epoxide, respectively. Yet treatment of *cis*-stilbene with the manganese(V) porphyrin (TPP)MnCl(O)·PhI gives a product that is predominantly the trans epoxide. Moreover, the iron analog, (TPP)FeCl(O)·PhI, fails to react with *trans*-stilbene at all. (Note, however, that several of the putative oxidants in Table 6.2 have not been isolated and their existence is surmised from indirect evidence.)

REACTIONS WITH NUCLEOPHILES   249

TABLE 6.2 Stereochemistry of Olefin Epoxidation with Various Oxometal Complexes

Oxidant	Olefin	Configuration	Epoxide %E	Epoxide %Z	Reference
(TPP)MnCl(O)	stilbene	E	100	0	176
		Z	62	38	
(TPP)CrCl(O)	stilbene	E	100	0	177
		Z	0	100	
(TPP)FeCl(O)	stilbene	E	N.R.	N.R.	178
		Z	0	100	
$[(bpy)_2(py)Ru(O)]^{2+}$	stilbene	E	100	0	179
		Z	5	95	
$[(salen)Cr(O)]^+$	stilbene	E	100	0	180
		Z	3	97	
$[(salen)Cr(O)]^+$	β-methylstyrene	E	100	0	180
		Z	2	98	
$CrO_2(NO_3)_2$	β-methylstyrene	E	100	0	180
		Z	1	99	
(TMP)Ru(O)$_2$	β-methylstyrene	E	99	1	181
		Z	4	96	

The lack of stereospecificity in the manganese system suggests that this epoxidation has a "stepwise radical nature" as shown in Eq. 66 [176]. Presumably, the triplet ground state for a high-spin d^2 oxomanganese(V) ion has affected the degree of concertedness in this case.

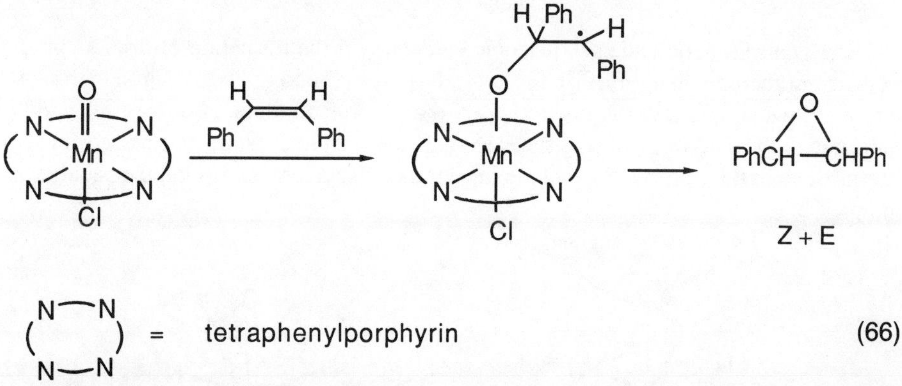

$$\text{(Mn=O complex)} \xrightarrow{\text{PhCH=CHPh}} \text{(Mn-O-CHPh-CHPh complex)} \longrightarrow \text{PhCH}\underset{O}{-}\text{CHPh}$$

$$\left(\begin{array}{c} N-N \\ N-N \end{array}\right) = \text{tetraphenylporphyrin} \tag{66}$$

The more common observation of stereoselective epoxidation has been interpreted in terms of oxametallacyclic intermediates. The first such proposal [153] invoked the intermediacy of chromium(VI) oxametallacycles in the oxidation of alkenes by chromyl chloride (Eq. 67) and this view has received some theoretical support [182]. Subsequently, it has been suggested that the active oxidation state in chromium-mediated epoxidations may in fact be chromium(V) [183]. An observation relevant to the issue of metallacyclic intermediates concerns the formation of side-product benzaldehyde when styrene is oxidized with [(salen)CrO]$^+$ cation in the presence of pyridine N-oxide (Eq. 68). Increasing the concentration of pyridine N-oxide increases the amount of benzaldehyde formed but has no effect on the rate of disappearance of styrene. This observation, taken together with labeling studies and other evidence, proves that an intermediate is formed in this system subsequent to the rate-limiting step for oxidation. However, it is not yet clear whether the intermediate is a chromium(V) oxametallacycle or a chromium(III) adduct of styrene oxide [180].

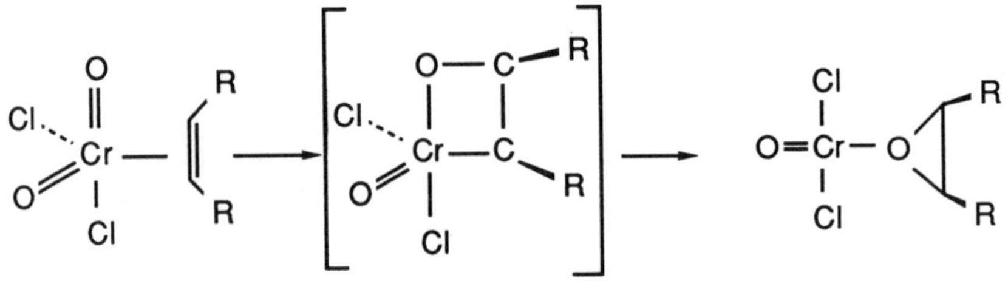

(67)

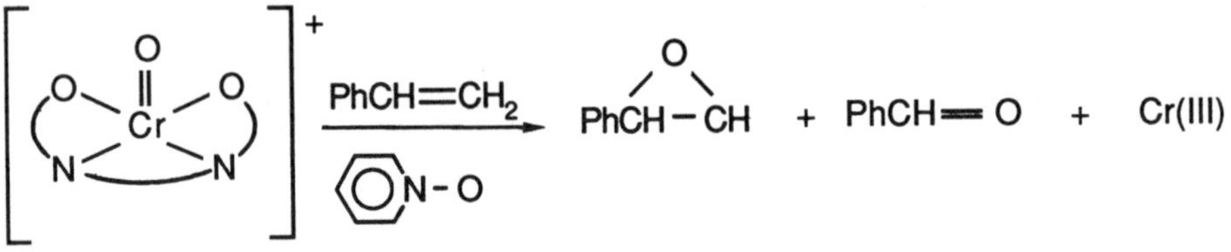

(68)

O⌒N⌒N⌒O = N,N'-ethylenebis(salicylideneaminato)(2-)

There is also kinetic and spectroscopic evidence for the formation of discrete intermediates in the epoxidation of olefins by oxo–porphyrin complexes of iron and manganese. However, in these cases it appears that the intermediate is formed reversibly and prior to the rate-determining step [184,185]. Possible structures proposed for the intermediate in the iron system include the oxametallacyclobutane **1** and the peculiar π complex **2**.

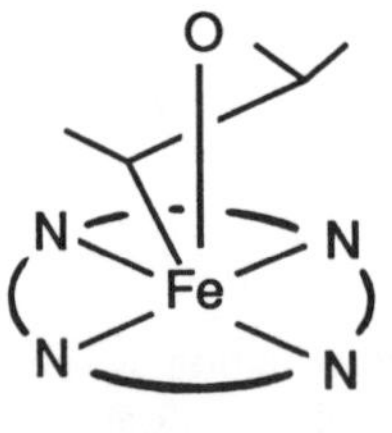

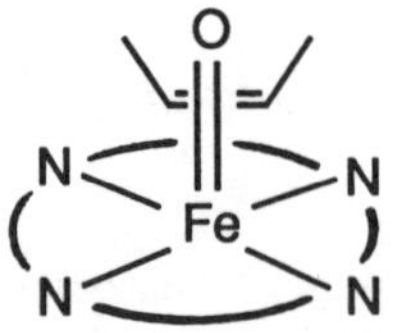

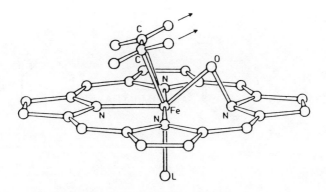

Figure 6.3 An alternative model for epoxidation by P-450 model systems which does not directly involve an oxoiron intermediate. (Adapted with permission from K.A. Jorgensen, *J. Am. Chem. Soc.* **1987**, *109*, 701. Copyright 1987 American Chemical Society.)

The enigmatic nature of this iron intermediate and the much higher reactivity of *cis*- versus *trans*-stilbene in the iron system are among the factors supporting the alternative mechanistic proposal of Jorgenson [186]. It was suggested that the active oxidant in P-450 and its model systems is not a terminal oxo complex at all but rather a structure in which the oxygen has inserted into an iron–nitrogen bond. This point of view is also supported by molecular orbital calculations and a structurally characterized nickel analog [187]. In this view, the structure of the intermediate olefin adduct is that shown in Figure 6.3. It is not clear how this type of structure would promote the facile hydroxylation of hydrocarbons observed in closely related systems. Moreover, it is evident that such a mechanism does not apply to the [(salen)CrO]$^+$ system where the structure of the active oxidant is well established [180].

Despite our incomplete understanding of the mechanism of epoxidation, recent years have witnessed several noteworthy technical achievements in this area. Groves and Myers have developed chiral iron porphyrin complexes which allow catalytic asymmetric epoxidations of terminal olefins with enantiomeric excesses of up to 51% [188]. Groves and Quinn have utilized the ruthenium tetramesitylporphyrin system to effect the catalytic epoxidation of olefins with air as stoichiometric oxidant [181]. The active form of the catalyst is believed to be the *trans*-(dioxo)ruthenium(VI) species. This stands in contrast to the [(bpy)$_2$(py)Ru(O)]$^{2+}$ system of Meyer [179] where the active oxidant is ruthenium(IV) and in particular to the osmium-based epoxidizing system of Che [132]. This last example utilizes an osmium complex with *N,N'*-bis-(2'-pyridinecarboxamide)-1,2-benzene ligands and iodosobenzene as stoichiometric oxidant. In this case, the monooxo osmium(IV) derivative epoxidizes cyclohexene; the dioxoosmium(VI) analog instead promotes allylic oxidation.

Finally we note that a single example of the "aza analog" of epoxidation has been reported [189]. Treatment of an acylimido manganese(V) intermediate with cyclooctene in Eq. 69 produced the corresponding aziridine in 90% yield.

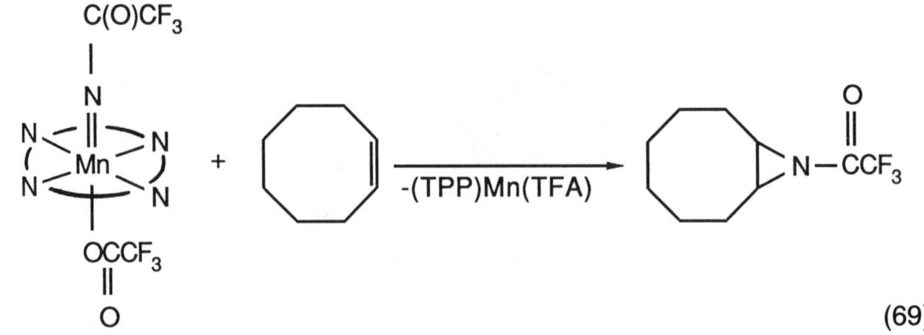

6.3.4.2 Cis Dihydroxylation and Related Reactions.

The synthetically useful cis dihydroxylation of alkenes to vicinal diols can be represented as in Eq. 70a, though we shall see this is a bit oversimplified. Both osmium tetroxide [190] and potassium permanganate [191] are commonly employed as the oxometal component in Eq. 70a. Evidence has also been presented [192] that ruthenium tetroxide also reacts with olefins to afford initially a metallate ester but this intermediate normally undergoes carbon–carbon bond cleavage according to Eq. 70b. The usual products are therefore aldehydes or ketones depending on the degree of substitution of the starting alkene [160,193]. A common feature in the reactions of OsO_4, RuO_4, and MnO_4^- is that trans olefins react faster than cis olefins [194]. On the other hand, reaction with permanganate is promoted by electron-withdrawing substituents on the olefin while reaction with OsO_4 proceeds most readily with electron-rich alkenes.

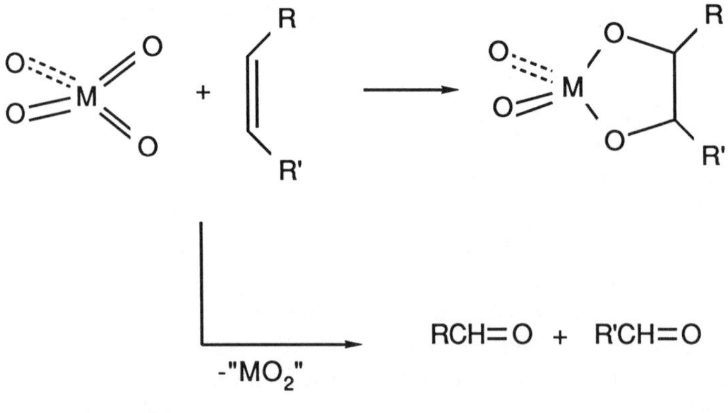

It has long been recognized [195] that the rate of reaction of osmium tetroxide with olefins is greatly increased in the presence of tertiary amines. The addition of pyridine in greater than stoichiometric ($>1:1$) amounts is especially effective, and under these conditions the kinetics are dominated by a term that is second order in pyridine [196]. It is also recognized that amines can coordinate to OsO_4 to form adducts $OsO_4 \cdot NR_3$; a 1:1 adduct of OsO_4 with quinuclidine has been characterized structurally [197]. Raman studies on OsO_4 in pyridine show only bands due to OsO_4, py, and $OsO_4 \cdot$py [198]. However, the isolated osmate esters from hydroxylations carried out in the presence of excess pyridine generally have two molecules of pyridine coordinated to osmium as in **3** [199]. Even when dihydroxylations are carried out in the absence of amine additives, the resultant osmate ester does not have the simple tetrahedral structure implied by Eq. 70. Instead, oxo-bridged dimers such as **4** are isolated [200].

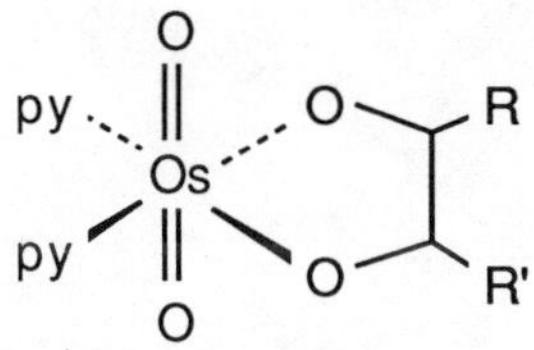

3 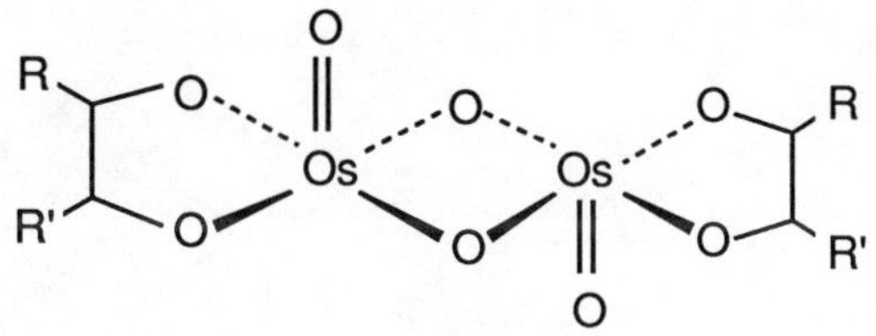   **4**

One explanation for these observations has been offered by Sharpless and coworkers [153,201]. As summarized in Eq. 71, coordination of one equivalent of amine ligand to an OsO_4-olefin complex triggers the formation of an oxametallacyclobutane intermediate. Attack of a second equivalent of amine then induces the formation of the final osmate ester. In connection with this work it was reported that OsO_4 in combination with an optically active amine could be used to convert alkenes to chiral diols, in some cases in >80% enantiomeric excess. These results could be rationalized in terms of Eq. 71.

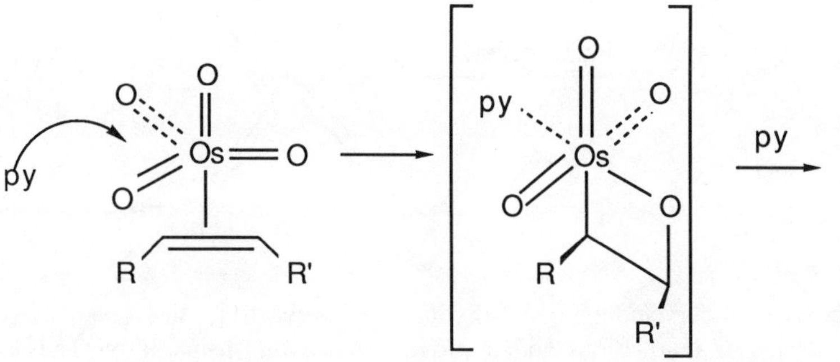

(71)

In the intervening years, several studies have appeared that provide some support for Eq. 71. These have included studies on the stereochemistry of oxidation of dienes by polyoxo manganese(VII) and chromium(VI) reagents [202,203]. Theoretical studies supporting the intervention of oxametallacyclobutane intermediates in many types of oxidations have appeared [182]. However, a report claiming the direct observation by NMR of such species [204] was promptly debunked [205]. (It is worthy of note that complexes containing an olefin coordinated to a d^0 metal oxo—or to any d^0 metal center—are unknown. However, oxotungsten(IV) complexes containing simple alkenes as ligands have been prepared [206].)

More recently Jorgenson and Hoffmann reported a frontier orbital study on the reaction of olefins with OsO_4 that supports a reasonable alternative mechanism [5]. They note that direct reaction of OsO_4 with an alkene would result in a highly disfavored tetrahedral d^2 complex. Accordingly, the role of the amine ligand is to generate an octahedral complex (Eq. 72) in which the equatorial oxygen atoms are both physically closer together and electronically activated. Both the HOMO and the LUMO of the adduct have the correct symmetry to interact with the π and π^* orbitals of an alkene; the reaction can be viewed as a concerted [3 + 2] cycloaddition.

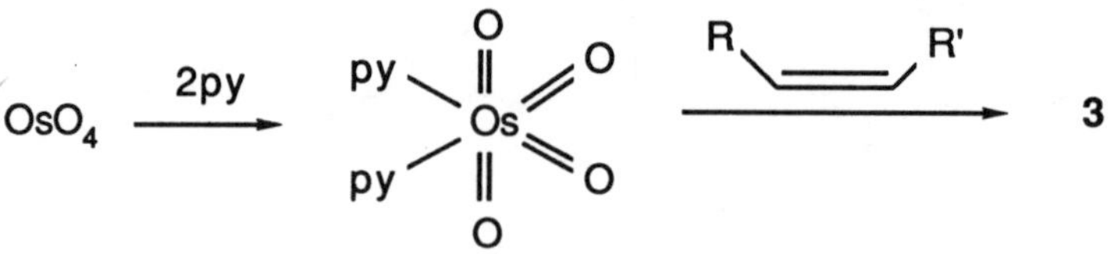

$$OsO_4 \xrightarrow{2py} \underset{py}{\overset{py}{\text{Os}}}(\text{O})_4 \xrightarrow{R\text{—}R'} 3 \qquad (72)$$

Providing support for this view, the rhenium(VII) complex Cp*ReO$_3$ has been shown to react with typical dienophiles to afford cyclic rhenium(V) products. For example, reaction with SO$_2$ affords an η^2–sulfate adduct [207] and diphenylketene reacts according to Eq. 73 [208]. The ready formation of these apparent [3 + 2] cycloadducts occurs despite the fact that oxorhenium complexes do not normally effect cis hydroxylation.

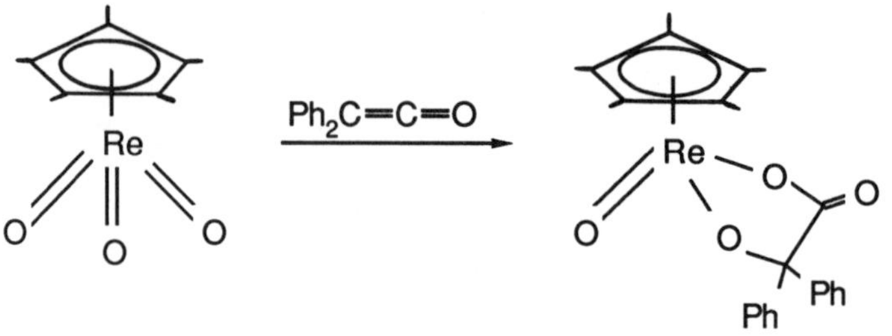

$$\text{Cp*ReO}_3 \xrightarrow{Ph_2C=C=O} \text{product} \qquad (73)$$

Our increased understanding of cis hydroxylation reactions has given rise to a remarkable series of innovations and improvements in this technology. The use of chiral diamine ligands like 5 and 6 in combination with OsO$_4$ now allows asymmetric cis hydroxylations of unsubstituted olefins with up to 90% enantiomeric excess [209,210]. Recognition of the importance of coordinating ligands in these reactions has led to significant improvements in the catalytic oxidation of olefins by ruthenium tetroxide [211]. It has also been realized that the reactivity of the permanganate anion is significantly enhanced in aprotic media where hydrogen bonding is eliminated. This first became evident in the phase-transfer studies of Starks [212] and was further underscored by Sam and Simmons who used crown ethers to solubilize KMnO$_4$ in benzene ("purple benzene") [213]. This observation has blossomed into a body of new chemistry which is described in a review aptly entitled "The Classical Permanganate Ion: Still a Novel Oxidant in Organic Chemistry" [214].

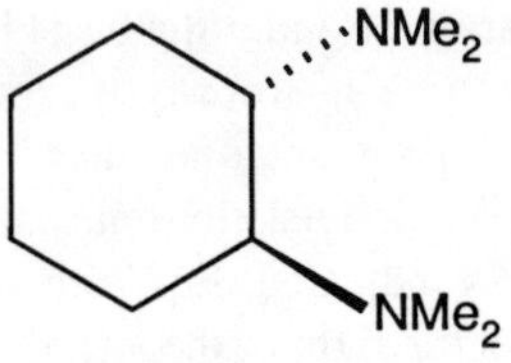

REACTIONS WITH NUCLEOPHILES 255

5

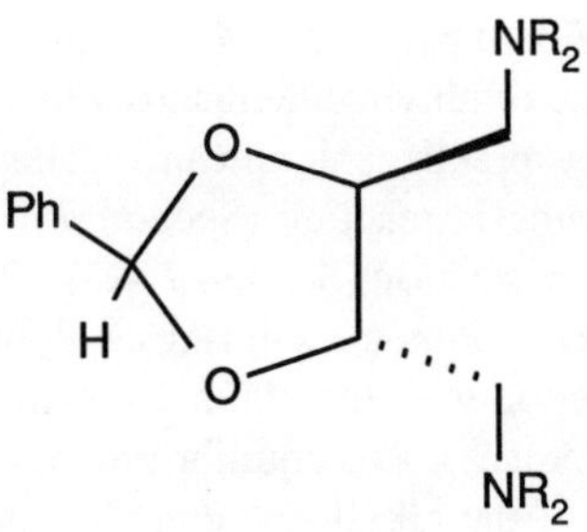

6

The development of the nitrogen analogs of osmium-mediated cis dihydroxylation is due to Sharpless and coworkers. In both the oxyamination reaction [9,215] (Eq. 74) and the diamination reaction [216] (Eq. 75), there is a pronounced tendency for C—N bond formation in preference to C—O bond formation. A frontier orbital explanation for this preference has been proposed [5]. Just as in the cis dihydroxylation reaction, Eq. 74 proceeds more readily with trans rather than cis olefins—and Eq. 75 occurs with trans olefins exclusively. Again, the oxyamination reaction is promoted by amine ligands, particularly quinuclidine [215], but the diamination reaction is apparently unaffected by strongly coordinating media. Of practical importance is the development of two efficient catalytic versions of the oxyamination reaction which use either chloramine-T or an N-chloro-N-argento carbamate as stoichiometric oxidant [217,218].

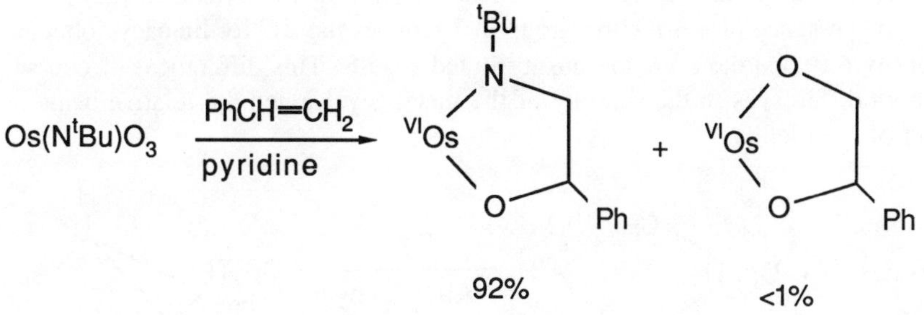

$$\text{Os}(N^t\text{Bu})O_3 \xrightarrow[\text{pyridine}]{\text{PhCH=CH}_2} \quad (74)$$

92%   <1%

$$\text{Os}(N^t\text{Bu})_2O_2 \xrightarrow[\text{CCl}_4]{\text{PhCH=CH}_2} \quad (75)$$

10%   73%

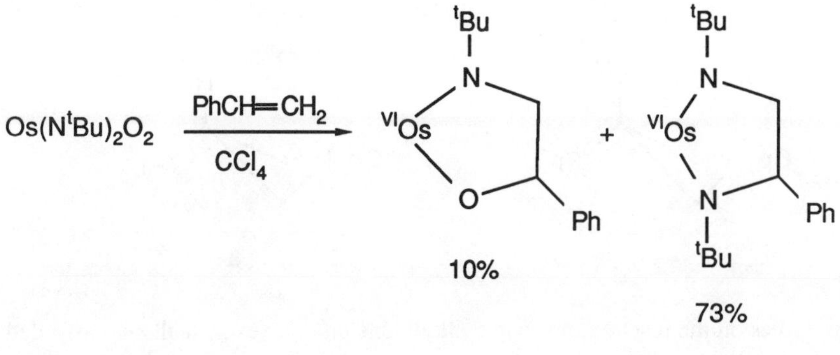

6.3.4.3 Formation of Metallacyclobutanes from Alkylidene Complexes.

The reaction of alkylidene ligands with olefins to afford metallacyclobutane complexes (Eq. 76) has practical significance. This transformation provides the basis for both the olefin metathesis reaction (Section 7.4.1) and the "Green–Rooney mechanism" for olefin polymerization (Section 7.4.3). (The Fischer carbene intermediates formed from organic diazo compounds during catalytic cyclopropanation reactions may also undergo such a reaction with the olefinic substrate [219].) As indicated, Eq. 76 is in general a reversible process. The equilibrium may lie either on the right or the left, depending on the nature of the alkylidene complex and the olefin.

$$R_2C = CR_2 + M = CR_2 \rightleftarrows \begin{array}{c} R_2C - CR_2 \\ | \quad | \\ M - CR2 \end{array} \quad (76)$$

R = H, alkyl, etc

The first such reaction to afford a discrete metallacyclobutane product was the reaction of the titanium methylidene complex in Eq. 77 with neohexene [76]. To effect this transformation it is necessary to precipitate the coordinated aluminum reagent by addition of a nitrogen base; 4-dimethylaminopyridine is especially effective in this regard. Strauss and Grubbs have reported that the structure of the olefinic component in reactions related to Eq. 77 has a significant impact on the thermodynamics [220]. For example, the presence of a β-methyl group in **7** reduces the ΔG for titanacyclobutane formation by 6.0 kcal/mole *vs.* the unsubstituted case **8**. This difference, of course, reflects both differences in the stability of the metallacycles and the relative heats of formation of the olefins.

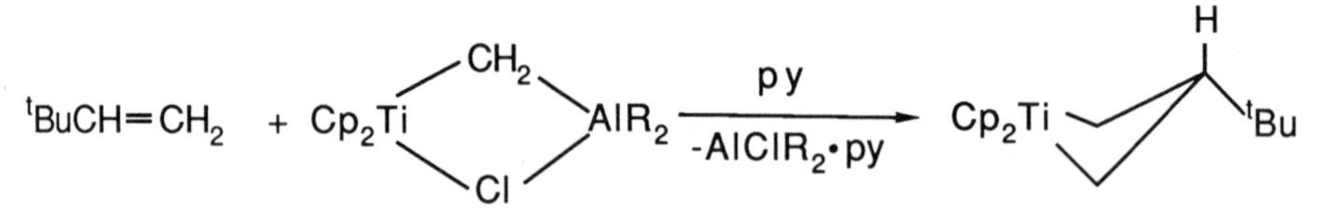

$$^tBuCH=CH_2 + Cp_2Ti\begin{array}{c}CH_2\\ \diagdown \\ Cl\end{array}AlR_2 \xrightarrow[-AlClR_2\cdot py]{py} Cp_2Ti\diagup\!\!\!\diagdown{}^tBu \quad (77)$$

7: $Cp_2Ti\diagup\!\!\!\diagdown{}^iPr$ with CH$_3$

8: $Cp_2Ti\diagup\!\!\!\diagdown{}^iPr$ with H

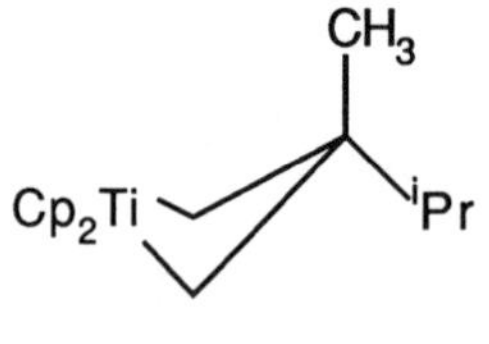

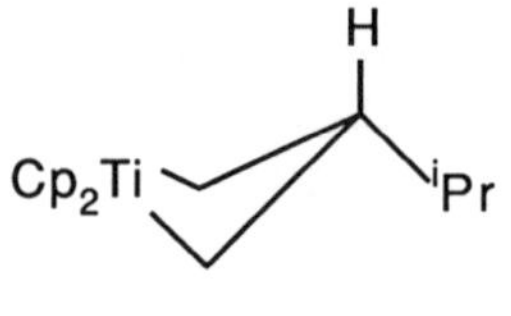

Early studies on the reaction of group 5 alkylidene complexes with alkenes afforded not metallacyclobutanes but a variety of rearrangement products derived therefrom. For example, the benzylidene ligand in Eq. 78 is recovered as allylbenzene, while the final tantalum product is a metallacyclopentane derived from two equivalents of ethylene [221]. A range of other alkylidene species was observed to react similarly, including niobium alkylidene derivatives [222] and tantalum bis(alkylidene) species [223]. In all cases the reaction is believed to proceed through a metallacyclobutane intermediate, which subsequently decomposes. (Interestingly, the metallacyclopentane products formed in this way are in several cases efficient catalysts for the tail-to-tail dimerization of alkenes [222,223].) The group 5 complexes which undergo such rearrangements invariably contained chloride ligands. Schrock suggested that such pathways would be disfavored by the use of hard, π donor ligands such as alkoxides [224]. This hypothesis not only proved correct but has provided the key for the design of highly active, Lewis acid-free olefin metathesis catalysts.

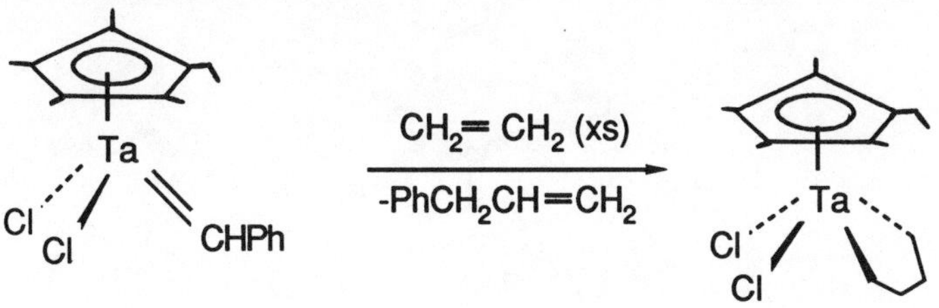

$$\text{Cp*TaCl}_2(\text{CHPh}) \xrightarrow[-\text{PhCH}_2\text{CH}=\text{CH}_2]{\text{CH}_2=\text{CH}_2 \text{ (xs)}} \text{Cp*TaCl}_2(\text{metallacyclopentane}) \quad (78)$$

Schrock has also suggested some factors that will promote Eq. 76 in general. The first is a low coordination number; the alkylidene complex should be either 4- or 5-coordinate. It is suggested that an olefin "almost certainly" must coordinate to the metal before a metallacyclobutane can form [224]. It is further proposed that the metal center should be electrophilic. Because of this factor it is suggested that in a given coordination environment tungsten will be be more reactive than molybdenum. The use of perfluoroalkoxide ligands has been proposed as one way to enhance the electrophilicity of a metal center. These factors have been harnessed by Schrock and coworkers to convert a range of alkylidene derivatives to stable metallacyclobutanes, as exemplified by Eqs. 79 and 80 [225,226]. In both cases "Ar" is 2,6-diisopropylphenyl. The structure of each product has been determined by X-ray crystallography.

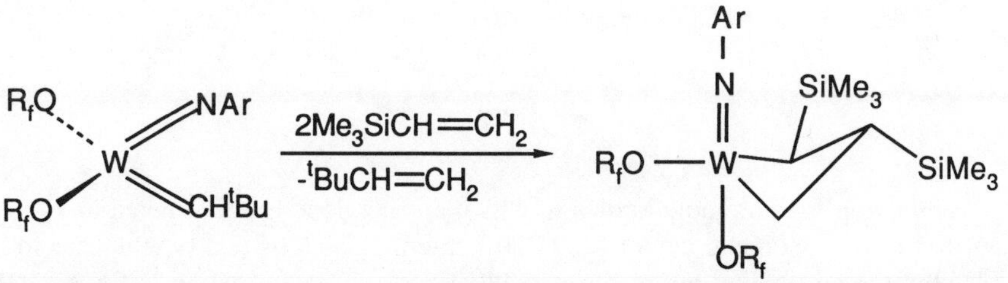

$$\text{W}(NAr)(CH^tBu)(OR_f)(R_fO) \xrightarrow[-^tBuCH=CH_2]{2Me_3SiCH=CH_2} \text{W metallacyclobutane with SiMe}_3 \text{ groups}$$

$R_f = -CMe(CF_3)_2$ \quad (79)

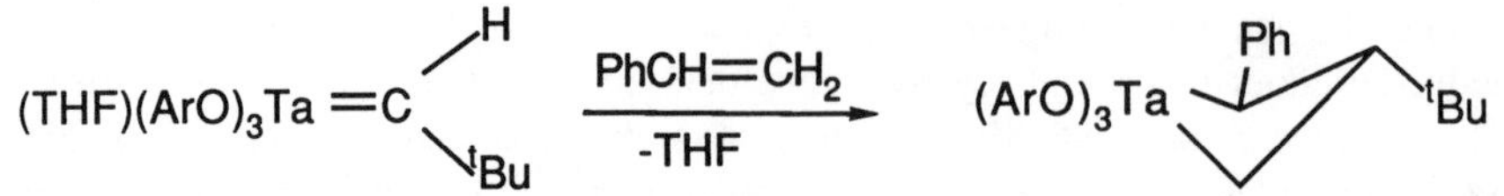

$$(THF)(ArO)_3Ta=C\begin{matrix}H\\ {}^tBu\end{matrix} \xrightarrow[-THF]{PhCH=CH_2} (ArO)_3Ta\underset{}{\overset{Ph}{\triangle}}{}^tBu \quad (80)$$

Theoretical studies on Eq. 76 have been reported. Goddard and Rappe have carried out *ab initio* calculations on the interaction of ethylene with model compounds $MO(CH_2)Cl_2$ where M = Cr, Mo, or W. They find that formation of a metallacyclobutane is energetically favorable in all cases with a ΔG of −18 to −24 kcal mole^{-1} [182]. Eisenstein and Hoffmann have stressed that for Eq. 76 to be facile it is crucial that the orientation of the carbene and olefin be as shown in structure **9** [313].

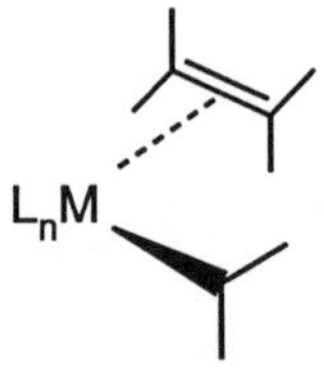

9

6.3.5 Reactions with Acetylenes

The simple tetraoxo species OsO_4, RuO_4, and MnO_4^- all react with alkynes [190, 193, 214]. With any of these oxidants it is possible, by careful control of reaction conditions, to isolate the α-diketone as the principal organic product. However, only in the case of osmium have the metal-containing products of such a reaction been isolated and characterized. In the presence of amine bases (L = pyridine or isoquinoline) a variety of acetylenes were found to react with two equivalents of osmium tetroxide to afford diamagnetic osmium(VI) adducts as shown in Eq. 81 [228].

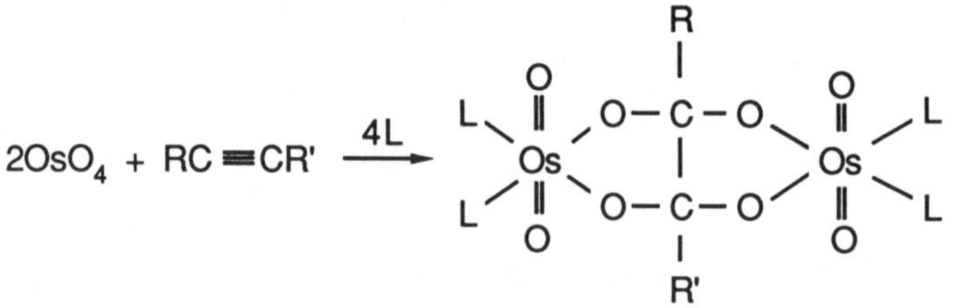

$$2OsO_4 + RC\equiv CR' \xrightarrow{4L} \text{[bis-osmium bridged diolate complex]} \quad (81)$$

Several complexes are now known in which an acetylene is coordinated to a lower valent (d^2 or d^4) oxometal center [229,230]. Intermediates of this type appear to be involved in the fascinating but mechanistically obscure transformation in Eq. 82 [69]. The identity of the product "rhenapyran" has been confirmed crystallographically.

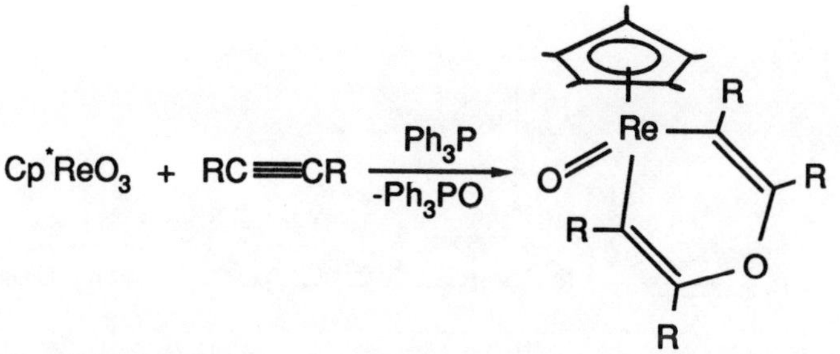

$$\text{Cp}^*\text{ReO}_3 + \text{RC}\equiv\text{CR} \xrightarrow[-\text{Ph}_3\text{PO}]{\text{Ph}_3\text{P}} \quad \text{[structure]} \quad (82)$$

Reactions of alkylidene complexes with acetylenes to afford metallacylobutenes have been reported. When the alkylidene is generated by cleavage of a titanacylobutane as in Eq. 83, the reaction can in some cases be first order in titanium complex and zeroth order in acetylene [94]. The products have been characterized structurally in cases where R = phenyl or trimethylsilyl [231,100]. The reaction of the tantalum neopentylidene complex in Eq. 84 with diphenylacetylene is likewise thought to proceed through a metallacyclobutene intermediate [232]. Equation 84 and also the recent characterization of an alkylidene complex containing a coordinated acetylene [233] appear to be relevant to the mechanism of coordination polymerization of acetylenes (Section 7.4.2).

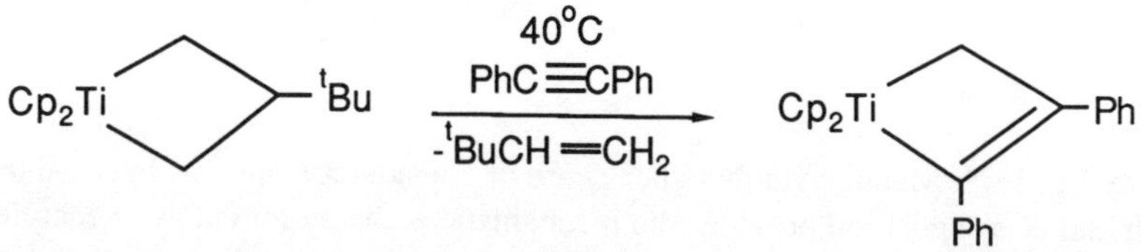

$$\text{Cp}_2\text{Ti}\text{—}^t\text{Bu} \xrightarrow[-^t\text{BuCH}=\text{CH}_2]{\substack{40°\text{C} \\ \text{PhC}\equiv\text{CPh}}} \text{Cp}_2\text{Ti}\text{—Ph, Ph} \quad (83)$$

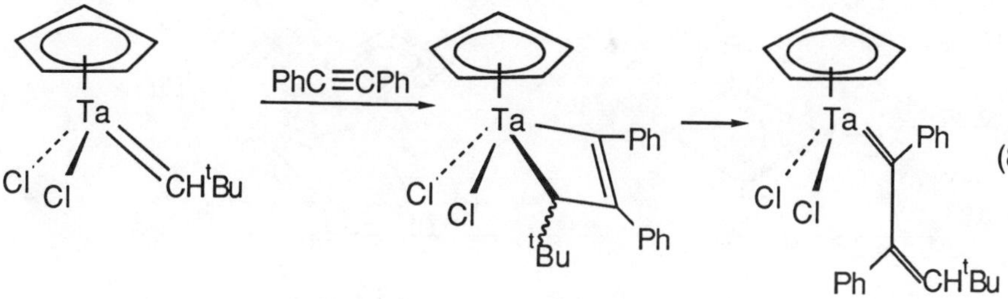

$$\text{[Ta complex]} \xrightarrow{\text{PhC}\equiv\text{CPh}} \text{[Ta metallacycle]} \longrightarrow \text{[Ta product]} \quad (84)$$

The reaction of an acetylene with an alkylidyne ligand can lead to a fascinating array of organometallic products as shown schematically in Figure 6.4. However, this diversity of products can be rationalized by a pathway commencing with formation of a metallacyclobutadiene via the equilibrium **A** ⇌ **B**. Metallacyclobutadienes may exhibit either a planar, delocalized structure exemplified by **10** or the puckered, localized structure **11**. The structure of several trigonal bipyramidal tungstacyclobutadienes corresponding to **10**, where X = chloride, hexafluoroisopropoxide, or 2,6-diisopropylphenoxide, have been determined crystallographically [234–236], as has the structure

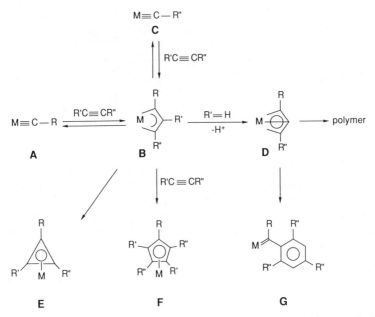

Figure 6.4 Known pathways for the reaction of acetylenes with transition metal alkylidene complexes.

of complex **11** [237]. Metallacyclobutadienes are of considerable interest because their formation and decomposition provides the mechanistic pathway for catalytic acetylene metathesis (Section 7.5).

Schrock and coworkers have identified several factors that can profoundly influence the position of equilibrium **A** ⇌ **B** (e.g. Eq. 85). In particular, the equilibrium will be shifted to the left (the metallacycle will be destabilized) by:

$$X_3W\equiv CEt\ +\ EtC\equiv CEt\ \rightleftharpoons\ X_3W\underset{Et}{\overset{Et}{\diamond}}Et \quad (85)$$

1. Replacing the metal with a less electrophilic (less Lewis acidic) one. For example, when X is the 2,6-diisopropylphenoxide ligand, the equilibrium in Eq. 85 heavily favors the tungstacyclobutadiene [236]. However, the corresponding molybdacyclobutadiene is essentially fully dissociated in solution [58].
2. Decreasing the electron-withdrawing power of the ancillary ligands. As one illustration, when X is hexafluoro-*t*-butoxide, the equilibrium in Eq. 85 overwhelmingly favors the tungstacycle [235]. Yet tungstacyclobutadienes are never observed in reactions of this type when X is *t*-butoxide [238].
3. Increasing the steric bulk of the ancillary ligands. For example, kinetic evidence suggests that cleavage of the the metallacyclobutadiene from Eq. 85 is essentially shut down when X is the relatively small hexafluoroisopropoxide ligand, inhibiting the normal "dissociative" mechanism for acetylene metathesis. Replacing the ancillary ligands with bulkier hexafluoro-*t*-butoxide allows normal acetylene methathesis via the reverse of Eq. 85 [235].
4. Addition of potentially coordinating ligands to the system. This is exemplified by the addition of dimethoxyethane to the system of Eq. 85 when X is hexafluoro-*t*-butoxide. The only tungsten species that can be isolated under these circumstances is the alkylidyne complex $W(CEt)(OR_f)_3(dme)$ [235].

If the metallacyclobutadiene **B** in Figure 6.4 comes apart in the opposite sense, it affords the stoichiometric metathesis product **C**. Normally one cannot apply this approach to cleanly convert one alkylidyne complex into another; the ready reversibility of $A \rightleftharpoons B \rightleftharpoons C$ demands that a grand mixture of all possible alkylidynes and acetylenes be generated. However, in selected cases such a transformation is possible. An example is Eq. 86 in which an existing alkylidyne ligand is replaced by an alkylidyne moiety containing a stablizing group R. Some stabilizing groups that have been used successfully to promote such stoichiometric metathesis reactions are phenyl, Me_3Si, NEt_2, CH_2CN, $C \equiv CEt$, and CN [58,238].

$$(^tBuO)_3W \equiv CMe + MeC \equiv CR \xrightarrow[-MeC \equiv CMe]{L} \begin{array}{c} R \\ | \\ C \\ ||| \\ ^tBuO - W \cdots O^tBu \\ | \quad \diagdown O^tBu \\ L \end{array}$$

L = pyridine, quinuclidine   (86)

When R' in Figure 6.4 is hydrogen, the product metallacyclobutadiene **B** will have a hydrogen atom in the β position which has been shown to be quite acidic. This proton can be removed with a mild base such as a trialkylamine. In the absence of basic additive, a portion of the metallacycle functions as an organometallic base and deprotonates the remainder. In either case the product is the "deprotio-metallacyclobutadiene" **D**. Examples of both molybdenum and tungsten complexes of this type have been structurally characterized [58,239]. Moreover, this intermediate apparently can react with additional acetylene to afford polymers or, under certain conditions, the oligomeric products **G** [240]. (Although **G** is drawn as a simple benzylidyne derivative for simplicity, an X-ray crystal structure in one case has shown it to possess a remarkable η^4-arene structure. Schrock has suggested that the simple benzylidyne species "probably would not exist as such.") Reactions of this type are invoked to explain the failure of catalytic metathesis of terminal acetylenes [241].

Rearrangement of the metallacyclobutadiene to the η^3-cyclopropenyl derivative **E** appears to be important when the coordination number of the complex exceeds five. For example, the structurally characterized analog $W[C_3(^tBu)Et_2](O_2CCH_3)_3$ contains three bidentate acetate ligands. If the cyclopropenyl ligand is considered to occupy a single coordination site, the coordination geometry in this species is very approximately pentagonal-bipyramidal [242]. It is conceivable that in some cases the cyclopropenyl derivative could form directly from an alkylidyne and an acetylene without the intermediacy of the metallacycle. However, in several instances it can be shown that adding strongly coordinating ligands to the metallacyclobutadiene does cause rearrangement to the cyclopropenyl species, for instance Eq. 87 [243].

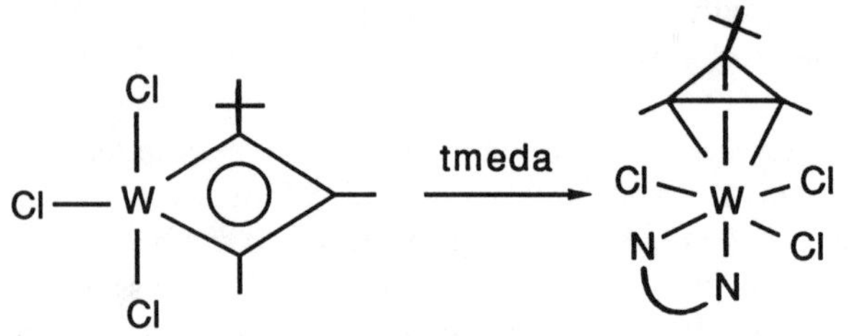

$$\text{(87)}$$

The final pathway in Figure 6.4, that resulting in η^5-cyclopentadienyl products **F**, requires the presence of excess acetylene. One transformation of this type that affords a structurally characterized product is shown in Eq. 88 [234].

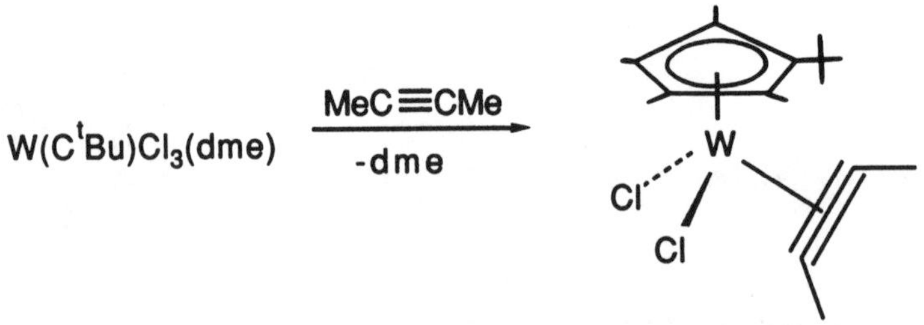

$$W(C^tBu)Cl_3(dme) \xrightarrow[-dme]{MeC \equiv CMe} \quad \text{(88)}$$

6.3.6 Reaction with Carbon Monoxide

A classical synthesis of $Os_3(CO)_{12}$ involves the treatment of a xylene solution of OsO_4 with carbon monoxide (128 bar, 175°C) [244]. However, nothing is known about the intermediates in this presumably complex and possibly heterogeneous reaction. An apparently simpler case has been described by Herrmann and coworkers [245]. Several $Cp^*Re(O)X_2$ derivatives react with CO under mild conditions (25°C, 50 bar) to afford the corresponding carbonyls as exemplified by Eq. 89.

REACTIONS WITH NUCLEOPHILES   263

$$\text{Cp*Re(O)Cl}_2 \xrightarrow[-CO_2]{3CO} \text{Cp*Re(CO)}_2\text{Cl} \quad (89)$$

Mayr found that treatment of the tungsten vinylalkylidene complex in Eq. 90 with sodium diethyldithiocarbamate ultimately affords the indicated vinylketene complex, which has been structurally characterized [87]. In fact, an unstable intermediate [W(C(Ph)CHCHMe)(dtc)$_2$(CO)$_2$] could be isolated which readily coupled to give the final product at 50°C.

$$\text{Br}_2\text{W(L)(CO)}_2(=C(Ph)CHCHMe) \xrightarrow[thf]{2\ Na(dtc)} \text{W(dtc)}_2(CO)_2(\text{vinylketene}) \quad (90)$$

L = 4-picoline

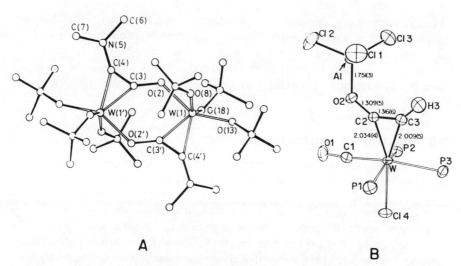

A   B

Figure 6.5 Two products derived from the reaction of tungsten alkylidynes with carbon monoxide: **(A)** structure of [(^{t}BuO)$_3$W (μ-OCCNMe$_2$)]$_2$ (from ref. 246 with permission of the Royal Society of Chemistry); **(B)** structure of W(η^2-HC≡COAlCl$_3$)(CO)(PMe$_3$)$_3$Cl (reprinted with permission from M.R. Churchill et al., *Organometallics* **1982**, *1*, 767; Copyright 1982 American Chemical Society).

There are two reported examples of coupling of an alkylidyne ligand with carbon monoxide. The first involves the tungsten(VI) complex (^{t}BuO)$_3$W(CNMe$_2$) and results in a dimeric product [(^{t}BuO)$_3$W(μ-OCCNMe$_2$)]$_2$ [246]. Its molecular structure is shown in Figure 6.5. Also shown in Figure 6.5 is the product from the reaction of W(CH)(PMe$_3$)$_4$Cl with CO in the presence of AlCl$_3$ [55,247]. In each case the product is that expected from attack of CO on the W–C triple bond consistent with the conceptual model proposed in Section 6.1. Another similarity between these two reactions is indirect evidence that each requires Lewis acid promotion. Addition of pyridine in the first example or omitting the AlCl$_3$ in the second case completely stops the coupling reaction.

6.3.7 Reactions with M≡M Triple Bonds

The metal–metal multiple bond in complexes such as (RO)$_3$M≡M(OR)$_3$, M = Mo or W, is an electron-rich unsaturation that in some respects resembles the C≡C bond of an acetylene [248]. Chisholm and coworkers have studied the addition of complexes containing multiply bonded ligands to such species. An early example involved the addition of (i-PrO)$_4$W(O) to afford trinuclear μ_3 oxo clusters according to Eq. 91 [249]. Attempts to replace the oxometal reactant in Eq. 91 with an organoimido complex were unsuccessful [249]. However, the group 6 nitrides (t-BuO)$_3$M≡N, M = Mo or W, do apparently undergo addition reactions analogous to Eq. 91 [250,251].

$$(^iPrO)_3Mo \equiv Mo(O^iPr)_3 \ + \ (^iPrO)_4W \equiv O \longrightarrow Mo_2W(O)(O^iPr)_{10} \quad (91)$$

More recently, it was shown that the alkylidyne complex (t-BuO)$_3$W(≡CMe) will react analogously to afford triangulo clusters M$_3$(CMe)(OR)$_9$ containing a μ_3-alkylidyne moiety [252]. Although it lies somewhat outside the scope of this book, the interested reader is also referred to an extensive body of work by Stone and coworkers concerning the addition of Fischer-type carbenes to metal–metal multiple bonds [253].

6.3.8 Reaction with Hydrocarbons

The C—H bonds of saturated hydrocarbons can react with some high-valent oxometal species to afford oxygenated organic products. Ruthenium tetroxide will slowly oxidize cycloalkanes [254–256] as will oxochromium(VI) reagents and permanganate under a variety of conditions [191,214]. Trifluoroacetic acid is a particularly effective solvent for the oxidation of hydrocarbons with permanganate [257]. Some oxoiron [178,258] and oxomanganese [176,259] containing porphyrins will also hydroxylate alkanes. Suggs and Ytuarte [260] have shown that chromyl trifluoroacetate is remarkably reactive toward alkanes. Reaction with n-hexane is complete in seconds at −50°C!

Despite the fairly extensive literature on the organic aspects of these reactions, few mechanistic details are known. It is, however, becoming evident that a mechanistic dichotomy exists. On the one hand there is evidence in some cases for a bona fide electrophilic attack on the carbon–hydrogen bond, formally tantamount to an "oxenoid" insertion. On the other hand, other cases are better understood in terms of a stepwise mechanism involving homolysis of the C—H bond followed by a separate step in which the hydrocarbon radical undergoes C—O bond formation.

The most compelling evidence supporting some type of direct insertion into the C—H bond of alkanes is based on stereochemical studies. Equation 92 is one of several examples when oxidation of a tertiary center in an alkane has been shown to proceed with at least partial retention of configuration at the 3° carbon [261–263]. (Biological hydroxylations involving P-450 enzymes are also highly stereoselective and at one time were thought to be examples of direct electrophilic attack. More recent experimental evidence indicates that these are stepwise processes but that constraints provided by the enzyme cavity prevent racemization of the intermediate alkyl radical; see Section 7.1.3.)

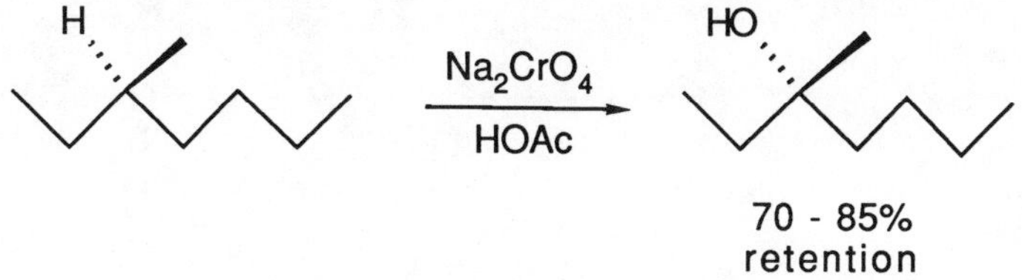

$$\text{(92)}$$

70 - 85%
retention

Observations such as Eq. 92 led Sharpless to propose a mechanism involving addition of the C—H bond across the metal–oxygen multiple bond to give an organometallic intermediate [153]. The C—O bond then resulted from a subsequent 1,2-shift as shown in Eq. 93. No evidence either rejecting or supporting this proposal has been forthcoming. However, in considering the mechanism of alkane hydroxylation, it should be recalled that other "oxenoid" reagents are known that lack a transition metal–oxygen multiple bond but can still effect stereospecific hydroxylation. For example, the reaction of *cis*- and *trans*-1,2-dimethyl cyclohexanes with trifluoroperacetic acid affords the corresponding alcohols with 100% retention of configuration [264].

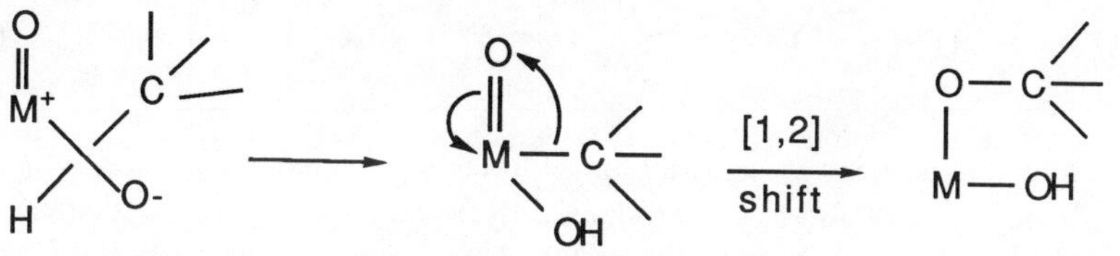

$$\text{(93)}$$

Evidence supporting the alternative two-step homolytic mechanism in Eq. 94 has been summarized elsewhere [191,214]. Particularly careful studies on alkane oxidation by oxomanganese(V) porphyrins clearly support such a mechanism [176,259]. It is worth noting that d^0 oxometals in general appear to be much more efficient at effecting Eq. 94b than Eq. 94a. This is entirely reasonable; the oxidation potential of organic radicals is considerably lower than that of the corresponding hydrocarbons. Consequently, the rate of oxidation of alkyl radicals, for example by permanganate ion, is diffusion controlled [265]. There are indications in some systems that the agent responsible for Eq. 94a is not the starting d^0 oxometal complex but, for example, some paramagnetic intermediate derived from its partial decomposition. To cite one example, chromyl chloride is reported to be stable in solvent methylcyclohexane. However, upon addition of a small amount of olefin (1% relative to $CrCl_2O_2$) oxidation of the alkane proceeds to completion [314].

$$R^2-\underset{R^3}{\underset{|}{\overset{R^1}{\overset{|}{C}}}}-H \quad + \quad Rad\cdot \quad \longrightarrow \quad R^2-\underset{R^3}{\underset{|}{\overset{R^1}{\overset{|}{C}}}}\cdot \quad \quad (94a)$$

$$R^2-\underset{R^3}{\underset{|}{\overset{R^1}{\overset{|}{C}}}}\cdot \quad + \quad ^nM=O \quad \longrightarrow \quad ^{(n-1)}M-O-\underset{R^3}{\underset{|}{\overset{R^1}{\overset{|}{C}}}}-R^2 \quad (94b)$$

The reactions of hydrocarbons containing allylic or benzylic hydrogen with oxo–metal reagents are considerably more facile than those involving alkanes. Indeed, selective allylic oxidation using oxochromium(VI) reagents is an important transformation in organic synthesis [266]. Even the oxoruthenium(IV) cation $[(trpy)(bpy)Ru(O)]^{2+}$, which is unreactive toward cyclohexane, will oxidize toluene to benzoate anion [267]. A limitation in using highly oxidizing oxometal species for benzylic oxidations is competitive electrophilic attack on the arene ring. Ruthenium tetroxide completely oxidizes arenes as shown in Eq. 95; in some cases the reaction can be quite violent [160,211]. Similar reactions occur with permanganate in highly acidic media where the very electrophilic MnO_3^+ ion is said to be formed [257].

$$R-C_6H_5 \quad \xrightarrow[-CO_2, H_2O]{RuO_4} \quad RCO_2H \quad (95)$$

An alternative two-electron pathway for hydrocarbon oxidation has been delineated for the benzylic oxidation of aromatic hydrocarbons such as cumene by $[Ru(trpy)(bipy)(O)]^{2+}$. In this case, although hydride is apparently transferred from the hydrocarbon to the oxo ligand, C—O bond formation proceeds exclusively by external nucleophilic attack of solvent water [268]. Meyer has described this process as "solvent assisted hydride transfer." It would be interesting to know the stereochemistry of this process.

The nitrogen analog of alkane hydroxylation is known. Breslow has reported that treatment of cyclohexane with (tosyliminoiodo)benzene, TsN=IPh, in the presence of manganese(III)- or iron(III)-tetraphenylporphyrin chloride affords N-cyclohexyltoluene-p-sulfonamide in amounts reflecting several catalytic turnovers [269]. It is conceivable that such reactions involve intermediate imido species as shown in Eq. 96 by analogy to P-450 model systems. Subsequent efforts to carry out this chemistry using purified P-450 isozyme as the metalloporphyrin component instead afforded cyclohexanol as the organic product [270]. More recently it has been shown that simple metal salts such as iron(II) chloride and rhodium(II) acetate could also promote apparent "nitrenoid" insertions into C—H bonds [271,272]. A possibly related reaction is the apparent intramolecular example shown in Eq. 97. Collins has suggested that this reaction may involve a diimido-osmium(VIII) intermediate wherein one phenylimido nitrogen atom inserts into an *ortho* C—H bond of the other [273].

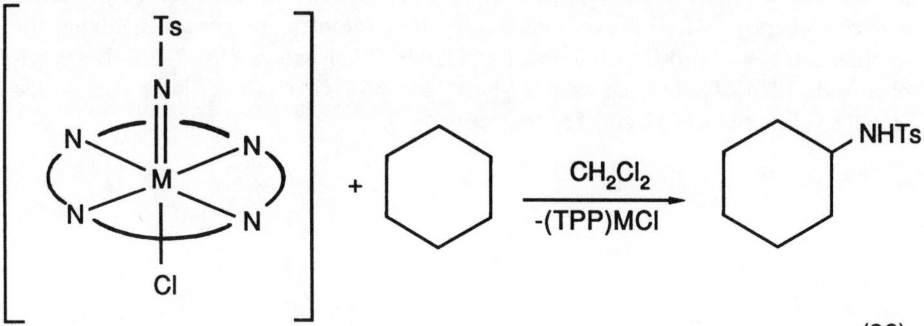

(96)

M = Mn, Fe; Ts = $SO_2C_6H_4Me$

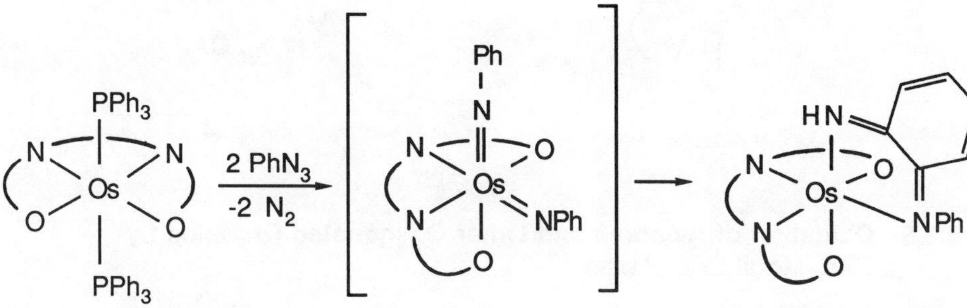

(97)

O⌒N⌒N⌒O = o-phenylenediaminebis(o-hydroxybenzamide) (4-)

A reaction that assuredly proceeds by a two-step mechanism is the oxidation of toluene to benzylidene-*t*-butylamine shown in Eq. 98. Nugent and Chan [85] showed that this reaction proceeds only in the presence of a stoichiometric source of radicals such as dibenzoyl peroxide (DBPO in Eq. 98). Initiator-derived phenyl radicals readily abstract benzylic hydrogen from toluene; the resultant benzyl radicals are efficiently trapped by the chromium(VI) imido reagent. The stoichiometry of the reaction suggests that the resulting chromium(V) amide may be further oxidized in a reaction requiring a second equivalent of chromium(VI) reagent.

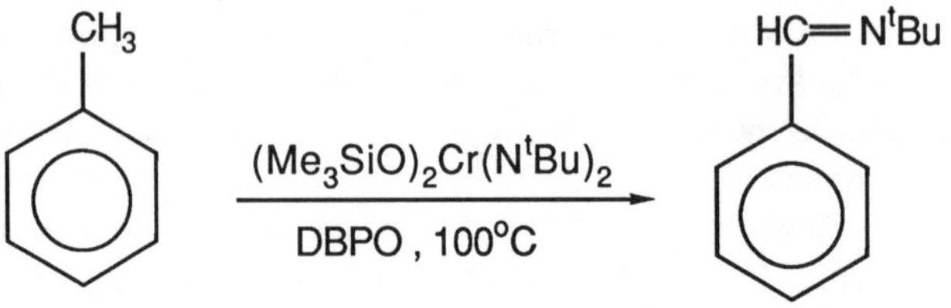

$$\text{toluene} \xrightarrow[\text{DBPO, 100°C}]{(Me_3SiO)_2Cr(N^tBu)_2} \text{PhCH=N}^t\text{Bu} \quad (98)$$

Intermolecular insertion of an early transition metal alkylidene or alkylidyne into a hydrocarbon C—H bond has yet to be demonstrated. Perhaps the closest precedent is the addition of H_2 across the W≡C bond of $W(C^tBu)Cl_3(PMe_3)_2$ reported by Schrock and coworkers [274]. However, intramolecular examples are known involving the intermediates $W(C^tBu)(OAr)_2Cl$ and $Ta(CH_2)Me(OAr)_2$ where OAr = 2,6-di-*t*-butylphenoxide [236,51]. In each case a *t*-butyl group of OAr is metallated to give the metallacyclic products **12** and **13**, respectively.

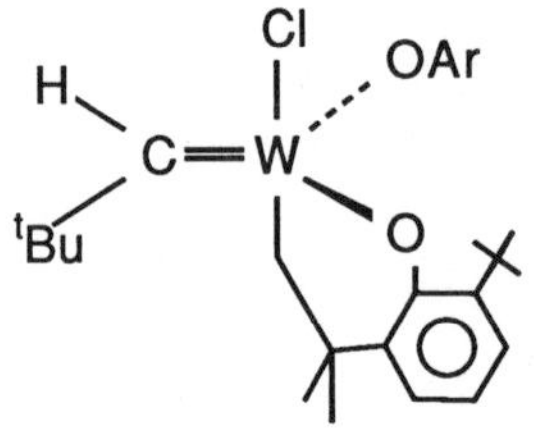

12

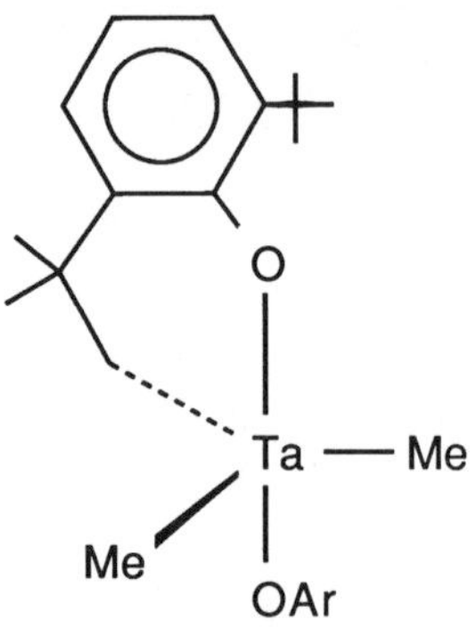

13

6.3.9 Oxidation of Alcohols and Other Oxygenated Organics by Oxo–Metal Complexes

Secondary alcohols react with a variety of d^0 oxometal reagents including RuO_4, MnO_4^-, and oxochromium(VI) to afford the corresponding ketone (Eq. 99). Oxochromium(VI) and (acidic or basic) permanganate are also useful for oxidation of primary alcohols to aldehydes (Eq. 100). Some longstanding problems in these reactions have included insufficient selectivity for certain applications and the requirement for large excesses of oxidant. These difficulties are partially circumvented by the introduction of two "improved" oxochromium(VI) reagents pyridinium chlorochromate, $(PyH)(CrO_3Cl)$, and pyridinium dichromate, $(PyH)_2(Cr_2O_7)$ [275,276]. These reagents, "PCC" and "PDC," are now used frequently for alcohol oxidation in organic synthesis, although perhaps not as often as the alternative "Swern oxidation." Another potentially interesting reagent, the easily prepared and organic-soluble triphenylsilyl chromate has received little study to date [277]. (From the standpoint of *inor-*

ganic synthesis, we can also note that reduction of OsO$_4$ with ethanol provides the favored route to potassium osmate K$_2$[Os(O)$_2$(OH)$_4$] [278].)

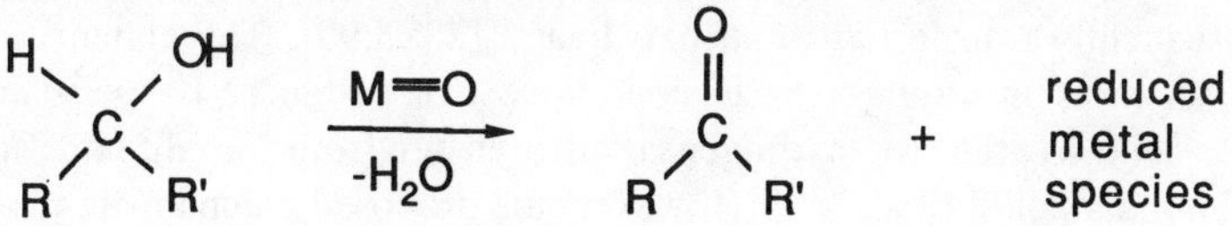

$$\underset{R}{\overset{H}{\diagdown}}\underset{R'}{\overset{OH}{\diagup}}C \xrightarrow[-H_2O]{M=O} \underset{R}{\overset{O}{\diagdown}}\underset{R'}{\overset{\|}{\diagup}}C \; + \; \text{reduced metal species} \qquad (99)$$

$$RCH_2OH \xrightarrow[-H_2O]{M=O} RCH=O \; + \; \text{reduced metal species} \qquad (100)$$

Ruthenium tetroxide and neutral permanganate cannot normally be used for the oxidation of primary alcohols to aldehydes; with these oxidants, further oxidation of the aldehyde product to the carboxylic acid is rapid relative to Eq. 100. Indeed even the oxidants that are useful for Eqs. 99 and 100 appear to oxidize aldehydes at some finite rate. However, an early report of aldehyde oxidation by weakly oxidizing oxomolybdenum(VI) reagents such as *cis*-MoO$_2$(ethyl-L-cysteinate)$_2$ [279] could be reproduced in subsequent studies only upon photochemical irradiation [145,280].

Some pioneering investigations have provided a degree of mechanistic insight into alcohol oxidations using "simple" oxidants such as permanganate and oxochromium(VI) [191,281]. Such reactions invariably show k_H/k_D isotope effects significantly

TABLE 6.3 Summary of Kinetic Studies on Oxidation of Alcohols by Oxometal Complexes

Oxidant	Substrate	$\Delta H^\ddagger$ (kcal/mol)	$\Delta S^\ddagger$ (eu)	k_H/k_D[a]	Ref.
[(trpy)(bipy)Ru(O)]$^{2+}$	^{i}PrOH	9	−34	5.2	282
[(bipy)$_2$(py)Ru(O)]$^{2+}$	PhCH$_2$OH	5.7	−38	50	283
[MnO$_4$]$^-$ [b]	PhCH(OH)Ph	5.7	−38	6.6	284
[MnO$_4$]$^-$ [b]	PhCH(OH)CF$_3$	9.1	−24	16	285
RuO$_4$	^{i}PrOH			4.6	286
[CrO$_4$]$^{2-}$	^{i}PrOH			~7[c]	287[d]
[CrO$_4$]$^{2-}$	cyclo-C$_4$H$_7$OH			8.9	288

[a] k_H/k_D for the α C—H bond.
[b] In basic solution.
[c] Estimated from results on 55% deuterated alcohol.
[d] See also ref. 289.

greater than unity upon substituting α C—H with deuterium (see Table 6.3). In the case of oxochromium(VI) and oxovanadium(V) there is kinetic evidence for the formation of discrete, inner-sphere metal–alcoholate complexes wherein hydride (or hydrogen atom) is subsequently transferred to an oxo ligand [288,290]. The enhanced rate of oxidation of alcohols in strongly basic conditions (for instance, by permanganate) reflects more facile hydride (or hydrogen atom) transfer from the alkoxide anion as compared with the alcohol [284,285]. However, the proposed mechanistic models are generally unsatisfying to the inorganic chemist. One reason is that there is virtually no information on the nature and structure of the metal-containing intermediates formed during the course of the reaction. Moreover, these partially reduced metal species often participate in (and occasionally dominate) the oxidation of the organic substrate [291,288]. Fortunately, recent work with monooxoruthenium oxidants is providing some insight into the fundamental mechanistic steps by which alcohol oxidations can take place.

Polypyridyl oxoruthenium(IV) complexes such as $[(bpy)_2(py)Ru(O)]^{2+}$ are nearly ideal oxidants for mechanistic studies. The complexes and their one- and two-electron reduction products are well characterized spectroscopically and electrochemically. These compounds are substitutionally inert; even exchange of the oxo ligand with solvent water does not occur at any significant rate. Meyer has shown [282,283] that these compounds will oxidize a variety of alcohols at rates increasing along the series methyl < primary < secondary < allylic ~ benzylic. The reactions are first order in alcohol and oxidant; upon substituting C—H by deuterium, large isotope effects are observed including a remarkable $k_H/k_D = 50 \pm 3$ in the case of benzyl alcohol (see Table 6.3). Most significantly, stopped-flow spectroscopic studies unambiguously show that a two-electron (hydride transfer or C—H insertion) mechanism is operant.

In these studies, a ruthenium(III) intermediate, $[(bpy)_2(py)Ru(OH)]^{2+}$, is formed by synproportionation of the ruthenium(II) product $[(bpy)_2(py)Ru(OH_2)]^{2+}$ with the ruthenium(IV) starting material. This ruthenium(III) species can also oxidize alcohols. However, the rate constant for this competing one-electron pathway is typically three orders of magnitude slower, making it unimportant under most conditions. It is interesting to note that changes in the ligation of oxoruthenium(IV) can divert the reaction toward formation of stable ruthenium(III) alkoxides [292] or slow the reaction so that further oxidation to ruthenium(V) is required before alcohol oxidation can take place [293].

6.4 "α CLEAVAGE" REACTIONS

In Section 3.2 we have already discussed the cleavage of the α bond of imido, NX, and alkylidene ligands as a route to the corresponding nitrido and alkylidyne complexes. We have also noted that, under appropriate conditions, a hydrazido complex underwent N—N bond cleavage to produce an imido complex. In this section we will confine our comments to some examples from the hydrazido literature in which N—N bond cleavage occurs as indicated by ammonia formation but wherein no isolable imido or nitrido complexes have been identified. The related topic of enzymatic nitrogen fixation will be discussed in Section 7.3.

TABLE 6.4 Yields of Ammonia from N–N Cleavage in Hydrazido(2−) Complexes

Complex[b]	Reagent	Solvent	Yield[a] NH_3	Yield[a] N_2H_4	Yield[a] N_2	Ref.
$W(NNH_2)I_2L_3$	H_2SO_4	MeOH	1.88	0.04	0.0	294
$W(NNH_2)Cl_2L_3$	H_2SO_4	MeOH	1.26	0.12	—[c]	294
$W(NNH_2)Cl_2L_3$	HCl	DME	0.47	0.50	—[c]	47
$Mo(NNH_2)Cl_2L_3$	H_2SO_4	MeOH	0.61	0.0	0.44	294
$Mo(NNH_2)Cl_2L_3$	HCl	DME	0.24	0.52	—[c]	47
$[Mo(NNH_2)(quin)L_3]I$	H_2SO_4	MeOH	0.55	0.0	0.17	294
$[W(NNH_2)(quin)L_3]I$	H_2SO_4	MeOH	0.0	0.39	—[c]	294
$[W(NNH_2)(H)Cl_2L_3]I$	H_2SO_4	MeOH	0.97	0.42	—[c]	46
$[W(NNH_2)(quin)L_3]I$	KOH	H_2O	0.03	0.43	0.45	294
$W(NNH_2)Cl_2L_3$	KOH	H_2O	1.40	0.14	—[c]	294
$W(NNH_2)Br_2L_3$	$NaBH_4$	THF	1.3	0.0	—[c]	294

[a] Moles per metal atom.
[b] L = dimethylphenylphosphine; quin = 8-hydroxyquinolinate.
[c] Not determined.

As originally reported by Chatt and coworkers [42], tungsten and molybdenum hydrazido complexes undergo N—N cleavage to afford ammonia upon treatment with mineral acids. A selection of results from such reactions is summarized in Table 6.4. For the molybdenum complexes only, a portion of the hydrazido nitrogen is typically recovered as dinitrogen; an idealized stoichiometry of 1 mole of ammonia and 0.5 mole of N_2 per molybdenum atom has been suggested. Table 6.4 also shows that hydrazine is often a significant side-product in these reactions, with the relative yields of NH_3 and N_2H_4 being a sensitive function of the reaction conditions. For example, Hidai and coworkers found that simply replacing solvent THF with dimethoxyethane markedly increases the extent of the side-reaction leading to hydrazine [47]. The use of hydridometal carbonyls as the acid component in these reactions has also been investigated but generally results in lower yield [295].

The ancillary ligands in these studies are usually monodentate phosphines. Complexes containing chelating diphosphines such as dppe do not yield ammonia upon treatment with acid [296]. (In contrast, tungsten hydrazido complexes supported by diphos ligands have been shown to yield ammonia upon electrochemical reduction [297].) George and coworkers have shown that a tridentate phosphine can be useful in studies of hydrazide cleavage [298]. A mixture of isomeric exo and endo hydrazido complexes is generated *in situ* by initial protonation of the corresponding bis(dinitrogen) complex according to Eq. 101; subsequent loss of the mondentate phosphine is crucial to the completion of the reaction. This system has two interesting features. First, it appears that the "exo" hydrazide (phenyl group and hydrazido ligand mutually trans) formed in Eq. 101 is much more reactive in ammonia formation than "endo" hydrazide. Second, it is found that hydrolysis of this reaction during its early stages produces high yields of hydrazine, but at long reaction times ammonia is observed as the exclusive product of nitrogen reduction. This resembles the behavior exhibited by nitrogenase and "suggests a chemical similarity between two systems." (See Figure 7.2).

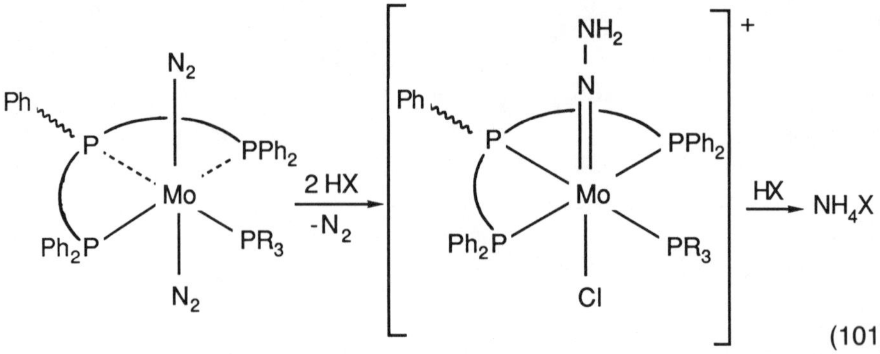

$$\text{Ph} \cdots \text{P}\overset{N_2}{\underset{N_2}{\overset{|}{\underset{|}{\text{Mo}}}}}\overset{PPh_2}{\underset{PR_3}{\cdots}} \xrightarrow[-N_2]{2\text{ HX}} \left[\text{Ph} \cdots \text{P}\overset{NH_2}{\underset{Cl}{\overset{|}{\underset{|}{\text{Mo}}}}}\overset{PPh_2}{\underset{PR_3}{\cdots}} \right]^{+} \xrightarrow{\text{HX}} NH_4X \qquad (101)$$

At this time there appears to be a consensus [47,46] that the pathway producing side-product hydrazine in these reactions involves protonation of the metal to afford a hydridometal hydrazide complex. Indeed an X-ray crystal structure [47] has confirmed the formation of such an intermediate in Eq. 102. Whether this hydride intermediate also lies on the pathway to ammonia remains to be determined [47]. (Available theoretical studies unfortunately predate the recognition of the hydride pathway [44,299].)

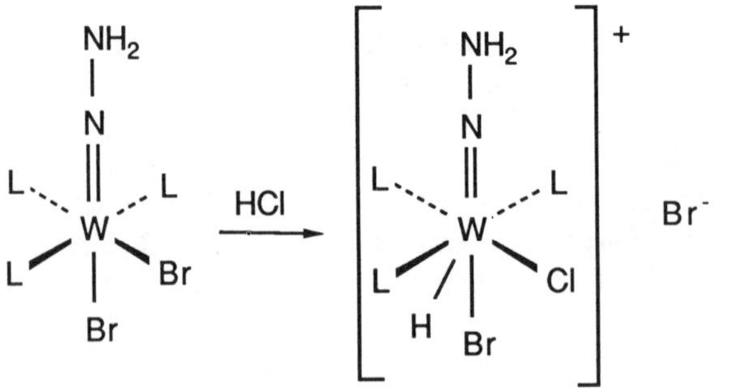

$$\overset{NH_2}{\underset{Br}{\overset{|}{\underset{|}{\text{L}\cdots\underset{L}{\overset{N}{\text{W}}}\cdots L}}}}\overset{}{\underset{Br}{}} \xrightarrow{\text{HCl}} \left[\overset{NH_2}{\underset{HBr}{\overset{|}{\underset{|}{\text{L}\cdots\underset{L/}{\overset{N}{\text{W}}}\cdots L}}}}\overset{}{\underset{Cl}{}} \right]^{+} \text{Br}^{-} \qquad (102)$$

L = PMePh$_3$; position of hydride not determined

It can also be seen in Table 6.4 that tungsten hydrazides can be cleaved to ammonia under basic conditions or upon treatment with borohydride. In contrast, molybdenum hydrazido(2−) complexes decompose in base with release of dinitrogen [294]. Finally, we note that certain *N,N*-dialkylhydrazido(2−) complexes have also been shown to undergo N—N cleavage upon treatment with acid [37] or lithium aluminum hydride [300].

6.5 MODIFICATION OF THE ORGANIC MOIETY IN ORGANOIMIDO AND ALKYLIDENE LIGANDS

A number of examples of α or β deprotonation reactions included in Section 6.3.1 might equally well have been included under this heading, but in general this area has been little explored. Nevertheless this is a potentially interesting area of chemistry. We back this assertion with the pair of examples in Eqs. 103 and 104. In the first example, both the "organic" and the "inorganic" chlorine in the imidotungsten(VI) starting material was replaced by fluorine upon treatment with silver fluoride in acetonitrile [301]. The latter reaction illustrates the electrochemical coupling of diazomethane ligands reported by Pickett [302].

$$AsPh_4[Cl_5W \equiv NCCl_3] \xrightarrow[-8\ AgCl]{8\ AgF} AsPh_4[F_5W \equiv NCF_3] \quad (103)$$

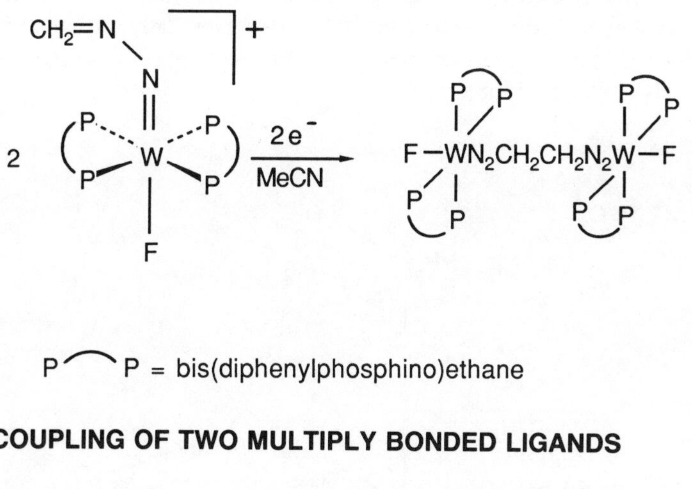

$$2\ \begin{pmatrix} CH_2=N \\ \| \\ N \\ P_{\cdots}|_{\cdots}P \\ P^{\diagup}W^{\diagdown}P \\ | \\ F \end{pmatrix}^+ \xrightarrow{2e^-}_{MeCN} F-WN_2CH_2CH_2N_2W-F \quad (104)$$

P⌒P = bis(diphenylphosphino)ethane

6.6 COUPLING OF TWO MULTIPLY BONDED LIGANDS

It has long been known that permanganate, upon UV irradiation, releases molecular oxygen. Recent studies indicate that the reaction proceeds according to Eq. 105, that is, with intramolecular coupling of two oxo ligands to form a peroxomanganese(V) intermediate [303]. This relatively long-lived intermediate was shown to be a more reactive oxidant than permanganate itself. Tyler and coworkers have demonstrated the intermolecular coupling of two oxo ligands [304]. They noted that Cp$_2$Mo(O) upon UV irradiation produces molecular oxygen. Moreover, Liebelt and Dehnicke have provided evidence for the coupling of two nitrido ligands to give a bridging hydrazido structure [305]. However, to date reactions of this type have been observed principally for alkylidene and alkylidyne ligands.

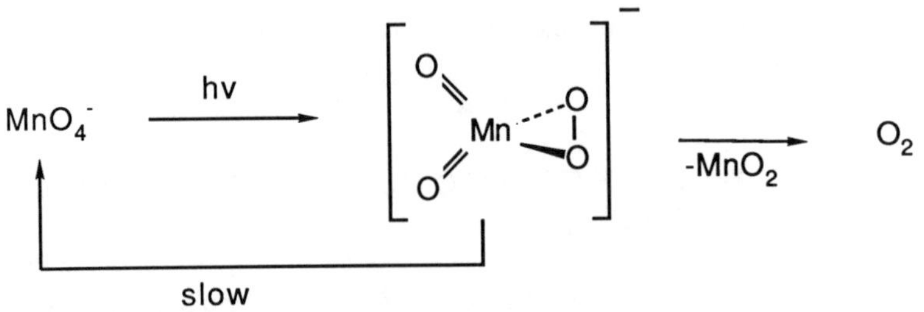

$$\text{MnO}_4^- \xrightarrow{h\nu} \left[\begin{array}{c} \text{O} \\ \text{Mn} \\ \text{O} \end{array} \right]^- \xrightarrow{-\text{MnO}_2} \text{O}_2 \qquad (105)$$

slow

Thermal decomposition of the tantalum methylidene complex $Cp_2Ta(CH_2)(CH_3)$ affords the ethylene complex $Cp_2Ta(CH_2CH_2)(CH_3)$ in what is necessarily an intermolecular reaction [77]. The yield of the ethylene complex is nearly quantitative when the reaction is carried out under ethylene. The proposed mechanism involves formation of a methylidene-bridged dimer which decomposes with formation of one equivalent of a methyltantalum(III) complex; the latter can be trapped by excess ethylene as shown in Eq. 106. A less clear-cut example involves the formation of ethylene in 58% yield during the decomposition of a putative $Cl_4W=CH_2$ intermediate [306]. Schrock has suggested that reactions related to Eq. 106 may be an important chain-termination step in olefin metathesis [227]. This pathway is believed to be most facile for methylidene complexes. Nevertheless, evidence has been presented that benzylidene complexes may also decompose in analogous fashion [307]. (However, only traces of ethane or ethylene are formed upon thermal decomposition of a zirconium(IV) methylidene complex, $Cp_2Zr(CH_2)(PPh_2Me)$ [308].)

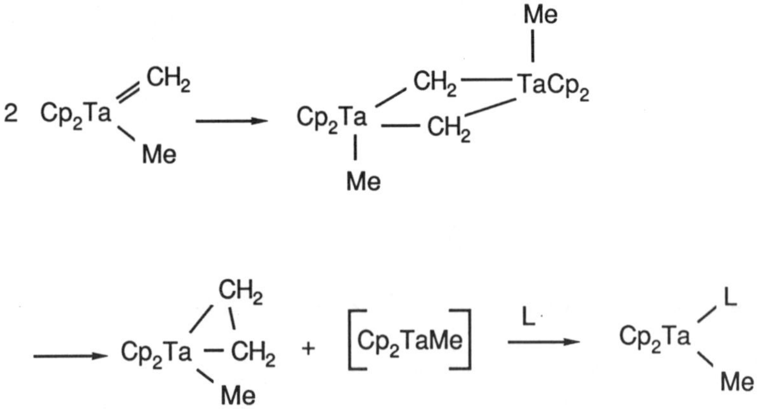

$$2\ Cp_2Ta\!\!\begin{array}{c}\diagup CH_2 \\ \diagdown Me\end{array} \longrightarrow Cp_2Ta\!\!\begin{array}{c}\diagup CH_2—TaCp_2 \\ \diagdown CH_2 \\ | \\ Me\end{array}\!\!\!\!\overset{Me}{|} \qquad (106)$$

$$\longrightarrow Cp_2Ta\!\!\begin{array}{c}CH_2 \\ \diagup \diagdown \\ —CH_2 \\ \diagdown Me\end{array} + [Cp_2TaMe] \xrightarrow{L} Cp_2Ta\!\!\begin{array}{c}\diagup L \\ \diagdown Me\end{array}$$

Chisholm and coworkers originally used ^{13}C-labeling studies to demonstrate that the complex $(t\text{-BuO})_3W(CH)$ in the presence of pyridine is in equilibrium with the acetylene-bridged dimer $W_2(O\text{-}t\text{-Bu})_6(\mu\text{-}C_2H_2)(py)$ [309]. Subsequently, it was shown that the μ-alkyne species in this type of equilibrium could be trapped, for example, by the addition of carbon monoxide in Eq. 107. The reaction of $(i\text{-PrO})_3W(CNEt_2)$ with CO proceeds further. Two molecules of CO are taken up per tungsten dimer and the isolated product contains the acetylene bound to a single metal center [246].

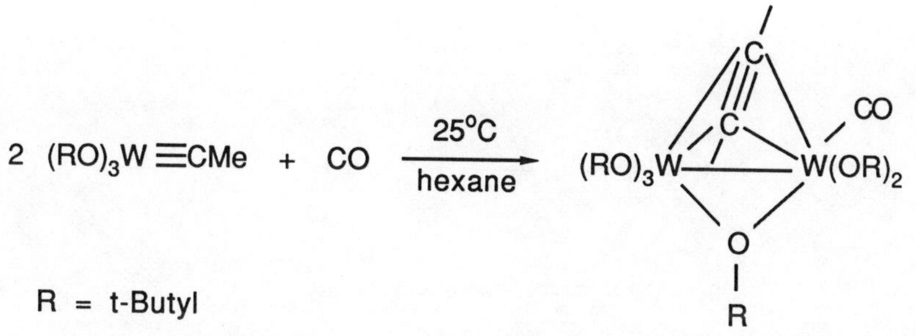

$$2\ (RO)_3W \equiv CMe\ +\ CO\ \xrightarrow[\text{hexane}]{25°C}\ \text{[dimer complex]} \quad (107)$$

R = t-Butyl

A theoretical study on the coupling of two alkylidene or alkylidyne ligands has been reported [310]. In the intramolecular case where two alkylidyne ligands are coupled to a coordinated acetylene at a single metal center, the reaction was predicted to be facile. In contrast, the coupling of two alkylidene ligands will be subject to both orientational and electronic constraints. For example, with an electron count of d^4 (counting a methylidene ligand as neutral) the coupling reaction was shown to be forbidden. However, it was stressed that the barrier to such coupling reactions could be lowered, for example by (1) using a more electronegative metal, (2) introducing acceptor ligands on the metal, (3) incorporating donor substituents on the methylidene moiety, or (4) substituting a more electropositive atom for the methylidene carbon.

Mayr has provided an example of the coupling of two alkylidynes on a single tungsten center [311]. The starting complex in Eq. 108 was treated with triphenylphosphine at −78°C. This should in principle give rise to a bis(alkylidyne) intermediate. In reality the indicated acetylene complex is the only observed product.

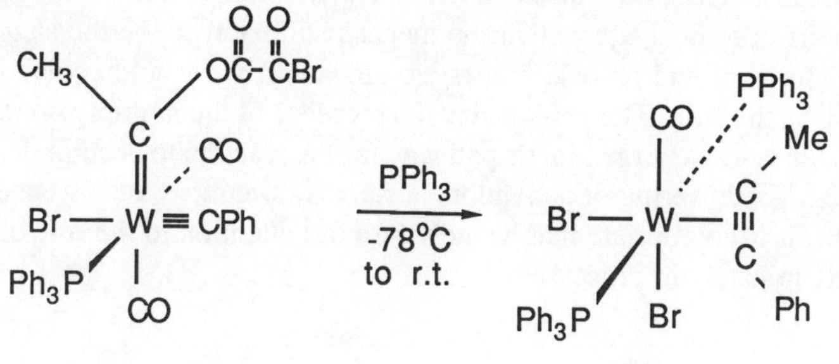

$$\xrightarrow[\text{-78°C to r.t.}]{PPh_3} \quad (108)$$

6.7 CONNECTIONS

The authors settled on the current organization of the chemistry of multiple bonded ligands along the lines of Chapters 3 and 6 only with some misgivings. For every system of organization, something is gained and something is lost. The present format has one major shortcoming. It does not stress the remarkable frequency with which a particular multiple bond forming reaction in Chapter 3 represents the exact reverse of one of the ligand transformations in Chapter 6. Rather than disrupt the flow of the chapters by constantly harping on this point, we attempt to make amends by including

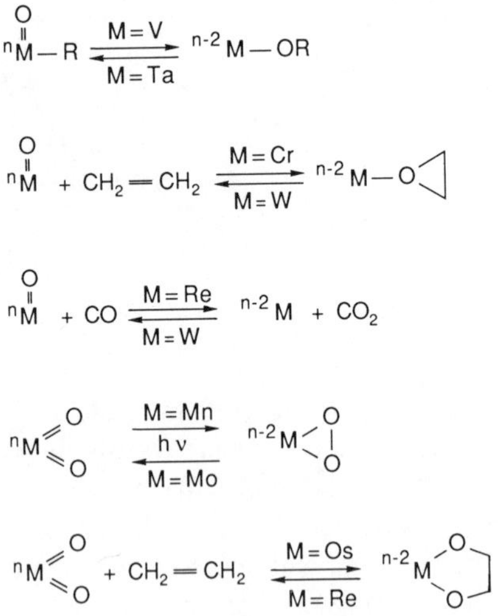

Figure 6.6 A few cases where both the forward and reverse of a chemical transformation is known, using the example of oxo complexes. For examples proceeding left to right, the metal is in the d^0 oxidation state.

Figure 6.6. This figure summarizes a few cases of this phenomenon using the example of oxo chemistry; the alert reader will no doubt spot many others.

It can be seen that the direction of these reactions can be reversed by changing the nature of the metal and its oxidation state. Of course, the observation of these correlated pairs of reactions does not require that the mechanistic pathway be the same or even related for the forward and reverse processes. However, we would contend that this may frequently be the case. The point is that the presence of the multiply bonded ligand appears to provide a low-energy kinetic pathway for the reaction to occur independently of the direction. The lowering of activation barriers is, of course, the essence of catalysis. Therefore, it is appropriate that we now turn our attention to the role of multiply bonded ligands in catalytic processes.

REFERENCES

1. Block, T.F.; Fenske, R.F.; Casey, C.P. *J. Am. Chem. Soc.* **1976**, *98*, 441-443.
2. Kostic, N.M.; Fenske, R.F. *J. Am. Chem. Soc.* **1981**, *103*, 4677-4685.
3. Kostic, N.M.; Fenske, R.F. *Organometallics* **1982**, *1*, 489-496.
4. Ushio, J.; Nakatsuji, H.; Yonezawa, T. *J. Am. Chem. Soc.* **1984**, *106*, 5892-5901.
5. Jørgensen, K.A.; Hoffmann, R. *J. Am. Chem. Soc.* **1986**, *108*, 1867-1876.
6. Thomas, I.M. *Can. J. Chem.* **1961**, *39*, 1386-1388.
7. Clifford, A.F.; Kobayashi, C.S. Abstracts, 130th National Meeting of the American Chemical Society, Atlantic City, NJ, Sept. 1956, p.50R.
8. Clifford, A.F.; Kobayashi, C.S. *Inorg. Synth.* **1960**, *6*, 204-208.
9. Sharpless, K.B.; Patrick, D.W.; Truesdale, L.K.; Biller, S.A. *J. Am. Chem. Soc.* **1975**, *97*, 2305-2307.
10. Henderson, R.A. *J. Chem. Soc., Dalton Trans.* **1983**, 51-60.
11. Shapley, P.A.B.; Own, Z.-Y.; Huffman, J.C. *Organometallics* **1986**, *5*, 1269-1271.
12. Maatta, E.A.; Wentworth, R.A.D. *Inorg. Chem.* **1979**, *18*, 2409-2413.
13. Henderson, R.A.; Davies, G.; Dilworth, J.R.; Thorneley, R.N.F. *J. Chem. Soc., Dalton Trans.* **1981**, 40-50.
14. Goddard, R.J.; Hoffmann, R.; Jemmis, E.D. *J. Am. Chem. Soc.* **1980**, *102*, 7667-7676.
15. Chen, G.J.-J.; McDonald, J.W.; Bravard, D.C.; Newton, W.E. *Inorg. Chem.* **1985**, *24*, 2327-2333.
16. Green, M.L.H.; Lynch, A.H.; Swanwick, M.G. *J. Chem. Soc., Dalton Trans.* **1972**, 1445-1447.
17. Bristow, S.; Enemark, J.H.; Garner, C.D.; Minelli, M.; Morris, G.A.; Ortega, R.B. *Inorg. Chem.* **1985**, *24*, 4070-4077.
18. Liebeskind, L.S.; Sharpless, K.B.; Wilson, R.D.; Ibers, J.A. *J. Am. Chem. Soc.* **1978**, *100*, 7061-7063.
19. Muccigrosso, D.A.; Jacobson, S.E.; Apgar, P.A.; Mares, F. *J. Am. Chem. Soc.* **1978**, *100*, 7863-7865.
20. Feinstein-Jaffe, I.; Dewan, J.C.; Schrock, R.R. *Organometallics* **1985**, *4*, 1189-1193.
21. Moore, F.W.; Larson, M.L. *Inorg. Chem.* **1967**, *6*, 998-1003.
22. Connor, J.A.; Ebsworth, E.A.V. *Adv. Inorg. Chem. Radiochem.* **1964**, *6*, 279-381.
23. Schrauzer, G.N.; Schlemper, E.O.; Liu, N.H.; Wang, Q.; Rubin, K.; Zhang, X.; Long, X.; Chin, C.S. *Organometallics* **1986**, *5*, 2452-2456.
24. Griffith, W.P. *J. Chem. Soc.* **1965**, 3694-3697.
25. Chatt, J.; Choukroun, R.; Dilworth, J.R.; Hyde, J.; Vella, P.; Zubieta, J. *Trans. Met. Chem.* **1979**, *4*, 59-63.
26. Ueyama, N.; Zaima, H.; Okada, H.; Nakamura, A. *Inorg. Chim. Acta* **1984**, *89*, 19-23.

27. Weller, F.; Müller, U.; Weiher, U.; Dehnicke, K. *Z. Anorg. Allg. Chem.* **1980**, *460*, 191–199.
28. Noble, M.E.; Folting, K.; Huffman, J.C.; Wentworth, R.A.D. *Inorg. Chem.* **1983**, *22*, 3671–3676.
29. Buslaev, Yu.A.; Kokunov, Yu.V.; Gustyakova, M.P.; Chubar, Yu.D.; Moiseev, I.I. *Dokl. Akad. Nauk SSSR* **1977**, *233*, 357–360.
30. Osborne, J.H.; Rheingold, A.L.; Trogler, W.C. *J. Am. Chem. Soc.* **1985**, *107*, 7945–7952.
31. Milas, N.A.; Iliopulos, M.I. *J. Am. Chem. Soc.* **1959**, *81*, 6089.
32. Maatta, E.A.; Haymore, B.L.; Wentworth, R.A.D. *Inorg. Chem.* **1980**, *19*, 1055–1059.
33. Edwards, D.S.; Biondi, L.V.; Ziller, J.W.; Churchill, M.R.; Schrock, R.R. *Organometallics* **1983**, *2*, 1505–1513.
34. Cotton, F.A.; Hall, W.T. *J. Am. Chem. Soc.* **1979**, *101*, 5094–5095.
35. Bradley, D.C.; Thomas, I.M. *Proc. Chem. Soc., London* **1959**, 225–226.
36. Chiu, K.W.; Wong, W.-K.; Wilkinson, G.; Galas, A.M.R.; Hursthouse, M.B. *Polyhedron* **1982**, *1*, 31–36.
37. Chatt, J.; Crichton, B.A.L.; Dilworth, J.R.; Dahlstrom, P.; Zubieta, J.A. *J. Chem. Soc., Dalton Trans.* **1982**, 1041–1047.
38. Chatt, J.; Dilworth, J.R.; Richards, R.L. *Chem. Rev.* **1978**, *78*, 589–625.
39. Hidai, M. in "Molybdenum Enzymes" Spiro, T.G., ed. Wiley, New York, 1985.
40. Chatt, J.; Fakley, M.E.; Richards, R.L.; Hanson, I.R.; Hughes, D.L. *J. Organomet. Chem.* **1979**, *170*, C6–C8.
41. Rocklage, S.M.; Schrock, R.R. *J. Am. Chem. Soc.* **1982**, *104*, 3077–3081.
42. Chatt, J.; Pearman, A.J.; Richards, R.L. *J. Organomet. Chem.* **1975**, *101*, C45–C47.
43. Takahashi, T.; Mizobe, Y.; Sato, M.; Uchida, Y.; Hidai, M. *J. Am. Chem. Soc.* **1979**, *101*, 3405–3407.
44. Yamabe, T.; Hori, K.; Fukui, K. *Inorg. Chem.* **1982**, *21*, 2816–2818.
45. Chatt, J.; Dilworth, J.R.; Dahlstrom, P.L.; Zubieta, J. *J. Chem. Soc., Chem. Comm.* **1980**, 786–787.
46. Chatt, J.; Fakley, M.E.; Hitchcock, P.B.; Richards, R.L.; Luong-Thi, N.T. *J. Chem. Soc., Dalton Trans.* **1982**, 345–352.
47. Takahashi, T.; Mizobe, Y.; Sato, M.; Uchida, Y.; Hidai, M. *J. Am. Chem. Soc.* **1980**, *102*, 7461–7467.
48. Kauffmann, T.; Fiegenbaum, P.; Wieschollek, R. *Angew. Chem., Int. Ed. Engl.* **1984**, *23*, 531–532.
49. Hill, A.F.; Roper, W.R.; Waters, J.M.; Wright, A.H. *J. Am. Chem. Soc.* **1983**, *105*, 5939–5940.
50. Schrock, R.R.; Fellmann, J.D. *J. Am. Chem. Soc.* **1978**, *100*, 3359–3370.
51. Chamberlain, L.R.; Rothwell, I.P.; Huffman, J.C. *J. Am. Chem. Soc.* **1986**, *108*, 1502–1509.
52. Fellmann, J.D.; Schrock, R.R.; Traficante, D.D. *Organometallics* **1982**, *1*, 481–484.
53. Schrock, R.R.; Clark, D.N.; Sancho, J.; Wengrovius, J.H.; Rocklage, S.M.; Pedersen, S.F. *Organometallics* **1982**, *1*, 1645–1651.

54. Holmes, S.J.; Clark, D.N.; Turner, H.W.; Schrock, R.R. *J. Am. Chem. Soc.* **1982**, *104*, 6322–6329.
55. Holmes, S.J.; Schrock, R.R.; Churchill, M.R.; Wasserman, H.J. *Organometallics* **1984**, *3*, 476–484.
56. Aristov, N.; Armentrout, P.B. *J. Am. Chem. Soc.* **1984**, *106*, 4065–4066.
57. Freudenberger, J.H.; Schrock, R.R. *Organometallics* **1985**, *4*, 1937–1944.
58. McCullough, L.G.; Schrock, R.R.; Dewan, J.C.; Murdzek, J.C. *J. Am. Chem. Soc.* **1985**, *107*, 5987–5998.
59. Fanwick, P.E.; Ogilvy, A.E.; Rothwell, I.P. *Organometallics* **1987**, *6*, 73–80.
60. Feinstein-Jaffe, I.; Pedersen, S.F.; Schrock, R.R. *J. Am. Chem. Soc.* **1983**, *105*, 7176–7177.
61. Stravropoulos, P.; Edwards, P.G.; Wilkinson, G.; Motevalli, M.; Malik, K.M.A.; Hursthouse, M.B. *J. Chem. Soc., Dalton Trans.* **1985**, 2167–2175.
62. Sergienko, V.S.; Porai-Koshits, M.A. *Koord. Khim.* **1982**, *8*, 251.
63. Hess, H.; Hartung, H. *Z. Anorg. Allg. Chem.* **1966**, *344*, 157–166.
64. Mathew, M.; Carty, A.J.; Palenik, G.J. *J. Am. Chem. Soc.* **1970**, *92*, 3197–3198.
65. Kress, J.; Wesolek, M.; Le Ny, J.-P.; Osborn, J.A. *J. Chem. Soc., Chem. Comm.* **1981**, 1039–1040.
66. Nakamura, S.; Dedieu, A. *Nouv. J. Chim.* **1982**, *6*, 23–30.
67. Lappert, M.F.; Prokai, B. *J. Chem. Soc. A* **1967**, 129–131.
68. Levason, W.; Ogden, J.S.; Rest, A.J.; Turff, J.W. *J. Chem. Soc., Dalton Trans.* **1982**, 1877–1878.
69. DeBoer, E.J.M.; DeWith, J.; Orpen, A.G. *J. Am. Chem. Soc.* **1986**, *108*, 8271–8273.
70. Herrmann, W.A.; Voss, E.; Küsthardt, U.; Herdtweck, E. *J. Organomet. Chem.* **1985**, *294*, C37–C40.
71. Chatt, J.; Heaton, B.T. *J. Chem. Soc. A* **1971**, 705–707.
72. Kafitz, W.; Weller, F.; Dehnicke, K. *Z. Anorg. Allg. Chem.* **1982**, *490*, 175–181.
73. Forsellini, E.; Casellato, U.; Graziani, R.; Magon, L. *Acta Cryst.* **1982**, *B38*, 3081–3083.
74. Dantona, R.; Schweda, E.; Strähle, J. *Z. Naturforsch.* **1984**, *39B*, 733–735.
75. Tebbe, F.N.; Parshall, G.W.; Reddy, G.S. *J. Am. Chem. Soc.* **1978**, *100*, 3611–3613.
76. Howard, T.R.; Lee, J.R.; Grubbs, R.H. *J. Am. Chem. Soc.* **1980**, *102*, 6876–6878.
77. Schrock, R.R.; Sharp, P.R. *J. Am. Chem. Soc.* **1978**, *100*, 2389–2399.
78. Sharp, P.R.; Holmes, S.J.; Schrock, R.R.; Churchill, M.R.; Wasserman, H.J. *J. Am. Chem. Soc.* **1981**, *103*, 965–966.
79. Schmidt, M.; Schmidbauer, H. *Inorg. Synth.* **1967**, *9*, 149–151.
80. Nugent, W.A. *Inorg. Chem.* **1983**, *22*, 965–969.
81. Sheldrick, G.M.; Sheldrick, W.S. *J. Chem. Soc. A* **1969**, 2160.
82. Feil, S.E.; Tyree, Jr., S.Y.; Collier, Jr., F.N. *Inorg. Synth.* **1967**, *9*, 123–126.
83. Hecht, H.; Jander, G.; Schlapmann, H. *Z. Anorg. Allg. Chem.* **1947**, *254*, 255–264.

84. Gambarotta, S.; Chiesi-Villa, A.; Guastini, C. *J. Organomet. Chem.* **1984**, *270*, C49–C52.
85. Chan, D.M.-T.; Nugent, W.A. *Inorg. Chem.* **1985**, *24*, 1422–1424.
86. Parshall, G.W.; Nugent, W.A.; Chan, D.M.-T.; Tam, W. *Pure Appl. Chem.* **1985**, *57*, 1809–1818.
87. Mayr, A.; Asaro, M.F.; Glines, T.J. *J. Am. Chem. Soc.* **1987**, *109*, 2215–2216.
88. Bevan, P.C.; Chatt, J.; Hidai, M.; Leigh, G.J. *J. Organomet. Chem.* **1978**, *160*, 165–176.
89. Hidai, M.; Mizobe, Y.; Uchida, Y. *J. Am. Chem. Soc.* **1976**, *98*, 7824–7825.
90. Mizobe, Y.; Uchida, Y.; Hidai, M. *Bull. Chem. Soc. Jpn.* **1980**, *53*, 1781–1782.
91. Schrock, R.R. *J. Am. Chem. Soc.* **1976**, *98*, 5399–5400.
92. Wittig, G. *J. Organomet. Chem.* **1975**, *100*, 279–287.
93. Pine, S.H.; Zahler, R.; Evans, D.A.; Grubbs, R.H. *J. Am. Chem. Soc.* **1980**, *102*, 3270–3272.
94. Brown-Wensley, K.A.; Buchwald, S.L.; Cannizzo, L.; Clawson, L.; Ho, S.; Meinhardt, D.; Stille, J.R.; Straus, D.; Grubbs, R.H. *Pure Appl. Chem.* **1983**, *55*, 1733–1744.
95. Cannizzo, L.F.; Grubbs, R.H. *J. Org. Chem.* **1985**, *50*, 2386–2387.
96. Clawson, L.; Buchwald, S.L.; Grubbs, R.H. *Tetrahedron Lett.* **1984**, 5733–5736.
97. RajanBabu, T.V.; Reddy, G.S. *J. Org. Chem.* **1986**, *51*, 5458–5461.
98. Stille, J.R.; Grubbs, R.H. *J. Am. Chem. Soc.* **1983**, *105*, 1664–1665.
99. Hartner, Jr., F.W.; Schwartz, J.; Clift, S.M. *J. Am. Chem. Soc.* **1983**, *105*, 640–641.
100. McKinney, R.J.; Tulip, T.H.; Thorn, D.L.; Coolbaugh, T.S.; Tebbe; F.N. *J. Am. Chem. Soc.* **1981**, *103*, 5584–5586.
101. Kauffmann, T.; Ennen, B.; Sander, J.; Wieschollek, R. *Angew. Chem., Int. Ed. Engl.* **1983**, *22*, 244–245.
102. Kauffmann, T.; Abeln, R.; Welke, S.; Wingbermuehle, D. *Angew. Chem., Int. Ed. Engl.* **1986**, *25*, 909–910.
103. Kauffmann, T.; Kieper, G. *Angew. Chem., Int. Ed. Engl.* **1984**, *23*, 532–533.
104. Kauffmann, T.; Enk, M.; Kaschube, W.; Toliopoulos, E.; Wingbermuehle, D. *Angew. Chem., Int. Ed. Engl.* **1986**, *25*, 910–911.
105. Kauffmann, T.; Möller, T.; Rennefeld, H.; Welke, S.; Wieschollek, R. *Angew. Chem., Int. Ed. Engl.* **1985**, *24*, 348–350.
106. Buchwald, S.L.; Grubbs, R.H. *J. Am. Chem. Soc.* **1983**, *105*, 5490–5491.
107. Gilliom, L.R.; Grubbs, R.H. *Organometallics* **1986**, *5*, 721–724.
108. Yoshida, T. *Chem. Lett.* **1982**, 429–432.
109. Clift, S.M.; Schwartz, J. *J. Am. Chem. Soc.* **1984**, *106*, 8300–8301.
110. Aguero, A.; Kress, J.; Osborn, J.A. *J. Chem. Soc., Chem. Comm.* **1986**, 531–533.
111. Freudenberger, J.H.; Schrock, R.R. *Organometallics* **1986**, *5*, 398–400.
112. Weiss, K.; Schubert, U.; Schrock, R.R. *Organometallics* **1986**, *5*, 397–398.

113. Chisholm, M.H.; Heppert, J.A.; Huffman, J.C.; Streib, W.E. *J. Chem. Soc., Chem. Comm.* **1985**, 1771–1773.
114. LaMonica, G.; Cenini, S. *Inorg. Chim. Acta* **1978**, *29*, 183–187.
115. LaMonica, G.; Cenini, S. *J. Chem. Soc., Dalton Trans.* **1980**, 1145–1149.
116. Bishop, M.W.; Chatt, J.; Dilworth, J.R. *J. Chem. Soc., Dalton Trans.* **1979**, 1–5.
117. Hursthouse, M.B.; Motevalli, M. *J. Chem. Soc., Dalton Trans.* **1979**, 1362–1366.
118. Baldas, J.; Bonnyman, J.; Mackay, M.F.; Williams, G.A. *Aust. J. Chem.* **1984**, *37*, 751–759.
119. Kaden, L.; Lorenz, B.; Kirmse, R.; Stach, J.; Abram, U. *Z. Chem.* **1985**, *25*, 29–30.
120. Anhaus, J.; Siddiqi, Z.A.; Roesky, H.W.; Bats, J.W.; Elerman, Y. *Z. Naturforsch.* **1985**, *40B*, 740–744.
121. Chatt, J.; Dosser, R.J.; King, F.; Leigh, G.J. *J. Chem. Soc., Dalton Trans.* **1976**, 2435–2443.
122. Adcock, P.A.; Keene, F.R.; Smythe, R.S.; Snow, M.R. *Inorg. Chem.* **1984**, *23*, 2336–2343.
123. Lane, J.D.; Henderson, R.A. *J. Chem. Soc., Dalton Trans.* **1987**, 197–200.
124. Lane, J.D.; Henderson, R.A. *J. Chem. Soc., Dalton Trans.* **1986**, 2155–2163.
125. Nicholls, D.; Wilkinson, D.N. *J. Chem. Soc. A* **1970**, 1103–1104.
126. Srinivasan, K.; Kochi, J.K. *Inorg. Chem.* **1985**, *24*, 4671–4679.
127. Johnson, J.W.; Brody, J.F.; Ansell, G.B.; Zentz, S. *Inorg. Chem.* **1984**, *23*, 2415–2418.
128. Rybak, W.K.; Ziółdowski, J.J. *Polyhedron* **1983**, *2*, 541–542.
129. Dilworth, J.R.; Neaves, B.D.; Hutchinson, J.P.; Zubieta, J.A. *Inorg. Chim. Acta* **1982**, *65*, L223–L224.
130. Che, C.-M.; Wong, K.-Y.; Mak, T.C.W. *J. Chem. Soc., Chem. Comm.* **1985**, 546–548.
131. Salmon, D.J.; Walton, R.A. *Inorg. Chem.* **1978**, *17*, 2379–2382.
132. Che, C.-M.; Cheng, W.-K.; Mak, T.C.W. *J. Chem. Soc., Chem. Comm.* **1986**, 200–202.
133. Hentges, S.G.; Sharpless, K.B.; Tulip, T.H. unpublished results.
134. Hanzlik, R.P.; Williamson, D. *J. Am. Chem. Soc.* **1976**, *98*, 6570–6573.
135. Bortolini, O.; Meunier, B. *J. Chem. Soc., Chem. Comm.* **1983**, 1364–1366.
136. Chin, D.-H.; LaMar, G.N.; Balch, A.L. *J. Am. Chem. Soc.* **1980**, *102*, 5945–5947.
137. Ledon, H.; Bonnet, M.; Lallemand, J.-Y. *J. Chem. Soc., Chem. Comm.* **1979**, 702–704.
138. Barral, R.; Bocard, C.; Sérée de Roch, I.; Sajus, L. *Tetrahedron Lett.* **1972**, 1693–1696.
139. Holm, R.H.; Berg, J.M. *Acc. Chem. Res.* **1986**, *19*, 363–370. Holm, R.H. *Chem. Rev.* **1987**, 1401–1449.
140. Harlan, E.W.; Berg, J.M.; Holm, R.H. *J. Am. Chem. Soc.* **1986**, *108*, 6992–7000.
141. Chen, G.J.-J.; McDonald, J.W.; Newton, W.E. *Inorg. Chem.* **1976**, *15*, 2612–2615.

142. Nakamura, A.; Nakayama, M.; Sugihashi, K.; Otsuka, S. *Inorg. Chem.* **1979**, *18*, 394–400.
143. Pickett, C.; Kumar, S.; Vella, P.A.; Zubieta, J. *Inorg. Chem.* **1982**, *21*, 908–916.
144. Deli, J.; Speier, G. *Trans. Met. Chem.* **1981**, *6*, 227–229.
145. Speier, G. *Inorg. Chim. Acta* **1979**, *32*, 139–141.
146. Ueyama, N.; Yano, M.; Miyashita, H.; Nakamura, A.; Kamachi, M.; Nozakura, S.-I. *J. Chem. Soc., Dalton Trans.* **1984**, 1447–1451.
147. Devore, D.D.; Maatta, E.A. *Inorg. Chem.* **1985**, *24*, 2846–2849.
148. Watt, G.D.; McDonald, J.W.; Newton, W.E. *J. Less-Common Met.* **1977**, *54*, 415–423.
149. Durant, R.; Garner, C.D.; Hyde, M.R.; Mabbs, F.E. *J. Chem. Soc., Dalton Trans.* **1977**, 955–956.
150. Durant, R.; Garner, C.D.; Hyde, M.R.; Mabbs, F.E.; Parsons, J.R.; Richens, D. *J. Less-Common Met.* **1977**, *54*, 459–464.
151. Reynolds, M.S.; Berg, J.M.; Holm, R.H. *Inorg. Chem.* **1984**, *23*, 3057–3062.
152. Moyer, B.A.; Sipe, B.K.; Meyer, T.J. *Inorg. Chem.* **1981**, *20*, 1475–1480.
153. Sharpless, K.B.; Teranishi, A.Y.; Bäckvall, J.-E. *J. Am. Chem. Soc.* **1977**, *99*, 3120–3128.
154. Marmion, M.E.; Takeuchi, K.J. *J. Am. Chem. Soc.* **1986**, *108*, 510–511.
155. Djerassi, C.; Engle, R.R. *J. Am. Chem. Soc.* **1953**, *75*, 3838–3840.
156. Henbest, H.B.; Khan, S.A. *J. Chem. Soc., Chem. Comm.* **1968**, 1036.
157. Ando, W.; Tajima, R.; Takata, T. *Tetrahedron Lett.* **1982**, 1685–1688.
158. Takata, T.; Yamazaki, M.; Fujimori, K.; Kim, Y.H.; Oae, S.; Iyanagi, T. *Chem. Lett.* **1980**, 1441–1444.
159. Thorn, D.L.; Tulip, T.H. *J. Am. Chem. Soc.* **1981**, *103*, 5984–5986.
160. Courtney, J.L.; Swansborough, K.F. *Rev. Pure Appl. Chem.* **1972**, *22*, 47–54.
161. Pawson, D.; Griffith, W.P. *J. Chem. Soc., Dalton Trans.* **1975**, 417–423.
162. Schmidt, I; Kynast, U.; Hanich, J.; Dehnicke, K. *Z. Naturforsch.* **1984**, *39B*, 1248–1251.
163. Dehnicke, K.; Prinz, H.; Kafitz, W.; Kujanek, R. *Liebigs Ann. Chem.* **1981**, 20–27.
164. Buchler, J.W.; Dreher, C.; Lay, K.-L. *Chem. Ber.* **1984**, *117*, 2261–2274.
165. Chatt, J.; Dilworth, J.R. *J. Indian Chem. Soc.* **1977**, *54*, 13–18.
166. Hayes, J.C.; Pearson, G.D.N.; Cooper, N.J. *J. Am. Chem. Soc.* **1981**, *103*, 4648–4650.
167. Cooper, N.J.; Green, M.L.H. *J. Chem. Soc., Chem. Comm.* **1974**, 761–762.
168. Reichle, W.T.; Carrick, W.L. *J. Organomet. Chem.* **1970**, *24*, 419–426.
169. Nugent, W.A.; Harlow, R.L. *J. Am. Chem. Soc.* **1980**, *102*, 1759–1760.
170. Seidel, W.; Kreisel, G. *Z. Chem.* **1982**, *22*, 113.
171. Hursthouse, M.B.; Motevalli, M.; Sullivan, A.C.; Wilkinson, G. *J. Chem. Soc., Chem. Comm.* **1986**, 1398–1399.
172. Jernakoff, P.; Cooper, N.J. *Organometallics* **1986**, *5*, 747–751.

173. Muetterties, E.L.; Band, E. *J. Am. Chem. Soc.* **1980**, *102*, 6572-6574.
174. Chiu, K.W.; Jones, R.A.; Wilkinson, G.; Galas, A.M.R.; Hursthouse, M.B.; Malik, K.M.A. *J. Chem. Soc., Dalton Trans.* **1981**, 1204-1211.
175. Fischer, E.O.; Frank, A. *Chem. Ber.* **1978**, *111*, 3740-3744.
176. Groves, J.T.; Kruper, W.J.; Haushalter, R.C. *J. Am. Chem. Soc.* **1980**, *102*, 6375-6377.
177. Groves, J.T.; Kruper, W.J. *J. Am. Chem. Soc.* **1979**, *101*, 7613-7615.
178. Groves, J.T.; Nemo, T.E.; Myers, R.S. *J. Am. Chem. Soc.* **1979**, *101*, 1032-1033.
179. Dobson, J.C.; Seok, W.K.; Meyer, T.J. *Inorg. Chem.* **1986**, *25*, 1513-1514.
180. Samsel, E.G.; Srinivasan, K.; Kochi, J.K. *J. Am. Chem. Soc.* **1985**, *107*, 7606-7617.
181. Groves, J.T.; Quinn, R. *J. Am. Chem. Soc.* **1985**, *107*, 5790-5792.
182. Rappé, A.K.; Goddard, III, W.A. *J. Am. Chem. Soc.* **1982**, *104*, 448-456.
183. Miyaura, N.; Kochi, J.K. *J. Am. Chem. Soc.* **1983**, *105*, 2368-2378.
184. Collman, J.P.; Brauman, J.I.; Meunier, B.; Raybuck, S.A.; Kodadek, T. *Proc. Natl. Acad. Sci.* **1984**, *81*, 3245-3248.
185. Groves, J.T.; Watanabe, Y. *J. Am. Chem. Soc.* **1986**, *108*, 507-508.
186. Jørgenson, K.A. *J. Am. Chem. Soc.* **1987**, *109*, 698-705.
187. Balch, A.L.; Chan, Y.-W.; Olmstead, M.M. *J. Am. Chem. Soc.* **1985**, *107*, 6510-6514.
188. Groves, J.T.; Myers, R.S. *J. Am. Chem. Soc.* **1983**, *105*, 5791-5796.
189. Groves, J.T.; Takahashi, T. *J. Am. Chem. Soc.* **1983**, *105*, 2073-2074.
190. Schröder, M. *Chem. Rev.* **1980**, *80*, 187-213.
191. Stewart, R. in "Oxidation in Organic Chemistry, Part A" Wiberg, K.B., ed. Academic, New York, 1965, pp 1-68.
192. Sharpless, K.B.; Akashi, K. *J. Am. Chem. Soc.* **1976**, *98*, 1986-1987.
193. Lee, D.G.; Van den Engh, M. in "Oxidation in Organic Chemistry, Part B" Trahanovsky, W.S., ed. Academic, New York, 1973, pp 177-227.
194. Sharpless, K.B.; Williams, D.R. *Tetrahedron Lett.* **1975**, 3045-3046.
195. Criegee, R.; Marahand, B.; Wannowius, H. *Justus Liebigs Ann. Chem.* **1942**, *550*, 99-133.
196. Clark, R.L.; Behrman, E.J. *Inorg. Chem.* **1975**, *14*, 1425-1426.
197. Griffith, W.P.; Skapski, A.C.; Woode, K.A.; Wright, M.J. *Inorg. Chim. Acta* **1978**, *31*, L413-L414.
198. Griffith, W.P.; Rossetti, R. *J. Chem. Soc., Dalton Trans.* **1972**, 1449-1453.
199. Cartwright, B.A.; Griffith, W.P.; Schröder, M.; Skapski, A.C. *Inorg. Chim. Acta* **1981**, *53*, L129-L130.
200. Phillips, F.L.; Skapski, A.C. *J. Chem. Soc., Dalton Trans.* **1975**, 2586-2590.
201. Hentges, S.G.; Sharpless, K.B. *J. Am. Chem. Soc.* **1980**, *102*, 4263-4265.

202. Walba, D.M.; Stoudt, G.S. *Tetrahedron Lett.* **1982**, 727–730.
203. Walba, D.M.; Wand, M.D.; Wilkes, M.C. *J. Am. Chem. Soc.* **1979**, *101*, 4396–4397.
204. Schröder, M.; Constable, E.C. *J. Chem. Soc., Chem. Comm.* **1982**, 734–736.
205. Casey, C.P. *J. Chem. Soc., Chem. Comm.* **1983**, 126–127.
206. Su, F.-M.; Cooper, C.; Geib, S.J.; Rheingold, A.L.; Mayer, J.M. *J. Am. Chem. Soc.* **1986**, *108*, 3545–3547.
207. Herrmann, W.A.; Jung, K.; Schaefer, A.; Kneuper, H.J. *Angew. Chem. Int. Ed. Engl.* **1987**, *26*, 464–465.
208. Herrmann, W.A.; Küsthardt, U.; Ziegler, M.L.; Zahn, T. *Angew. Chem., Int. Ed. Engl.* **1985**, *24*, 860–861.
209. Tokles, M.; Snyder, J.K. *Tetrahedron Lett.* **1986**, 3951–3954.
210. Yamada, T.; Narasaka, K. *Chem. Lett.* **1986**, 131–134.
211. Carlsen, P.H.J.; Katsuki, T.; Martin, V.S.; Sharpless, K.B. *J. Org. Chem.* **1981**, *46*, 3936–3938.
212. Starks, C.M. *J. Am. Chem. Soc.* **1971**, *93*, 195–199.
213. Sam, D.J.; Simmons, H.E. *J. Am. Chem. Soc.* **1972**, *94*, 4024–4025.
214. Fatiadi, A.J. *Synthesis* **1987**, 85–127.
215. Hentges, S.G.; Sharpless, K.B. *J. Org. Chem.* **1980**, *45*, 2257–2259.
216. Chong, A.O.; Oshima, K.; Sharpless, K.B. *J. Am. Chem. Soc.* **1977**, *99*, 3420–3426.
217. Herranz, E.; Biller, S.A.; Sharpless, K.B. *J. Am. Chem. Soc.* **1978**, *100*, 3596–3598.
218. Herranz, E.; Sharpless, K.B. *J. Org. Chem.* **1978**, *43*, 2544–2548.
219. Doyle, M.P. *Chem. Rev.* **1986**, *86*, 919–939.
220. Straus, D.A.; Grubbs, R.H. *Organometallics* **1982**, *1*, 1658–1661.
221. Messerle, L.W.; Jennische, P.; Schrock, R.R.; Stucky, G. *J. Am. Chem. Soc.* **1980**, *102*, 6744–6752.
222. McLain, S.J.; Sancho, J.; Schrock, R.R. *J. Am. Chem. Soc.* **1980**, *102*, 5610–5618.
223. Fellmann, J.D.; Schrock, R.R.; Rupprecht, G.A. *J. Am. Chem. Soc.* **1981**, *103*, 5752–5758.
224. Schrock, R.; Rocklage, S.; Wengrovius, J.; Rupprecht, G.; Fellmann, J. *J. Mol. Catal.* **1980**, *8*, 73–83.
225. Schaverien, C.J.; Dewan, J.C.; Schrock, R.R. *J. Am. Chem. Soc.* **1986**, *108*, 2771–2773.
226. Wallace, K.C.; Dewan, J.C.; Schrock, R.R. *Organometallics* **1986**, *5*, 2162–2164.
227. Rocklage, S.M.; Fellmann, J.D.; Rupprecht, G.A.; Messerle, L.W.; Schrock, R.R. *J. Am. Chem. Soc.* **1981**, *103*, 1440–1447.
228. Schröder, M.; Griffith, W.P. *J. Chem. Soc., Dalton Trans.* **1978**, 1599–1602.
229. Schneider, P.W.; Bravard, D.C.; McDonald, J.W.; Newton, W.E. *J. Am. Chem. Soc.* **1972**, *94*, 8640–8641.
230. Mayer, J.M.; Thorn, D.L.; Tulip, T.H. *J. Am. Chem. Soc.* **1985**, *107*, 7454–7462.
231. Tebbe, F.N.; Harlow, R.L. *J. Am. Chem. Soc.* **1980**, *102*, 6149–6151.
232. Wood, C.D.; McLain, S.J.; Schrock, R.R. *J. Am. Chem. Soc.* **1979**, *101*, 3210–3222.

233. Mayr, A.; Lee, K.S.; Kjelsberg, M.A.; Van Engen, D. *J. Am. Chem. Soc.* **1986**, *108*, 6079–6080.
234. Pedersen, S.F.; Schrock, R.R.; Churchill, M.R.; Wasserman, M.J. *J. Am. Chem. Soc.* **1982**, *104*, 6808–6809.
235. Freudenberger, J.H.; Schrock, R.R.; Churchill, M.R.; Rheingold, A.L.; Ziller, J.W. *Organometallics* **1984**, *3*, 1563–1573.
236. Churchill, M.R.; Ziller, J.W.; Freudenberger, J.H.; Schrock, R.R. *Organometallics* **1984**, *3*, 1554–1562.
237. Churchill, M.R.; Ziller, J.W.; McCullough, L.; Pedersen, S.F.; Schrock, R.R. *Organometallics* **1983**, *2*, 1046–1048.
238. Listemann, M.L.; Schrock, R.R. *Organometallics* **1985**, *4*, 74–83.
239. McCullough, L.G.; Listemann, M.L.; Schrock, R.R.; Churchill, M.R.; Ziller, J.W. *J. Am. Chem. Soc.* **1983**, *105*, 6729–6730.
240. Strutz, H.; Dewan, J.C.; Schrock, R.R. *J. Am. Chem. Soc.* **1985**, *107*, 5999–6005.
241. Freudenberger, J.H.; Schrock, R.R. *Organometallics* **1986**, *5*, 1411–1417.
242. Schrock, R.R.; Murdzek, J.S.; Freudenberger, J.H.; Churchill, M.R.; Ziller, J.W. *Organometallics* **1986**, *5*, 25–33.
243. Schrock, R.R.; Pedersen, S.F.; Churchill, M.R.; Ziller, J.W. *Organometallics* **1984**, *3*, 1574–1583.
244. Bradford, C.W.; Nyholm, R.S. *J. Chem. Soc., Chem. Comm.* **1967**, 384–385.
245. Herrmann, W.A.; Küsthardt, U.; Schäfer, A.; Herdtweck, E. *Angew. Chem., Int. Ed. Engl.* **1986**, *25*, 817–818.
246. Chisholm, M.H.; Huffman, J.C.; Marchant, N.S. *J. Chem. Soc., Chem. Comm.* **1986**, 717–718.
247. Churchill, M.R.; Wasserman, H.J.; Holmes, S.J.; Schrock, R.R. *Organometallics* **1982**, *1*, 766–768.
248. Cotton, F.A.; Walton, R.A. "Multiple Bonds Between Metal Atoms" J. Wiley & Sons, New York, 1982.
249. Chisholm, M.H.; Folting, K.; Huffman, J.C.; Kober, E.M. *Inorg. Chem.* **1985**, *24*, 241–245.
250. Chisholm, M.H.; Folting, K.; Huffman, J.C.; Leonelli, J.; Marchant, N.S.; Smith, C.A.; Taylor, L.C.E. *J. Am. Chem. Soc.* **1985**, *107*, 3722–3724.
251. Chisholm, M.H.; Hoffman, D.M.; Huffman, J.C. *Inorg. Chem.* **1985**, *24*, 796–797.
252. Chisholm, M.H.; Folting, K.; Heppert, J.A.; Hoffman, D.M.; Huffman, J.C. *J. Am. Chem. Soc.* **1985**, *107*, 1234–1241.
253. Stone, F.G.A. *Angew. Chem., Int. Ed. Engl.* **1984**, *23*, 89–99.
254. Spitzer, U.A.; Lee, D.G. *J. Org. Chem.* **1975**, *40*, 2539–2540.
255. Carlsen, P.H.J. *Synth. Commun.* **1987**, *17*, 19–23.
256. Bakke, J.M.; Lundquist, M. *Acta Chem. Scand., Ser. B.* **1986**, *B40*, 430–433.
257. Stewart, R.; Spitzer, U.A. *Can. J. Chem.* **1978**, *56*, 1273–1279.
258. Mansuy, D.; Bartoli, J.-F.; Chottard, J.-C.; Lange, M. *Angew. Chem., Int. Ed. Engl.* **1980**, *19*, 909–910.
259. Hill, C.L.; Schardt, B.C. *J. Am. Chem. Soc.* **1980**, *102*, 6374–6375.
260. Suggs, J.W.; Ytuarte, L. *Tetrahedron Lett.* **1986**, 437–440.
261. Wiberg, K.B.; Foster, G. *J. Am. Chem. Soc.* **1961**, *83*, 423–429.
262. Wiberg, K.B.; Fox, A.S. *J. Am. Chem. Soc.* **1963**, *85*, 3487–3491.

263. Schleyer, P.V.R.; Nicholas, R.D. Abstracts of the 140th National Meeting of the American Chemical Society Chicago, Il, Sept. **1971**, abstract no. 75Q.
264. Hamilton, G.A.; Giacin, J.R.; Hellman, T.M.; Snook, M.E.; Weller, J.W. *Ann. New York Acad. Sci.* **1973**, *212*, 4–12.
265. Steenken, S.; Neta, P. *J. Am. Chem. Soc.* **1982**, *104*, 1244–1248.
266. Cainelli, G. "Chromium Oxidations in Organic Chemistry" Springer-Verlag, New York, 1984.
267. Moyer, B.A.; Thompson, M.S.; Meyer, T.J. *J. Am. Chem. Soc.* **1980**, *102*, 2310–2312.
268. Thompson, M.S.; Meyer, T.J. *J. Am. Chem. Soc.* **1982**, *104*, 5070–5076.
269. Breslow, R.; Gellman, S.H. *J. Chem. Soc., Chem. Comm.* **1982**, 1400–1401.
270. White, R.E.; McCarthy, M.-B. *J. Am. Chem. Soc.* **1984**, *106*, 4922–4926.
271. Barton, D.H.R.; Hay-Motherwell, R.S.; Motherwell, W.B. *J. Chem. Soc., Perkin Trans. I* **1983**, 445–447.
272. Breslow, R.; Gellman, S.H. *J. Am. Chem. Soc.* **1983**, *105*, 6728–6729.
273. Barner, C.J.; Collins, T.J.; Mapes, B.E.; Santarsiero, B.D. *Inorg. Chem.* **1986**, *25*, 4322–4323.
274. Wengrovius, J.H.; Schrock, R.R.; Churchill, M.R.; Wasserman, H.J. *J. Am. Chem. Soc.* **1982**, *104*, 1739–1740.
275. Corey, E.J.; Suggs, J.W. *Tetrahedron Lett.* **1975**, 2647–2650.
276. Corey, E.J.; Schmidt, G. *Tetrahedron Lett.* **1979**, 399–402.
277. Holecek, J.; Handlir, K.; Klikorka, J.; Nadvornik, M. *J. Prakt. Chem.* **1982**, *324*, 345–348.
278. Lott, K.A.K.; Symons, M.C.R. *J. Chem. Soc.* **1960**, 973–976.
279. Spence, J.T.; Kronek, P. *J. Less-Common Met.* **1974**, *36*, 465–474.
280. Garner, C.D.; Durant, R.; Mabbs, F.E. *Inorg. Chim. Acta* **1977**, *24*, L29–L30.
281. Wiberg, K.B. in "Oxidation in Organic Chemistry" Wiberg, K.B., ed. Academic, New York, 1965.
282. Thompson, M.S.; Meyer, T.J. *J. Am. Chem. Soc.* **1982**, *104*, 4106–4115.
283. Roecker, L.; Meyer, T.J. *J. Am. Chem. Soc.* **1987**, *109*, 746–754.
284. Stewart, R. *J. Am. Chem. Soc.* **1957**, *79*, 3057–3061.
285. Stewart, R.; Van der Linden, R. *Disc. Faraday Soc.* **1960**, *29*, 211–218.
286. Lee, D.G.; Van den Engh, M. *Can. J. Chem.* **1972**, *50*, 2000–2009.
287. Westheimer, F.H.; Nicolaides, N. *J. Am. Chem. Soc.* **1949**, *71*, 25–28.
288. Rocek, J.; Radkowsky, A.E. *J. Am. Chem. Soc.* **1973**, *95*, 7123–7124.
289. Kaplan, L. *J. Am. Chem. Soc.* **1955**, *77*, 5469–5471.
290. Littler, J.S.; Waters, W.A. *J. Chem. Soc.* **1959**, 4046–4052.
291. Hasan, F.; Rocek, J. *J. Am. Chem. Soc.* **1976**, *98*, 6574–6578.
292. Aoyagi, K.; Nagao, H.; Yukawa, Y.; Ogura, M.; Kuwayama, A.; Howell, F.S.; Mukaida, M.; Kakihana, H. *Chem. Lett.* **1986**, 2135–2138.
293. Wong, K.-Y.; Che, C.-M.; Anson, F.C. *Inorg. Chem.* **1987**, *26*, 737–741.

294 Anderson, S.N.; Fakley, M.E.; Richards, R.L.; Chatt, J. *J. Chem. Soc., Dalton Trans.* **1981**, 1973–1980.
295 Nishihara, H.; Mori, T.; Tsurita, Y.; Nakano, K.; Saito, T.; Sasaki, Y. *J. Am. Chem. Soc.* **1982**, *104*, 4367–4372.
296 Chatt, J.; Pearman, A.J.; Richards, R.L. *J. Chem. Soc., Dalton Trans.* **1977**, 1852–1860.
297 Pickett, C.J.; Ryder, K.S.; Talarmin, J. *J. Chem. Soc., Dalton Trans.* **1986**, 1453–1457.
298 Baumann, J.A.; Bossard, G.E.; George, T.A.; Howell, D.B.; Koczon, L.M.; Lester, R.K.; Noddings, C.M. *Inorg. Chem.* **1985**, *24*, 3568–3578.
299 DuBois, D.L.; Hoffmann, R. *Nouv. J. Chim.* **1977**, *1*, 479–492.
300 Bevan, P.C.; Chatt, J.; Leigh, G.J.; Leelamani, E.G. *J. Organomet. Chem.* **1977**, *139*, C59–C62.
301 Massa, W.; Hermann, S.; Dehnicke, K. *Z. Anorg. Allg. Chem.* **1982**, *493*, 33–40.
302 Pickett, C.J.; Tolhurst, J.E.; Copenhauer, A.; George, T.A.; Lester, R.K. *J. Chem. Soc., Chem. Comm.* **1982**, 1071–1072.
303 Lee, D.G.; Moylan, C.R.; Hayashi, T.; Brauman, J.I. *J. Am. Chem. Soc.* **1987**, *109*, 3003–3010.
304 Silavwe, N.D.; Chiang, M.Y.; Tyler, D.R. *Inorg. Chem.* **1985**, *24*, 4219–4221.
305 Liebelt, W.; Dehnicke, K. *Z. Naturforsch.* **1979**, *34B*, 7–9.
306 Dolgoplosk, B.A.; Oreshkin, I.A.; Makovetsky, K.L.; Tinyakova, E.I.; Ostrovskaya, I.Ya.; Kershenbaum, I.L.; Chernenko, G.M. *J. Organomet. Chem.* **1977**, *128*, 339–344.
307 Rupprecht, G.A.; Messerle, L.W.; Fellmann, J.D.; Schrock, R.R. *J. Am. Chem. Soc.* **1980**, *102*, 6236–6244.
308 Schwartz, J.; Gell, K.I. *J. Organomet. Chem.* **1980**, *184*, C1–C2.
309 Chisholm, M.H.; Folting, K.; Hoffman, D.M.; Huffman, J.C. *J. Am. Chem. Soc.* **1984**, *106*, 6794–6805
310 Wilker, C.N.; Hoffmann, R.; Eisenstein, O. *Nouv. J. Chim.* **1983**, *7*, 535–544.
311 McDermott, G.A.; Mayr, A. *J. Am. Chem. Soc.* **1987**, *109*, 580–582.
312 Roecker, L.; Dobson, J.C.; Vining, W.J.; Meyer, T.J. *Inorg. Chem.* **1987**, *26*, 779–781.
313 Eisenstein, O.; Hoffmann, R.; Rossi, A.R. *J. Am. Chem. Soc.* **1981**, *103*, 5582–5584.
314 Tillotson, A.; Houston, B. *J. Am. Chem. Soc.* **1951**, *73*, 221–222.

CHAPTER 7

ROLE OF METAL–LIGAND MULTIPLE BONDS IN CATALYSIS

This chapter deals with the role of multiply bonded ligands in catalytic processes. The subject matter encompasses examples of heterogeneous, homogeneous, and enzymatic catalysis. Obviously we can not hope to provide a comprehensive review of the current knowledge of these processes—many of the individual reactions are themselves the topics of several books and/or review articles. Instead we shall focus on two particular points that are relevant to our principal subject. First, we will summarize existing evidence for the involvement of multiply bonded species in each of the catalytic transformations. Second, we will discuss the results of mechanistic studies that may provide insight into the particular function of metal–ligand multiple bonding in promoting this chemistry.

7.1 OXOMETAL SPECIES IN CATALYSIS

7.1.1 Heterogeneous Oxidations

A heterogeneous oxidation reaction of industrial importance is the selective oxidation of propylene to acrolein in Eq. 1. This process and the closely related ammoxidation of propylene (Section 7.2.1) have been the subject of extensive mechanistic investigations; work in this area through early 1978 has been reviewed [1]. As indicated in Eq. 1, selective propylene oxidation is promoted by so-called bismuth molybdate catalysts. It should be noted that contemporary commercial catalysts contain not only bismuth molybdate but also a variety of empirical transition metal additives whose function is not well understood. However, bismuth molybdate itself remains a favored catalyst for mechanistic studies.

$$CH_3CH=CH_2 + O_2 \xrightarrow[400°C]{Bi_2O_3/MoO_3} CH_2=CHCH=O + H_2O \qquad (1)$$

Such catalysts contain multiply bonded oxo species. Three distinct "bismuth molybdate" phases are known, referred to as α, β, and γ. These correspond to the compositions $Bi_2Mo_3O_{12}$, $Bi_2Mo_2O_9$, and Bi_2MoO_6, respectiveiy. The bulk structures of the α and γ phases are known [2,3] while the structure of the β phase remains somewhat controversial [1]. Both long and short Mo—O bonds are present (1.72 and 1.87 Å for the tetrahedral α-molybdate; 1.75 and 2.24Å for the octahedral γ-molybdate). The shorter Mo—O distances are clearly indicative of multiple bond character. Perhaps more relevant to catalysis are the studies of Trifiro and coworkers using reflectance UV and IR spectroscopy [4]. Their results indicate the presence of multiply bonded terminal oxo groups on the catalyst surface. Both octahedral and tetrahedral oxomolybdenum species are said to be present.

Moreover, there can be little doubt that lattice oxygen is involved in the oxidation process. According to the Mars–van Krevelen model [5] such catalytic oxidations are believed to proceed by initial oxidation of the substrate followed by reoxidation of the reduced form of the catalyst by dioxygen (Eqs. 2 and 3). Such a model can be confirmed in the case of propylene oxidation by studies using ^{18}O-labeled catalyst or ^{18}O-labeled dioxygen [6,7]. Moreover, it has been shown that stoichiometric propylene oxidation can be carried out in the absence of oxygen, essentially using bismuth molybdate as a reagent [8].

$$\text{substrate} + M_{ox} \rightarrow \text{product} + M_{red} \qquad (2)$$

$$M_{red} + O_2 \rightarrow M_{ox} \qquad (3)$$

The proposal that multiply bonded terminal oxo groups are directly involved in propylene oxidation was set forth by Trifiro and Pasquon in 1968 [9]. They based their argument on a comparison of a series of metal oxide catalysts which do not contain multiply bonded oxygen and a series which do possess such multiple bonds. In general, the former series was active in the isotopic homomolecular exchange of O_2 and oxidized ammonia to N_2O, while CO, methanol, and propylene were all completely oxidized to CO_2. The latter series showed little activity for isotopic homomolecular exchange of O_2 or oxidation of CO; ammonia, methanol, and propylene were all selectively oxidized affording N_2, formaldehyde, and acrolein, respectively. They concluded that "the dehydrogenating power of the catalysts of the second class is directly connected with the metal–oxygen double bond." In the intervening years, this notion has been widely embraced by the heterogeneous catalysis community, although, in our opinion, there has been no definitive demonstration of its validity.

Another widely held tenet is that the rate-determining step in propylene oxidation is C—H bond cleavage to afford a symmetrical allyl intermediate—most frequently asserted to be an adsorbed allyl radical [1]. It has been shown that oxidation of propylene which is isotopically labeled at one end with ^{2}H or ^{14}C produces acrolein in which the label is scrambled between the 1 and 3 positions [10]. However, caution appears to be required in interpreting these results. Subsequent studies on the oxidation of allyl alcohol-1,1-d_2 and -3,3-d_2 over bismuth molybdate also result in statistical scrambling of the isotopic label in the product acrolein [11]. Apparently molybdenum allyloxide species on the catalyst undergo rapid interconversion of the 1 and 3 positions—not a suprising result in view of the facile homogeneous allylic rearrangements, which we discuss in Section 7.1.2.

It is frequently suggested that allyl radicals are generated at bismuth sites in the bismuth molybdate catalyst. This suggestion seems sensible on several scores. When a mixture of propylene and oxygen is passed over bismuth oxide at 475°C, catalytic dehydrodimerization to 1,5-hexadiene takes place [12]. This is the expected product of allyl radical dimerization and in fact Lunsford has demonstrated the formation of allyl radicals under these conditions using low-temperature matrix isolation and ESR [13]. As early as 1973 Trifiro pointed out the apparent analogy between heterogeneous propylene oxidation and allylic oxidations involving selenium dioxide [14]. In the intervening years it has been shown that the latter reactions are essentially ene reactions involving the ^{IV}Se=O double bond [15]. Evidence has been presented that surface cation vacancies in propylene oxidation catalysts generate "higher bond order" bismuth–oxygen sites which promote allyl formation [16]. Most significant in this regard are recent experiments in which Bi^{3+} was deposited selectively on the (100) and (001) faces of molybdenum oxide platelets. At low coverages it could be shown that the yield of acrolein was directly proportional to the surface concentration of Bi^{3+} [17].

In the subsequent step it appears that the adsorbed allyl radicals are further oxidized with oxomolybdenum(VI) acting as an "electron sink." Allyl radicals as open shell species are fairly potent reducing agents. Even typical alkyl radicals will reduce permanganate with a rate constant $>10^9 \ M^{-1} \ s^{-1}$, in the diffusion controlled range [18]. Likewise, surface molybdate species apparently provide the necessary "hole" to convert adsorbed allyl radicals to adsorbed allyl cations. When allyl radicals generated by passing propylene over bismuth oxide subsequently are passed over MoO_3 placed immediately downstream, the radicals are efficiently scavenged [13]. Grasselli and coworkers studied the products formed from authentic allyl radicals (generated by thermal extrusion of dinitrogen from 3,3'-azopropene) and molybdenum trioxide at 320°C [19]. Under sufficiently dilute conditions acrolein was formed in greater than 50% yield based on N_2.

Additional indirect evidence supports the intermediacy of allyl cations. This includes the scrambling of labeled allyl alcohol under propylene oxidation conditions as noted above [12]. Furthermore, replacing propylene with 1-butene gives high yields of 1,3-butadiene, best rationalized as the product of deprotonation of crotyl cation (Eq. 4). Indeed, treatment of MoO_3 with an allyl cation equivalent, allyl iodide, at 320°C is said to afford acrolein in 100% yield [20]. However, it is not clear whether allyl iodide is acting as an alkylating agent or radical source under these conditions.

$$CH_3CH_2CH{=}CH_2 + 0.5 \ O_2 \xrightarrow[350°C]{Bi_2O_3/MoO_3} CH_2{=}CHCH{=}CH_2 + H_2O \quad (4)$$

Finally, it appears that allyl cations are trapped by lattice oxygen to form a surface allyloxide moiety. The expected fate of an alkoxide at such elevated temperatures would be to lose β hydrogen. This additional C—H bond cleavage gives the final product, acrolein.

In summary, terminal oxo groups on the catalyst surface are necessary components of propylene oxidation catalysts. At minimum they stabilize the requisite high oxidation states and impart coordinative unsaturation in the surface molybdenum atoms. There is some suggestion that they additionally promote C—H bond cleavage and provide nucleophilic sites to trap allyl cations. However, unambiguous insight into these mechanistic details will have to await additional studies.

Perhaps the best hope for truly detailed mechanistic insight into a heterogeneous oxidation involves the oxidation of methanol. In commercial practice this chemistry often utilizes an iron molybdate catalyst (Eq. 5). ^{2}H Isotope effects confirm that the reaction proceeds via rate limiting C—H bond cleavage [21]. Moreover, the model reaction between crystalline MoO_3 and methanol has proven amenable to studies using the powerful techniques of contemporary surface science, including spectroscopic methods (XPS and UPS) and temperature-programmed desorption studies [22]. It was shown that:

1. Dissociative chemisorption of methanol, that is, formation of surface methoxide groups, occurs with concurrent loss of H_2O.
2. The sites for dissociative chemisorption are localized on non-[010] faces only. (Note that dioxomolybdenum species are present on all of the surfaces of MoO_3. No Mo—O bonds need be broken forming the [010] face. Long bridging Mo—O bonds must be broken to form the remaining faces freeing potential coordination sites.)
3. Both dissociative adsorption and oxidation activity can be blocked by pretreatment with Lewis bases and therefore probably involve Lewis acid sites. However, it is not known whether terminal oxomolybdenum species are involved in cleavage of the methyl C—H bond.

$$MeOH + 0.5\ O_2 \xrightarrow{Fe_2(MoO_4)_3} CH_2{=}O + H_2O \quad (5)$$

A final oxidation of commercial interest appears much more complicated. The oxidation of simple C_4 hydrocarbons (butane and butenes) to maleic anhydride typically employs a vanadium phosphorus oxide catalyst [194]. Limited insight into this complex transformation is available. Nevertheless, X-ray diffraction studies on active catalysts have revealed the presence of a vanadium(IV) pyrophosphate phase $(VO)_2P_2O_7$. The isolated phase was shown to be an active catalyst [23]. An X-ray crystal structure of this material indicates that the structure is built of double chains of VO_6 octahedra—again consistent with a catalyst surface comprised of terminal oxo–vanadium units [24].

7.1.2 Rearrangement of Allylic and Propargylic Alcohols

The catalytic rearrangement of propargylic and allylic alcohols is promoted by soluble oxometal complexes [25]. These reactions are exemplified by the rearrangement of dehydrolinalool to citral (Eq. 6) and the equilibration of linalool with geraniol/nerol (Eq. 7). Commercial applications of this chemistry have recently been reviewed [26]. The propargyl alcohol rearrangement has also been applied to organic synthesis [27,28].

$$\text{dehydrolinalool} \xrightarrow[\text{catalyst}]{160°C} \text{citral} \quad (6)$$

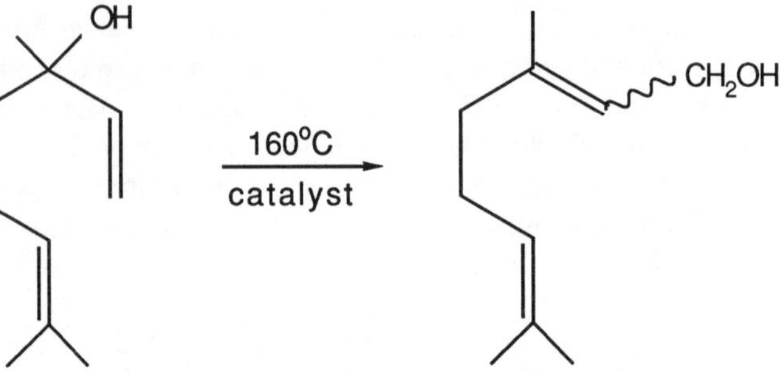

[Reaction (7): allylic alcohol rearrangement at 160°C with catalyst]

(7)

Both propargylic and allylic rearrangements are promoted by oxovanadium(V) catalysts [29,30], including triphenylsilyl vanadate, **1**. Allylic alcohol rearrangement is also catalyzed by tungsten(VI) catalysts such as **2** [31]. Soviet workers have favored the use of polymeric silyl vanadates [32].

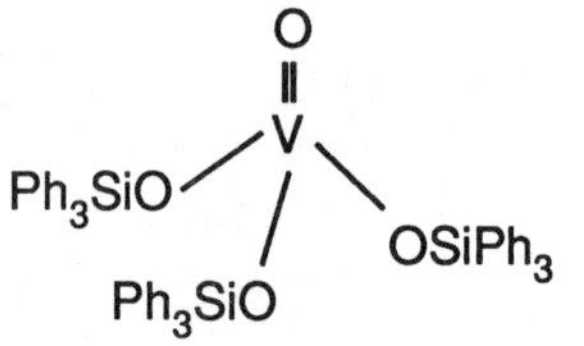

1: O=V(OSiPh$_3$)$_3$

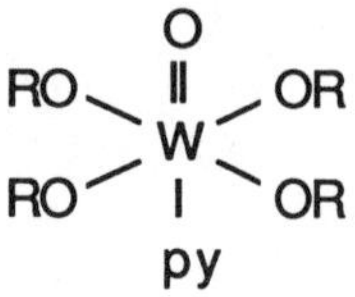

2: O=W(OR)$_4$·py

These rearrangements are widely presumed to proceed via a cyclic mechanism as shown in Eqs. 8 and 9. Several facts are consistent with this proposal: These pathways represent the transition metal analog of known electrocyclic transformations in organic chemistry, that is, the Claisen rearrangement. (Indeed the original invention by Chabardes [25] was based on this line of reasoning.) The reaction is kinetically zero order in excess substrate, supporting the intermediacy of the metallate ester [33]. Again consistent with the "metalla-Claisen" model, the reaction proceeds best in nonpolar hydrocarbon solvents. Because rearrangement of propargyl alcohols initially gives the thermodynamically disfavored Z aldehyde, it is proposed that subsequent tautomerism of the allenyl moiety from Eq. 8 occurs in the metallate ester rather than the free allenyl alcohol [34].

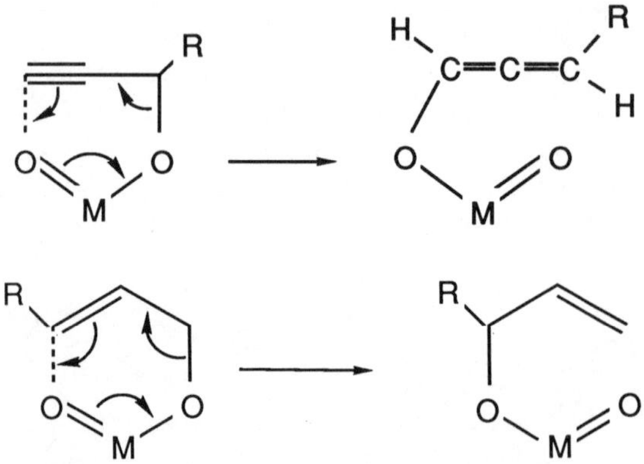

(8)

(9)

The concerted pathway in Eq. 9 suggests that the rearrangement should be stereospecific. No test of this premise has been made using typical (high temperature) catalysts. However, Takai and coworkers [35] have recently introduced new Mo and V catalysts which effect the rearrangement of primary and secondary allyl alcohols to the tertiary isomers under mild conditions. Using a catalyst derived from a 1:1 mixture of $VO(acac)_2$ and bis(trimethylsilyl) peroxide the optically active alcohol in Eq. 10 rearranged with significant retention of configuration.

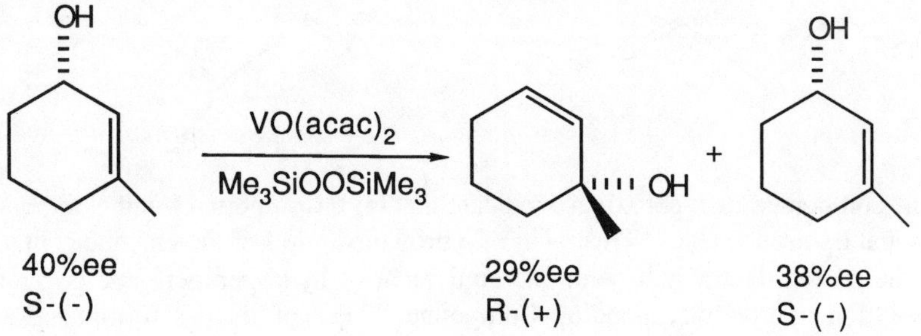

$$\text{(structures)} \quad (10)$$

40%ee   29%ee   38%ee
S-(−)   R-(+)   S-(−)

This result clearly supports the metalla-Claisen mechanism of Eq. 9 under these particular conditions. These results may be contrasted with those for the allylation of imido ligands to be discussed in Section 7.2.1. Some caution seems indicated in generalizing the results of Eq. 10 to other systems.

7.1.3 Enzymatic Catalysis by the Heme-Containing Peroxidases and Oxygenases

The peroxidases [36] are enzymes that catalyze the oxidation of inorganic or organic substrates according to Eq. 11 where the stoichiometric oxidant ROOH may be H_2O_2, an alkyl hydroperoxide, or a peracid. Most peroxidases contain ferriprotoporphyrin IX as their prosthetic group. Perhaps the most widely studied of these enzymes is the readily available horseradish peroxidase.

$$ROOH + AH_2 \xrightarrow{\text{enzyme}} ROH + A + H_2O \qquad (11)$$

Many peroxidase-catalyzed oxidations proceed by the sequence summarized in Eqs. 12–15, where Enz represents the native enzyme:

$$Enz + ROOH \rightarrow \text{Compound I} + ROH \qquad (12)$$

$$\text{Compound I} + AH_2 \rightarrow \text{Compound II} + AH\cdot \qquad (13)$$

$$\text{Compound II} + AH_2 \rightarrow Enz + AH\cdot \qquad (14)$$

$$AH\cdot \rightarrow \text{Nonradical products} \qquad (15)$$

Horseradish peroxidase exhibits this type of catalytic cycle with a range of reducing substrates AH_2 including phenylbutazone (Eq. 16), p-cresol, and N,N-dimethylaniline [37–39]. In some other cases Compound I undergoes direct two-electron reduction to the native enzyme without formation of Compound II. An example is the enzyme catalase whose function is to destroy hydrogen peroxide in peroxisomes [40]. In this case the substrate AH_2 is a second equivalent of hydrogen peroxide which undergoes direct two-electron oxidation to dioxygen.

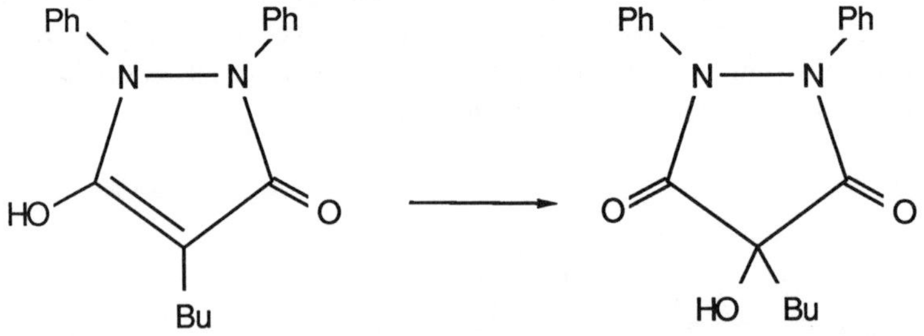

$$\text{(16)}$$

Studies on horseradish peroxidase indicate that both Compound I and Compound II are terminal oxoiron species. Treatment of a brown solution of the enzyme containing iron in the 3+ oxidation state with one equivalent of hydrogen peroxide generates a green solution of the corresponding Compound I. The product is formulated as an oxoiron(IV) complex in which the porphyrin ring has additionally been oxidized to the radical cation. This description is supported by Mossbauer [41], EPR [41], ENDOR [42], NMR [43], and MCD [44] studies. Upon treatment with ferrocyanide, Compound I undergoes one-electron reduction to the red Compound II. Mossbauer and EPR data show that this is again an oxoiron(IV) species but that the porphyrin cation radical has been reduced [41]. EXAFS studies indicate that the iron–oxygen distance in both Compounds I and II is about 1.64 Å, consistent with double-bond character [45].

Enzymes of the oxygenase family differ from the peroxidases in that they utilize molecular oxygen as stoichiometric oxidant. Among the heme-containing oxygenases, cytochrome P-450 is by far the most widely studied [46]. We have previously noted the unique capability of this enzyme for selective hydroxylation of unactivated hydrocarbons under physiological conditions (for example, Eq. 17).

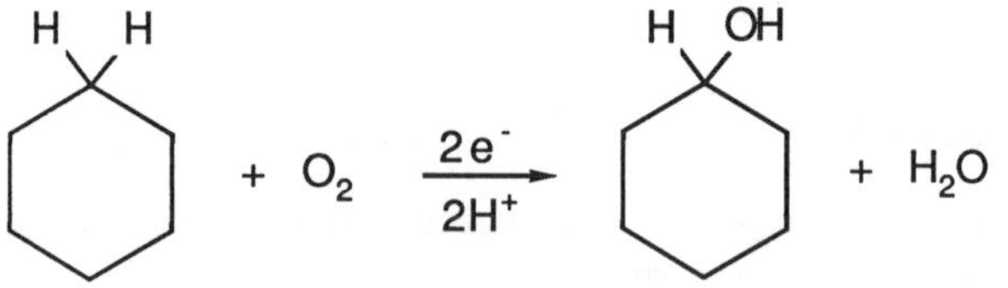

$$\text{(17)}$$

Catalysis by cytochrome P-450 is believed to resemble that involving horseradish peroxidase, at least to the extent that the activated form of the enzyme is an oxoiron species which is two oxidizing equivalents above the iron(III) resting state [47]. (Note, however, that horseradish peroxidase, in contrast to cytochrome P-450, generally does not catalyze oxygen-transfer reactions. Recent results indicate that oxygen-transfer reactions do not occur because the oxo ligand and the substrate are physically separated by a protein-imposed barrier in horseradish peroxidase [48].) Generation of this oxo species from native (low-spin) ferric P-450 apparently involves five distinct steps:

1. Binding of the substrate to generate a high-spin ferric complex.

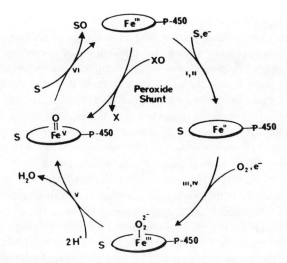

Figure 7.1 Catalytic cycle of cytochrome P-450. (From ref. 47 with permission of Plenum Publishing Corp.)

2. One-electron reduction of the iron to the +2 oxidation state.
3. Binding of dioxygen to give a iron(III) superoxide intermediate.
4. A second one-electron reduction which yields an iron(III) peroxo (O_2^{2-}) species.
5. Formal heterolysis of the O—O bond affording the reactive oxidant $[Fe=O]^{3+}$ and a molecule of water.

These steps are incorporated into the catalytic cycle [47] shown in Figure 7.1.

Another similarity to peroxidase is that the overall two-electron oxidation in P-450 mediated hydroxylation is believed to occur as a pair of discrete one-electron steps. The stepwise mechanism is supported by the stereochemical scrambling that is observed with certain substrates [49–51]. For example, oxidation of optically active ethylbenzene-1-d_1 with P-450 from rabbit liver microsomes (Eq. 18) gave a partially racemized product [52]. The observation of large intrinsic isotope effects in the range k_H/k_D = 10–13 further militates against concerted oxenoid insertion [53–55]. For example, the intramolecular isotope effect for ω-hydroxylation of [1,1,1-d_3]-n-octane is 9.8; 15% of this is due to the secondary isotope effect leaving a primary k_H/k_D = 7.6 ± 0.3 [56]. (To avoid kinetic complications it is essential that isotope effects in such enzymatic systems be determined via intramolecular competition.)

Such results suggest that initial C—H cleavage in P-450 results in formation of an alkyl radical. However, subsequent trapping of this radical appears to be extremely fast. Ordinary "radical clocks" such as the opening of the cyclopropylmethyl radical (1×10^8 s^{-1}) cannot compete. Experiments using the highly strained probe bicyclo[2.1.0]pentane indicate that the radical pair formed in P-450 catalyzed hydroxylation collapses at a rate in excess of 1×10^9 s^{-1} [57].

Given this radical mechanism it is clear that the high stereospecificity normally observed in such hydroxylations must arise from interactions between the substrate and surrounding protein. Our understanding of this phenomenon has been greatly enhanced by the availability of high-resolution crystal structures of bacterial cytochrome P-450$_{cam}$ in both the substrate-free and substrate-bound forms [58,59]. This enzyme effects the stereospecific hydroxylation of camphor to 5-exo-hydroxycamphor. Substrate specificity is imposed by hydrogen bonding between the carbonyl oxygen of camphor and a strategically placed tyrosine residue and also by van der Waals attraction between the methyl groups of camphor and hydrophobic portions of the protein. In the activated enzyme the 5-exo hydrogen will be held proximate to the oxo oxygen as in structure **3**.

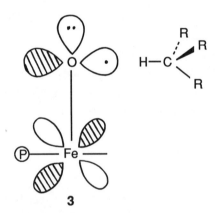

3

A final comment pertains to the so-called "peroxide shunt" in Figure 7.1. This arrow represents the important discovery that P-450 hydroxylations can be carried out without the involvement of molecular oxygen by the addition of exogenous oxygen sources such as ROOH [60], PhIO [61], or NaIO$_4$ [62]. This finding opened the door for extensive studies using metalloporphyrin models for cytochrome P-450. These model studies, which were discussed in Sections 6.3.4.1 and 6.3.8, have significantly contributed to our understanding of P-450 activity. McMurry and Groves have published a comprehensive review covering these model systems [47].

7.1.4 The Molybdenum "Oxo-Transferase" Enzymes

"Oxo-transferase" is a term coined by Holm [63] to describe a widely distributed family of molybdenum redox enzymes that promote the addition of an oxygen atom to, or its removal from, a substrate. These include hydroxylases such as xanthine oxidase, aldehyde oxidase, and sulfite oxidase. Also included are formate dehydrogenase and reductases such as nitrate reductase [65].

The presence of oxo ligands in these enzymes is indicated by ^{17}O EPR effects in the partially reduced enzymes and especially by X-ray absorption fine-structure (EXAFS) studies [66]. Data on the number of oxo ligands and Mo—O bond length for several enzymes are shown in Table 7.1 All of these enzymes additionally contain one or more sulfur ligands as indicated by Mo—S distances in the 2.15 to 2.86 Å range. This fact is consistent with model studies that have uncovered the unique ability of sulfur-containing ligands to moderate the reduction potential of molybdenum [70].

It is now generally believed that these enzymes function, at least in some cases, by direct transfer of the oxo ligand to the organic substrate. This model is supported by experiments [71] involving the enzymatic oxidation of xanthine to uric acid with ^{18}O-labeled nicotinamide N-oxide as stoichiometric oxidant (Eq. 19). Some 67% of the uric acid produced contained the ^{18}O label. An interpretation of these results has been given by Holm and Berg [63]: The amine oxide transfers its oxygen to molybdenum(IV) to give oxomolybdenum(VI); this in turn transfers the oxo moiety to substrate. (The lack of complete incorporation may reflect exchange of an oxo ligand with unlabeled water and/or another pathway. Analogous oxidations carried out in labeled water gave ~25% labeled product.)

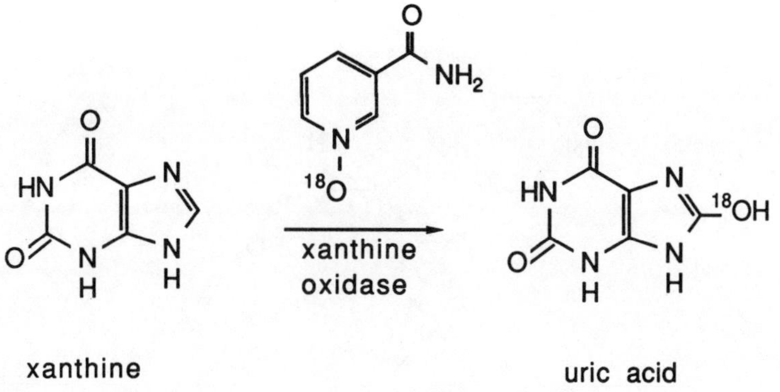

$$\text{xanthine} \xrightarrow{\text{xanthine oxidase}} \text{uric acid} \quad (19)$$

xanthine ⟶ uric acid

Model studies have lent significant support to this oxo transfer model [70]. In particular, it has been possible to reduce the natural substrate d-biotin d-(S-oxide) to biotin using discrete oxomolybdenum complexes as enzyme mimics. Equally important is the demonstration that thiols can be employed as "physiologically relevant" stoichiometric reductants. Model studies relevant to oxo transferase activity have been reviewed [63].

A development that portends significant advances in this area is the isolation of the dissociable molybdenum cofactor "Mo-co." This substance has the ability to generate enzyme activity when added to Mo-co-deficient apoproteins [72]. Degradation studies indicate that this cofactor has the minimum structure **4** [73].

TABLE 7.1 Metal–Oxygen Distances in Some Oxomolybdoenzymes by EXAFS

Protein	Source	Oxidation State	Number of Oxo's	Distance (Å)	Ref.
Xanthine Oxidase	Bovine Milk	ox.	1 or 2	1.75	67
Xanthine Dehydrogenase	Chicken Liver	ox.	1	1.70	68
		red.	1	1.68	68
Sulfite Oxidase	Chicken Liver	ox.	2	1.68	68
		red.	1	1.69	68
Nitrate Reductase	Chorella vulgaris	ox.	2	1.72	69
		red.	1	1.67	69
Mo-protein	Desulfovibrio gigas	ox.	2	1.68	66
		red.	1	1.68	66
Formate Dehydrogenase	Clostridium Pasteurianum	ox.	3	1.74	66

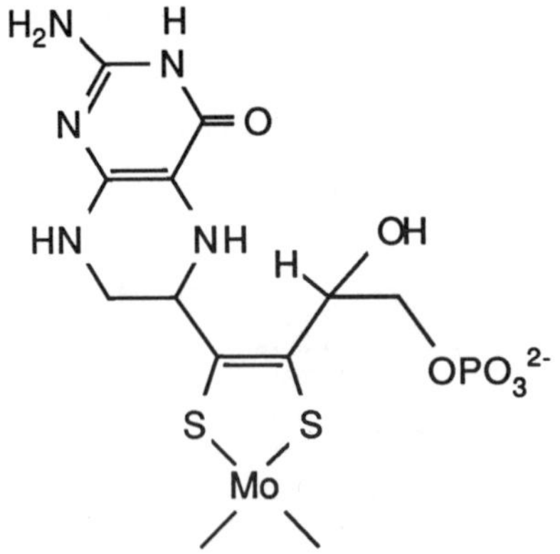

4

7.1.5 Other Homogeneous Oxidations Involving Oxometal Intermediates

A variety of catalytic homogeneous oxidations are mediated by oxometal species. In each example enumerated below the corresponding stoichiometric oxidation utilizing the isolated oxometal reagent can be studied and compared with the catalytic process. These studies indicate that the role of the stoichiometric oxidant is to regenerate the high-valent oxometal continuously *in situ* by pathways we have discussed in Chapter 3. Reaction with the substrate then proceeds as described in Chapter 6.

The cis-hydroxylations of olefins involving OsO_4 [74], as well as a variety of RuO_4-mediated oxidations [75,76], can be carried out in catalytic fashion. Newer versions of the OsO_4 reactions using t-butyl hydroperoxide [77] or N-methylmorpholine N-oxide [78] as stoichiometric oxidant have the advantage of avoiding overoxidation to the α-ketol. In the case of RuO_4-catalyzed oxidations, hypochlorite or periodate are the stoichiometric oxidants of choice. Sharpless and coworkers have optimized the procedure for the periodate-based reactions [79]. A solvent system containing both acetonitrile and carbon tetrachloride was shown to be advantageous.

In Chapter 6 we described some of the stoichiometric oxidations that can be achieved using Meyer's oxoruthenium(IV) system $[(trpy)(bipy)RuO]^{2+}$. It has subsequently been shown that these reactions can be carried out catalytically under phase-transfer conditions using hypochlorite as stoichiometric oxidant [80]. Although there remains some question about the active form of ruthenium under these catalytic conditions, an electochemical procedure has been reported that clearly involves a Ru^{2+}/Ru^{4+} cycle [81]. This electrocatalytic procedure allows the net electrochemical oxidation of alcohols, aldehydes, and unsaturated hydrocarbons under mild conditions. Moreover, Meyer has pointed out the potential for a tantalizing future development: If a reversible, high current-density oxygen electrode could be developed and coupled with these reactions, the resulting cells could be made to operate spontaneously, in effect becoming electrochemical synthesis fuel cells [81].

A long-standing goal of homogeneous catalysis has been to carry out oxidations of organic substrates using only molecular oxygen. In the case of olefin epoxidation, this goal has now been achieved by Groves and coworkers [82]. The active ruthenium species is a $trans$-dioxoruthenium(VI) porphyrin complex, $(TMP)Ru(O)_2$. The generation of this species by aerobic oxidation of the ruthenium(II) porphyrin was discussed in Section 3.2.5. The catalytic epoxidation proceeds with retention of olefin configuration; yields corresponding to 45 catalyst turnovers were reported.

7.2 IMIDOMETAL SPECIES IN CATALYSIS

7.2.1 Ammoxidation of Propylene and Methylarenes

When ammonia is added to the feed, the same bismuth molybdate catalysts that effect acrolein synthesis (Section 7.1.1) can be used for the manufacture of acrylonitrile (Eq. 20) [1]. Using this technology, 8 billion pounds of acrylonitrile are produced worldwide each year. A related process is the ammoxidation of methylarenes to aryl nitriles (Eq. 21). A vanadium oxide catalyst is more commonly employed for this latter transformation [83].

$$CH_2\!\!=\!\!CHCH_3 + 1.5\ O_2 + NH_3 \xrightarrow[\text{catalyst}]{450°C} CH_2\!\!=\!\!CHC\!\!\equiv\!\!N + 3H_2O \quad (20)$$

$$ArCH_3 + 1.5\ O_2 + NH_3 \xrightarrow[\text{catalyst}]{450°C} ArC\!\!\equiv\!\!N + 3H_2O \quad (21)$$

In Section 7.1.1 we saw that propylene oxidation catalysts could be regarded as efficient engines for converting propylene to allyl radicals and for their subsequent one-electron oxidation to the allyl cation oxidation state. In general this picture can be applied directly to propylene ammoxidation as well [1]. The principal bone of contention in this area is whether the ultimate nucleophiles are molybdenum imido species or some other species such as adsorbed ammonia. The imido model has been advocated by Grasselli and coworkers at SOHIO [19] and has subsequently been embraced by many other workers in the field [84,85].

The SOHIO group proposes the diimido structure **5** for the active sites. C—H cleavage occurs at the adjacent bismuth, and the allyl moiety is then trapped by imido nitrogen. The proposal is based principally on analogy to acrolein synthesis, where it is assumed that terminal oxo groups on molybdenum are the ultimate nucleophiles. (Note that, despite some heroic studies on acrolein synthesis [86], this point remains unresolved.) Others have observed that in flow systems, when a pulse of ammonia is applied to the catalyst, a pulse of displaced water vapor is detected in the effluent. If a pulse of propylene is injected after 3 seconds, acrylonitrile is formed cleanly; however, injection of propylene after 60 seconds gives only acrolein [84]. This indicates that at the reaction temperature and in the absence of oxygen, the ammonia is being oxidized to nitrogen oxides. While consistent with the formation of imido species, such observations do not require imido intermediates.

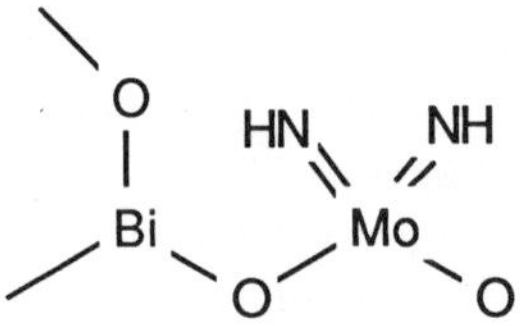

5

Observations on model systems indicate that if imido sites exist, their proposed reaction with allyl or benzyl radicals are chemically reasonable. For example, the diimido–chromium(VI) complex in Eq. 22 reacts with toluene in the presence of a stoichiometric radical source to afford the imine, benzylidene-t-butylamine, in up to 50% yield [87]. Imine was also formed in experiments using the corresponding diimido–molybdenum(VI), but the yield fell to 7%.

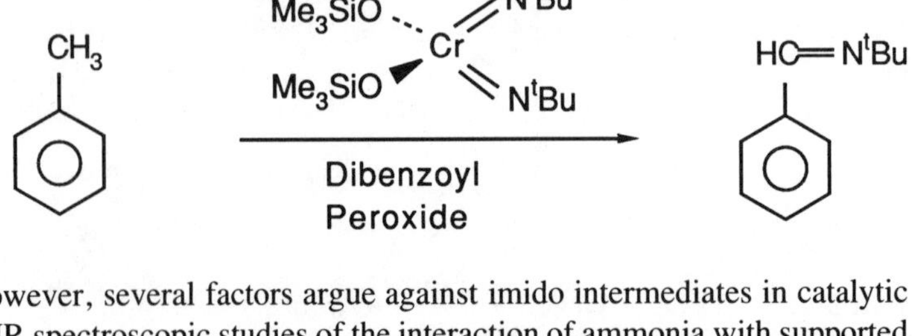

$$\text{PhCH}_3 + (\text{Me}_3\text{SiO})_2\text{Cr}(\text{N}^t\text{Bu})_2 \xrightarrow{\text{Dibenzoyl Peroxide}} \text{PhCH=N}^t\text{Bu} \quad (22)$$

However, several factors argue against imido intermediates in catalytic ammoxidation. IR spectroscopic studies of the interaction of ammonia with supported molybdates indicate the presence of only ammonium ions and adsorbed NH_3 [88]. From our discussion in Chapter 6 it is evident that molybdenum-bound imido ligands are hydrolytically unstable with respect to the corresponding oxos. While a monoimido molybdenum species could be kinetically sampled, the proposal that significant concentrations of diimido sites like **5** exist on the catalyst surface seems to us unlikely. Moreover, it appears unnecessary to invoke imido intermediates: Once the allyl cation oxidation state has been reached (whether as an adsorbed carbocation or a molybdate ester), the allyl moiety should be trapped by any available nucleophilic nitrogen including ammonia itself. Note that antimonate-based ammoxidation catalysts in which terminal imido ligands are unquestionably absent function quite adequately [1,89].

Kinetic studies by the SOHIO group have suggested that under some circumstances during ammoxidation the allyl moiety intially undergoes C—O bond formation to afford a surface allyloxide moiety [89]. Under such circumstances the catalyst is said to tranfer the allyl moiety to imido nitrogen in a separate subsequent step, either by heterolysis to regenerate the allyl cation or by the electrocyclic process shown in Eq. 23. The ammoxidation of allyl alcohol is cited in support of such allyl transfer [11]. (The oxidation of labeled allyl amine shows that no further migration of the allyl moiety occurs after C—N bond formation [90].) Model studies using tungsten(VI) imido complexes [87] indicate that such migrations can in fact occur under mild conditions (100–140°C) and favor an ionic pathway for this rearrangement.

$$\text{(23)}$$

7.2.2 Imido Intermediates in Homogeneous Catalysis

Earlier (Section 6.3.4.2) we discussed the stoichiometric vicinal oxyamination of olefins by osmium(VIII) complexes including $Os(O)_3(N^tBu)$. The Sharpless group has developed several catalytic versions of this reaction in which chloramine-T [91,92] or N-chloro-N-argentocarbamates [93] serve as stoichiometric oxidant. It is believed that these oxidizing agents continuously reoxidize osmium to an imidoosmium(VIII) which then adds to the olefinic substrate. Although this pathway has never been demonstrated, it appears reasonable. Chloramine-T was shown to be competent to oxidize other reduced osmium species to an isolable imidoosmium(VIII) compound (see Eq. 49 in Chapter 3).

A somewhat mysterious reaction is the electrocatalytic tosylamidation reported by Breslow and coworkers [94]. Cyclohexenyl hydroperoxide was found to react with toluenesulfonyl amide at an applied potential of 1.8 V according to Eq. 24. The remarkable observation is that this reaction takes place only with a vanadium anode, leading to the suggestion that it specifically involves vanadium tosylimido sites on the electrode surface. (It is interesting to speculate whether there might be a relationship between this reaction and the facile allylic rearrangements promoted by oxovanadium(IV) in the presence of peroxide (Eq. 10) and/or with the known electrochemical synthesis of oxovanadium(IV) complexes from a vanadium anode [95].) Further studies would be of interest.

$$\text{C}_6\text{H}_{11}\text{-OOH} \xrightarrow[-2e^-]{\text{ArSO}_2\text{NH}_2} \text{C}_6\text{H}_{11}\text{-NHSO}_2\text{Ar} \qquad (24)$$

Weiss and coworkers [96] have utilized imido complexes to catalyze the exchange of imide groups in carbodiimides (Eq. 25). This "aza analog" of olefin metathesis appears to proceed by a four-center mechanism of the type discussed in Section 3.2.6.

$$\text{RN}=\text{C}=\text{NR} + \text{R'N}=\text{C}=\text{NR'} \xrightleftharpoons[]{\text{Cl}_4\text{W}=\text{NR}} 2\text{RN}=\text{C}=\text{NR'} \qquad (25)$$

7.3 HYDRAZIDO INTERMEDIATES IN ENZYMATIC NITROGEN FIXATION

Nitrogen-fixing microorganisms utilize the enzyme nitrogenase [97,98] to promote Eq. 26. Nitrogenase can be separated into two metalloproteins, the larger of which (mw 220,000) contains two atoms of molybdenum and 28–32 atoms of iron. The smaller protein (mw 68,000) contains four atoms of iron in a Fe_4S_4 ferredoxin-type structure. The binding and reduction of dinitrogen is believed to occur at molybdenum, whereas the iron clusters serve as a sort of electron reservoir. No crystal structure is yet available for nitrogenase. EXAFS studies indicate the presence of four or five sulfur ligands around the molybdenum as well as two or three iron atoms in the second coordination sphere [99–101].

$$N_2 + 8H^+ + 8e^- \xrightarrow{\text{nitrogenase}} 2NH_3 + H_2 \qquad (26)$$

There is growing agreement that molybdenum hydrazido species are key intermediates for Eq. 26. However, it remains uncertain whether the hydrazides are terminal Mo=NNH$_2$ compounds or bridging hydrazido(4−) structures Mo=NN=Mo. Evidence for a hydrazido intermediate includes studies in which the functioning enzyme is destroyed by the addition of either acid or base, whereupon free hydrazine is produced [102]. As conversion of dinitrogen to ammonia proceeds, the presumed hydrazido intermediate is observed to increase in concentration, pass through a maximum, and finally diminish as shown in Figure 7.2. Also shown in Figure 7.2 is the similar course of events observed by George and coworkers [103] using the model system Mo(N$_2$)$_2$(PPh$_3$)(triphos). The model system is known to involve discrete hydazido(2−) intermediates (see Section 6.4).

Some indirect evidence supports the hydrazido(2−) model. For example, an iron–molybdenum cofactor "FeMo–co" has been isolated from the large protein which

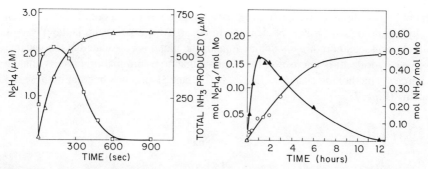

Figure 7.2 *(Left)* Formation of hydrazine and ammonia upon quenching nitrogenase at various stages of the catalytic reduction of N_2 by *K.pneumoniae*. (Reprinted by permission from *Nature* **1978**, *272*, 557. Copyright © 1978 Macmillan Magazines Limited.) *(Right)* A similar plot of hydrazine and ammonia formation during stoichiometric N_2 reduction by the model system *trans*-[Mo(N$_2$)$_2$(triphos)(PPh$_3$)]. (Reprinted with permission from J.A. Baumann et al. *Inorg. Chem.* **1985**, *24*, 3573. Copyright 1985 American Chemical Society.)

appears to contain a single molybdenum atom. The cofactor [104] is the only known molybdenum material capable of activating the large protein of mutant bacteria which contain no molybdenum. Moreover, the large protein consists of two matching subunits, suggesting that the molybdenum atom in each subunit may be significantly separated from its counterpart. (Of course, the molybdenum atoms could be paired on the boundary between subunits.) Proponents of the hydrazido(2−) model can also point to the recently developed system for electrocatalytic nitrogen fixation in which hydrazido(2−) intermediates are demonstrably present [105].

The bridging hydrazido(4−) model has long been advocated by Shilov and coworkers [106]. This school of thought has received fresh impetus from the recent synthesis of model M=NN=M complexes from dinitrogen [107,108] and the demonstration that such complexes afford hydrazine in high yield upon hydrolysis [109]. As expected, this experimental work was quickly followed by new theoretical studies supporting the viability of hydrazido(4−) intermediates in enzymatic nitrogen fixation [110,111].

The minority opinion that enzymatic nitrogen fixation proceeds via the reduction of side-bound dinitrogen has been advocated by G.N. Schrauzer. He has designed catalyst systems to fit this model that are claimed [112] to reduce dinitrogen to ammonia in 0.06–0.10% yield based on molybdenum (5 days, 140 atm N_2). A chemically worrisome aspect of this work is the assertion that highly reactive diimide (HN=NH) is "accumulated during the first 40 min" of reaction [113]. Several published [114,115] and unpublished attempts to reproduce Schrauzer's results have been unsuccessful.

Despite evidence for the involvement of hydrazido molybdenum intermediates in enzymatic nitrogen fixation and the model studies summarized in Section 6.4, it is not currently understood how such species promote N—N bond cleavage. A striking feature of the chemical (as opposed to electrochemical) model systems is the fact that N—N cleavage can take place when molybdenum is coordinated with four monodentate phosphines or the combination of one monodentate and one tridentate phosphine but not by two diphosphines. This suggests a requirement for a coordination site cis to the hydrazido ligand for N—N cleavage to occur. Also noteworthy is the observation that, after dissociation of a phosphine ligand, the metal is protonated at the vacant site to give a hydride. It is interesting to speculate that an α elimination process analogous to Bercaw's methoxide cleavage (Eq. 39 in Chapter 3) might be operant in these model systems (Eq. 27):

$$\begin{array}{c}\overset{+}{N}H_3\\|\\N\\||\\Mo-H\end{array} \rightleftharpoons \begin{array}{c}H\diagdown\overset{+}{N}H_3\\N\\|\\Mo-\square\end{array} \rightarrow \begin{array}{c}H\\|\\N\\||\\Mo-NH_3\end{array} \qquad (27)$$

7.4 ALKYLIDENE INTERMEDIATES IN CATALYSIS

7.4.1 Olefin Metathesis

"Olefin metathesis" [116,117] is the name given to the catalytic transalkylidenation of olefins, a simple example being the equilibration of propylene with ethylene and 2-butene in Eq. 28. A related reaction—in fact, a subset of olefin metathesis reactions—is the ring opening polymerization of cyclic olefins exemplified by Eq. 29. These are both reactions with industrial roots. The first ring opening polymerization was patented by Du Pont in 1957 [118] and the first report of the methathesis of acyclic olefins was provided by workers at Phillips in 1964 [119]. The most active catalysts for olefin metathesis are based on molybdenum, tungsten, or rhenium. In industrial practice heterogeneous catalysts are preferred but homogeneous catalysts can be advantageous for ring opening polymerization.

$$2CH_3CH=CH_2 \underset{}{\overset{catalyst}{\rightleftharpoons}} H_2C=CH_2 + CH_3CH=CHCH_3 \qquad (28)$$

$$\text{cyclopentene} \overset{catalyst}{\longrightarrow} (=CH(CH_2)_3CH=)_n \qquad (29)$$

The first mechanisms proposed for olefin metathesis were based on "pairwise" processes involving metal-bound cyclobutane [120] or related [121] structures. However, by 1970 Herisson and Chauvin [122] proposed that the reaction proceeds via alkylidene and metallacyclobutane intermediates according to Eq. 30. This proposal is now universally accepted (for most metals) on the strength of a series of elegant mechanistic studies. These included (1) studies on cross-metathesis reactions which show that the products are incompatible with pairwise exchange [123–126]; (2) studies on the stereochemistry of metathesis of internal olefins [127–131]; and (3) the demonstration that discrete metal alkylidene complexes could serve as highly active catalysts for olefin metathesis [132,133]. In fact, one system has been identified where it is possible simultaneously to observe tungsten alkylidene and tungstacyclobutane intermediates during the course of metathesis [134]. Additional support comes from studies on the molecular weight distribution of the products of ring opening metathesis. A detailed and lucid explication of all of these studies can be found in ref. 116.

$$\begin{array}{c} R \\ H \end{array} C = CH_2 \atop + \quad M = CHR' \quad \rightleftarrows \quad \begin{array}{c} RHC - CH_2 \\ | \quad | \\ M - CHR' \end{array} \quad \rightleftarrows \quad \begin{array}{c} RHC \\ \| \\ M \end{array} + \begin{array}{c} CH_2 \\ \| \\ C \\ / \; \backslash \\ H \quad R' \end{array} \quad (30)$$

Upton and Rappe have provided a theoretical basis for the low barriers in $2\pi + 2\pi$ reactions involving transition metals, of which olefin metathesis is an example [135]. It was suggested that a concerted 2 + 2 reaction between olefin and alkylidene is possible because of the second angular node present in d orbitals. (We should note however, that the orbital restriction on concerted 2 + 2 reactions only strictly applies to the symmetrical case, for example, olefin plus olefin. As the system departs from ideal symmetry the restriction is relaxed.)

It was also regarding olefin metathesis that Goddard and Rappe first suggested that "spectator oxo" groups may play a key role in catalysis [136]. It was proposed that, in addition to an alkylidene ligand, active metathesis catalysts should possess a second multiply bonded ligand such as an oxo group. The increased strength of the triple-bonded oxo group in the metallacyclic intermediate versus the double-bonded oxo of the alkylidene complex (Fig. 7.3) was said to provide a significant driving force for metallacycle formation. In contrast, the addition of ethylene to $Cl_4Mo(CH_2)$, a model compound with no spectator oxo, was calculated to be endothermic by 15 kcal/mole.

The original communications [137,138] on the spectator oxo effect can be confusing to the beginning student of catalysis. We should stress that the stabilization of one intermediate in a catalytic cycle as has occurred in Figure 7.3 is not a useful way to promote catalysis. One ideally wants the intermediates to be energetically proximate. In the full paper on this work [136] it is proposed that the presence of a Lewis acid in typical catalyst systems may alter the energetics of Figure 7.3 by promoting decomposition of the metallacycle. It was also predicted that the smaller spectator effect of an imido ligand might prove advantageous in Lewis acid-free systems. As we shall see below, this proposal proved quite prophetic.

Typical homogeneous catalysts for olefin metathesis are generated by treatment of a transition metal chloride or oxychloride with a main-group alkyl. In some cases an innocent main-group organometallic (e.g., Me_4Sn) appears simply to alkylate the transition metal [139]. However, in particular when alkylaluminum reagents are used as the main group component, the Lewis acidic aluminum seems to play a direct role in catalysis by a mechanism that is not fully understood. Several different roles for the Lewis acid have been suggested. By coordinating to a spectator oxo group the Lewis acid has been suggested to (1) promote initial α elimination to generate the active alkylidene catalyst [140], (2) promote electrophilic attack of the alkylidene on the

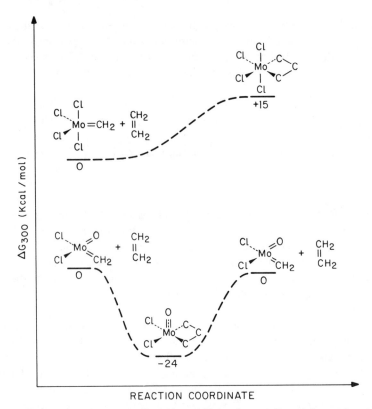

Figure 7.3 Effect of a spectator oxo ligand in stabilizing the metallacyclobutane intermediate in the degenerate methathesis of ethylene. (Adapted with permission from A.K. Rappé and W.A. Goddard, III, *J. Am. Chem. Soc.* **1982**, *104*, 449. Copyright 1982 American Chemical Society.

olefin [140,141], or, as proposed by Goddard and Rappe, (3) decrease the activation energy for metallacycle decomposition [136]. In other cases the Lewis acid has been suggested to (4) remove an anionic ligand to afford a highly active cationic catalyst [142] or (5) bind to the alkylidene moiety to protect it sterically and electronically [143].

Given such uncertainties, small wonder that the preparation of a well-characterized, highly active, Lewis acid-free olefin metathesis catalyst became something of a Holy Grail in this field. Schrock has chronicled this quest in his review "On the Trail of Metathesis Catalysts" [144]. The search has now culminated in the synthesis of tungsten [145] and molybdenum [146] catalysts (**6**) which meet these criteria. The successful approach combines a low-coordination number with hard, sterically encumbered alkoxide ligands as well as a sterically encumbered "spectator" imido ligand.

For the industrial chemist, the real excitement of catalysts **6** and of the discrete titanium-based catalysts developed by Grubbs and coworkers (see Eq. 41 in Chapter 3) lies in their potential applications in ring-opening polymerization. Such catalysts can provide polymers that are monodisperse (i.e., having very narrow molecular weight distributions). The polymers they produce are "living" (i.e., the active catalyst is still attached to the polymer chain), which allows the incorporation of blocks of different polymers to achieve interesting combinations of properties. Especially compounds **6** allow the polymerization of monomers containing some functional groups. All of these possibilities have now been demonstrated in the laboratory [147,148].

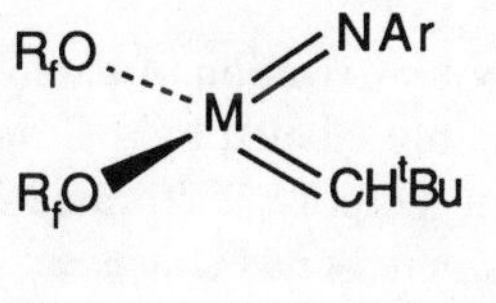

$R_f = -CMe(CF_3)_2$

Ar = 2,6-diisopropylphenyl

M = Mo, W

6

There clearly remain interesting discoveries to be made. The recent demonstration by Anslyn and Grubbs [149] that olefin metathesis by titanium metallacycles may in some cases proceed by an associative ("S_N2-like") mechanism is indicative of the surprises that await us. One question of interest is whether ring-opening polymerization promoted by low-valent group 8 metals (Os,Ir,Ru) occurs via a carbene mechanism related to that now established for early transition metals. Some striking differences have been observed; for example, cases where a rhenium(V) catalyst produces all cis polymer while a ruthenium(III) catalyst gives exclusively trans polymer [150]. In such cases "chauvinism" to the usual mechanism does not seem warranted.

7.4.2 Acetylene Polymerization

Alkyne polymerization is promoted by a variety of molybdenum and tungsten catalysts including $MoCl_5$ and WCl_6. Masuda and coworkers [151] proposed in 1975 that these reactions proceed via the intermediacy of metal alkylidene species. Reaction of this alkylidene with the acetylene according to Eq. 31 was suggested to afford a metallacyclobutene intermediate; opening the four-membered ring in the opposite sense gives a new alkylidene that can continue the process. (It is interesting to note that this proposal predates by several years the Green–Rooney mechanism for olefin polymerization that we discuss in Section 7.4.3.)

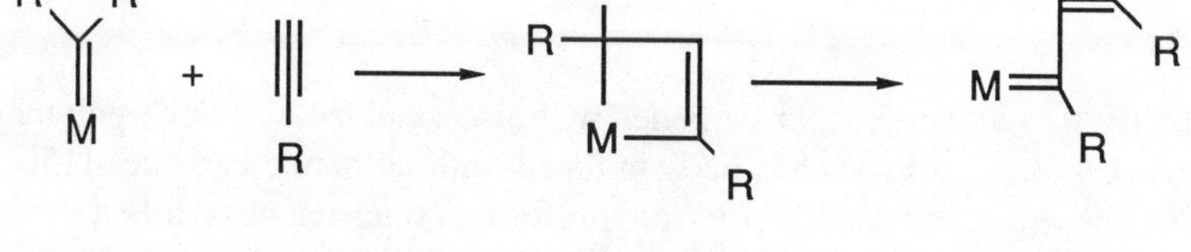

(31)

Several subsequent studies lend solid support to the Masuda mechanism.

1. A model transformation consistent with Eq. 31 has been reported.

CpTa(CH^{t}Bu)Cl$_2$ reacts with diphenylacetylene quantitatively to form a new alkylidene complex CpTa[C(Ph)—C(Ph)=CH^{t}Bu]Cl$_2$ [152].

2. The finding that the living polymer is an active catalyst for ring-opening polymerization of cycloalkenes supports the contention that the growing polyacetylene is bound to the metal via an alkylidene linkage. Katz and coworkers characterized the polymer produced when cyclopentene is added to a mixture of WCl$_6$ and phenylacetylene. Poly(phenylacetylene) was found to be attached as a block at the start of all polypentenamer chains [153]

3. In another elegant study, Katz examined the polymerization of phenylacetylene containing a small percentage of monomer doubly labeled by ^{13}C on the triple bond [154]. Nutation ^{13}C NMR confirmed that the polymer formed in Eq. 32b contains pairs of ^{13}C atoms joined by C—C single bonds consistent with polymerization via Eq. 31. In contrast, the conventional titanium-based catalyst in Eq. 32a gave a polymer in which ^{13}C's were separated by C=C double bonds as expected for an insertion-type mechanism.

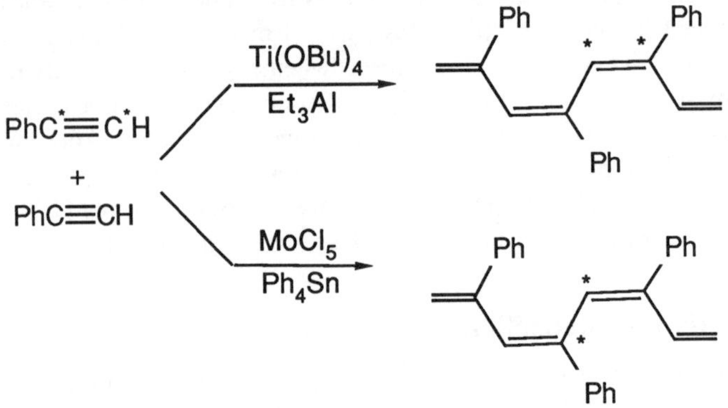

$$PhC^*{\equiv}C^*H + PhC{\equiv}CH \xrightarrow{\text{Ti(OBu)}_4 / \text{Et}_3\text{Al}} \quad (32a)$$

$$\xrightarrow{\text{MoCl}_5 / \text{Ph}_4\text{Sn}} \quad (32b)$$

Schrock [155] has presented evidence that metallacyclobutadienes, like other high-valent alkylidenes, can initiate acetylene polymerization according to Eq. 31. In this case the polyacetylene will be propagated as a metallacycle as shown in Eq. 33:

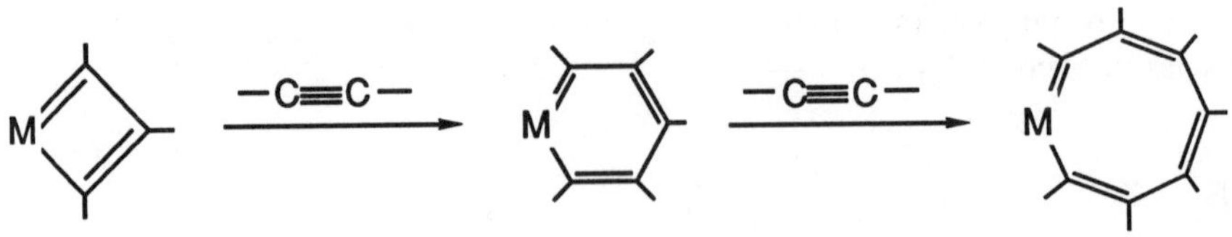

$$(33)$$

Interestingly low-valent Fischer carbenes will also catalyze acetylene polymerization. Polymerization of alkynes has been initiated with authentic carbenes [156,157] and also with carbenes generated *in situ* from reaction of tungsten hexacarbonyl with an acetylene [158,159] or from low-valent carbynes [160]. However, generalization of the mechanistic scheme of Eq. 31 to the low-valent systems does not appear warranted at this time. We note that preliminary studies of the type shown in Eq. 32 were carried out using Fischer type carbenes as catalysts. The composition of the polymers was essentially the same as when the titanium-containing mixture was the catalyst [154].

7.4.3 Alkylidene Intermediates in Ziegler–Natta Polymerization of Olefins

The polymerization of ethylene and α-olefins at low pressure is promoted by the combination of an early transition metal compound and a main group organometallic [161,162]. For example, typical Ziegler–Natta catalysts can be prepared from titanium(III) chloride and trialkylaluminum reagents. The newer Phillips catalyst is based on low-valent chromium. In industrial practice such catalysts are invariably heterogeneous. However, homogeneous model systems are available for laboratory studies. Considering the tremendous industrial importance of this chemistry, our understanding of Ziegler–Natta catalysis at the molecular level is remarkably incomplete.

The conventional rationale for Ziegler–Natta polymerization is provided by the Cossee–Arlman mechanism (Eq. 34) [163]. In this model the key C—C bond forming step involves insertion of a coordinated olefin into a transition metal–carbon bond. That such direct insertions can occur has been confirmed by labeling studies on an organocobalt system [164]. Moreover, Watson has reported an organolutetium-based model system in which it is possible to observe directly chain growth resulting from sequential insertion of propylene units [165]. In this case, direct insertion seems to be required due to the inaccessibility of higher oxidation states.

$$M-R \xrightarrow{H_2C=CH_2} \underset{\underset{M-R}{|}}{H_2C=CH_2} \longrightarrow M-CH_2CH_2R \qquad (34)$$

An alternative mechanism for olefin polymerization based on alkylidene intermediates was proposed by Ivin, Rooney, Green, and coworkers [166]. These workers noted the dearth of examples of insertion of a coordinated olefin into a transition metal alkyl and further pointed out that the same or similar catalysts often promote both Ziegler–Natta and ring-opening polymerization of appropriate substrates. At the heart of their proposal is the equilibrium between a low-valent transition metal alkyl and the alkylidene hydride arising by α elimination (Eq. 35). Polymer chain growth then takes place via a metallacyclobutane intermediate as shown in Eq. 36.

$$M-CH_2R \rightleftharpoons \overset{H}{\underset{|}{M}}=CHR \qquad (35)$$

$$\overset{H}{\underset{|}{M}}=CHR \xrightarrow{H_2C=CH_2} \underset{\underset{H_2C-CH_2}{|\quad\;|}}{\overset{H}{\underset{|}{M}}-CHR} \longrightarrow M-CH_2CH_2CH_2R \qquad (36)$$

Distinguishing between this "Green–Rooney" mechanism and the Cossee–Arlman model is not a simple exercise. (Note that neither the absence of H—D scrambling during polymerization of labeled olefins [167] nor studies on polymer stereoregularity will allow one to make such a distinction. A study claiming the latter approach [168] was roundly debunked by Casey [169].) Several IR spectroscopic studies [170–172] are claimed to provide evidence for alkylidene species on the surface of heterogeneous (Ti or Cr) olefin polymerization catalysts. However, the interpretation of the spectroscopic data is by no means obvious.

Several studies have been reported which, when taken together, disfavor the Green–Rooney model in the case of titanium catalysts.

1. Phenyl titanium derivatives which lack α hydrogen can nevertheless be used to initiate polymerization of propylene. The phenyl group is incorporated as the terminal group on the polymer chain [173].

2. Copolymerization studies on C_2H_4/C_2D_4 mixtures indicate an isotope effect for the propagation step of $K_H/K_D = 1.04 \pm 0.03$. This argues against hydrogen transfer during the rate-determining step of propagation [174].

3. An additional elegant study was carried out as a check against the possibility of kinetic masking of the key C—C bond-forming step in the previous study. This study confirmed the absence of an isotope effect on the *stereochemistry* of olefin insertion [175].

Schrock and coworkers have presented evidence that the tantalum catalyst $Ta(CH^tBu)H(PMe_3)_3I_2$ may polymerize ethylene according to Eq. 36 [176]. Living polymers produced with this catalyst were shown by NMR to possess the alkylidene structure $Ta[CH(CH_2CH_2)_n{}^tBu]H(PMe_3)_3I_2$. Magnetization transfer studies on the catalyst and on the living polymer provide evidence for rapid reversible α hydrogen elimination. Also consistent with the Green–Rooney formulation, ethylene polymerization takes place even in the presence of excess PMe_3. In contrast, $Ta(C_2H_4)Et(PMe_3)_2I_2$, a conceivable intermediate for Cossee–Arlman polymerization, does not polymerize ethylene.

It appears that under appropriate circumstances either the Green–Rooney or Cossee–Arlman mechanism can be viable. It remains for organometallic chemists to delineate the scope of each process.

7.5 ALKYLIDYNE COMPLEXES IN ACETYLENE METATHESIS

It has been known for some time that the metathesis of internal acetylenes (Eq. 37) is promoted by ill-defined catalysts generated from molybdenum hexacarbonyl and either phenols or fluoroalcohols [177–181]. (Heterogeneous acetylene metathesis is also known but requires extremely high temperatures—typically 400°C [182, 183].) ^{13}C-labeling studies using the $Mo(CO)_6$/PhOH system confirm that acetylene metathesis proceeds with cleavage of the C—C triple bond [184].

$$RC\equiv CR + R'C\equiv CR' \xrightleftharpoons{\text{catalyst}} 2RC\equiv CR' \qquad (37)$$

As early as 1975 Katz proposed that the reversible reaction between an acetylene and an alkylidyne complex to afford a metallacyclobutadiene (Eq. 38) constitutes the mechanistic pathway for acetylene metathesis [185]. This proposal is now supported by a wealth of experimental evidence.

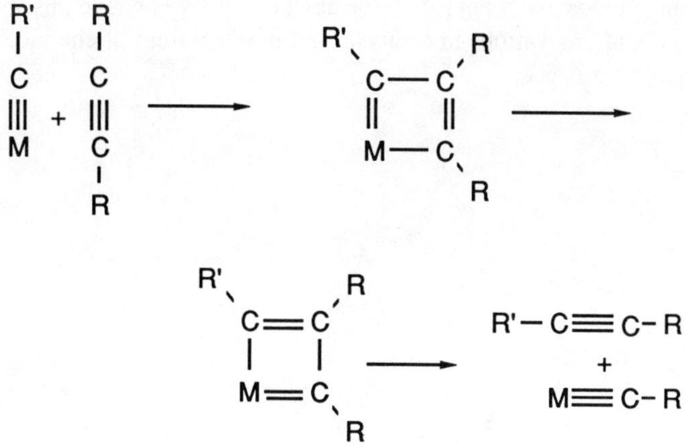

$$ \qquad (38)$$

Schrock and coworkers demonstrated that an authentic tungsten alkylidyne complex (^{t}BuO)$_3$W≡C^{t}Bu would catalyze the metathesis of internal acetylenes in a reaction that is first order in tungsten and first order in acetylene [186, 187]. The Schrock group subsequently isolated and structurally characterized a series of metallacyclobutadiene complexes, all prepared by the interaction of alkylidyne complexes with acetylenes [188–191]. By resorting to sterically bulky 2,6-diisopropylphenoxide ligands, it was possible to prepare a tungstacyclobutadiene W(C$_3$Et$_3$)(OAr)$_3$ which is an active catalyst for acetylene metathesis. Significantly, this complex catalyzes metathesis at a rate independent of acetylene concentration, indicating rate-limiting loss of acetylene from the metallacycle [190]. (Interestingly, related tungstacylobutadienes containing fluoroalkoxy ligands under some conditions appear to promote metathesis by an associative mechanism [191].)

Delineation of the various pathways available when alkylidynes react with acetylenes, which we earlier summarized in Figure 6.4, provides additional insight into acetylene metathesis. In particular, explanations have been offered for the failure of terminal acetylenes to undergo methathesis [192] and for the ultimate deactivation of the catalyst [189]. The crowning achievement in this area has been the rational design of trialkoxymolybdenum alkylidyne catalysts which are many orders of magnitude more active than the original Mo(CO)$_6$-based systems [193]. Indeed, Schrock has proposed that minute quantities of d^0 alkylidynes from the interaction of Mo(CO)$_6$ with phenols or fluoroalcohols may be the active catalysts in the older systems [193].

7.6 CONCLUDING REMARKS

It is appropriate that we conclude our discussion of multiply bonded ligands in catalysis with the example of acetylene metathesis. In the space of a decade this reaction has been transformed from a rather inefficient "black box" to a viable and well understood piece of chemistry. This development has paralleled the growth in understanding of metal–ligand multiple bonds.

As we have tried to illustrate throughout this book, the pace of new discoveries and new insight in this field actually appears to be quickening. We believe this should lead to many more rational innovations in catalysis and in other areas of chemical technology. It is an exciting prospect.

REFERENCES

1. Keulks, G.W.; Krenzke, L.D.; Notermann, T.M. *Adv. Catal.* **1978**, *27*, 183-225.
2. van den Elzen, A.F.; Rieck, G.D. *Acta Cryst.* **1973**, *B29*, 2433-2436.
3. van den Elzen, A.F.; Rieck, G.D. *Acta Cryst.* **1973**, *B29*, 2436-2438.
4. Mitchell, P.C.H.; Trifiro, F. *J. Chem. Soc. A* **1970**, 3183-3188.
5. Mars, P.; van Krevelen, D.W. *Chem. Eng. Sci., Spec. Suppl.* **1954**, *3*, 41-59.
6. Keulks, G.W. *J. Catal.* **1970**, *19*, 232-235.
7. Wragg, R.D.; Ashmore, P.G.; Hockey, J.A. *J. Catal.* **1971**, *22*, 49-53.
8. Peacock, J.M.; Parker, A.J.; Ashmore, P.G.; Hockey, J.A. *J. Catal.* **1969**, *15*, 398-406.
9. Trifiro, F.; Pasquon, I. *J. Catal.* **1968**, *12*, 412-416.
10. Adams, C.R.; Jennings, T.J. *J. Catal.* **1963**, *2*, 63-68. Sachtler, W.M.H. *Rec. Trav. Chim. Pays-Bas* **1963**, *82*, 243-245.
11. Burrington, J.D.; Kartisek, C.T.; Grasselli, R.K. *J. Catal.* **1980**, *63*, 235-254.
12. Swift, H.E.; Bozik, J.E.; Ondrey, J.A. *J. Catal.* **1971**, *21*, 212-224.
13. Martir, W.; Lunsford, J.H. *J. Am. Chem. Soc.* **1981**, *103*, 3728-3732.
14. Trifiro, F. *Chim. Ind. (Milan)* **1974**, *56*, 835-839.
15. Arigoni, D.; Vasella, A.; Sharpless, K.B.; Jensen, H.P. *J. Am. Chem. Soc.* **1973**, *95*, 7917-7919.
16. De Rossi, S.; LoJacono, M.; Porta, P.; Valigi, M.; Gazzoli, D.; Minelli, G.; Anichini, A. *J. Catal.* **1986**, *100*, 95-102.
17. Bruckman, K.; Haber, J.; Wiltowski, T. *J. Catal.* **1987**, *106*, 188-201.
18. Steenken, S.; Neta, P. *J. Am. Chem. Soc.* **1982**, *104*, 1244-1248.
19. Burrington, J.D.; Grasselli, R.K. *J. Catal.* **1979**, *59*, 79-99.
20. Grzybowska, B.; Haber, J.; Janas, J. *J. Catal.* **1977**, *49*, 150-163.
21. Machiels, C.J.; Sleight, A.W. *J. Catal.* **1982**, *76*, 238-239.
22. Ohuchi, F.; Firment, L.E.; Chowdhry, U.; Ferretti, A. *J. Vac. Sci. Technol. A* **1982**, *2*, 1022-1023. Farneth, W.E.; Ohuchi, F.; Staley, R.H.; Chowdhry, U.; Sleight, A.W. *J. Phys. Chem.* **1985**, *89*, 2493-2497. Compare: Tatibouet, J.M.; Germain, J.E. *J. Catal.* **1981**, *72*, 375-378.
23. Johnson, J.W.; Johnston, D.C.; Jacobsen, A.J.; Brody, J.F. *J. Am. Chem. Soc.* **1984**, *106*, 8123-8128.
24. Gorbunova, Yu.A.; Linde, S.A. *Sov. Phys. Dokl.* **1979**, *24*, 138-140.
25. Chabardes, P.; Querou, Y., Brit. patent 1,204,754 (1970); *Chem. Abstr.* **1970**, *72*, 43923g.
26. Parshall, G.W.; Nugent, W.A. *Chemtech* **1988**, *18*, in press.

27. Olson, G.L.; Cheung, H.-C.; Morgan, K.D.; Borer, R.; Saucy, G. *Helv. Chim. Acta* **1976**, *59*, 567–585.
28. Olson, G.L.; Morgan, K.D.; Saucy, G. *Synthesis* **1976**, 25–26.
29. Chabardes, P.; Kuntz, E.; Varagnat, J. *Tetrahedron* **1977**, *33*, 1775–1783.
30. Pauling, H.; Andrews, D.A.; Hindley, N.C. *Helv. Chim. Acta* **1976**, *59*, 1233–1243.
31. Hosogai, T.; Fujita, Y.; Ninagawa, Y.; Nishida, T. *Chem. Lett.* **1982**, 357–360.
32. Erman, M.B.; Aul'chenko, I.S.; Kheifits, L.A.; Dulova, V.G.; Novikov, Yu. N.; Vol'pin, M.E. *Tetrahedron Lett.* **1976**, 2981–2984.
33. Erman, M.B.; Aul'chenko, I.S.; Kheifits, L.A.; Dulova, V.G.; Novikov, Yu. N.; Vol'pin, M.E. *J. Org. Chem. (U.S.S.R.)* **1976**, *12*, 931–938.
34. Gulyi, S.E.; Erman, M.B.; Novikov, N.A.; Aul'chenko, I.S.; Vol'pin, M.E. *J. Org. Chem. (U.S.S.R.)* **1983**, *19*, 715–721.
35. Matsubara, S.; Okazoe, T.; Oshima, K.; Takai, K.; Nozaki, H. *Bull. Chem. Soc., Japan* **1985**, *58*, 844–849.
36. Saunders, B.C.; Holmes-Seidel, A.G.; Stark, B.P. "Peroxidases" Butterworths, London, 1964. Dunford, H.B. *Adv. Inorg. Biochem.* **1982**, *4*, 41–68.
37. Portoghese, P.S.; Svanborg, K.; Samuelsson, B. *Biochem. Biophys. Res. Comm.* **1975**, *63*, 748–755.
38. Hewson, W.D.; Dunford, H.B. *J. Biol. Chem.* **1976**, *251*, 6043–6052.
39. Kedderis, G.L.; Hollenberg, P.F. *J. Biol. Chem.* **1983**, *258*, 8129–8138.
40. Schonbaum, G.R.; Chance, B. in: "The Enzymes" Vol. 13, Academic, New York, 1976, pp. 345–362.
41. Schulz, C.E.; Rutter, R.; Sage, J.T.; Debrunner, P.G.; Hager, L.P. *Biochem.* **1984**, *23*, 4743–4754.
42. Roberts, J.E.; Hoffman, B.M.; Rutter, R.; Hager, L.P. *J. Am. Chem. Soc.* **1981**, *103*, 7654–7656.
43. LaMar, G.N.; deRopp, J.S.; Smith, K.M.; Langry, K.C. *J. Biol. Chem.* **1981**, *256*, 237–243.
44. Browlett, W.R.; Gasyna, Z.; Stillman, M. *J. Biochem. Biophys. Res. Comm.* **1983**, *112*, 515–520.
45. Penner-Hahn, J.E.; Eble, K.S.; McMurry, T.J.; Renner, M.; Balch, A.L.; Groves, J.T.; Dawson, J.H.; Hodgson, K.O. *J. Am. Chem. Soc.* **1986**, *108*, 7819–7825.
46. Ortiz de Montellano, P.R., ed. "Cytochrome P-450: Structure, Mechanism, and Biochemistry" Plenum, New York, 1986.
47. McMurry, T.J.; Groves, J.T., in ref. 46, pp. 1–28.
48. Ator, M.A.; Ortiz de Montellano, P.R. *J. Biol. Chem.* **1987**, *262*, 1542–1551.
49. Groves, J.T.; McClusky, G.A.; White, R.E.; Coon, M. *J. Biochem. Biophys. Res. Comm.* **1978**, *81*, 154–160.
50. Gelb, M.H.; Heimbrook, D.C.; Malkonen, P.; Sligar, S.G. *Biochemistry* **1982**, *21*, 370–377.
51. Groves, J.T.; Subramanian, D.V. *J. Am. Chem. Soc.* **1984**, *106*, 2177–2181.
52. White, R.E.; Miller, J.P.; Favreau, L.V.; Bhattacharyya, A. *J. Am. Chem. Soc.* **1986**, *108*, 6024–6031.

53. Hjelmelend, L.M.; Aronow, L.; Trudell, J.R. *Biochem. Biophys. Res. Comm.* **1977**, *76*, 541–549.
54. Foster, A.B.; Jarman, M.; Stevens, J.D.; Thomas, P.; Westwood, J.H. *Chem. Biol. Interact.* **1974**, *9*, 327–340.
55. Miwa, G.T.; Walsh, J.S.; Lu, A.Y.H. *J. Biol. Chem.* **1984**, *259*, 3000–3004.
56. Jones, J.P.; Trager, W.F. *J. Am. Chem. Soc.* **1987**, *109*, 2171–2173.
57. Ortiz de Montellano, P.R.; Stearns, R.A. *J. Am. Chem. Soc.* **1987**, *109*, 3415–3420.
58. Poulos, T.L.; Finzel, B.C.; Howard, A.J. *Biochemistry* **1986**, *25*, 5314–5322.
59. Poulos, T.L.; Finzel, B.C.; Gunsalus, I.C.; Wagner, G.C.; Krautt, J. *J. Biol. Chem.* **1985**, *260*, 16122–16130.
60. Hrcay, E.G.; Gustafsson, J.-A.; Ingelman-Sundberg, M.; Ernster, L. *FEBS Lett.* **1975**, *56*, 161–165.
61. Lichtenberger, F.; Nastainczyk, W.; Ullrich, V. *Biochem. Biophys. Res. Comm.* **1976**, *70*, 939–946.
62. Hrcay, E.G.; Gustafsson, J.; Ingelman-Sundberg, M.; Ernster, L. *Biochem. Biophys. Res. Comm.* **1975**, *66*, 209–216.
63. Holm, R.H.; Berg, J.M. *Acc. Chem. Res.* **1986**, *19*, 363–370.
64. Hille, R.; Massey, V. in Spiro, T.G., ed. "Molybdenum Enzymes" Wiley-Interscience, New York, 1985, pp. 443–518.
65. Adams, M.W.W.; Mortenson, L.E., in ref. 64, pp. 519–593.
66. Cramer, S.P. *Adv. Inorg. Bioinorg. Mech.* **1980**, *2*, 259–316.
67. Bordas, J.; Bray, R.C.; Garner, C.D.; Gutteridge, S.; Hasnain, S.S. *Biochem. J.* **1980**, *191*, 499–508.
68. Cramer, S.P.; Wahl, R.; Rajagopalan, K.V. *J. Am. Chem. Soc.* **1981**, *103*, 7721–7727.
69. Cramer, S.P.; Solomonson, L.P.; Adams, M.W.W.; Mortenson, L.E. *J. Am. Chem. Soc.* **1984**, *106*, 1467–1471.
70. Harlan, E.W.; Berg, J.M.; Holm, R.H. *J. Am. Chem. Soc.* **1986**, *108*, 6992–7000.
71. Murray, K.N.; Watson, J.G.; Chaykin, S. *J. Biol. Chem.* **1966**, *241*, 4798–4801.
72. Cramer, S.P.; Stiefel, E.I., in ref. 64, pp. 411–441.
73. Johnson, J.L.; Rajagopalan, K.V. *Proc. Natl. Acad. Sci.* **1982** *79*, 6856–6860. Johnson, J.R.; Hainline, B.E.; Rajagopalan, K.V.; Arison, B.H. *J. Biol. Chem.* **1984**, *259*, 5414–5422.
74. Schröder, M. *Chem. Rev.* **1980**, *80*, 187–213.
75. Courtney, J.L.; Swansborough, K.F. *Rev. Pure Appl. Chem.* **1972**, *22*, 47–54.
76. Lee, D.G.; van den Engh, M. in "Oxidation in Organic Chemistry, Part B" Trahanovsky, W.S., ed. Academic, New York, 1973.
77. Sharpless, K.B.; Akashi, K. *J. Am. Chem. Soc.* **1976**, *98*, 1986–1987.
78. VanRheenan, V.; Kelly, R.C.; Cha, D.Y. *Tetrahedron Lett.* **1976**, 1973–1976.
79. Carlsen, P.H.J.; Katsuki, T.; Martin, V.S.; Sharpless, K.B. *J. Org. Chem.* **1981**, *46*, 3936–3938.
80. Dobson, J.C.; Seok, W.K.; Meyer, T.J. *Inorg. Chem.* **1986**, *25*, 1513–1514.

81. Moyer, B.A.; Thompson, M.S.; Meyer, T.J. J. Am. Chem. Soc. **1980**, *102*, 2310-2312.
82. Groves, J.T.; Quinn, R. J. Am. Chem. Soc. **1985**, *107*, 5790-5792.
83. Busca, G.; Cavani, F.; Trifiro, F. J. Catal. **1987**, *106*, 471-482.
84. Bruckman, K.; Haber, J.; Wiltkowski, T. J. Catal. **1987**, *106*, 188-201.
85. Andersson, A. J. Catal. **1986**, *100*, 414-428.
86. Otsuba, T.; Miura, H.; Morikawa, Y.; Shirasaki, T. J. Catal. **1975**, *36*, 240-243.
87. Chan, D.M.-T.; Nugent, W.A. Inorg. Chem. **1985**, *24*, 1422-1424.
88. Alsdorf, E.; Hanke, W.; Schnable, K.-H.; Schreier, E. J. Catal. **1986**, *98*, 82-87.
89. Burrington, J.D.; Kartisek, C.T.; Grasselli, R.K. J. Catal., **1984**, *87*, 363-380.
90. Burrington, J.D.; Kartisek, C.T.; Grasselli, R.K. J. Catal. **1982**, *75*, 225-232.
91. Sharpless, K.B.; Chong, A.O.; Oshima, K. J. Org. Chem. **1976**, *41*, 177-179.
92. Herranz, E.; Sharpless, K.B. J. Org. Chem. **1978**, *43*, 2544-2548.
93. Herranz, E.; Biller, S.A.; Sharpless, K.B. J. Am. Chem. Soc. **1978**, *100*, 3596-3598.
94. Breslow, R.; Kluttz, R.Q.; Khanna, P.L. Tetrahedron Lett. **1979**, 3273-3274.
95. Casey, A.T.; Vecchio, A.M. Trans. Met. Chem. **1986**, *11*, 366-368.
96. Meisel, I.; Hertel, G.; Weiss, K. J. Mol. Catal. **1986**, *36*, 159-162.
97. Chatt, J.; da Camera Pina, L.M.; Richards, R.L. eds. "New Trends in the Chemistry of Nitrogen Fixation" Academic Press, London, 1980.
98. Spiro, T.G. ed. "Molybdenum Enzymes" Wiley-Interscience, New York, 1985.
99. Cramer, S.P.; Hodgson, K.O.; Gillum, W.O.; Mortenson, L.E. J. Am. Chem. Soc. **1978**, *100*, 3398-3407.
100. Cramer, S.P.; Gillum, W.O.; Hodgson, K.O.; Mortenson, L.E.; Stiefel, E.I.; Chisnell, J.R.; Brill, W.J.; Shah, V.K. J. Am. Chem. Soc. **1978**, *100*, 3814-3819.
101. Flank, A.M.; Weininger, M.; Mortenson, K.E.; Cramer, S.P. J. Am. Chem. Soc. **1986**, *108*, 1049-1055.
102. Thorneley, R.N.F.; Eady, R.R.; Lowe, D.J. Nature **1978**, *272*, 557-558.
103. Baumann, J.A.; Bossard, G.E.; George, T.A.; Howell, D.B.; Koczon, L.M.; Lester, R.K.; Noddings, C.M. Inorg. Chem. **1985**, *24*, 3568-3578.
104. Stiefel, E.I.; Cramer, S.P., in ref. 98, chapter 2, pp. 89-116.
105. Pickett, C.J.; Ryder, K.S.; Talarmin, J. J. Chem. Soc., Dalton Trans. **1986**, 1453-1457.
106. Shilov, A.E., in ref. 97, chapter 5, pp. 123-150.
107. Rocklage, S.M.; Turner, H.W.; Fellmann, J.D.; Schrock, R.R. Organometallics **1982**, *1*, 703-707.
108. Murray, R.C.; Schrock, R.R. J. Am. Chem. Soc. **1985**, *107*, 4557-4558.
109. Rocklage, S.M.; Schrock, R.R. J. Am. Chem. Soc. **1982**, *104*, 3077-3081.

110. Powell, C.B.; Hall, M.B. *Inorg. Chem.* **1984**, *23*, 4619-4627.
111. Rappé, A.K. *Inorg. Chem.* **1984**, *23*, 995-996.
112. Schrauzer, G.N.; Schlesinger, G.; Doemeny, P.A. *J. Am. Chem. Soc.* **1971**, *93*, 1803-1804.
113. Schrauzer, G.N.; Kiefer, G.W.; Tano, K.; Doemeny, P.A. *J. Am. Chem. Soc.* **1974**, *96*, 641-653.
114. Werner, D.; Russel, S.A.; Evans, H.J. *Proc. Natl. Acad. Sci.* **1973**, *70*, 339-342.
115. Vorontsova, T.A.; Krushch, A.P.; Shilov, A.E. *Kinet. Catal.* **1975**, *16*, 1403-1404.
116. Ivin, K.J. "Olefin Metathesis" Academic Press, London, 1983.
117. Dragutan, V.; Balaban, A.T.; Dimonie, M. "Olefin Metathesis and Ring Opening Polymerization of Cyclo-Olefins" Wiley-Interscience, Chichester, 1985.
118. Eleuterio, H.S., U.S. patent 3074918; filed June 20, 1957, issued Jan. 22, 1963. *Chem. Abstr.* **1961**, *55*, 16005.
119. Banks, R.L.; Bailey, G.C. *Ind. Eng. Chem., Prod. Res. Devel.* **1964**, *3*, 170-173.
120. Bradshaw, C.P.C.; Howman, E.J.; Turner, L. *J. Catal.* **1967**, *7*, 269-276.
121. Lewandos, G.B.; Pettit, R. *Tetrahedron Lett.* **1971**, 789-793.
122. Herrison, J.-P.; Chauvin, Y. *Makromol. Chem.* **1970**, *141*, 161-176.
123. Katz, T.J.; Rothchild, R. *J. Am. Chem. Soc.* **1976**, *98*, 2519-2526.
124. Grubbs, R.H.; Carr, D.D.; Hoppin, C.; Burk, P.L. *J. Am. Chem. Soc.* **1976**, *98*, 3478-3483.
125. Katz, T.J.; McGinnis, J. *J. Am. Chem. Soc.* **1977**, *99*, 1903-1912.
126. Grubbs, R.H.; Hoppin, C.R. *J. Am. Chem. Soc.* **1979**, *101*, 1499-1508.
127. Katz. T.J.; Hersh, W.H. *Tetrahedron Lett.* **1977**, 585-588.
128. Ofstead, E.A.; Lawrence, J.P.; Senyek, M.L.; Calderon, N. *J. Mol. Catal.* **1980**, *8*, 227-242.
129. Bilhou, J.L.; Basset, J.M. *J. Organomet. Chem.* **1977**, *132*, 395-407.
130. Tanaka, K.; Miyahara, K.; Tanaka, K.-I. *Chem. Lett.* **1980**, 623-626.
131. Garnier, F.; Krausz, P.; Dubois, J.-E. *J. Organomet. Chem.* **1979**, *170*, 195-201.
132. Rocklage, S.M.; Fellmann, J.D.; Rupprecht, G.A.; Messerle, L.W.; Schrock, R.R. *J. Am. Chem. Soc.* **1981**, *103*, 1440-1447.
133. Wengrovius, J.H.; Schrock, R.R.; Churchill, M.R.; Missert, J.R.; Youngs, W.J. *J. Am. Chem. Soc.* **1980**, *102*, 4515-4516.
134. Kress, J.; Osborn, J.A.; Greene, R.M.E.; Ivin, K.J.; Rooney, J.J. *J. Am. Chem. Soc.* **1987**, *109*, 899-901.
135. Upton, T.H.; Rappé, A.K. *J. Am. Chem. Soc.* **1985**, *107*, 1206-1218.
136. Rappé, A.K.; Goddard, III, W.A. *J. Am. Chem. Soc.* **1982**, *104*, 448-456.
137. Rappé, A.K.; Goddard, III, W.A. *J. Am. Chem. Soc.* **1980**, *102*, 5114-5115.
138. Rappé, A.K.; Goddard, III, W.A. *Nature* **1980**, *285*, 311-312.
139. Herrmann, W.A.; Felixberger, J.K.; Herdtweck, E.; Schafer, A.; Okuda, J. *Angew. Chem., Int. Ed. Engl.* **1987**, *26*, 466-467.
140. Kress, J.; Wesolek, M.; Le Ny, J.-P.; Osborn, J.A. *J. Chem. Soc., Chem. Comm.* **1981**, 1039-1040.

141. Nakamura, S.; Dedieu, A. *Nouv. J. Chim.* **1982**, *6*, 23-30.
142. Kress, J.; Osborn, J.A. *J. Am. Chem. Soc.* **1983**, *105*, 6346-6347.
143. Schrock, R.R.; Sharp, P.R. *J. Am. Chem. Soc.* **1978**, *100*, 2389-2399.
144. Schrock, R.R. *J. Organomet. Chem.* **1986**, *300*, 249-262.
145. Schaverien, C.J.; Dewan, J.C.; Schrock, R.R. *J. Am. Chem. Soc.* **1986**, *108*, 2771-2773.
146. Murdzek, J.S.; Schrock, R.R. *Organometallics* **1987**, *6*, 1373-1374.
147. Grubbs, R.H.; Gilliom, L.R.; *J. Am. Chem. Soc.* **1986**, *108*, 733-742. U.S. patent 4,607,112; filed March 28, 1984, issued August 19, 1986.
148. Schrock, R.R.; Feldman, J.; Grubbs, R.H.; Cannizzo, L. *Macromolecules*, **1987**, *20*, 1169-1172.
149. Anslyn, E.V.; Grubbs, R.H. *J. Am. Chem. Soc.* **1987**, *109*, 4880-4890.
150. Hamilton, J.G.; Ivin, K.J.; McCann, G.M.; Rooney, J.J. *Makromol. Chem.* **1985**, *186*, 1477-1494.
151. Masuda, T.; Susaki, N.; Higashimura, T. *Macromolecules* **1975**, *8*, 717-721.
152. Wood, C.D.; McLain, S.J.; Schrock, R.R. *J. Am. Chem. Soc.* **1979**, *101*, 3210-3222.
153. Han, C.-C.; Katz, T.J. *Organometallics* **1985**, *4*, 2186-2195.
154. Katz, T.J.; Hacker, S.M.; Kendrick, R.D.; Yannoni, C.S. *J. Am. Chem. Soc.* **1985**, *107*, 2182-2183.
155. Strutz, H.; Dewan, J.C.; Schrock, R.R. *J. Am. Chem. Soc.* **1985**, *107*, 5999-6005.
156. Katz, T.J.; Lee, S.J. *J. Am. Chem. Soc.* **1980**, *102*, 422-424.
157. Katz, T.J.; Lee, S.J.; Nair, M.; Savage, E.B. *J. Am. Chem. Soc.* **1980**, *102*, 7940-7942.
158. Landon, S.J.; Shulman, P.M.; Geoffroy, G.L. *J. Am. Chem. Soc.* **1985**, *107*, 6739-6740.
159. Parlier, A.; Rudler, H. *J. Chem. Soc., Chem .Comm.* **1986**, 514-515.
160. Katz, T.J.; Ho, T.H.; Shih, N.-Y.; Ying, Y.-C.; Stuart, V.I.W. *J. Am. Chem. Soc.* **1984**, *106* 2659-2668.
161. Boor, J., Jr. "Ziegler-Natta Catalysts and Polymerizations" Academic, New York, 1979.
162. Keii, T.; Soga, K., eds. "Catalytic Polymerization of Olefins" Elsevier, Amsterdam, 1986.
163. Cossee, P.; *J. Catal.* **1964**, *3*, 80-88. Arlman, E.J. *J. Catal.* **1964**, *3*, 89-98. Arlman, E.J.; Cossee, P., *J. Catal.* **1964**, *3*, 99-104.
164. Evitt, E.R.; Bergman, R.C. *J. Am. Chem. Soc.* **1980**, *102*, 7003-7011.
165. Watson, P.L. *J. Am. Chem. Soc.* **1982**, *103*, 337-339.
166. Ivin, K.J.; Rooney, J.J.; Stewart, C.D.; Green, M.L.H.; Mahtab, R. *J. Chem. Soc., Chem. Comm.* **1978**, 604-606.
167. McDaniel, M.P.; Cantor, D.M. *J. Polym. Sci., Polym. Chem. Ed.* **1983**, *21*, 1217-1221.
168. Zambelli, A.; Locatelli, P.; Sacchi, M.C.; Rigamonti, E. *Macromolecules* **1980**, *13*, 798-790.
169. Casey, C.P. *Macromolecules* **1981**, *14*, 464-465.

170. Ghiotti, G.; Garrone, E.; Coluccia, S.; Morterra, C.; Zecchina, A. *J. Chem. Soc., Chem. Comm.* **1979**, 1032-1033.
171. Al-Mashta, F.; Davanzo, C.U.; Sheppard, N. *J. Chem. Soc., Chem. Comm.* **1983**, 1258-1259.
172. Davanzo, C.U.; Sheppard, N.; Al-Mashta, F. *Spectrochim. Acta* **1985**, *41A*, 263-269.
173. Locatelli, P.; Sacchi, M.C.; Tritto, I.; Zannoni, G.; Zambelli, A.; Piscitelli, V. *Macromolecules* **1985**, *18*, 627-630.
174. Soto, J.; Steigerwald, M.L.; Grubbs, R.H. *J. Am. Chem. Soc.* **1982**, *104*, 4479-4480.
175. Clawson, L.; Soto, J.; Buchwald, S.L.; Steigerwald, M.L.; Grubbs, R.H. *J. Am. Chem. Soc.* **1985**, *107*, 3377-3378.
176. Turner, H.W.; Schrock, R.R.; Fellman, J.D.; Holmes, S.J. *J. Am. Chem. Soc.* **1983**, *105*, 4942-4950.
177. Mortreux, A.; Delgrange, J.C.; Blanchard, M.; Lubochinsky, B. *J. Mol. Catal.* **1977**, *2*, 73-82.
178. Devarajan, S.; Walton, D.R.M.; Leigh, G.J. *J. Organomet. Chem.* **1979**, *181*, 99-104.
179. Bencheik, A.; Petit, M.; Mortreux, A.; Petit, F. *J. Mol. Catal.* **1982**, *15*, 93-101.
180. Villemin, D.; Cadiot, P. *Tetrahedron Lett.* **1982**, 5139-5140.
181. Petit, M.; Mortreux, A.; Petit, F. *J. Chem. Soc., Chem. Comm.* **1982**, 1385-1386.
182. Pannella, F.; Banks, R.L.; Bailey, G.C. *J. Chem. Soc., Chem. Comm.* **1968**, 1548-1549.
183. Moulijn, J.A.; Reitsma, H.J.; Boelhouwer, C. *J. Catal.* **1972**, *25*, 434-436.
184. Leigh, G.J.; Rahman, M.T.; Walton, D.R.M. *J. Chem. Soc., Chem. Comm.* **1982**, 541-542.
185. Katz, T.J.; McGinnis, J. *J. Am. Chem. Soc.* **1975**, *97*, 1592-1594.
186. Wengrovius, J.H.; Sancho, J.; Schrock, R.R. *J. Am. Chem. Soc.* **1981**, *103*, 3932-3934.
187. Sancho, J.; Schrock, R.R. *J. Mol. Catal.* **1982**, *15*, 75-79.
188. Churchill, M.R.; Ziller, J.W.; McCullough, L.; Pedersen, S.F.; Schrock, R.R. *Organometallics* **1983**, *2*, 1046-1048.
189. Pedersen, S.F.; Schrock, R.R.; Churchill, M.R.; Wasserman, M.J. *J. Am. Chem. Soc.* **1982**, *104*, 6808-6809.
190. Churchill, M.R.; Ziller, J.W.; Freudenberger, J.H.; Schrock, R.R. *Organometallics* **1984**, *3*, 1554-1562.
191. Freudenberger, J.H.; Schrock, R.R.; Churchill, M.R.; Rheingold, A.L.; Ziller, J.W. *Organometallics* **1984**, *3*, 1563-1573.
192. McCullough, L.G.; Listemann, M.L.; Schrock, R.R.; Churchill, M.R.; Ziller, J.W. *J. Am. Chem. Soc.* **1983**, *105*, 6729-6730.
193. McCullough, L.G.; Schrock, R.R. *J. Am. Chem. Soc.* **1984**, *106*, 4067-4068. McCullough, L.G.; Schrock, R.R.; Dewan, J.C.; Murdzek, J.C. *J. Am. Chem. Soc.* **1985**, *107*, 5987-5998.
194. Centi, G.; Trifirò, F.; Ebner, J.R.; Franchetti, V.M. *Chem. Rev.* **1988**, *88*, 55-80.

INDEX

Ab initio calculations, 39–41, 89, 220, 258
Acetylene:
  from coupling of alkylidynes, 274–275
  metathesis of, 15, 93, 260, 262, 310–312
    mechanism, 311
  polymerization of, 14, 259, 262, 307–309
  reactions:
    with alkylidene, 259
    with alkylidyne, 259–262
    C–C cleavage, 15, 86
    with oxo, 258
Acid chloride, 236
Acrolein manufacture, 288
Acrylonitrile manufacture, 299
Alcohols:
  allylic rearrangement of, 291–293
  oxidation of, 268–270
    isotope effect in, 269–270
  propargylic, rearrangement of, 291–293
Aldehyde oxidase, 297. *See also* Oxo transferase enzymes
Aldehydes:
  oxidation of, 269
  Wittig-like reactions, 87, 135, 234, 239
Alkane, *see* Hydrocarbons
Alkene, *see* Olefin
Alkoxide ligand, 3
Alkylation, 90, 92–93, 232–234
  of alkylidene, 233
  of alkylidyne, 233
  in hydrazido synthesis, 93–94
  of imido, 222, 233, 301
  of nitrido, 92
  of oxo, 232
Alkylidene, Schrock type, 3, 12–14. *See also* Carbene ligand, Fischer-type
  in acetylene polymerization, 307–308
  applications, 14
  with β hydrogens, 59, 90
  bond angles, M–C–R, 155
    calculations on, 43–44
    and C–H coupling constant, 134, 136

    and C–H stretch, 127
    and ^{1}H NMR, 132
  bonding, 11–12, 24–25
    ab initio calculations, 39–40
  bond length, M–C, 148–152
  bond strengths, 32–33
  C–H coupling constant, 132, 134–136
  C–H stretching frequency, 127
  ^{13}C NMR, 132, 134–136
  coupling, to form olefins, 274
  cyclic, 97–98
  distorted structure, 155
    calculations on, 43–44
    and C–H coupling constant, 134, 136
    and C–H stretch, 127
    and ^{1}H NMR, 132
    calculations on, 43–44
  donor/acceptor *vs.* double bond, 39–40
  history, 12–13
  ^{1}H NMR, 132
  hyperconjugation in, 43–44
  occurrence, 12, 145–147
  in olefin metathesis, 14, 90, 304–307
  in olefin polymerization, 304–307, 309–310
  in organic synthesis, 12, 14, 235–239
  reactions, *see also* Olefin metathesis; Tebbe–Grubbs reagent
    with acetylenes, 97, 236, 259, 307–308
    with acid chlorides, 236
    with aldehydes, 87, 235
    alkylation, 233
    1,2-alkyl migration, 247–248
    α elimination, 58–59
    with amides, 235
    calculations on, 220, 258, 305
    with carbon dioxide, 235
    with carbon monoxide, 263
    with C–H bonds, 268
    coupling to form olefins, 274
    with esters, 14, 87, 235–238
    with imines, 88–89, 238
    with ketones, 87, 235–239

321

Alkylidene, Schrock type, reactions
(*Continued*)
   with ketones, 87, 235–239
   with Lewis acids, 231
   metallacycle formation, 67, 236–238, 256–258, 304–310
   migration of alkyl groups to, 247–248
   with nitriles, 97
   nucleophilic *vs.* electrophilic, 28–29
   olefin metathesis, 14, 90, 256, 304–307
   with olefins, 67, 236–238, 256–258, 305
   with phosphines, 245
   protonation, 59, 226–227
  in ring-opening polymerization, 304–307
  structures, table of, 186–190
  substituents, 25
   effect on bond length, 151
  synthesis:
   from alkylidene and acetylene, 97
   from alkylidyne, 59, 61, 91, 228, 268
   by cleavage of ketones, 71
   by cleavage of metal–carbon bond, 64
   by cleavage of olefins, 86
   by cleavage of Si—C bond, 61
   by dealkylation, 67
   by deprotonation, 52–61
   from diazoalkanes, 79
   from metallacyclobutanes, 67
   by olefin metathesis, 90
   from oxo, 89
   by protonation, 59, 61, 91
   from Wittig reagents, 80
  transfer from Ta to W, 64
  trans influence, 157
  in Ziegler–Natta polymerization, 14, 309–310

Alkylidyne, Schrock type, 3, 15–16. *See also* Carbyne ligand, Fischer-type
  in acetylene metathesis, 310–311
  applications, 15
  bond angle, M—C—R, 24, 154–155
  bonding, 24
  bond length, M—C, 148–149, 151–152
  bridging, 3, 15, 228
   ^{13}C NMR, 136
  ^{13}C NMR, 135–136
  coupling to form acetylenes, 274–275
  d^4 complex, 30, 154
  history, 15
  occurrence, 15, 145–147
  in organic synthesis, 15
  photochemistry, 24, 33–35
  reactions, *see also* Acetylene, metathesis of
   with acetylenes, 15, 93, 259–262
   with aldehydes, 87, 239
   alkylation, 233
   alkyl migration to, 248
   with amides, 239
   with carbon monoxide, 239, 263–264
   with C—H bonds, 268
   coupling to acetylenes, 274–275
   with esters, 239
   with H$_2$, 268
   hydride transfer to, 248
   with isocyanates, 239
   with isonitriles, 239
   with ketones, 87, 239
   with Lewis acids, 231–232
   to form metallacycles, 93, 259–262, 311
   migrations to, 248
   with M≡M triple bond, 264
   with nitriles, 15, 89, 239
   nucleophilic vs. electrophilic, 28–29
   with phosphines, 245
   protonation, 59, 61, 91, 228–229
   Wittig-like, 15, 87
  structures, table of, 186–190
  substituent, 25–26
   effect on bond distance, 151
  synthesis:
   by acetylene metathesis, 261
   from carbyne ligands, 15, 71
   by cleavage of acetylenes, 15, 86
   by cleavage of nitriles, 67, 79
   by deprotonation, 52–61
  trans influence, 157
  vibrational spectra, 123
Alkylimido, *see* Imido ligand
1,2-Alkyl migration, 66, 246–248, 265
Alkyl radicals, oxidation of, 266, 267, 295–296, 300
Alkyne, *see* Acetylene
Allyl alcohol, oxidation, 289
  scrambling of 1- and 3- positions, 289
Allyl cation, in allylic oxidation, 290, 300–301
Allylic alcohols, catalytic rearrangement, 291–293, 301
Allylic oxidation, 251, 266, 288–291, 299–301
Allylic rearrangement, 291–293, 301, 302
Allyl radicals, in propylene oxidation, 289, 300
Allyl transfer, 233, 301
Amide, 235, 239
Amido ligand, 3. *See also* Imido ligand reactions; Imido ligand synthesis
  methyleneamido, 240
Amine, oxidation of, 244
Amine-*N*-oxide:
  from oxo, 244
  in oxo synthesis, 75, 297, 299

INDEX   **323**

Ammonia, 270–273, 299–301
Ammoxidation, 233, 299–301
Ancillary ligands:
  and multiple bond length, 150
  and reactivity, 220, 222
    metallacyclobutadienes, 261
    protonation, 266
    sulfur ligands and O atom transfer, 241–242, 297
Angular distortions, 156–158
  and trans influence, 156–157
Arene oxidation, 266. *See also* Hydrocarbons
Arsine oxide deoxygenation, 80
Arylimido, *see* Imido ligand
Azides, 76–78
Aziridene, from imido, 251
Azoalkanes, N–N cleavage of, 85

Bending modes, in IR and Raman, 112
  of bridging nitrides, 121–122
  in tetraoxo complexes, 121
Bent imido, *see* Imido ligand, bent *vs.* linear
Benzylic oxidation, 266, 299–300
Bicyclo[2.1.0]pentane radical clock, 296
Biotin S-oxide, 297
Bis(cyclopentadienyl) complexes:
  imido:
    electronic structure, 42
    stretching frequency, 125
  multiple bond lengths, 152
  oxo, stretching frequency, 117
    photolysis to give O_2, 273
Bismuth molybdate, 288–290, 299–301
Bond angle:
  in alkylidene, M—C—R, 155
  in alkylidyne, M—C—R, 154–155
  in imido, M—N—R, 154–155
Bond dipole, change in, 112
Bond dissociation energy, 32–33, 40–41
Bond distance, *see* Bond length
Bonding:
  comparison between multiply bonded ligands, 23, 24, 26–28
  comparison with π-acid ligands, 26
  competition between multiply-bonded ligands, 23, 40–41, 125–126
  effect of substituents, 25–26
Bond length, 148–154
  ancillary ligand, effect of, 152
  and bond order, 150
  ligand comparison, 148–150
  in molybdenum complexes, 149
  and ^{17}O NMR, 127–129
  unusually long, 152–154

Bond order:
  alkylidene and carbene ligands, 25–26
  and bond length, 150
  in dioxo, diimido complexes, 35–36
  and metal–oxo stretching frequencies, 117, 118
  and ^{17}O NMR, 127–129
  oxo, nitrido, imido ligands, 23, 40–41
  in trioxo complexes, 36
Bond strengths, metal–ligand multiple, 32–33, 40–41
Bridging *vs.* terminal ligand, 29–32, 147
Butane oxidation to maleic anhydride, 291
Butene oxidation:
  to butadiene, 290
  to maleic anhydride, 291

Camphor, hydroxylation of, 296
Carbene ligand, Fischer-type, 3, 11–12. *See also* Alkylidene, Schrock-type
  acetylene polymerization initiator, 308
  applications, 12
  bonding, 11–12, 25–26
    ab initio calculations, 39–40, 220
  bond length, 151
  comparison to alkylidene, 11–12, 24–25, 39–40, 151
  history, 11
  in organic synthesis, 12
  singlet *vs.* triplet, 39–40
  synthesis by cleavage of activated olefins, 86
Carbido ligand, bridging, 15
Carbodiimide:
  cleavage of, 67
  metathesis of, 302
Carbon dioxide, cleavage of, 67, 235
Carbon monoxide, 67, 239, 262–264
Carbonyl ligand stretching frequency, 27
Carbonyl olefination reaction, *see* Kauffmann methylenating reagents; Tebbe–Grubbs reagent; Wittig-like reactions
Carbyne ligand, Fischer-type, 3, 15. *See also* Alkylidyne, Schrock-type
  bonding, 25–26
    ab initio calculations, 220
  bond length, 151
  ^{13}C NMR, 135–136
  conversion to alkylidyne, 15, 71
  structures, table, 186–190
  synthesis, 71
  vibrational spectrum, 123
Catalase enzymes, 294
Catalysts, heterogeneous, *see* Heterogeneous catalysis

C–glucosides, 236
Charge control, of reactivity, 4, 28–29, 243
Chauvin mechanism for olefin metathesis, 304
C—H bond, see Hydrocarbons
Chloramine-T, in imido synthesis, 70, 255, 301
N-Chloro-N-argento carbamate, 255, 301
Chromium(VI) reagents, 4
  chromate/dichromate equilibrium, 129
  imidochromium(VI), 267, 300
  reactions:
    with alcohols, 268–270
    with benzyl radicals, 300
    with hydrocarbons, 264–266
Cis ligand:
  displacement from multiple bond, 157
  and stretching frequency, 116, 118
Citral, manufacture of, 291
^{14}C labeling, propylene oxidation, 289
Claisen rearrangement, 292–293, 301
^{13}C NMR:
  acetylene polymerization, study of, 308
  alkylidene, 132, 135–136
  alkylidyne, 135–136
  and imido reactivity, 133–135
  t-butylimido, 133–135
Cofactor, enzyme:
  "FeMo–Co" in Nitrogenase, 302–303
  "Mo-Co" in Oxo transferases, 297
Comparison between multiply bonded ligands, see Periodic trends
Competition between multiply-bonded ligands, 23, 40–41, 59–61, 125–126
Coordination geometry, see Octahedral complex; Square Pyramidal complexes; Tetrahedral complexes; Trigonal bipyramidal structures
  of d^1 complexes, 147
  of diimido complexes, 35–36
  of dioxo complexes, 35–36
  distortions of, 155–158
  of trioxo complexes, 36
Cossee–Arlman mechanism, 14, 309–310
Coupling constant:
  C-H, alkylidene, 132, 134–136
  C-H, alkylidyne, 135
  NH, imido, 130
Coupling of two multiply bonded ligands, 273–275
  calculations on, 275
Coverage in this volume, 1
Cresol, 293
[2 + 2] Cycloaddition reactions, see Olefin metathesis; Wittig-like reactions
[3 + 2] Cycloaddition reactions, 254

Cyclopentadienyl ligand, synthesis from alkylidyne, 260–262. See also Bis(cyclopentadienyl) complexes
Cyclopropanation, 12
Cyclopropenyl ligand, from alkylidyne, 260–262
Cyclopropylmethyl, radical clock, 296
Cytochrome P-450, 5, 81–83, 265, 267, 294–296. See also Iron(IV)-oxo
  crystal structure of cytochrome P-450$_{cam}$, 296
  model systems, 241, 248–251, 265–267, 296
  peroxide shunt, 296

Dealkylation:
  of imido, 65
  in multiple bond synthesis, 64–69
Dehydrolinalool, rearrangement, 291
d electron count:
  and bond length, 150–151
  d^1, μ-oxo structures, 147
  effect on multiple bonding, 22, 35, 54, 147
  higher than d^2, 6, 8, 22, 30, 35, 37–39, 41–42, 147, 296
    and angular distortions, 157
    and stretching frequencies, 117
  and occurrence of multiple bonds, 147
  and trans influence, 156
Deprotonation:
  α, to form a multiply bonded ligand, 53–61
  by external base, 56, 240–241
  factors promoting, 53–54
  steric effects, 57–58
  of NH imido, 53–54
  of CHR alkylidene, 58
Deuterium labeling:
  methanol oxidation, 291
  propylene oxidation, 289
Diamination of olefins, 8–9, 255
Diazenido ligand, 25, 93, 95
Diazoalkane:
  complex, 79, 234–235
  coupling, 273
  in carbene synthesis, 79
  from hydrazido, 10, 234–235
N,N-Dichlorophenylsulfonamide, 70
Dihydroxylation of olefins, 4, 252–255. See also Diamination of olefins; Oxyamination of olefins
  asymmetric, 254
  mechanism, 253–254
Diimide, putative intermediate, 303
Diimido complexes, coordination geometry, 35–36
α-diketone synthesis, 258

Dinitrogen oxide, 74
Diol synthesis, 252–254
Dioxo complexes:
    bond angle, O—M—O, 157–258
    bond length, M—O, 149–150
    bond order, 35–36
    cis vs. trans, 35–36, 157
    [3+2] cycloaddition to, 254
    d^1, structures, 147
    stretching frequencies, 118–119
Distortional isomers, 152–154
Distribution of compounds, 145–147
Ditungsten hexaalkoxide, 15, 86
DuPont Co., 12–13, 304

Effective Atomic Number (EAN) rule, 23
Electrocatalytic:
    nitrogen fixation, 303
    oxidations, 299, 301
Electronic materials, 5, 7
Electronic spectroscopy, 34–35
    oxo-halide complexes, 34–35
    permangante, 37
    vanadyl ion, 34
    vibrational structure, 35, 117
Electron Spin Resonance, see EPR
Enzymes, see Catalase enzymes; Cytochrome P-450; Nitrogenase; Oxo-transferase enzymes; Oxygenase enzymes; Peroxidase enzymes
Episulfides, 240
Epoxidation, 248–251
    asymmetric, 251
    aza analog, 251
    catalytic, 83, 251, 299
    mechanism, 249–251
    stereochemistry, 248–249
Epoxides:
    formation from metal-oxo, 83, 248–251
    deoxygenation to form oxo, 64, 68
    catalytic, 68–69
EPR:
    of d^1 complexes, 34
    ^{17}O effect on, 297
Esters, 14, 87, 235–239
EXAFS, 294, 297–298, 302
Excited states, electronic, 35, 117

Ferrate, FeO_4^{2-}, 4
Fischer Carbene ligand, see Carbene ligand, Fischer-type
Fischer Carbyne ligand, see Carbyne ligand, Fischer-type
Fluoride ligand, 3

Formaldehyde ligand, 66, 89
Formaldehyde manufacture, 291
Formamidines, in imido synthesis, 88
Formate dehydrogenase, 297. See also Oxo-transferase enzymes
Four center mechanisms, see Acetylene, metathesis of; Olefin; Wittig-like reactions
Frontier orbital control, 4, 28–29, 220, 244, 253, 255

Gas phase reactions, 89–90
Geraniol/nerol manufacture, 291
Green–Rooney mechanism, 14, 256, 307, 309–310

^2H, see Deuterium labeling
Heterogeneous catalysis:
    metathesis:
        of acetylenes, 310
        of olefins, 304
    oxidation:
        ammoxidation, 299–301
        bismuth molybdate catalyst, 288–290, 299–301
        of butane, butenes, 291
        comparison of oxide catalysts, 289
        imido ligand in, 9, 299–301
        iron molybdate catalyst, 291
        of methanol, 291
        molybdenum trioxide catalyst, 291
        of propylene, 288–290, 299–301
        surface vacancies in, 290
        vanadium oxide for ammoxidation, 299
        vandyl pyrophosphate catalyst, 291
    polymerization:
        of acetylenes, 307
        of olefins, 304, 309–310
Heteropolyanions, see Polyoxoanions
^1H NMR:
    alkylidene ligands, 132
    alkylimido ligands, 131–132
Homogeneous catalysis:
    in acetylene metathesis, 310–312
    in acetylene polymerization, 307–309
    alkylidene ligand in, 303–310
    alkylidyne ligand in, 310–312
    imido ligand in, 9, 301–302
    in olefin metathesis, 303, 306–307
    in olefin polymerization, 303–310
    oxidation reactions, 298–299, 301
    oxo ligand in, 5, 291–293, 298–299
Homopolyanions, see Polyoxoanions
Horseradish peroxidase, 293–294
    electronic structure of FeO unit, 41

326 INDEX

Hydrazido, 2, 9–11. *See also* Hydrazido(4-)
  bond angle, M—N—N, 154
  bonding, 25–26, 151
    ab initio calculation, 40
  bond length, unusually long, 153
  bridging, 2–3
  and nitrogen fixation, 9–10, 302–304
  ^{15}N NMR, 131
  occurrence, 10, 145–147
  in organic synthesis, 10
  reactions:
    with aldehydes, ketones, 234–235
    deprotonation, 241
    to form N_2, 271–272
    with hydride donors, 271–273
    hydride migration to, 266, 272
    N—N cleavage, 270–273, 302–304
    protonation, 76, 226, 271–273, 303–304
  structures, table of, 179–186
  synthesis:
    by alkylation of N_2 ligand, 93–95
    by cleavage of Si–N bond, 62
    from diazonium cation, 95
    from hydrazine and metal oxo, 55
    by protonation of N_2 ligand, 93–95, 302–304
Hydrazido(4-), 3, 10. *See also* Hydrazido
  from coupling of two nitrides, 273
  IR spectrum, 123
  in nitrogen fixation, 302–303
  ^{15}N NMR, 131
  reactions:
    with acetone, 235
    protonation, 303
  synthesis:
    from alkylidenes, 88
    from N_2, 303
Hydrazines:
  from hydrazido, 271, 302–303
  in nitrido and imido synthesis, 76–79, 85
  in nitrogen fixation, 302–304
  reaction with oxo, 60
Hydride, *see* Deprotonation; Protonation
  intermediate in N—N bond cleavage, 272, 304
  migration to alkylidene, alkylidyne, 58–61, 228, 248
  migration to hydrazido, 226, 272
  solvent assisted transfer to oxo, 266, 270
Hydrocarbons, *see* Acetylene; Arene oxidation; Methyl arene, oxidation of; Olefin
  reaction with alkylidene, 268
  reaction with alkylidyne, 268
  reaction with imido, 266–267, 299–301
    radical intermediates, 300
  reaction with oxo, 264–266, 295–296
    radical intermediates, 265–266, 295–296
    stereochemistry, 265, 295
β-hydrogen:
  alkylidene stability and, 59
  in metallacyclobutanes, 67
Hydrogen (H_2) addition to alkylidyne, 268
α-hydrogen elimination, 58–59, 309
Hydrogen peroxide, 84, 224, 293–296
Hydrogen transfer to form a multiply bonded ligand, 52–61
  intramolecular, 59–61
Hydroperoxides, *see* Peroxides
Hydroxylation of hydrocarbons, 264–267, 294–296
  stereochemistry, 265, 295
Hypochlorite, in oxo synthesis, 69, 299

Imido ligand, 2, 8–9. *See also* NX-type ligands
  in ammoxidation, 299–301
  applications, 9
  bent *vs.* linear, 22–23, 24, 154–155
    calculation, 42
  bond angle, M—N—R, 154–155
  bonding, 22–23, 42–43
  bond length, M—N, 148–151
  bond order, 23
  bridging, 2–3
  t-butylimido, ^{13}C NMR, 133–135
  ^{13}C NMR of *t*-butylimido, 133–135
  diimido ligand, bridging, 96, 97
  in enzymatic reactions, 304
  in heterogeneous catalysis, 9, 299–301
  history, 8
  ^{1}H NMR, 131–132
  in homogeneous catalysis, 9, 301–302
  ^{15}N NMR, 129–131
  nucleophilic attack at, 29, 42
  occurrence, 8, 145–147
  in olefin metathesis, 9, 305–307
  in organic synthesis, 9, 255, 301–302
  organometallic derivatives, 9
  pentachlorethyl imido ligand, 96
  radical reactions of, 300
  reactions:
    with aldehydes and ketones, 87, 135, 234
    alkylation, 233, 301
    allyl transfer to, 301
    with carbocations, 300–301
    with carbodiimide, 302
    Claisen rearrangement, 301
    correlation with ^{13}C NMR, 134–135

dealkylation, 65, 221
deprotonation of NCH$_2$R, 240
deprotonation of NH, 53–54
diamination of olefins, 8–9, 255
  with hydrocarbons, 266–267
  migration of alkyl to, 246–247
  of the organic moiety, 273
  with O$_2$, 239
  with olefins, 251, 255
  with organometallic reagents, 246–247
  oxyamination of olefins, 9, 255, 301
  protonation of 60, 221–222, 224–226
  with radicals, 300
"spectator" imido, 305–306. *See also* 40–41
structure of, 22–23
structures, table of, 179–186
substituent, 25–26
  and bond distance, 151
synthesis:
  by alkylation of nitrides, 92
  from alkylidene, 88–89
  from azides, 76, 78
  from azoalkanes, 85
  from *t*-butyl sulfur diimine, 73
  from carbodiimide, 67–68
  from chloramine-T, 70, 255, 301
  by cleavage of Cl–N bond, 70, 255, 301
  by cleavage of Si–N bond, 62–63
  by dealkylation, 65–66
  by deprotonation, 55–57, 60
  from hydrazido complexes, 76–79, 304
  from hydrazines, 76–79, 85
  from imines, 88–89, 96–97
  via η^2-imine intermediate, 66
  from isocyanate, 67, 88
  from nitrides, 90–92
  by nitrile dimerization, 96, 97
  by nitrile insertion, 96, 97, 98
  from nitrosobenzene, 71, 75–76
  from oxo, 88, 234
  from phosphinimines, 88
  by protonation of nitrides, 53, 90–91
  from (tosyliminoiodo)benzene, 266–267
  from triazine complex, 78
theoretical studies, 41–42
trans influence, 156–157
Imine:
  formation, 300
  as ligand, imido precursor, 66
  reactions of, 88–89, 96–97, 238
  Me$_3$SiN=CHR dimerization, 97
Iodosobenzene, in oxo synthesis, 69–70, 248, 296
  (tosyliminoiodo)benzene, 266–267

Iron–Molybdenum Cofactor, 302–303
Iron(IV)-oxo, *see* Cytochrome P-450; Horseradish peroxidase
  characterization, 294
  electronic structure, 41
  IR, 117
  triplet ground state in, 41
IR spectra, *see* Vibrational spectra
Isocyanates, 67–68, 88, 239
Isonitriles, 239
Isopolyanions, *see* Polyoxoanions
Isotope effect:
  in alcohol oxidation, 269–270
  in ethylene polymerization, 310
  in hydrocarbon hydroxylation, 295
Isotopic substitution:
  in acetylene polymerization, 308
  in enzymatic reactions, 295, 297
  imido, 123–125
  nitrido, 115, 122
  oxo, 114–115
  in propylene oxidation, 289, 301

Kauffmann methylenating reagents, 237
Ketazine, dimethyl, 235
Ketene:
  complex, from alkylidene, 263
  reaction with dioxo complex, 254
Ketones, 68, 71, 87, 234–239

β-lactam synthesis, 12
Lewis acid:
  binding to multiple bond, 229–232
    effect on nitride stretch, 121
    effect on oxo stretch, 230
  in olefin metathesis, 230, 231, 305–306
Ligand comparison, *see* Periodic trends
Ligand field descriptions, 33–39
  octahedral complexes, 33–36
  square pyramidal complexes, 36
  tetrahedral complexes, 36–38
Linalool, rearrangement, 291
"Living" polymer, 307–308

Main group elements, multiple bond formation, 31
Maleic anhydride production, 291
Mars–van Krevelen model, 289
Metal, effect on reactivity of multiply-bonded ligand, 29–32, 221, 276
Metalla-Claisen rearrangement, 292–293, 301
Metallacyclobutadienes, 93, 259–262
  in acetylene metathesis, 311
  acetylene polymerization initiator, 308

Metallacyclobutadienes (*Continued*)
  deprotonated (β), 261
  factors affecting stability, 261
Metallacyclobutanes, 67, 236–238, 256–258, 304–307
  calculations on, 258, 305
  in olefin metathesis, 256, 304–307
  in olefin polymerization, 309–310
  oxametallacyclobutane, 249–250, 253
  rearrangement of, 256–257
Metallacyclobutenes, 67, 259, 307
Metal–ligand multiple bonds, *see also individual ligands;* Periodic trends
  bond angle in, 154–155
  bonding, 21ff
    d electron count and, 22
  bond distance, 148–152
  bond order, 23, 25–26, 35–36, 40–41, 117–118, 127–129, 150
  bond strengths, 32–33, 40–41
  vs. bridged structures, 29–32, 147
  comparison with π-acceptor ligands, 26, 156
  competition between, 23, 40–41, 59–61, 125–126
  coupling of two, 273–275
  distribution by metal, 145–147
  formation:
    by addition to the β atom, 93–98
    by alkylation of α atom, 90, 92–93
    by cleavage of α carbon, 64–69, 86
    by cleavage of α halogen, 69–70
    by cleavage of α hydrogen, 52–61
    by cleavage of α nitrogen, 73–79, 85
    by cleavage of α oxygen, 70–72, 81–85
    by cleavage of α phosphorus, arsenic, etc., 79–80
    by cleavage of α silicon, 61–64
    by cleavage of α sulfur, 72–73
    by protonation of α atom, 90–91
    by Wittig-like reactions, 86–90, 234–235, 238–239
  reactions, 220–279. *See also* Reactivity of multiply-bonded ligands
    with acetylenes, 258–262
    alcohol oxidation, 268–270
    alkylation, 92–93, 232–234
    1,2-alkyl migration to, 246–248, 265
    with amines, 244
    with carbon dioxide, 235
    with carbon monoxide, 262–264
    coupling, 273–275
    [2+2] cycloadditions, *see* Wittig-like reactions
    [3+2] cycloaddition, 254
    dealkylation, 65, 221
    deprotonation, 53–58, 222, 240–241
    with episulfides, 240
    gas phase, 89–90
    hydrocarbon oxidation, 264–268
    with Lewis acids, 229–234
    with M≡M triple bonds, 264
    with O_2, 239
    with organic carbonyls, 14, 87, 234–239
    with organometallic reagents, 246–248
    with olefins, 248–258
    with phosphines, 79, 241–246
    protonation, 58–61, 90–91, 221–222, 223–229
    with sulfides, sulfoxides, 244
    with sulfur, S_2Cl_2, S_4N_4, 240
    Wittig-like, 10, 14, 15, 86–90, 234–239
  rearrangements involving, 66, 246–248
  replacement of one by another, *see* Wittig-like reactions
  transfer between metals, 81
  vibrational spectra, 112–127
Metal-ligand stretching frequencies, *see* Stretching frequencies
Metallooxaziridine, 75–76
Metal–metal multiple bonds, 31
  reaction with:
    acetylenes, 15, 86
    alkylidyne complexes, 264
    nitrido complexes, 264
    nitriles, 67, 79
    oxo complexes, 264
Metathesis, *see* Acetylene; Olefin
Methanol, oxidation of, 291
Methyl arene, ammoxidation of, 299–300
Methylenating reagents, *see* Kauffmann methylenating reagents; Tebbe-Grubbs reagent; Wittig-like reactions
Methylene, *see* Alkylidene, Schrock-type
Methyleneamido complex, 240
Methylimido osmium, instability, 240
Migration:
  of alkyl to oxo, imido, alkylidene, alkylidyne, 246–248, 265
  of hydride, 58–61, 228, 248. *See also* Deprotonation; Protonation
Molybdate catalysts, 5, 288–291, 299–301
Molybdenum enzymes, *see* Nitrogenase; Oxo-transferase enzymes
Molybdenum-95 NMR, 136
Molybdenum-oxo, oxygen atom transfer, 241–243. *See also* Oxo-transferase enzymes
Monodisperse polymer, 307

Naphthoquinone synthesis, 12
Nitrate ion, N–O cleavage in, 74
Nitrate reductase, 297. *See also* Oxo-
 transferase enzymes
Nitrene, *see* Imido ligand
Nitrene extrusion, 75–76
Nitrenoid C–H insertion, 267
Nitrido ligand, 2, 6–8. *See also* Nitrido ligand,
  bridging
 applications, 7–8
 bonding, 22, 36
  ab initio calculation, 40
 bond length, M–N, 148–151
  unusually long, 153
 bond order, 23
 in electronic materials, 7
 history, 6–7
 ^{15}N NMR, 130–131
 occurrence, 7, 145–147
 organometallic derivatives, 7
 reactions:
  alkylation, 90, 92
  with ArSCl, 92
  coupling to hydrazido, 273
  with episulfides, 240
  with Lewis acids, 230
  with M≡M triple bond, 264
  with NCl$_3$, 92
  with (NSCl)$_3$, 93
  nucleophilic attack at, 29, 245
  with phosphines, 245
  with PhSO$_2$Cl, 92
  protonation, 53–54, 90, 224
  with sulfur, S$_2$Cl$_2$, S$_4$N$_4$, 240
 stretching frequencies, 113–118
  effect of Lewis acid, 121
 structures, table of, 179–186
 synthesis:
  from alkylidynes, 89
  from azides, 76–77
  by cleavage of nitriles, 89
  by cleavage of nitrosyl, 71
  by cleavage of S–N bond, 73
  by dealkylation, 64–65, 67
  from hydrazines, 85
  by proton transfer, 54
 transfer between metals, 81
 trans influence, 156–157
Nitrido ligand, bridging, 2–3, 6, 8, 81, 230
 metal nitride materials, 7
 oligomers and polymers containing, 8, 70
  theoretical studies, 42
 reactions:
  protonation, 224

 structures, table of, 179–186
 synthesis:
  by cleavage of Cl—N bond, 70, 81
  by cleavage of Si—N bond, 8, 62
 vibrational spectra, 121–122
Nitriles, reactions:
 with alkylidene, 97
 with alkylidyne, 15, 67, 79, 89, 239
 cleavage of, 15, 67, 79, 89, 239
 coupling of, 96, 98
 insertions of, 96, 97
Nitrite ion, N—O cleavage in, 74–75
Nitrobenzene, N–O cleavage, 75
Nitrogenase, 9–10, 76, 302–304
Nitrogen monoxide, in oxo synthesis, 73–74
Nitrogen (N$_2$), from hydrazido, 271
Nitrogen (N$_2$) complex, *see* Hydrazido(4-)
 protonation, alkylation, 93–95, 302
Nitrogen (N$_2$) fixation, 9–10, 224, 270, 302–
  303. *See also* Nitrogenase
 theoretical studies, 42, 272, 303
Nitrosobenzene, in oxo and imido synthesis,
  71, 75–76
Nitrosyl ligand, cleavage of, 71
Nitrous oxide, 74
NMR, *see* ^{13}C NMR; ^{1}H NMR; ^{14}N NMR; ^{15}N
 NMR; ^{17}O NMR
 of metal atom, 136
^{14}N NMR, 129–130
^{15}N NMR, 130–131
N-oxide, *see* Amine-*N*-oxide
NX-type ligands, 2, 10–11. *See also* Imido
 ligand
 cyclic, 11, 93
 N-halogen, 70
  halogen exchange in, 70
 synthesis:
  by cleavage of halogen-N bond, 70
  from nitrido + ArSCl, 92
  from nitrido + NCl$_3$, 92
  from nitrido + (NSCl)$_3$, 93
  from nitrido + PhSO$_2$Cl, 92

Occurrence of metal–ligand multiple bonds,
 30–31, 145–147. *See also individual
 ligands*
Octahedral complex, 147
 bonding in, 33–36
 distortion of, 36, 156–158
  trans influence, 156–157
  angular distortions, 157–158
^{18}O labeling:
 propylene oxidation, 289
 xanthine oxidation, 297

Olefin, *see also* Olefin metathesis; Olefin polymerization
  from alkylidene coupling, 274
  C—C cleavage, 86, 89
  dihydroxylation, 252–255
  epoxidation, 83, 248–252
  metallacycles from, 67, 236, 256–258
Olefin metathesis, 90, 256, 257, 304–307
  associative mechanism, 307
  aza analog, 302
  calculations on, 258, 305
  catalyst decomposition, 248, 274
  imido catalyst for, 9, 307
  Lewis acid in, 230, 231, 305–306
  mechanism, 304–305
Olefin polymerization:
  Cossee–Arlman mechanism, 14, 309–310
  Green–Rooney mechanism, 14, 256, 307, 309–310
  ring opening, 14, 304–307, 308, 309
  Ziegler–Natta, 14, 309–310
^{17}O NMR, 127–129
Organic synthesis:
  alkylidene ligand in, 12, 14, 235–239
  alkylidyne ligand in, 15
  hydrazido ligand in, 10
  imido ligand in, 9, 255, 301–302
  oxo ligand in, 4–5, 252–256, 258, 266, 291
Osmium–imido–oxo complexes:
  bond lengths, 28
  diamination by, 255
  hydrolytic stability, 31–32, 221
  instability of methylimido, 240
  IR, 125–126
  oxyamination by, 255, 301
Osmium tetroxide, OsO$_4$, 4, 31, 262
  catalytic oxidations, 75, 299, 301
  complexes with amines, olefins, 252–254
  dihydroxylation by, 252–254
  Os—O bond strength, 32
  reaction with acetylenes, 258
  reaction with alcohols, 269
Oxametallacyclobutane, 249–250, 253
Oxenoid reaction with C—H bonds, 265, 295
Oxidation state, *see* d electron count
Oxo-halide complexes, UV-Visible spectra, 34–35
Oxo ligand, 2, 3–6
  applications, 3–6
  bonding, 22–23
    ab initio calculations, 40–41
  bond length, M–O, 148–151
    and bond order, 150–151
    and oxidation state, 150–151
    unusually long, 152–154

bond order, 23, 150–151
bond strength, 32, 40
bridging, 2, 229–230
  favored for d^1, 147
  ^{17}O NMR, 127–129
  stretching frequency, 113, 116–117
d^4, d^5 complexes, 6, 22, 30, 35, 37–39, 147, 296
distortional isomers, 152–154
in electronic materials, 5
in enzymatic processes, 293–298
halide complexes, UV-Visible spectra, 34–35
in heterogeneous catalysis, 5, 288–291
history, 3–4
in homogeneous catalysis, 5, 291–293, 298–299
mechanisms of reactions, 4
nucleophilic attack at, 29
occurrence, 4, 145–147
^{17}O NMR, 127–129
in organic synthesis, 4–5, 252–256, 258, 266, 291
organometallic derivatives, 5–6
photochemistry, 35, 84, 269, 273
radical reactions of, 249, 264–266, 289–290, 295–296
reactions, *see also* Oxo transferase enzymes; Oxygen atom transfer
  with acetylenes, 258
  alcohol oxidation, 268–270, 291
  alcohol rearrangement, 291–293
  aldehyde oxidation, 269
  alkylation, 232
  alkyl migration to, 246–247, 265
  with amines, 31–32, 56, 60, 244
  aniline, dimethyl, oxidation, 293
  arene oxidation, 266
  with carbon monoxide, 262
  Claisen rearrangement, 292–293, 301
  coupling to O$_2$, 273
  coupling to peroxo, 273
  cresol oxidation, 293
  [3 + 2] cycloadditions, 254
  deoxygenation, 240–245
  with dialkylsulfides, 244
  epoxidation of olefins, 248–251
  exchange with H$_2^{18}$O, 53
  to form dihalide, 230
  gas phase, 89–90
  with H$_2$O$_2$, 224
  hydride transfer to, 266, 270
  with hydrocarbons, 264–266
  with isocyanates, 88
  with ketenes, 254
  with Lewis acids, 229–230

migration of alkyl group to, 246–247, 265
with M≡M triple bond, 264
with olefins, 248–254, 288–291
with organometallic reagents, 246–247
phenylbutazone oxidation, 293
with phosphines, 79, 241–244
with phosphinimines, 88
protonation of, 59–60, 223–224
with sulfoxides, sulfides, 244
with thiols, 72, 297
with Wittig reagents, 89
significance, 3–6
"spectator" oxo, 40–41, 305
stretching frequencies:
of d^4 oxo complexes, 117
monooxo complexes, 115–118
polyoxo complexes, 118–121
structures, table of, 159–179
synthesis:
from aldehydes and ketones, 68, 87, 134–135, 234–239
from alkylidene, 87, 235
alkylidyne, 87, 239
from amine-N-oxides, 75
from arsine oxides, 79–80
from CO and CO_2, 67
by cleavage of Si–O bond, 61
by cleavage of sulfoxides, 72
by dealkylation, 64–69
from dinitrogen oxide, 74
from epoxides, 68–69
from hydrazido, 235
from hypochlorite, 69
from imido, 87, 134–135, 234
from iodosobenzene, 69–70, 296
from nitrate, 74
from nitrite, 74–75
from nitrobenzene, 75–76
from nitrogen monoxide, 73–75
from nitrosobenzene, 75–76
from nitrous oxide, 74
from O_2, 81–84, 241
from peracids, 70–71
from periodate, 69–70, 299
from peroxides, 84–85
from phosphine oxides, 79–80
by proton transfer, 52–56, 58–59
from sulfoxides, 72
transfer between metals, 81
trans influence, 156–157
Oxo transfer, see Oxo ligand, reactions; Oxo ligand, synthesis
Oxo-transferase enzymes, 5, 72, 74, 83, 241, 296–298

model studies, 297
aldehyde oxidation, 269
molybdenum cofactor, 297
Oxyamination of olefins, 9, 221, 255, 301
catalytic, 255, 301
Oxygen (O_2):
cleavage to form oxo, 81–84, 241
from coupling of oxo ligands, 273
reaction with imido, 239
Oxygenase enzymes, 294–296. See also Cytochrome P-450
Oxygen atom transfer, see Oxo ligand, reactions; Oxo, ligand, synthesis; Oxo transferase enzymes
kinetics and mechanism, 243–244
thermodynamics, 242

Pairwise mechanism for olefin metathesis, 304
Peracid, in oxo synthesis, 70–71
Periodate, in oxo synthesis, 69, 296, 299
Periodic trends:
comparison with π-acceptor ligands, 26, 156
ligand comparisons, 23–24, 26–29, 59–60
bond lengths in oxo–imido complexes, 28
by IR, 125–126
trans influence, 156–157
in ligand reactivity, 28–29, 221–223
by metal, 29–32, 221–223, 276
in stretching frequencies, 117–118, 120–121
Permanganate, MnO_4^-:
electronic structure, 36–37
origin of color, 37
photolysis to give O_2, 273
reactions, 254
with acetylenes, 258
with alcohols, aldehydes, 268–270
with alkyl radicals, 265
with arenes, 266
with hydrocarbons, 264
with olefins, 252
Peroxidase enzymes, 293–294. See also Horseradish Peroxidase
Peroxides, 293, 301
formation of oxo ligand, 84, 296, 299
and peroxidase enzymes, 293–294
reaction with oxo ligand, 224
Peroxo complexes:
from coupling of oxo ligands, 273
O—O cleavage, 82–84
from oxo and H_2O_2, 224
as oxygen donor, 84–85
μ-peroxo dimer, 82
Perrhenate ion, silylation, 232
Pertechnetate ion, silylation, 232

Phillips catalyst, 309
Phillips Petroleum Co., 304
Phosphine oxide deoxygenation, 79–80
Phosphines:
 attack at alkylidene, 245
 attack at alkylidyne, 245
 attack at nitrido, 245
 deoxygenation of oxo complexes, 79–80, 241–244
 promotion of α elimination, 57–58
Phosphinimidate ligand, R_3PN, 245
Phosphinimine, 88
Photochemistry, 24, 34–35, 84
π bond, metal–ligand, 21–22
π donating ability, 23–24, 26–28, 125–126
π donation, and ligand electronegativity, 27
Polymerization, see Acetylene; Olefin; Ring opening polymerization; Ziegler–Natta polymerization
Polyoxoanions, 2, 6, 31
Propargylic alcohols, catalytic rearrangement, 291–293
Propylene:
 heterogeneous ammoxidation of, 299–301
 heterogeneous oxidation of, 288–291
Protonation, see also Deprotonation
 of alkylidene, 59, 226–227
 of alkylidyne, 58–59, 61, 90–91, 228–229
 of hydrazido, 226, 303
 hydridometal intermediate, 226, 272, 304
 of imido, 59, 224–226
 of nitrido, 53–54, 90–91, 224
 of oxo, 58–60, 223–224
Publication activity, 1
Pyridine, oxidation, 244
Pyridine-N-oxide, see Amine-N-oxide

Radical reactions, 249, 264–266, 289–290, 294–296, 300
 radical clocks, 296
Radiopharmaceuticals, 7
Raman spectra, see also Vibrational spectra
 of bridging nitrides, 121–122
 of polyoxo complexes, 118–121
 water as solvent for, 113
Reactions, see Metal–ligand multiple bonds, reactions
Reactivity of multiply-bonded ligands, see also Metal–ligand multiple bonds
 charge control, 4, 28–29, 243
 electrophilic vs. nucleophilic, 28–29, 221
 frontier orbital control, 4, 28–29, 220, 244, 253, 255
 metal, influence of, 29–32, 221, 276

oxidation state, effect of, 30–32, 222, 276
 periodic trend by ligand, 28–29
Rearrangements involving multiple bonds, 66, 246–248
Ring opening polymerization, 14, 304–307, 308, 309
Ruthenium(VI) dioxo porphyrin:
 catalytic epoxidation, 251, 299
 formation from O_2, 83
Ruthenium(IV) oxo:
 bonding in, 42
 in catalytic oxidation, 299
 formation of, 54–55
 IR, 117
 reaction with alcohols, 269–270
 reaction with C—H bonds, 266
 reaction with phosphines, 243–244
Ruthenium tetroxide, 4
 catalytic oxidations, 299
 reactions:
  with acetylenes, 258
  with alcohols, aldehydes, 268–269
  with arenes, 266
  with hydrocarbons, 264–265
Ru-O bond strength, 32

S_2Cl_2, 240
Selection rules, IR and Raman spectra, 112, 118, 121
Selenium dioxide, allylic oxidation by, 289
Seven coordinate complexes, 148
Silicon group:
 addition of, 232
 cleavage of, 61–64, 97
Silylation of oxo, 232
S_4N_4, 240
SOHIO, 300–301
Solid state compounds:
 metal nitrides, 7
 metal oxides, 5, 30–31, 156
"Spectator" imido, 305–306
"Spectator" oxo effect, 40–41, 305
Square pyramidal complexes, 147–148
 bonding, 36
Stretching frequencies, see also Vibrational spectra
 alkylidene, C—H stretch, 127
 alkylidyne, 123
 bridging nitrido, 121–122
 bridging oxo, 113
 carbyne, 123
 cis ligand, effect of, 116, 118
 comparison of ligands by, 27, 125–126
 in electronic excited states, 35, 117

INDEX 333

hydrazido(2−), N—H stretch, 127
hydrazido(4−), 123
imido, 123–125
   N—H stretch, 126–127
   isotopic shifts, 113, 114–115, 122, 123–125
nitrido, terminal, 112–121
NX ligands, 124–125
oxo, 112–121
   bridging oxo, 113, 116–117
   monooxo, 114–118
   polyoxo, 118–121
   range of, for monooxo and nitrido (Table), 116
   trans ligand, effect of, 115
Structural tables, of complexes with:
   metal–carbon multiple bonds, 186–190
   metal–nitrogen multiple bonds, 179–186
   metal–oxo bonds, 159–179
Structures, see Coordination geometry; Structural tables
Substituents:
   and bond length, 151
   effect on bonding, 25–26
Sugars, 1-methylene, 236
Sulfide, dialkyl, 72, 244
Sulfido ligand, 3
Sulfite oxidase, 297. See also Oxo transferase enzymes
Sulfone, 244
Sulfoxides:
   from oxo, 244
   in oxo synthesis, 72, 297
Sulfur, 240
Sulfur diimine, in imido synthesis, 73
Sulfur monochloride, 240
Sulfynylamines, in imido synthesis, 88
Syntheses, see Metal–ligand multiple bonds, formation of; individual ligands

Tebbe–Grubbs reagent, 12, 64, 231, 235–238, 256
Tetrahedral complexes, 148
   bonding, 36–39
Tetraoxo complexes, see Osmium Tetroxide; Permanganate MnO_4^-; Ruthenium tetroxide
   bonding, 36–37
   ^{17}O NMR, 128–129
   vibrational spectra, 118–121, 125
Thermodynamic measurements, 32, 242
Thiols, reduction of oxo, 72, 297
Thionitrosyl, 73, 240
Titanacyclobutane, 67, 236–237, 256. See also Metallacyclobutanes; Tebbe–Grubbs

Titanium methylene, see Tebbe–Grubbs
Tosylamidation, electrocatalytic, 301
(Tosyliminoiodo)benzene, 266–267
Trans influence, 148, 156–157
   and angular distortions, 156–158
Trans ligand, effect on IR, 115
Trialkylphosphine, see Phosphines
Triazine complex, as imido precursor, 78
Trigonal bipyramidal structures, 148
Trimethylsilyl, see Silicon group
Trioxo complexes:
   bond length, M—O, 150
   bond order, 36
   coordination geometry, 36
   stretching frequencies, 118, 120
Triphenylantimony oxide, 80
Triphenylarsenic oxide, 80
Triphenylphosphine, see Phosphines
Triphenylphosphine oxide, see Phosphine oxide
Tungsten–tungsten triple bond, see Metal–metal multiple bonds

Uranyl ion, trans structure, 36
Uric acid, 297
UV-visible spectroscopy, 35–36, 37

Vanadium anode, 55, 301–302
Vanadium-51 NMR, 130, 136
Vanadium—oxo, IR, 117
Vanadyl ion, origin of color, 34
Vanadyl pyrophosphate catalyst, 291
Vibrational spectra, 112–127
   alkylidene, C—H stretch, 127
   alkylidyne, 123
   bending modes, 112
      bridging nitrides, 122
      polyoxo complexes, 121
   bridging nitrido, 121–122
   bridging oxo, 113
   carbyne, 123
   comparison of ligands by, 27, 125–126
   coupling of modes, 124–125
   hydrazido(2−), N—H stretch, 127
   hydrazido(4−), 123
   imido, 123–125
      N—H stretch, 126–127
   isotopic shifts, 113, 114–115, 122, 123–125
   matrix isolation, 113
   molybdenum oxide tetrafluoride, 113
   nitrido, terminal, 112–121
   NX ligands, 124–125
   oxo, 112–121
      bridging oxo, 113, 116–117

Vibrational spectra, oxo (*Continued*)
   monooxo, 114–118
   polyoxo, 118–121
   selection rules for, 112, 118, 121
Vinylidene, 26, 151
   structures, table of, 186–190

Wittig-like reactions, 10, 14, 15, 86–90, 134–135, 234–239, 302
   calculations on, 305
   and ^{13}C NMR of *t*-butylimido, 134–135

Wittig reagent (phosphorane), 80, 89

Xanthine oxidase, 297. *See also* Oxo transferase enzymes
X-ray crystallography, 145, 159. *See also* Structural tables, of complexes with

Ziegler–Natta polymerization, 14, 309–310
Zirconium alkylidene, 238